Emerging Perspectives on the Mobile Content Evolution

Juan Miguel Aguado
University of Murcia, Spain

Claudio Feijóo
Technical University of Madrid, Spain & Tongji University, China

Inmaculada J. Martínez
University of Murcia, Spain

A volume in the Advances in Multimedia and Interactive Technologies (AMIT) Book Series

Information Science REFERENCE
An Imprint of IGI Global

Managing Director: Lindsay Johnston
Managing Editor: Keith Greenberg
Director of Intellectual Property & Contracts: Jan Travers
Acquisitions Editor: Kayla Wolfe
Production Editor: Christina Henning
Development Editor: Janine Haughton
Cover Design: Jason Mull

Published in the United States of America by
Information Science Reference (an imprint of IGI Global)
701 E. Chocolate Avenue
Hershey PA, USA 17033
Tel: 717-533-8845
Fax: 717-533-8661
E-mail: cust@igi-global.com
Web site: http://www.igi-global.com

Library of Congress Cataloging-in-Publication Data

Emerging perspectives on the mobile content evolution / Juan Miguel Aguado, Claudio Feijoo, Inmaculada J. Martinez, editors.
pages cm
Includes bibliographical references and index.
ISBN 978-1-4666-8838-4 (hardcover) -- ISBN 978-1-4666-8839-1 (ebook) 1. Mobile apps. 2. Mobile computing. I. Aguado, Juan Miguel, 1977-
QA76.59.E46755 2016
004--dc23

2015022715

This book is published in the IGI Global book series Advances in Multimedia and Interactive Technologies (AMIT) (ISSN: 2327-929X; eISSN: 2327-9303)

British Cataloguing in Publication Data
A Cataloguing in Publication record for this book is available from the British Library.

Advances in Multimedia and Interactive Technologies (AMIT) Book Series

Joel J.P.C. Rodrigues
Instituto de Telecomunicações, University of Beira Interior, Portugal

ISSN: 2327-929X
EISSN: 2327-9303

Mission

Traditional forms of media communications are continuously being challenged. The emergence of user-friendly web-based applications such as social media and Web 2.0 has expanded into everyday society, providing an interactive structure to media content such as images, audio, video, and text.

The **Advances in Multimedia and Interactive Technologies (AMIT) Book Series** investigates the relationship between multimedia technology and the usability of web applications. This series aims to highlight evolving research on interactive communication systems, tools, applications, and techniques to provide researchers, practitioners, and students of information technology, communication science, media studies, and many more with a comprehensive examination of these multimedia technology trends.

Coverage

- Web Technologies
- Digital Technology
- Digital Images
- Internet technologies
- Gaming Media
- Multimedia Services
- Digital Communications
- Multimedia Streaming
- Multimedia technology
- Digital Watermarking

IGI Global is currently accepting manuscripts for publication within this series. To submit a proposal for a volume in this series, please contact our Acquisition Editors at Acquisitions@igi-global.com or visit: http://www.igi-global.com/publish/.

The Advances in Multimedia and Interactive Technologies (AMIT) Book Series (ISSN 2327-929X) is published by IGI Global, 701 E. Chocolate Avenue, Hershey, PA 17033-1240, USA, www.igi-global.com. This series is composed of titles available for purchase individually; each title is edited to be contextually exclusive from any other title within the series. For pricing and ordering information please visit http://www.igi-global.com/book-series/advances-multimedia-interactive-technologies/73683. Postmaster: Send all address changes to above address.

Titles in this Series

For a list of additional titles in this series, please visit: www.igi-global.com

Emerging Research on Networked Multimedia Communication Systems
Dimitris Kanellopoulos (University of Patras, Greece)
Information Science Reference • copyright 2016 • 449pp • H/C (ISBN: 9781466688506) • US $200.00 (our price)

Emerging Research and Trends in Gamification
Harsha Gangadharbatla (University of Colorado Boulder, USA) and Donna Z. Davis (University of Oregon, USA)
Information Science Reference • copyright 2016 • 456pp • H/C (ISBN: 9781466686519) • US $215.00 (our price)

Experimental Multimedia Systems for Interactivity and Strategic Innovation
Ioannis Deliyannis (Ionian University, Greece) Petros Kostagiolas (Ionian University, Greece) and Christina Banou (Ionian University, Greece)
Information Science Reference • copyright 2016 • 379pp • H/C (ISBN: 9781466686595) • US $195.00 (our price)

Design Strategies and Innovations in Multimedia Presentations
Shalin Hai-Jew (Kansas State University, USA)
Information Science Reference • copyright 2015 • 590pp • H/C (ISBN: 9781466686960) • US $225.00 (our price)

Cases on the Societal Effects of Persuasive Games
Dana Ruggiero (Bath Spa University, UK)
Information Science Reference • copyright 2014 • 345pp • H/C (ISBN: 9781466662063) • US $205.00 (our price)

Video Surveillance Techniques and Technologies
Vesna Zeljkovic (New York Institute of Technology, Nanjing Campus, China)
Information Science Reference • copyright 2014 • 369pp • H/C (ISBN: 9781466648968) • US $215.00 (our price)

Techniques and Principles in Three-Dimensional Imaging An Introductory Approach
Martin Richardson (De Montfort University, UK)
Information Science Reference • copyright 2014 • 324pp • H/C (ISBN: 9781466649323) • US $200.00 (our price)

Computational Solutions for Knowledge, Art, and Entertainment Information Exchange Beyond Text
Anna Ursyn (University of Northern Colorado, USA)
Information Science Reference • copyright 2014 • 511pp • H/C (ISBN: 9781466646278) • US $180.00 (our price)

Perceptions of Knowledge Visualization Explaining Concepts through Meaningful Images
Anna Ursyn (University of Northern Colorado, USA)
Information Science Reference • copyright 2014 • 418pp • H/C (ISBN: 9781466647039) • US $180.00 (our price)

Table of Contents

Detailed Table of Contents

This chapter aims at providing a comprehensive overview of the mobile landscape. It bridges the global trends affecting the mobile industry as a whole, with the creation and distribution of contents, emphasizing the new patterns of production and distribution. It shed light on the drivers of the changes taking place. The first part of this chapter identifies and sums up the main trends in a global landscape, namely the role of mobile communications (devices, networks) combined with the rise of the "app economy", the spread of "Big Data", and the move toward a multiscreen / cross-media paradigm. It provides a synthesis of available data. The first section tracks the evolution of the global market and the emerging geography with the rise of Asia as a leading player in the mobile world. The second section analyses the smartphone phenomenon. The third section deals with the rise of the "app economy". The fourth section reviews the new modes of data driven management, stemming from "Big Data". The fifth section follows the move a multi-screen world. The second part investigates some of the impacts of this fast evolving environment on the production of an array of mobile contents (portable movies, video, video games…) tracking the way content industries, while standing at the nexus of technology and content, are rapidly changing. The section documents the way mobile industry is indeed an innovation platform for new services. The section delineates the growing role of the consumer in a mobile world, its various modes of involvement in the production and distribution of digital content. We conclude with some elements about the new challenges brought by these trends, by the entry of new players in both the content and the mobile field.

It is commonly accepted that entrepreneurship is a key activity in the current economic context, and entrepreneurs create value in form of new businesses ventures, and generate new jobs by mobilizing economic resources, which in combination has a positive impact in terms of increasing productivity and economic growth, among other effects. In this chapter, the authors acknowledge the importance of innovation in the mobile industry driven by entrepreneurs and firms as a source of disruptive changes that have completely reshaped the world, by presenting the key elements of current mobile industry and the emerging areas with the highest potential to attract entrepreneurs in the upcoming years.

Personal information is a key intangible asset for companies. In particular, for those companies that collect personal data to attract advertisers through being a channel to reach their target markets, i.e., through offering targeted advertising. A universe of always located and always connected potential consumers enhance the potential of behavioural targeting, which is seen as the business model on which the success of new mobile content, applications and services relies. This chapter aims to explain the implications of the use of behavioural targeting within the mobile environment. It provides an overview of the exploitation of personal data, a comprehensive description of the behavioural targeting ecosystem, and a view of the promises and pitfalls associated to the rise of mobility.

This chapter explores the innovation landscape for the mobile content ecosystem building upon the hypothesis of the disruption potential of wearables. Google Glass is taken as the paradigm, as it offers direct access to network, becoming an eventual alternative to the smartphone. Equipment manufacturers lost traction as innovators right after the arrival of the iPhone in 2007, which brought a key innovation to the ecosystem. Observing the strategies of the different players operating in the supply side, two innovation scenarios are proposed for the debate: a sustaining-innovation based scenario, and a disruptive-innovation one. The main findings admit a clear disruption potential in wearables, mostly focused on the power of glasses as a driver to create a new wave of content, oriented towards augmented-reality features. However, it is still uncertain if these new devices can trigger a new era of innovation expanding its effects towards the whole ecosystem.

Music industry is arguably the foremost example of the advantages and disadvantages of having been exposed early on to digitalization. But not only has music been a pioneer in the area of digitalization, but it has also been one of the first sectors in the content industry to be massively "mobilized". It is therefore an area where lessons can be learnt for other mobile media and content market segments. From this framework, this chapter examines four main stages in the evolution of mobile music, investigates in some detail the emerging business models in the domain and concludes with the main challenges and opportunities for mobile music.

The multimedia features brought by the rise of the mobile access to internet are tightly linked to the massive access of the public to mobile search. To date, there is limited knowledge on how distinct groups make use of mobile search in their everyday life. Attempting to fill that void, the purpose of this chapter is to describe and explain how mobile search behaviours have gained traction among distinct groups in Sweden from 2010 to 2012. The findings evidence a growing uptake among all groups over time. Smartphone ownership and age score the highest odds ratios, revealing smartphone users and 16–29 year olds as most inclined to use mobile search. In addition, a high education and use of a personal or company subscription show significant results.

mHealth is an emerging field that comprises the delivery of health through mobile technologies, being part of a broader concept: eHealth. With the rapid advances and adoption of mobile technologies and applications, mHealth opens new opportunities to deliver healthcare and wellness promotion through personal mobiles. This will transform the current delivery of services linked to particular places to a delocalization and personalization of services delivery. Although the adoption and integration of mobile technologies is high across the globe, the particular adoption of mHealth remains low and scattered due to some remaining challenges. This chapter presents the current picture of mHealth and it later discusses the opportunities both at market level and with regards to healthcare delivery. Then, it presents future research directions as big data; and it closes with conclusions on this challenging emerging field.

"friendly" to mobile devices. The idea of "friendly" can range from simple accessibility requirements to a dynamic redesign. Included in this discussion is a summary of the advantages and disadvantages of current techniques. Examples of different techniques are presented and illustrated along with the system framework necessary to implement them.

Researchers generally have not been able to keep up with the rapid pace of mobile app innovation in many ways. One of those holes is with typology. Not being able to clearly describe mobile app artifacts, as a way to start a discussion about them, is a key factor holding researchers back. This chapter splinters off a promising but relatively underdeveloped genre of mobile apps, the locative ones that juxtapose interactive digital media with particular physical environments, for closer examination and better labeling. This piece quickly provides theoretical, historical, and contextual overviews to help situate locative apps as a genre of importance. It then provides informal frameworks, as thought exercises, for examining and understanding apps through a locative perspective. Two cases, also led by the author, will then be taken through this process to illustrate what can be learned from it, followed by reflections and suggestions for related future areas of inquiry.

The spread of mobile devices and digital networks generate new perspectives for the circulation of news. The development of mobile communication technologies and the migration of habits consolidated in personal computers promote the expansion of news aggregators for smartphones and tablets. This work seeks to inquire the operational logic of applications designed to aggregate journalistic content. An immersive analysis shows that the general operations of these apps are based on three dimensions: integration between news and online social media, always-on recommendation systems, and parameterization of outputs. It is argued that the approach between aggregators and social media makes the news extrapolate its information domain and start to circulate as symbolic reference for online interaction. Finally, this work posits that the operational logic of aggregators neglect fundamental aspects of new technologies related to locative information.

The aim of this chapter is to examine in depth the potential for ecological awareness that emerging artistic and creative narrative practices involving mobility may possess. In order to do so, this chapter will analyze the current situation in reference to several projects that exemplify this potential. Our time sees the coming of the Creative Activist in a practice that uses art and media to scale up the ecological message to a global level, to reach a lot of people and to move them.

This chapter discusses the new social configurations society is undergoing on the basis of media emergence. Media are embedded in the arousal of communication and information transmission becoming the form, the infrastructure and the institution for the social and culture. This chapter focuses on mobile communication, having as central goal to debate on the processes of mediatization and mediation of society, as well as on the processes of belonging and social cohesion. Data from mobile internet adoption and use will be discussed in the light of the above mentioned theoretical approaches. An empirical case study will also be approached and results will provide contributions for the understanding of this type of technology adoption processes and the increasing importance of mobility in cultural and social practices, promoting an exciting discussion on the centrality of media nowadays and the current transformation processes society is undergoing.

Preface

A MOBILE REVOLUTION INSIDE A DIGITAL REVOLUTION

The Mobile Revolution: The Latest ICT Innovation Engine

Mobile technology is revolutionizing the world. In less than a decade it has changed our culture, our societies, and our economies. From old feature phones to new smartphones, tablets and wearables, mobile devices, and the services they enable, have literally turned upside down our social relations, our personal memories, our professional performance and even our very understanding of space and time (Katz, 2006).

Mobile has become the fastest diffusing technology in history. It has outstripped the adoption rate of other relevant technology in the century before: It took 38 years for radio broadcasting to reach 50 million users. Landline telephone needed over 75 years to gather that amount of users. Television had to wait for more than 13 years to have a 50 million audience, and even the World Wide Web took almost 4 years to reach that number of users. Apple sold 50 million iPhone 3G in just 2 years and 9 months (from its original launch on June 29, 2007 to April 8, 2010) (McKinsey Global Institute, 2012). Today –barely 8 years after- more than 2 million smartphones are sold each day and mobile devices (smartphones and tablets) are close to represent half of the global consumer electronics industry (Evans, 2014). But this is not only about scale and scope. It is also about acceleration: in less that ten years we have turned from 1.6 billion people renewing hardware every 5 years to 4 billion users buying equipment every two years (Evans, 2014).

Mobile is a revolution inside the digital revolution. In a sense, mobile is to the Internct what the print was to writing and literacy: it expands and transforms the already existing possibilities by making them accessible and ubiquitous. To understand the scale of that assumption, the reader may think for instance of cloud computing with and without mobile technologies in the background or just take a walk in any metropolis of the developing world where mobile Internet is the dominating means to access the net.

It is not therefore accidental that much of the importance of mobile technologies lies in reshaping the whole ICT industry ecosystem. In fact, the evolution of mobile communications has much to do with the very understanding of ICTs as an ecosystem -a complex net of interdependent players and environmental factors that internally generates its own order- rather than a plain and structured market sector. Mobile is taking the Internet revolution to its ultimate consequences, expanding connectivity across time and space, from persons to relationships (Web 2.0), from relationships to things and contexts (IoT) and back to persons through massive data management (Big Data).

Mobile has pushed innovation in almost all levels of the ecosystem (De Prato, Sanz and Simon, 2014): it has contributed to create new products (smartphones, tablets, wearables…), new logics (personal, ubiquitous, frictionless access, context awareness, IoT capable built-in sensors…), new players (app developers, platforms, aggregators...) and new interdependences amongst them (multi-sided markets).

The Central Nature of Platforms and Users

Fransman (2014) explains the ICT ecosystem as a symbiotic interaction amongst players organised into three layers: The first layer includes ICT equipment providers (who create the elements that make up ICT networks and enable ICT services); the second layer is that of network operators (telecoms, cable and satellite operators, and broadcasters). The third layer is comprised of platform, content and applications providers, which create content, applications and services that run over the networks provided by equipment providers and network operators. The recent evolution of the ICT ecosystem has been marked by the increasing influence of the players in the third layer.

Platforms –and very especially mobile platforms, such as Apple's and Google's- have acquired a central role in re-shaping the ICT ecosystem. This has evolved in the last decade from a network centred structure, in which network operators and equipment providers had a privileged position, to a content and platform centred structure, in which software developers (especially OS suppliers) influence the relations amongst the different mobile players. Peter Ballon (2009) refers to that shift from the telecom paradigm to the software paradigm with the term 'platformisation'. Cross-network externalities facilitated through this platformisation have become the driving force behind a surge of innovation in this leading ICT ecosystem.

There is however a particular player that has become especially relevant in the (mobile) digital environment: users, as these are the ground for the social and cultural relevance of innovative technology related meaning practices. So it is users –and their relational and emotional context- that provide with sense what the market provides with value. Giving some light to the connections between these two spheres –the socio-cultural and the economic- of the mobile content evolution is indeed one of the purposes that guide this book.

Consequently, users conform a fourth layer in Fransman's model (2014). They are not only a key element in many of layer 3 value creation processes, being the target for most of the players involved. Also the players from layers 2 and 1 maintain relevant interactions with users. In fact, the increasing influence of platforms and content/application providers lies in their privileged access to final users, which has become an appeal for players in the other layers of the ecosystem (De Prato, Sanz and Simon, 2014; Aguado and Martinez, 2014).

Because of their capacity to determine social and meaning practices, users occupy a very relevant place in the innovation creation networks within the new digital ecosystem. The answer to the so-called 'Consumer Demand Paradox' condenses both the innovative impact of mobile and the deep links existing between social practices and technological innovation. The consumer demand paradox frames the question on "how consumers can express a demand for something that does not yet exist and which they cannot conceive of since they lack of the underlying technologies and designs" (Fransman, 2014:33). The well-known Steve Jobs line claiming that "people don't know what they want until we show them", or the famous quotation by Henry Ford, "If I'd asked customers what they wanted, they would have told me, 'A faster horse!'" plainly summarize the nature of the consumer demand paradox. Consumers have the power to attribute social meaning but they cannot anticipate it since it emerges as a result of the social practices they get involved in with technology (Aguado & Martínez, 2010). And also the economic value of that social meaning emerges as a result of the appropriation of technology in social practices. Foreseeing or anticipating new meanings, new practices and their value is thus a risky –but nevertheless disruptive- business.

If mobile platforms around content and application providers have managed to create a very ecosystem within the global ICT ecosystem (Feijóo et al., 2012), it has been possible to a given extent thanks to their ability to produce a user-centred viable emergent innovation system as a specific 'mobile' answer to the consumer demand paradox (Fransman, 2014). On the basis of their control over the distribution channels and the reciprocal access of the different players (app developers, service providers, content creators, users…), mobile platform providers such as Apple and Google create and run a system of emergent innovation based upon the principles of evolution, namely self-steered selection over increasing population and diversity. In fact, the layered approach to the mobile ecosystem and platforms has been successfully applied to a number of upcoming mobile sub-domains to explain their status and dynamics, namely, mobile search (Gomez-Barroso et al., 2010), mobile television (Aguado et al., 2012), or mobile gaming (Feijóo et al., 2012).

The way this emergent innovation processes work is relevant not only because it is based to a large extent on content distribution and related services. It is also important because it shows an effective way to deal with cost-worthiness uncertainty as perceived by users, which is one key issue to be addressed in the context of media digitization and new viable business models. As Fransman (2014:33) puts it,

Given that there is an important degree of novelty in all new apps consumer-users first have to experience the app and learn what it can do before they can form a conclusion about how useful it is and, therefore, how much they are willing to pay. Business models that involve a combination of free apps plus paid-for additional features facilitate this experience-learning process by in effect subsidising the learning period. Naturally, each individual app provider starts by creating a value-creating conjecture in which he or she hypothesises about what will add value for consumer-users. The belief behind the individual app may turn out to be right, or, alternatively, it may prove to be wrong.

Platforms like Apple's iOS/iTunes/AppStore or Google's Android/Chrome/Google Play provide via cross-network externalities significant incentives for app developers (up to 70% of the revenue generated from the sale of the app or advertising platform services for those who prefer ad based revenues). This makes possible a large –and rapidly increasing- population of apps with a considerable amount of variety. Consumers/users select from this variety according to their expectations. Trial formats, like lite versions or freemium models, do not only subsidize learning process. They help users to cope with the uncertainty about the effective value they pay for. Platforms facilitate the matching of effective selection and users' expectations by creating a whole system for the management of visibility, which includes category management, search protocols, ranking systems and validation networks (like reviews, users rating, etc.). By this they enable specific content and app marketing environment that –together with their control over distribution channels- makes platform owners relevant intermediaries in the digital media system (De Prato, Sanz and Simon, 2014; Aguado and Martinez, 2014).

Media and Digital Content after the Mobile Revolution

Mobile ecosystem is in fact a key driver in the digitization process of media industries. Content is the king of mobile screens, and apps are the prevalent interface through which content is accessed (Aguado and Martinez, 2014). According to a recent report (Khalaf, 2014), 86% of mobile Internet use is through mobile apps, leaving only a 14% to mobile browsers. A 71% of the total app-related mobile Internet use is dealing with digital content and media, be it user-made (28% for social networks like Facebook, Twitter or Wechat) or channelled from media industries (32% games; 8% video entertainment, like YouTube, Netflix or legacy media apps; and 3% news).

Content consumption has become a leading activity in mobile devices. This has turned content into a relevant asset in the mobile ecosystem but also has turned mobile into a relevant channel for media industries. Mobile content has become thus a core aspect of the innovation processes in both the ICT ecosystem and the media system.

However, the idea of content itself is also subjected to transformation in the mobile ecosystem. As with previous media forms, a given technology pairs with new forms of industry, new businesses and new social consumption practices. It is a similar process to that of cinema during the first half of XX century: technology enables the channel, channel enables the industry and the right coupling of technology, channels and industry enables the social and cultural practice to start taking form while being transformed by them. In that process the very idea of content becomes constantly reshaped.

One first approach to that transformation of mobile content may look at the formal and functional aspects of mobile content. Taking into account the degree of creativity involved and the extent to which mobile specific technologies and capabilities are involved, Feijóo et al. (2009) classify mobile content into four categories:

1. Adapted content comprises existing information (coming from different media) that has been adjusted in order to be displayed and used on a mobile device.
2. Repurposed content refers to those creative contents that –coming from other media- have been processed and adapted to mobility. Mobile versions of desktop or console games as well as mobisodes from TV series are examples of this category.
3. Augmented content includes those kinds of mobile content that –being conceived in the frame of other existing media- make use of mobile technology functional capacities. Mobile versions of web content including location aware functionalities are an example.
4. Mobile specific contents are those contents conceived and created to be mobile, taking advantage of mobile technology capacities and addressed to the specific mobile frames of use.

From this perspective mobile content has evolved from the mere adaptation of content formats coming from other media (television, CD players, game consoles) to a progressive consolidation mobile specific formats, syntax and uses. This involves integrating and experimenting with new meaning factors, like movement, location or other information coming from built-in sensors. It also implies new ways of interaction between content and users and amongst users around content, which means in the end new interfaces for content consumption. And that content innovation is to be understood within new ways of merging meaning access through co-ordinated but different media and interfaces (transmedia narratives, multiscreening).

As it happened with cinema, television or Web video in last century, through that process mobile devices have become a specific new media with its own semantics and its own social meaning frames. Describing mobile as the seventh mass media, Ahonen (2007) reminded already in the dawn of the smartphone era that mobile is not a miniaturized little version of television, nor "the dumb little cousin of the Internet". Mobile media have also contributed in turn to change existing media and meaning practices around them to such extent that old media (cinema, television, news, video games, music…) cannot think about themselves without taking mobile as a relevant variable (De Prato, Sanz and Simon, 2014).

Furthermore, as it will be developed in some of the chapters in this book, mobile focused distribution channels (apps stores) and mobile specific ubiquitous access interface for content consumption (mobile apps) are changing the very nature of media content from a passive conception (content is something to be viewed, read or listened to) to an active and social one (content is something to act upon, transforming, recommending or commenting it, and to use it to relate to others). This (mobile) shift in the social meaning of digital content consumption has obviously roots in the technological and economic possibilities of mobile ecosystem players and elements, but is not limited to them. Quite on the contrary, it influences the techno-economic implications of the mobile revolution by merging social interaction dynamics into the digital communications and the content consumption sphere.

A Key Field of Study

Understanding the trends and shapes of mobile content evolution constitutes a major challenge and a strategic advantage in a wide spectrum of business platforms and industry sectors nowadays. Its development impacts all the players in the digital content and advertising value chain (carriers, OEMs, aggregators –including legacy media-, software developers, users…) as well as it concerns important digital related policy issues regarding spectrum, privacy and transparency regulation, freedom of expression, and intellectual property rights management (Feijóo et al. 2012).

This book discusses the disruptive nature of mobile media by underlying three convergent evolutionary trends of the digital ecosystem:

1. Thc central nature of digital content distribution in the evolution of mainstream digital companies (as a counterpart for the adapt-or-perish dilemma of legacy media in the increasingly mobile digital environment),
2. The relevant role digital and mobile content plays in the new consumption habits that develop around networked performance of identity, and
3. The increasing importance of personal information management (profiles) as an economic asset in the context of Big Data ubiquitous Internet, and its implications for media and content industries, fostering new business models, new content formats and new forms of relationships with audiences.

These three trends mark a territory of conflict and innovation that concerns the redefinition of the advertising system, the pervasiveness of the SOLOMO (social-located-mobile) golden rule in format an product strategy design in the sphere of media and new regulatory and policy challenges in the sphere of media and in the context of a Big Data driven mobile internet.

At the same time, the confluence of multi-screen content consumption scenarios and mobile devices' built-in or software enhanced capacities allow for new ways of telling, enjoying, understanding, and socially sharing narratives that come to put the very term 'content consumption' in brackets. Entertainment, content and advertising merge together in new forms of interactive dynamics and participatory scenarios. Also the conventional actions around digital content (creating, playing, managing and sharing) become increasingly merged (Aguado and Martínez, 2014) transforming thus the idea of content driven consumption into an increasingly social interaction driven value.

The evolution of mobile content in the last few years condenses therefore the innovative contributions to the ICT ecosystem and to the media industries carried out by mobile platform players. These innovations concern different aspects of both ecosystems: economic (business models, distribution channels, value creation), political (regulation and sector policies), semantic (new content formats and genres, new forms of interacting with cultural meanings through devices) or social (new forms of media and technology appropriation). These changes are still occurring, and they come along with extraordinary challenges –but also with unexpected opportunities- for legacy media industries.

The Structure of the Book

Taking this into account, the academic discussion about mobile content evolution has to be multidisciplinary. For that reason we have conceived this book as an open discussion in which researchers from different countries and disciplines (engineering, economy, sociology, arts, communication, design, etc.) converge. The reader will find thus three main groups of perspectives on the evolution of mobile content and its multiple consequences:

In the first section contributing authors address the debate on the economics and logics of the mobile content evolution. Chapters in this section deal with economic and structural aspects of the mobile content ecosystem, how the players are organizing in it and what the role of mobile content evolution is.

Chapter one (Global Trends in Mobile: A New Global Landscape for Supply and Demand) accounts for a broad picture of the economic evolution of mobile influenced media industries and their confluence with other IT players. Jean Paul Simon and Giuditta De Prato underline the changing conditions driven by the mobile revolution and the strategic relevance of the new distribution channels in the digital content value chain.

Sergio Ramos, Cristina Armuña, Alberto Arenal and Jesús Ferrandis consider in chapter two (Mobile Communications and the Entrepreneurial Revolution) the transformations in entrepreneurship and digital business driven by mobile technology and mobile platforms.

Chapter three (Behavioral Targeting in the Mobile Ecosystem) deals with the increasing importance of personal information and behavioural targeting as a key economic asset. José Luis Gómez-Barroso and Juan Ángel Ruiz-Rodríguez provide an up to date overview of the exploitation of personal data, a comprehensive description of the behavioural targeting ecosystem, and a view of the promises and pitfalls associated to the rise of mobility.

Of course, when talking about mobile content evolution and possible future scenarios, wearable technologies have their own place. These are the focus of chapter four (A Reflection on Wearables and Innovation in the Mobile Ecosystem: Two Possible Scenarios), where Andreu Castellet addresses the question whether the return of hardware-driven innovation after a decade of software-driven value creation.

Addressing specific mobile content sectors, like music (chapter five: Mobile Music: An Overview), allows identifying relevant experiences the recent evolution of which may contribute to shed some light into other areas of the mobile content ecosystem. According to Claudio Feijóo, Sergio Ramos and Victoria Bleda's proposal, the case of music is especially significant, since it is one of the cultural content sectors pioneering digital transformations and also especially impacted by mobile technologies and platforms in all the aspects of its characteristic value chain.

Lessons from the evolution of search practices and related behaviours are extracted by Andreu Castellet and Oscar Westlund in chapter six (Examining Mobile Search Adoption in Smartphone Users: The Swedish Experience in the Uptake Phase), contributing thus to frame mobile Internet specific aspects.

Chapter seven (From Personal to Mobile Healthcare, Challenges and Opportunities) in this section considers one of the emerging sectors with bigger potential relevance in the mobile application and content ecosystem. In the crossroad amongst lifestyle related consumption, education, service providing and research-addressed data gathering, mobile health conforms a crucial field of the coming mobile environment. Elena Villalba-Mora, Ignacio Peinado and Leocadio Rodríguez-Mañas present and discuss the challenges and opportunities of the mobile health, describe the background of personal healthcare applications and the current picture of mHealth around the globe.

Chapter eight (Mobile Big Data: a New Frontier of Innovation) accounts for a mobile driven characterization of the Big Data phenomenon and its influence in innovation and value creation. Shivom Aggarwal and Abbishek Nayak discuss the concept of Mobile Big Data, putting forth how Mobile Big Data is shaping society and business as well as possible influence of policy making.

In the second section of this book, the discussion focuses the semantics and pragmatics of mobile content evolution. Here, relevance is given to those processes of meaning construction and of social practice shaping around mobile content innovation.

In chapter nine (Driving media transformations: Mobile Content and Personal Information) Juan Miguel Aguado and Inmaculada J. Martínez consider the strategic role of mobile content as a source of social transformations (new content consumption scenarios), of content format evolution (towards participatory meaning frameworks) and business transformations (value chain expansion and personal data driven business models).

Christian Sonnenberg confronts in chapter ten (Mobile Content Adaptation: An Analysis of Techniques and Frameworks) the challenges of flexibility and adaptability that content design and access face in a massively fragmented and rapidly changing device ecosystem like the mobile one.

Location and connection to the situation of users have always been some of the most appealing –yet underdeveloped and to some extent unaware- features of mobile content applications. Brett Oppegaard addresses this lacuna in chapter 11 (Mobility Matters: Categorizing Locative Media Through Affordances and Constraints). To do so he applies specific theoretical approaches (affordances approach) and developes an appealing research line focused on classifying and analysing locative functionalities and performances.

Together with music, news media is another cultural industry singularly hit by the digital and mobile revolution. News is a media sector majorly affected by the changes in distribution channels and in creation processes, an also by the transformation of the users' role in the news media value chain (De Prato, Sanz and Simon, 2014). The new mobile-enhanced possibilities in distribution and content aggregation involve as well new players. News aggregators have a relevant place amongst them, and constitute the objecto of analysis in João Canavilhas and Ivan Satuf's chapter twelve (Who bring the news? Exploring the Logics of Content Aggregators Apps for Mobile Devices).

News media mobile-driven transformation is also the context for chapter thirteen (Mobile Journalism and Innovation: A Study on Content Formats of Autochthonous News Apps for Tablets). In this case, Marcos Palacios, Suzana Barbosa, Fernando Firmino da Silva and Rodrigo Cunha focus on analysing the new formats for news mobile applications, examining the specific affordances, structure, organization and types of narratives produced.

In chapter fourteen (The Role of the Profile and the Digital Identity on the Mobile Content) Ana Serrano puts ubiquitous personal information and profiling –which in the previous section were considered from the economic and structural perspectives- in the context of social and cultural practices, rising relevant questions regarding public/privacy balance and related social, economic and policy challenges.

If sections one and two are devoted to the economic, logic and logistic aspects of mobile content evolution, the third section in this book presents some specific experiences and performances of mobile content that contribute to illustrate –and some times also to challenge- what discussed and presented in the previous sections. Some of these performances and experiences discuss the challenges of designing and developing specific mobile content or the implications of mobile social networking experiences. The debate includes also contributions around multi-disciplinary insights like considering the links between ecology, art and mobile content, or investigating new forms of location and movement based narratives.

Enrique Castelló, Roi Méndez, Antia M. López, Julián Flores and Eduardo Sánchez-Vila present in chapter fifteen (Research, Development and Creativity in Ubiquitous Technologies in University: CIDUS Contribution) an account for the CIDUS platform for innovation and experimentation in digital content at the University of Santiago de Compostela (Spain). Their product innovation covers such diverse areas as telemedicine, tourism promotion, distribution of digital cinema in OV, the impact of the ecological footprint, cultural content generation and dissemination and enhancement of historical and artistic heritage.

The artist and researcher Fred Adam rises in chapter sixteen (Mobile Content and Walking Documentary: Teaching and Learning Science Step-by-Step with Smartphones) the challenges and potential of locative and moving driven narratives, connecting them to new strategies in mobile learning and new forms of documentaries.

Also artist and researcher, Verónica Perales contributes to chapter seventeen (The Message is the Medium: Ecology, Mobility and Emergent Storytelling) by exploring the links between transmedia storytelling and ecology. From the need of narrative complexity in understanding complex problems to the delimitation of ecology as a field for consciousness awakening digital art, Perales raises the appropriateness of location and movement aware narratives in dealing with problems like global warming.

Finally, in chapter eighteen (Mobile Media and Social Interaction - Mobile Services and Content as Drivers of Social Interaction) Manuel J. Damásio, Sara Henriques, Inês Teixeira-Botelho and Patrícia Dias approach the different ways in which contemporary mobile content, services and applications mold specific individual and collective experiences that have consequences for different aspects and forms of social cohesion, involvement and belonging.

We are well aware that the debate is far from finished in this volume, as it is the very evolution of mobile content or the increasing perspectives from which it is to be studied. Our aim is to contribute with it to better understand the centrality of mobile content in the recent and coming evolution of both the ICT ecosystem and the media industries.

Juan Miguel Aguado
University of Murcia, Spain

Claudio Feijóo
Technical University of Madrid, Spain & Tongji University, China

Inmaculada J. Martínez
University of Murcia, Spain

REFERENCES

Aguado, J. M., Feijóo, C., Martínez, I. J., & Roel, M. (2012). Mobile television, a paradigmatic case on the uncertainties and opportunities of the new media ecosystem. In A. Abruzzese, N. Barile, J. Gebhardt, J. Vincent, & L. Fortunati (Eds.), *The new television ecosystem* (pp. 141–171). Frankfurt am Main: Peter Lang.

Aguado, J. M., & Martínez, I. J. (2010, December). Liquid digital selves: Mobile Media Implicit Cultures, Social Relations and Identity Management. *Encyclopaideia: Journal of Phenomenology and Education, 15*(1), 63–88.

Aguado, J. M., & Martínez, I. J. (2014). The Relationship is the Medium: Understanding Media in a Mobile Age. In J. Katz (Ed.), *Living Inside Mobile Social Information* (pp. 77–108). Boston: Boston University Press.

Ahonen, T. (2007). Mobile the 7th mass media is to Internet what TV is to radio. Retrieved from http://communities-dominate.blogs.com/brands/2007/02/mobile_the_7th_.html

Ballon, P. (2009). The Platformisation of the European Mobile Industry. *Communications & Stratégies, 75*, 15–34.

Feijóo, C., Gómez Barroso, J. L., Aguado, J. M., & Ramos, S. (2012). Mobile gaming: Industry challenges and policy implications. *Telecommunications Policy, 36*(3), 212–221. doi:10.1016/j.telpol.2011.12.004

Feijóo, C., Maghiros, I., Abadie, F., & Gomez-Barroso, J. (2009). Exploring a heterogeneous and fragmented digital ecosystem: Mobile content. *Telematics and Informatics, 26*(3), 282–292. doi:10.1016/j.tele.2008.11.009

Fransman, M. (2014). Models of Innovation in Global ICT Firms: The Emerging Global Innovation Ecosystems. *EC-JRC Science and Policy Reports*. Retrieved from https://ec.europa.eu/jrc/sites/default/files/jrc90726.pdf?search

Gómez-Barroso, J.-L., Compañó, R., Feijóo, C., Bacigalupo, M., Westlund, O., Ramos, S., & García-Jiménez, M.-C. (2010). Prospects of mobile search. JRC Scientific and Technical Reports. Seville: Institute for Prospective Technological Studies. European Commission; Retrieved from ftp://ftp.jrc.es/pub/EURdoc/JRC56100.pdf

Katz, J. E. (2006). *Magic in the air: Mobile communication and the transformation of social life*. Piscataway, NJ: Transaction Publishers.

Khalaf, S. (2014). Apps Solidify Leadership Six Years into the Mobile Revolution. *Flurry Insights*. Retrieved from http://www.flurry.com/bid/109749/Apps-Solidify-Leadership-Six-Years-into-the-Mobile-Revolution#.VR1ExEat1NI

McKinsey Global Institute. (2012). *The Social Economy: Unlocking Value and Productivity Through Social Technologies*. Retrieved from http://www.mckinsey.com/insights/high_tech_telecoms_internet/the_social_economy

Trendforce (2015). Global Smartphone shipments in 2014 total 1167b. Retrieved from http://press.trendforce.com/press/20150120-1806.html

Acknowledgment

The content in this book and the academic debate that have made it possible are part of a research project supported by the Spanish Ministry of Economy with the reference CSO2013-47394-R addressed to explore the impact of mobile communications and ubiquitous personal data in media industries –including the advertising system- and users behavior. We are also indebted to all those who have made possible the editorial process and the final structure of the book.

Section 1

On the Logics and Economics of the Mobile Content Evolution

Chapter 1
Global Trends in Mobile:
A New Global Landscape for Supply And Demand

Giuditta de Prato
European Commission JRC IPTS, Spain

Jean Paul Simon
JPS Public Policy Consulting, Spain

ABSTRACT

This chapter aims at providing a comprehensive overview of the mobile landscape. It bridges the global trends affecting the mobile industry as a whole, with the creation and distribution of contents, emphasizing the new patterns of production and distribution. It shed light on the drivers of the changes taking place. The first part of this chapter identifies and sums up the main trends in a global landscape, namely the role of mobile communications (devices, networks) combined with the rise of the "app economy", the spread of "Big Data", and the move toward a multiscreen / cross-media paradigm. It provides a synthesis of available data. The first section tracks the evolution of the global market and the emerging geography with the rise of Asia as a leading player in the mobile world. The second section analyses the smartphone phenomenon. The third section deals with the rise of the "app economy". The fourth section reviews the new modes of data driven management, stemming from "Big Data". The fifth section follows the move a multi-screen world. The second part investigates some of the impacts of this fast evolving environment on the production of an array of mobile contents (portable movies, video, video games…) tracking the way content industries, while standing at the nexus of technology and content, are rapidly changing. The section documents the way mobile industry is indeed an innovation platform for new services. The section delineates the growing role of the consumer in a mobile world, its various modes of involvement in the production and distribution of digital content. We conclude with some elements about the new challenges brought by these trends, by the entry of new players in both the content and the mobile field.

DOI: 10.4018/978-1-4666-8838-4.ch001

Disclaimer: The views expressed are those of the authors and may not in any circumstances be regarded as stating an official position of the European Commission.

1. A MOBILE WORLD: SCALING UP[1-2]

In this section we set up the scene of the mobile environment focusing on some of the main trends that may open the way for new or renewed contents production and circulation. This section relies mostly in industry and consultancy data, as in such a fast changing environment official data even when accessible are in most cases too old to remain relevant to identify the trends. The figures may fluctuate from one source to another, but the direction looks clear enough. Therefore, the data used should be perceived as just signalling potential trends[3], but sufficient at the same time to give an overview of the evolution of the global mobile ecosystem.

An Impressive Growth of the Global Market

The World Bank stresses that "mobile telephony has been one of the most quickly adopted technologies of all time": it took 128 years before fixed telephone lines reached 1 billion users, mobile networks achieved this milestone in just over two decades (Minges, 2012, p. 115). The author adds: "even more astounding, mobile networks have roughly doubled in size every two years since 2002". Indeed, according to the ITU (2013, p. 1), there were, in 2013, almost as many mobile-cellular subscriptions as people in the world. Asia Pacific, the largest region in the global mobile market, accounted for almost half of all the world's unique subscribers at the end of 2013 (GSMA, Asia Pacific, 2014a: 10), as shown in Figure 1. At the end of 2003, there were a little over one billion unique subscribers, meaning that just under one in six people had subscribed to a mobile service. By the end of 2013 this figure had increased to 3.4 billion unique subscribers: equivalent to just under half of the global population. Globally there were 7 billion SIM connections at the end of 2014[4], with an average of 1.8 active SIM cards per unique subscriber (Mobile Economy 2015, p.2). The mobile services industry is one of the most globalized in the world (Minges, 2012, p. 125).

GSMA, the mobile trade association is tracking these evolutions, its research branch Wireless Intelligence (2013) forecasts 8.5 billion connections by 2017 with 50% operating under the new generation of mobile networks (3G[5]: 40%, 4G[6]: 10%). Even if the growth ratio will slow down, from a compound annual growth rate (CAGR) of 11.3% over the 2008-2013 period, to 4.2% growth rate for the next period (2014-2020), in 2020, the number of mobile connections will reach almost 9,2 billion (see Figure 1). Total operators revenues reached 1.15 in 2014 trillion US $ and will reach 1.4 trillion US $ in 2020 (GSMA, 2015, Mobile Economy, p. 4). If one considers the broader mobile ecosystem, it goes up to almost 2 trillion US $: network infrastructure (80 trillion), components (89 trillion), apps/content and advertising (205 trillion), device (436 trillion), and operators as noted a little under 1.2 trillion (GSMA, 2014b, p. 32).

This growth is almost mechanically accompanied by a growth in mobile traffic, now shifting to IP traffic. According to the Cisco Visual Networking Index (2014), for the first time in the history of the Internet, mobile and portable devices will generate more than half of global IP traffic by 2018[7]. No wonder the company may claim that (Cisco Visual Networking Index, 2013, p. 25): "Mobile data are well on its

Figure 1. Global mobile connections (2008-2020)
Source: GSMA. The Mobile Economy 2015 (6) © GSMA Intelligence 2015.

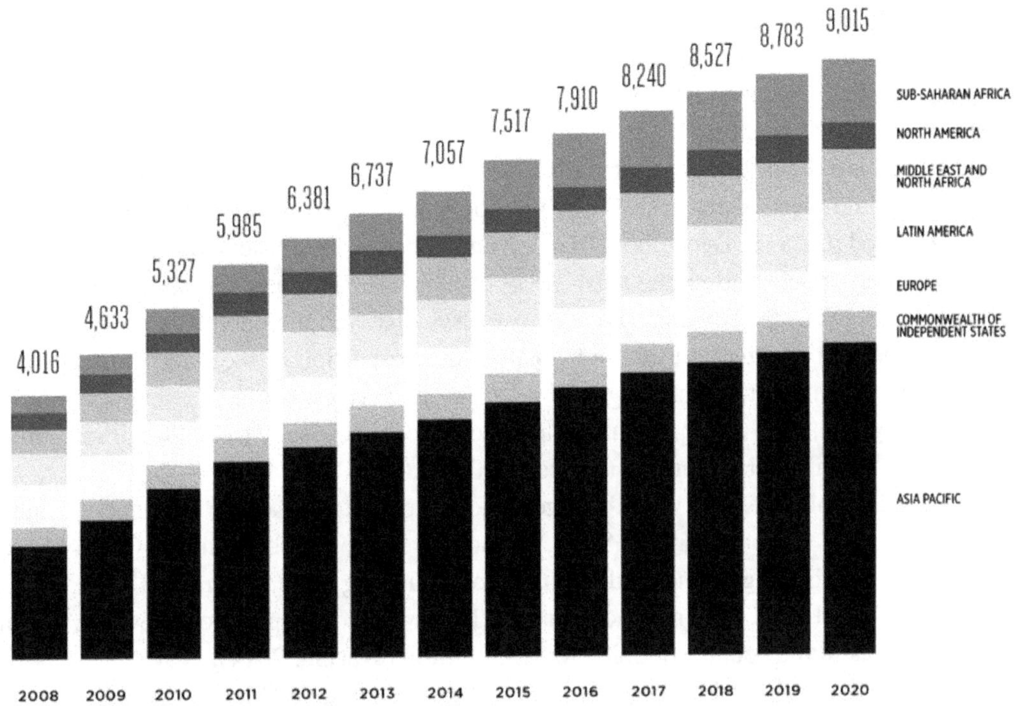

way to become necessities for most network users". Indeed, the shift to digital information is scaling up by several orders of magnitude in data volume every couple of years. Cisco (2014), forecasts global IP traffic for fixed and mobile connections between 2013 and 2018: 33 per cent of all IP traffic originated with non-PC devices in 2013, by 2018, however, that figure shoots up to 57 per cent, reaching an annual run-rate of 1.6 "zettabytes"[8] by 2018. Most of the growth will come from tablets (74 per cent CAGR in traffic, along with smartphones (64 per cent) and M2M modules (84 per cent). By 2018, there will be 8.2 billion handheld or personal mobile-ready devices and 2 billion machine-to-machine connections (e.g., GPS systems in cars, asset tracking systems in shipping and manufacturing sectors, or medical applications making patient records and health status more readily available, et al) (Cisco, 2014, p. 7).

Cisco attributes this impressive growth of mobile data traffic to a combination of trends, among which: the transitioning to smarter mobile devices, a move toward the Internet of Everything, video dominance, and the adoption of IPv6. These parameters allow for device capabilities to be combined with faster, higher bandwidth and more intelligent networks, paving the way to a wide adoption of advanced multimedia applications that contributes, in turn, to increased mobile and Wi-Fi traffic.

According to GSMA (2014b), the developed market average unique subscriber penetration figure is now 79%, which is around the "demographic ceiling" at which subscriber growth tends to stall. Europe with 84% is above this level (Mobile Economy, p. 10). In contrast, unique subscriber penetration on average in developing markets is only 41% this gives room for further growth. As stressed by GSMA (2014a) growth in the mobile internet will come from emerging markets, especially from Asia Pacific. Asian markets have been the main driver behind global growth.

The rapid migration towards 4G-LTE[9] is likely to drive a surge in data usage, with 4G users typically consuming twice as much data per month as other users. In some cases the introduction of 4G has led to an uplift in ARPU, however the impact on revenue varies widely depending on the market. Having reached a 100% population coverage and over half of total connections being 4G at the end of 2013 (compared to around a quarter in both Japan and the US), South Korea is the world's most advanced 4G market (GSMA 2014a, p. 20).

China and India are the fastest growing mobile (data) markets in the world. China was home to 630 million unique mobile subscribers in 2013. In its report on internet development in the country, the CNNIC (2014) stated that as of December 2013, the number of Internet users in China had reached 618 million and the Internet penetration rate was 45.8%, of which the number of mobile Internet users had reached 500 million and kept growing. The report stresses that "mobile phones are still the main driving force for the growth of internet users in China" as the proportion of the Internet users that access through mobile to the Internet is much higher than via other devices. China is going to a rapid shift to 4G (GSMA Intelligence, 2014a, p. 21). China Mobile claimed a 4G user base of 20 million (July 2014), and targets hitting 50 million by year-end (Hanford, 2014a).

In 2014, China Mobile was the world's largest mobile operator based on connections and annual mobile revenue (108.64 billion US$[10]) (GSMA, 2014d). Its domestic competitors, China Unicom and China Telecom, were climbing up the ladder in the global ranking, ranked third and tenth, respectively (Q2, 2014) (GSMA, 2014d). China now accounts for 15 per cent of Apple's total revenue – from 2011 to 2013 Apple's sales in the country doubled to $25.4 billion (Waring, 2014).

In the case of India, 69% of Internet users access via their mobile (McClelland, 2012). India had 405.5 unique mobile subscribers in 2013 (GSMA, 2014a, p. 5). India's Bharti Airtel was ranking four in the top ten mobile operators. Indonesia is the largest smartphone market in South Asia and is estimated to have 37 million Facebook users, second only to the United States, according to web statisticians Socialbakers (Ranjini Mei Hua & al, 2012: 15). As illustrated by Figure 2, the Asia Pacific and North America regions will account for almost two-thirds of global mobile traffic by 2018, with Asia Pacific accounting for over 42% leaving North America and Western Europe far behind.

Figure 2. Global mobile data traffic forecast by region (2013-2018) (ExaBytes (EB) per month)
Source: Cisco VNI (2014, p. 6), (figures in parentheses refer to traffic share in 2018)

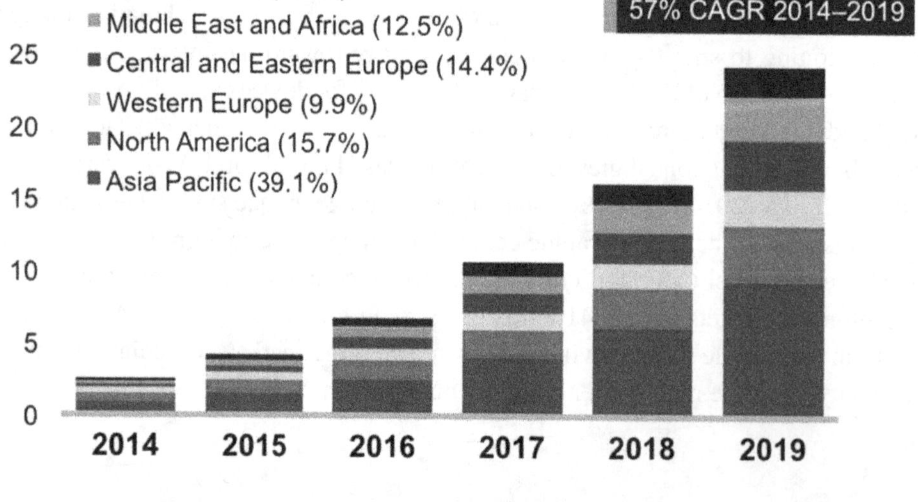

Overall, operators in the developing world generated around half a trillion dollars in revenue in 2012 ($488 billion) contributing 42% of global revenue. In 2012, mobile operators revenues in BRIC[11] countries (over US $250 billion out of US $1.16 trillion in 2012, Wireless Intelligence, 2013a) represented almost 22% of total global mobile revenue (up from 16% four years ago) having already surpassed North America in terms of sales during 2011 (Wireless Intelligence, 2013b). These growth markets have already turned into primary markets.

This impressive growth results from the increasing availability (and affordability) of mobile broadband, the increasing availability (and affordability) of smartphones, and now phablets, as well as the changing role of customers as avid demander –or even co-creators- of mobile content and applications (Feijóo, Maghiros & al, 2009). The result is a new circle between supply (network, devices) and demand led by the latter –the "app" economy.

Within this framework, the growth of traffic is now mainly driven by consumers (as opposed to business before) also active as "prosumers"[12] (Toffler, 1980). The growth is first media-led, with video being the driver. Smarter devices and updated networks[13] can accommodate mobile video content with a much higher bit rates than other mobile content types, thereby generating much of the mobile traffic growth through 2018 according to Cisco (2014: 13): a CAGR of 69 percent between 2013 and 2018, the highest growth rate of any mobile application (see Figure 3). In Europe, data from Vodafone showed that in the quarter to September 2013, 75% of the company's data traffic in Europe was already video and browsing (Mobile Economy, p. 18). According to the 2014 report from the China Internet Network Information Center (CCNIC), the number of China's Internet users who watched videos online or downloaded videos on mobile phone had reached 247 million, increasing by 112 million over the end of 2012, up 83.8%, ranking the first in the statistics about the growth of mobile phone application users.

Mobile cloud traffic follows a curve similar to video according to Cisco (2014, p. 14), up to 90% of the traffic in 2018 from 35% in 2013 (Mobile Economy, p. 29). Cloud applications allow mobile users to overcome the memory capacity and processing power limitations of mobile devices; reliance on cloud

Figure 3. Evolution of mobile data traffic, 2013-2018 (EB per month)
Source: Cisco VNI (2014, p. 14), (figures in parentheses refer to traffic share in 2018).

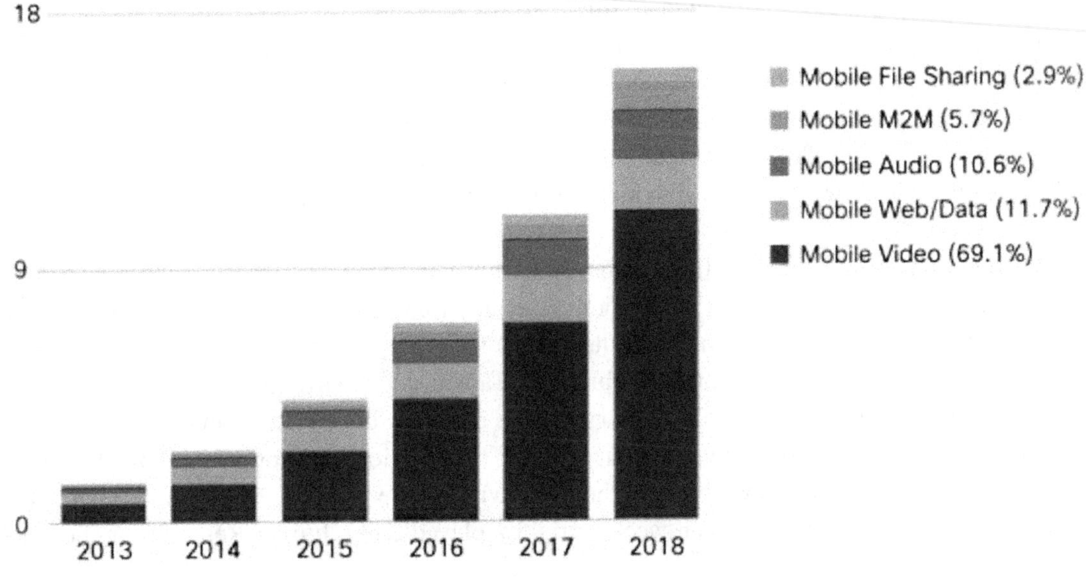

computing increases demand for the quantity of bandwidth as well (Feijóo, 2014). Applications, and especially content applications like games contribute decisively to this growth as they are increasingly played both from mobile devices and from the cloud – online gaming. They will also use increasingly more advanced graphics and controls, going from normal definition to high definition and ultimately 3D graphics, and including augmented reality elements (Feijóo, Gómez-Barroso, Aguado, & Ramos, 2012). In fact, a survey from Information Solutions (2011), with both UK and US online panellists, showed that when asked to identify which gaming-enabled device they played games on most often, 44% cited their phones, ahead of videogame consoles (21%) and computers (30%).

Smartphones, Phablets,[14] and Tablets: The New Hardware and Software Platforms

As part of the "virtuous" circle we alluded to, the growth in mobile broadband connections in many regions of the globe is being driven by rising smartphone penetration. The number of smartphones in use worldwide reached 1.457 billion at the end of 2013 from 237 million as of 2008 (Mobile Economy, p. 5). As of 2020, 81% of phones sold globally are forecasted to be smartphones (2.5 billion) from 26% in 2011 (400 million). According to data from Ericsson (quoted by GSMA, 2014b, p. 16), over 50% of all mobile phones sold in the second quarter of 2013 were smartphones, again half were in Asia Pacific. Going forward, new smartphone connections will largely come from the Asia-Pacific region, with just under 900 million new smartphones expected in the region in the period out to 2017. For the US markets the consultancy ComScore (2013a) indicates 125 million owners of smartphones and over 50 million for tablets in 2013.

Tablets are expected to reach 700 million in 2020, from 70 million in 2011, and 450 million in 2013. Tablets are considered as the "*fastest ramping mobile device in history*" (Morgan Stanley, 2012). A 51.8 per cent year-on-year growth was seen in 2013, Research house IDC predicts a lower but still healthy 12.1 per cent for 2014. The consultancy notes the growth of large-screen smartphones (phablets) (Costello, 2014a). The proportion of the installed base made up by phablets doubled in 2013, with 6 per cent of mobile users now having these devices (Ferguson, 2014a). In 2013, in the UK, tablets accounted for 8.1% of the website traffic, and smartphones 23.4% (UKOM, 2013).

The release of the Apple iPhone in late 2007 played a major role to trigger such a migration while mitigating the expected negative impact of the financial crisis in mobile telecoms, as data growth in mature markets accelerated (West and Mace, 2009). For instance, the appearance of the iPhone dramatically changed the circumstances of mobile gaming from the previous modest version of gaming in feature phones (Feijóo et al, 2012, 2014). According to Gartner, Apple which only sells smartphones, was the third largest player, selling 150.78 million units of its iPhones in 2013 (130.13 million smartphones in 2012), with a global market share of 8.3%.

If SMS and web browsing are still the dominant monthly activity on smartphone, according to Nielsen (mobiThinking, 2014), Apps reach 81% in South Korean 74% in Brazil, and 71% in China[15]. In South Korea around 80 percent of people use mobile screens for video viewing on a weekly basis (Ericson Mobility Report, 2014, p. 13) (see Part 2. Mobile TV) Tablets and smartphones are now being adopted as gaming devices for casual game players, driving the demand for wireless games; and becoming in turn one of the most dynamic segments. The survey from Information Solutions (2011) mentioned above showed that smartphone owners were by far the most avid mobile gamers: 93% of smartphone owners said that they played at least once each week, and 45% played daily. Jupiter Research predicts games sales for tablets up to 3.1 billion US$ in 2014 (SNJV, 2012). Gaming is currently the dominant use for tablets accounting for 67% of the time spent on that terminal (Michaud, 2013).

As a summary, the smartphones and tablets phenomenon allowed novel ways for users to interact with devices and screens, shifting the patterns of use toward a mobile Internet and all types of mobile media. It paved the way for the creation of new content and an array of new applications whose number has skyrocketed as discussed below.

The Rise of the App Economy in a Platform Ecosystem

Smartphones and the mobile internet saw the advent of applications for the consumer. As noted in 2011, by an industry player, Buongiorno (see Box 3), "the development of Apps accelerated the growth of an already rapidly expanding market for smartphones, revitalizing the innovative environment of digital entertainment, and creating a fundamental change in the value chain of mobile content" (Annual report 2011, p. 10). A simple figure displays the absolute success of the mobile applications: within five years (July 2008- October 2013) the number of apps available for download in the Apple platform has grown from 500 to more than 1,000,000 with 60 billion downloads in total. The market research company Vision Mobile (2014a) estimated that there are already over 100 billion downloads a year. App usage is fueling the growth of mobile, with over 50%, in the US, of the total digital media engagement being app, 60% of the digital media time spent taking place on mobile (ComScore, 2014).

App stores have changed –or rather disrupted- the value chain of software (distribution and pricing), and also the industries of music, books, and games in particular. The mobile app market is expected to be worth $27 billion in 2013 from revenue generated directly from apps according to Appnation, (Appnation, 2013). More broadly, by far the fastest growth rates in the global mobile ecosystem will come from the areas of applications, content and advertising, where annual revenue growth is expected to be 16% and to total US$ 576 billion by 2020 out of the predicted 2,897 trillion US $(GSMA, 2014b, p. 32).

Along these lines, the first '*State of the App Economy Report*' (Appnation, 2013) predicts that the largest contributor to this growth will be the app-enabled commerce, with revenue from downloads, in-app advertising and virtual goods complementing this (see Figure 4). The total size of the US 'app economy' is expected to hit US$ 151 billion by 2017, more than double the US$ 72 billion forecast for 2013 (Figure 4). Asia is "*the most lucrative app market in the world*" (Distimo quoted by Costello, 2014b), with 41% of total revenue generated globally in December 2013 coming from this region. In comparison, North America generated 31% of the total[16], while Europe accounted for 23%.

The blend of revenues coming from apps is also evolving. According to Vision Mobile (2014b), there is a chasm between former revenue models, such as pay-per-download and in-app advertising, and of the new, such as e-commerce and affiliate programs: "*E-commerce is the most lucrative revenue model at $2,750 in revenues on average per app / month, and also the most profitable; 52% of developers using e-Commerce revenue models report comfortable profits*". The value is progressively shifting from handset to apps, from respectively 82% and 18% as of 2012 to a 67%/ 33% split (half of handset market) (Vision Mobile 2014a).

App usage increased 115 per cent year-on-year in 2013, according to analytics firm Flurry (quoted by Ferguson, 2014a) with the success of messaging apps (including social and photo sharing titles) having a particularly significant impact. As noted by the same consultancy, some messaging apps, such as LINE[17] and KakaoTalk, (Box 1) started to evolve into platforms (see Box 1), rather than standalone apps with the inclusion of games and sticker stores. As of March 2014, LINE claimed 100 million games

Figure 4. The US app economy (2011-2017)
Source: Appnation, 2013.

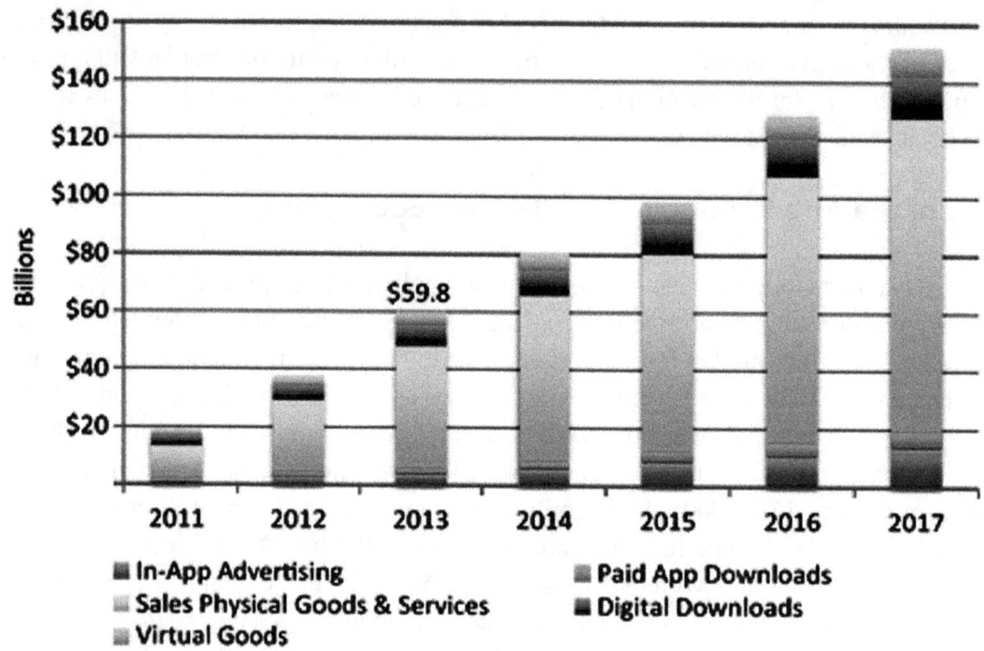

downloads via its messaging app. The strategic role of messaging apps came to the fore over the last months with the wave of exits and investments in the messaging app arena, with Facebook acquiring Whatsapp[18] for $19 billion and Rakuten acquiring Viber for $900 million[19]. Chinese Ecommerce giant Alibaba also made a $215 million investment in Tango[20] (Wee, 2014a).

In the US a sample[21] of 32 video news organizations studied by Pew Research reveals that the vast majority of these entities embraced mobile: fully 26 of the 32 stations had an app available in the iTunes Store, and 25 had an app available in the Google Play Store (News-Video-on-the-Web, 2014, p. 21). On smartphones, the 2014 Reuters Institute Digital News Report[22] finds that news apps, rather than mobile websites, are often the main way of accessing news: in the UK, almost 50% of smartphone users say they mainly use apps (Levy and Newman, 2014, p. 10). By the same token, F.Filloux (2014) states that by the end of 2014, many news media will collect around 50% of their page views via mobile devices. Filloux stresses that native apps lead the game at the expense of web apps and mobile, and claims that on mobile, news is now growing faster than game or social media.

Table 1. Rovio, a global company

Headquarters	Games Studio	Games Studio	Branch Office	Branch Office (Consumer Products)	Branch Office	Branch Office	Branch Office
Espoo Finland	Tampere, Finland	Stockholm, Sweden	Shanghai, China	Santa Monica, CA, USA	Seoul, South Korea	Tokyo, Japan	London, UK

Source: Rovio.com

Box 1. A global South Korean firm, KakaoTalk

KakaoTalk is a multi-platform texting app that allows iPhone, Android,, Bada OS, BlackBerry, Windows Phone, Nokia Asha and Personal Computer users free mobile, instant messaging application for smartphones with free text and free call features. The firm provides services in the arena of digital content, business marketing, commerce, and game platform optimized for mobile devices.

Launched on March 18, 2010, the app is currently available on iOS, Android. KakaoTalk has 140 million users and is available in 15 languages, with over 426 games on its games platform contributed by 217 game developers and partners. The app is also used by 93% of smartphone owners in South Korea where the company is headquartered. When the first smartphone was introduced in Korea in 2009 and eventually went mainstream, the founding team, J.B.Lee and B.Kim, saw an opportunity to build a world-class smartphone application.

In 2013, Kakao Corp. generated revenue of $203 million, netting $59 million in profits. 84 percent of the revenue was generated from commission through games, commerce, and digital content. The remaining revenue was generated from ads. Kakao now employs almost 700 employees. Total registered user count on KakaoTalk is at 140 million.

Kakao describes itself now as a mobile social platform. The company describes it mission in the following fashion: "to create and maintain a sound mobile ecosystem in which all our stakeholders including smartphone users, app developers and content providers can grow together". The South Korean company set up the ambitious goal of "thriving together with one million partners on our platform". In June 2014, the South Korean start-up launched Kakao Lab on the Android version of its popular messaging service, users will be able to access Kakao Lab through the Other menu in their KakaoTalk Settings and take part in testing out the new features designed by Kakao's developers.

Kakao Lab is another example of Kakao's continued dedication to providing the most enjoyable and user-friendly mobile experience possible. The features available in Kakao Lab will be fine-tuned based on user response

In August 2014, the company was named a 2015 Technology Pioneer by the World Economic Forum, making it the first Korean company to be recognized since 2001. Previous Technology Pioneers of the World Economic Forum include Google (2001), Mozilla (2007), Twitter (2010), Spotify (2011), Dropbox (2012) and AirBnB (2013).

Source: Company website, Wee (2014a, b), Internet sources.

Nevertheless, the top grossing iOS and Android apps generally came from games. In September 2012, games accounted for 64% of the combined total, a year later; this had increased to 77%. The iOS market (iPad, iPhone and iPod Touch) generates 89% of its revenues from mobile gaming[23] (Newzoo, 2012 quoted by SNJV, 2012). The Gaming category saw a 66 per cent growth during 2013, games being the most downloaded apps across the iPhone (33%), iPad (48%), from Google Play (37%) and the Amazon Appstore (a remarkable 63%) in 2013, according to a study from Distimo (Costello, 2014). App Store dominates in mobile game sales, but Google Play is catching up fast.

Besides, a host of Android-based tablet rivals have also expanded the market considerably, including Amazon's Kindle Fire and its successors, Google's Nexus line and Samsung Galaxy devices; many of them more affordable than Apple products. Thus, the installed base of mobile and portable devices used for gaming will grow from nearly 800 million in 2012 to more than 1.2 billion by 2014, with Android-powered devices seeing the strongest growth in market share at the expense of gaming-optimised hand-helds such as the Nintendo DS and Sony PlayStation Vita. According to IDC and App Annie (Ferguson, 2013), revenue in 2013 is on track to top US$ 12 billion, with in-app purchases generating the lion's share (51%), followed by paid app sales (44%) and advertising (5%).

The Next Wave: "Big Data"?[24]

The fast deployment of the new networks (4G, 5G) is key for the continuing growth of applications, it unlocks to the development of all kind of connected devices, including connected cars. The combination of the devices and the relevant networks paves the way for a fastest growth of the "Internet of Things" (OiT) that will generate huge amount of data. The data explosion is driven by the combination of exponentially expanding amount of data available (up to 7 "zettabytes" predicted for 2015), and the rapidly improving ability to process and analyse the data (Boston Consulting Group, 2012, p. 7).

Bain (2013a) describes big data, as "the mining and processing of petabytes[25] worth of information to gain insights into customer behavior, supply chain efficiency and many other aspects of business performance" (Bain2014a, p. 1). Mc Kinsey (2011, p. 1) states that "big data" in many sectors will range from a few dozen terabytes to multiple petabytes (thousands of terabytes). New tools to deal with the data (extract, load, and transform) are emerging[26].

The global market for "big data" was estimated at 6.3 billion US $ in 2012 and is expected to reach US $ 48.3 billion by 2018 (a CAGR of 40.5% from 2012 on to 2018) (Transparent Market Research, 2013). The basic components of "big data" include software, hardware, and storage, with software and service being the larger share. According to another specialised consultancy, IDC, the "big data" market will grow to $32.4 billion by 2020 (with an astonishing 212 billion devices connected up to the Internet of Things) (quoted by EMC2014). Atelier Parisbas holds that the "big data" phenomenon can be understood as the growing power of data in the economy of culture.

IT players, labelled "the digital dragons" by Atelier Parisbas (2013, p.78), beyond their power position, are intrinsically well placed to benefit from this shift, to make the most of new analytics, of "big data" and the cloud (Amazon is a leading provider of third party cloud computing services with AWS). This shift brings along some marketing innovation. New digital players also pioneer the use of data mining for compiling viewers' recommendations (Amazon, Netflix, Pandora, Zynga…) (Simon, 2015). Netflix produced its hit show, "House of Cards", after analysing the data from their consumer base (million "plays" per day, million searches, plus tags and other metadata) (Carr, 2013).

Platforms are poised to become an increasingly important point of aggregation in the advertising market with unique reach, customer data and measurability. By combining their traditional expertise with new techniques including data analysis and viral marketing, publishers/ content aggregators can retain their valuable position as curators in the digital space and play a key role in ensuring quality content finds its audience. Big data will be used online and with mobile to power the content recommendation engine. Content providers are setting up entities so supervise the analytics. For instance, videogames company Zynga relies on business unit organization, Google as well (Bain, 2013a). Amazon; LinkedIn created centres of excellence and Netflix fully centralised this activity. Zynga pioneered the use of "big data" by intimately linking game design and business models (Atelier Paribas, 2013, p. 81). Samsung uses "big data" to power the content recommendation engine on its newest smart TVs.

Besides, the pervasive integration of semiconductors, mobile communication, and "big data"/analytics may as well propel the Internet of Things (IoT) into the wider economy. However the next wave may still require some time for deployment, a 2012 survey from SAS (SAS, 2013, p. 1): "…found that 12 percent of organizations are currently implementing or executing a big data strategy". The "digital dragons" were arguably, built around big data from the beginning (Davenport and Dyché), they will act as catalysts within the mobile ecosystem.

However, "big data" raises concerns about the protection of privacy and other values as illustrated by a recent report for the second Obama administration (Podesta Privacy Report, 2014). At the same time, as underlined by R.Samarajiva big data and data analytics may have the potential of to improve the delivery of public services and thereby contribute to e-inclusion, specifically in the case of the developing-countries (Samarajiva, 2014, p. 2[27]). Indeed, bringing services to the bottom of the pyramid (Minges, 2013, p. 135, for India see Malik, 2013, p.30), the mobile revolution is changing the lives of millions of people in the developing world (UNCTAD, 2010, p. 26).

Figure 5. The distribution of app sessions by device[33]
Source: Flurry Analytics quoted by Ferguson (2014).

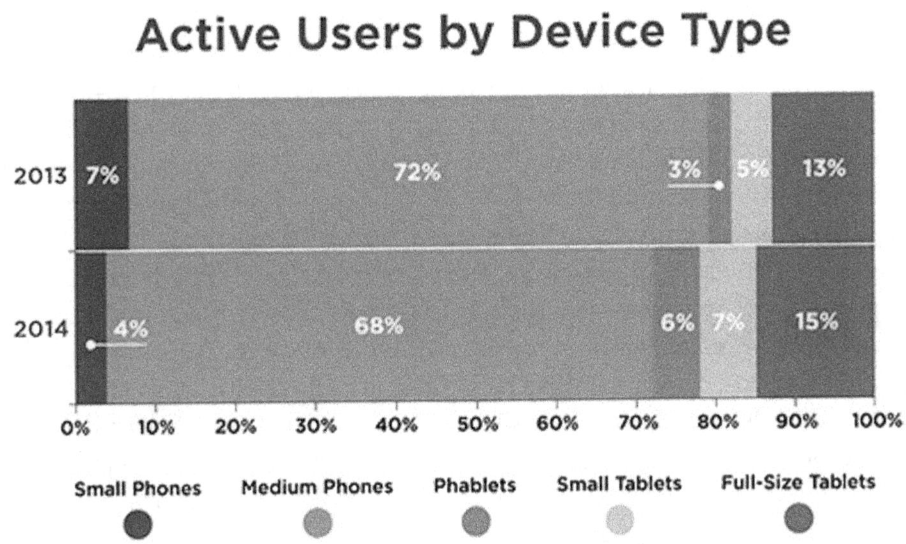

Moving to a Multiscreen/Multichannel World: A New Allocation of Screens for Media Consumption

Now, a variety of screens – long and wide and square, large and small, composed of grains, composed of pixels – compete for the attention of the customers. Media and content activities are moving toward a five screens world (Simon, 2012, 2014): TV, PC, game consoles, connected TVs and mobile devices (smartphones, phablets, and tablets see Figure 5 for the distribution of app sessions by device), an increasing array of digital media are now vying for the attention of consumers.

According to a 2012 Google study[28] for the US market: "Today 90% of our media consumption occurs in front of a screen", with an average of 4.4 hours per day (Google, 2012, p. 1, 9), that leaves a mere 10% for non-screen-based (radio, newspapers, and magazines). Adding, "this multi-screen behaviour is quickly becoming the norm", as consumers split and balance their time between the devices. The screens are used prevalently sequentially (mobile in the morning and commuting, PC at work, tablets peaking at night) (ComScore, 2013), but more and more often simultaneously (multitasking, complementary activities) (Nielsen, 2013, Pew, 2013) and the content industry starts to approach this scenario with a cross-media approach. The Google study claims that smartphones are the backbone of our daily media interactions with the highest number of user interactions per day (38%), noting that 60% are taking place at home. Tablets stand for 9%.

If TV, as put forward by Google, no longer commands full attention because of multitasking, it is a major catalyst for search (Google 2012: 30, 31). Besides, Ofcom's "Communications Market Report 2013" reveals that people are still coming together to watch TV in the living room[29]: 91% of UK adults view TV on the main set each week, up from 88% in 2002 (Ofcom 2014), while watching more TV than before, "people are streaming videos, firing off instant messages and updating their social media status" (Ofcom 2014). The UK regulators stresses as well that "these activities are mostly carried out using smartphones, with half of adults (51%) now owning these devices, almost double the proportion two years ago (27%)".

New entrants, especially social networks, are the engines of these changes toward multitasking, multi-screening behaviours. The Pew Internet "State of the News" 2014 report finds that half of Facebook and Twitter users get news on those sites as do 62% of reddit[30] users (Matsa and Mitchell, 2014), also stressing that news consumption varies widely across the sites. Still in the US, as of December 2012[31], more than three-quarters of adults owned a laptop or desktop computers and about 45% of adults owned a smartphone (Pew, 2013); and the number of adults tablet owners grew by about 50% since the summer of 2011, up to 31% (Pew, 2013). 62% of smartphone owners said they consume news on their device weekly, and 64% of tablet owners. (Pew 2013). In this increasingly multichannel scenario, 25% of US consumers play digital games every day, which makes it one of the most common daily activities, 29% read a newspaper and 14% read a magazine every day (Ericsson, 2013). Pew Internet (2014) finds still with the US case, that smartphone owners were more than twice as likely to watch online videos and online news videos as non-smartphone owners (News-Video-on-the-Web, 2014, p. 5). 88% of smartphone owners watch online videos, compared with 35% of those who do not have a smartphone.

According to a 2014 Reuters Report[32], stressing a rapid growth in both mobile and tablet use for news, the mobile applications are gaining pre-eminence for news over browser consumption: 47% of smartphone users say they use mainly apps for news, rising by 6 percentage in just one year, 38% of the news consumption is made via a browser (Levy and Newman, 2014, p. 8). Their sample indicates that almost four in ten (39%) of those who use digital devices for news now use more than two to access the news, and 12% are using more than three devices. The report confirms the pivotal role of smartphones as the main way of accessing online news: "on average a fifth (20%) of all users, with 10% using the tablet as their primary access point" (Levy and Newman, 2014, p. 9). The distribution of films on devices such as a mobile once seen with scepticism, "is increasingly the place where young people in particular are choosing to watch film" (M. Gubbins, 2014, p. 75) (See Figure 5).

Legacy players are also offering additional services on mobile. For instance, broadcasting organizations are distribution contents on the mobile channel, on top of to their on-line offers. Very often, they developed Apps on mobile as companion services as part of their so-called second screen strategy, like for instance BsB with Sky Player. But other broadcasters like the BBC with its celebrated iPlayer have already commissioned specific content for mobile as it becomes part of the overall BBC strategy to make the iPlayer easily accessible from all four screens, TV, mobile, tablet and computer. Its other UK competitor, Channel 4, claims that mobiles, and tablets together with connected TVs are its fastest growing platforms, accounting for more than two thirds of viewing to 4oD[34] in the first quarter of 2014 (Mc Donald, 2014). Hulu the US on-line channel of content companies did not have a mobile offering until three years ago, but now it represents 20 per cent of total usage (Hanford 2014b). According to Hanford (2014b), total usage is currently running at 500 million video views per month which conforms Cisco's predictions. The next section investigates some of the challenges for legacy players.

Other examples of the proliferation of screens, come from the games industry: 25% of all US gamers play on all screens (Warman, 2013a). Gamers a few years ago were mostly using console/TV/PC, but now they can allocate their gaming time across two additional screens in combination with the previously existing ones. This multiplication of screens at the same time enables ubiquitous or cross-media gaming (Michaud, 2013) across platforms. Therefore, it its highly significant of this new multiscreen world, that the specialised consultancy Newzoo proposed a new screen segmentation to better track the changes in the games markets: entertainment screen, computer screen, floating screen[35], and personal screen. As of the first semester of 2015, the distribution of revenues was the following: entertainment screen 45%, computer screen 24%, personal screen 16%, and floating screen 15% (Warman, 2015).

2. CHANGING PATTERNS OF PRODUCTION AND CREATION

In this second part we draw the consequences of these trends for the creation and production of mobile contents, first drawing the landscape about the role of users and their evolving relationships with content and service providers, and then concentrating on "portable movies" and video games (De Prato, 2014, De Prato & al, 2014). However it is to be noted, that if the previous section may look optimistic because of the growth envisaged, these evolutions are bringing disruptions to well established business models, and novel ways to monetize content are still a question mark for most players. We will come back to this point in the concluding part.

Mobile TV: Expectations Never Fulfilled, so Far?

After the optimistic section emphasizing growth, this section does open not with a success story but with a failure one. On the back of its ability to stream the content from the television on the move, the global mobile TV market was expected a significant growth over the last decade. Around 2007-8, newspapers gave a broad coverage of the then new mobile TV phenomenon[36], with much time devoted to the topic at conferences and events. However, the hype petered out as the spread turned out to be rather limited in scope, customers did not follow.

Mobile TV benefited from a strong and fastened standardisation support since 2007 (DVB-H), especially in the EU,[37] but has been since in a "*limbo*". It was expected that mobile TV could potentially be a market of up to 20 billion Euros by 2011, reaching 500 million customers worldwide[38]. Despite of attempting using major events (Olympics, Euro football 2008 in Switzerland) for commercial launches, the output remains modest with a limited response from EU markets (63 mobile TV platforms were available in the EU serving around 5 million users in 2010). Some modest success stories are to be found (Japan and South Korea[39]) but also notorious failures (BBC 2008, BT after less than a year in 2006). Nevertheless, in the wake of this early deployment some start-ups like Buongiorno (see Box 3) managed to devise a business model moving from mobile TV to apps. In a way, the App disruption came at the right moment to boost the content market.

Box 2. The Italian 2007 test bed

Mobile TV was launched in Italy, in June 2006, it reached 1.2 million mobile users in 2007: 60% of users view an average of 66 minutes per day of mobile TV programmes. It has been described DVB-H as the first large scale experiment of business integration between broadcasters and telcoes, although some more critical observers noted Italy started because telcoes were able to buy broadcasters. Nevertheless, it was perceived as a win-win for broadcasters and mobile operators.
Indeed, from a regulatory viewpoint, the Italian early start was made possible because authorised players were already holding authorisations for content distribution (DTT, cable, satellite) or a TV network operation for mobile TV networks.
Two models were to be found:

- A total vertical integration: H3G produces its own channel (la 3, la 3 Sports) and offers 12 channels(up to 20 in the medium term) as well as access to premium events. Two pricing structures: pre-paid from 3 euros per day to 59 for 6 months, an all included flat fee for 29/49 euros under a 2 years contracts (handset included),
- Partnerships between broadcasters and mobile operators: Tim and Vodafone with Mediaset. Vodafone has also a strong partnership with Sky. The MNO bills the customers and pays the broadcaster. Mediaset acts as a carrier. Mediaset channels are made available to all MNOs. TIM charges 5 euros per month for a subscription to 8 channels and football events. Content premium is charge at a price similar to the competing DTT channels. Vodafone under a strong agreement with Sky offers 9 channels pour a 9.90 euros monthly fee and started operating in December 2006.

Source: Authors based on the Mobile TV World Conference, 2007.

Box 3. A successful pioneer: Buongiorno

Buongiorno is one of the oldest mobile content companies around, developing and managing paid apps and services. Mauro Del Rio founded Buongiorno in Parma, Italy in 1999. The Italian successful start-up prides itself of having morph from the plain circulation of "good morning" SMS to a party of 20 friends into a "mobile content behemoth". In 1999, the company had revenues of 9 million euros and 59 employees, up to 429 in 2005, and to a turnover of 239 million in 2011. It claims it became profitable in 2005.

The company was acquired (over 90%) by the leading Japanese Mobile Operator NTT Docomo in 2012, for an amount estimated between 200 and 300 million euros through its Germany-based subsidiary, Docomo Deutschland. At that time the workforce reached 700 and the company boasted two billion customers in 57 countries across four continents.

Its services include a sprawling list of direct-to-consumer offerings and those it creates in partnership with carriers (connections with more than 130 telephone carriers) and others — cover gaming, music, casual content like wallpapers and ringtones, and mobile payments. The business activity of the Buongiorno Group is subdivided into the following segments: the core activity of the Group - value added services for mobile telephone users (B2C), services for tele-phone carriers and relationship marketing services for businesses (B2B), and New Biz & Shared Services, which refers to services such as Winga, Mobile-Payments (Cashlog), and Connected Devices. Buongiorno leaves telecom operators be in control of the billing. In 2011, the mobile entertainment market was effectively described by the word application (App) after the advent of the iPhone and various App Stores.

The history available on the company website shows that the company grew by mergers and acquisition: music store Vitaminic in 2003, 2004 acquisition of Gsmbox (adding mobile2TV voting capabilities); 2005 purchase of France's Freever, in 2007 it acquired Silicon Valley's Rocket Mobile, followed by HotSMS and FlyTXT (mobile marketing and loyalty), and acquired iTouch that same year. Their history shows at the same time that the company had been revisiting frequently its strategy. In 2007, Buongiorno was already addressing a target audience of 1 billion users in the EU, the US and Asia. They also draw in the evolution, from traditional broadcasting to PC with some form of content creation and mobile (content creation and PtoP). In the US they offered service first for Sprint, then Cingular and moved to South America. In 2011, Buongiorno refocused its business, concentrating on the B2C segment spinning off its B2B operations. Earlier, the company grasped the opportunity brought by the advent of the iPhone and various App Stores and the shift of the mobile entertainment market toward apps. The company became a leading app company.

Source: Company website, last consulted September 2014, Annual report 2011, Mobile TV World Conference, 2007.

This "*impasse*" situation stems from divergent technological options (broadcasting vs mobile communications[40]). Even if the EC was pushing for standards, for the adoption of DVB-H standards, the situation was a mess of competing standards globally, coming from different horizons and regions: broadcast (DAB, DVB families), mobile (OMA Bcast, UMTS, MediaFLO[41], Wimax[42]). DVB-H, DMB and MediaFLO all shared some similar flaws and were doomed to follow similar paths: technically they worked as mobile broadcast solutions using costly dedicated spectrum, by means of separate infrastructure that meant costly buildouts in order to achieve coverage levels that made services a viable proposition (Costello 2014c). The content producer was facing an extended complexity of layers of standards: modulation layer (DVB-H, MediaFLO, ISDB-T…, service layer (EPG, rights management, streaming, download), middleware layer (API) (interactive services, recording, sharing, replay, repurposing, games, trick modes).

It stems also from a lack of perceived value and usefulness from users. The unexpected behaviour of users (at home, after prime-time, long time view…) is another part of the explanation. Business models was an issue. It was also striking, at that time, that most players were looking at mobile TV with a rear glass: applying the business model of the last decade (pay TV). Only a few players user generated content and social networking as a way to bridge the gap between different patterns of use (mobile/ internet/ TV). Again the change came with Apps.

The regulatory regime was unclear in spite of some attempts by the EC[43]: they were pending regulatory issues such as spectrum allocation, digital dividend, privacy, interoperability and shared usage of networks. To sum it up, the main constraints were the following: too many standards, interoperability issues, batteries duration, indoor coverage limit, screen size at time where people are watching TV at home on huge screens or watching at home, costs (not tailored for young people).

The issue is now bouncing back with LTE Broadcast, although most operators remain very cautious, "*learning from mobile TV missteps*", taking into account that customers are not necessarily keen to pay for broadcast services (Costello, 2014c). South Korean carrier, SK Telecom introduced already additional LTE data services for its 4G network[44]. Operators like Telstra in Australia, Vodafone Group in Germany, Orange in France and Verizon in the US, are using pilots, backed by manufacturers like Ericsson, Huawei, and Qualcomm. They now point that "*the economics of LTE Broadcast are very different to its earlier mobile broadcast brethren*" (Costello, 2014c). Business models and revenues from advertising are still questions marks; as well as capacity on 4G networks. .

Producers/Produsers: A Diversification of Sources

New relationships between producers and consumers are emerging, not only on the marketing side as illustrated by Netflix or Amazon, through innovative ways of using and sharing recommendations. Amazon built part of its marketing strategy on the strength of these recommendations. As stressed by Pew (2014, p. 9) social and mobile developments are doing more than bringing consumers into the process, they are also changing the dynamics of the process itself. In the example of "The Guardian" customers are part of an "ecosystem of networked collaborators" becoming collaborators in their own right (Henley, 2014). The same holds for "The Huffington Post" relying more heavily on its readers.

Pew finds that half (50%) of social network users share or repost news stories, images or videos while nearly as many (46%) discuss news issues or events on social network sites (Pew 2014, p. 9). The research institute adds: "In addition to watching more digital videos, Americans are recording and sharing their own videos". Indeed, the institute reports than, as of July 2013, more than a third (36%) of U.S. adults reported recording videos on their cell-phones.

Besides, the dramatic fall of the costs audio-visual production tools contributes to the blurring of the boarders between professional and amateurs as stressed by Flichy (2010). For instance, Image Metrics, a leading provider of facial animation technology, allows self-representation and personalized content through its proprietary algorithms that make the recognition and real time animation of nuanced expressions and emotion possible with nothing more than a camera on a laptop, tablet or smartphone. "Everyone can now use professional tools - Protools, Ableton Live or simply desktop tools - to create their own music", notes Salmon (2013, p. 14), adding: " Self production remains trickier in the cinema sector, despite the rise of digital cameras for achieving quality films" (Salmon 2013, p. 14). Salmon quotes an ADAMI[45] study that reveals that 60% of the artists in their association made at least one recording via self-production between 2008 and 2010 (Salmon 2013, p. 15). Amazon is providing would-be musicians and writers with tools and even some funds to self-produce.

These evolutions lead to the consumption on the same device or in the same room of a new blend of various types of contents: truly "edited" contents, user generated contents (UGC) ranging from almost professional production to contents provided by "friends", transaction (utilitarian audio-visual like tele-shopping, tele-transaction, interactive advertising…), (Busson & al, 2011). Media consumption is gliding from push to pull. Not only users are posting pictures or circulating the pictures they took through their favourite social networks (Facebook) or specialised provider (Instagram), but they started circulating videos as well.

In the field of music, the DIY (Do it yourself) model (Bacache-Beauvallet & al, 2011; Byrne, 2012) adds another source of production, thereby increasing diversity, and offers some novel avenues for creativity: as David Byrne (singer of the US band "Talkin' Head") noted: "The totally DIY model is certainly not for everyone – but that's the point. Now there's choice"[46].

"Portable Movies"

What happened with music to a certain extent with DIY, then with pictures taken from mobile phones, may as well still grow with videos enabled by the introduction of more and more sophisticated cameras in mobile phones. The iPhones already allow for some kind of motion pictures; manufacturers are introducing new cameras that may not be different from the camera they produced for the professional sectors, at least for the lighter ones. Samsung just launched in April 2014 a "zoom"-branded "camera specialised smartphone"[47]. For instance, Jean-Luc Godard has created his own rudimentary 3D camera for the shoot-out, of two mobile phones fitted with lenses for his last film "Goodbye to Language" (2014).

Before the advent of digital technologies, the split between "amateurs" and professional films was clear cut even if "amateur" filmmaking was a very popular and widely spread hobby (Odin, 1995, 1999). YouTube, before being bought by Google, started with some amateurs' short videos; other sites like DailyMotion in the EU followed.

It does not mean that all the videos produced by users may become global hits, or even that the still to be perceived as amateurs are trying to become professional players. Most likely the vast majority of users producing contents will stick to some of the previous patterns of amateur film-making and just take the opportunity of the new channels of distribution to circulate their contents. However, just like in the case of music, where amateur musicians were also able to act as professionals or semi-professionals occasionally, as illustrated by various kinds of folk music (blues, country, flamenco…), this opens up a new space in between the realm of professionals and of amateurs, eventually allowing the scouting of new talents

Voci (2013) spotted some interesting examples, in a fast growing mobile market such as China, of amateurs circulating short videos shot with their mobile cameras. According to Voci (2014): "And thanks to portable and affordable digital technology, cinema today is also made up of a myriad of amateur videos that contribute creative and often dissonant perspectives on what China is". Voci analysed what she called "portable movies", giving examples of Chinese amateurs' short films: "Portable movies are generally freely circulating (i.e., shared rather than purchased) and almost exclusively targeting transient –online or on-phone– audiences. In particular, mobile-made and viewed movies often emphasize portability, playfulness, connectivity, and sharing as their main defining elements" (Fowler & Voci, 2011). Voci considers it is an attempt to develop an aesthetic of brevity away from a more traditional approach (Voci, 2010, p. 77-78), an innovation in art-form development to use Bakhshi and Throsby classification (2010, p. 16).

The production of these videos can be triggered by an array of factors. For instance, the huge viral success of the South Korean rap artist, "Psy", "Gangnam Style" posted on YouTube in 2012, gave birth to a lot of parodies, some originating without any doubt from professionals (including some "Angry Birds", "Minecraft" style), but other not. Parody on YouTube is a genre that is attracting amateurs and professional alike. YouTube offers a top ten of these musical parodies as well as parodies of films[48] produced by "clever, creative YouTubers" (Lazar, 2014)[49]. As noted by Erickson et al. (2013) parody is a highly significant consumer activity[50], driving consumers to post contents.

New Comers and Innovative Business Models into the Video Game Industry

In the video games industry, several firms actually rely on communities of users in order to design new products and services. Internet technologies are thus used as forums and social networks for providing consumers with the possibility not only to share information but also to actively contribute to design new content and functionalities for improving their products and services (Parmentier and Gandia, 2013). The communities of users are considered as new sources of co-innovation (Parmentier and Mangematin, 2014). Indeed, in the case of the video games industry: actual gamers can simultaneously be not only users but also developers, but "what is new is that the blurring of firm frontiers in digitized industries allows for fluid interactions between user communities and firms" (Parmentier and Mangematin, 2014, p. 40). For instance, from a technological viewpoint, Image Metrics introduced in 2012 "Zombie Wars", an iPad exclusive free digital comic app that lets users insert their images into a tale of a zombie outbreak and become the main characters: a new platform for creating 3D customizable avatars from 2D facial photos.

The fan culture for digital games is deeply embedded in shared practices and experiences among fan communities, and their active consumption contributes to the production: fan-programmers (generally known as "modders") play a role in the success of the PC digital game industry. According to A.Kerr (2014), community managers are becoming a growing category of the work force, at least in the Irish game industry.

The advent of the empowered Internet and mobile technology introduced new comers and disruptive and innovative business models into the video game industry. These new comers and their innovative scenarios are re-shaping the landscape of the traditional industry. Typically, during the 1st semester of 2013, the Chinese Internet service Portal and games company, Tencent[51] became n°1 overtaking Activision Blizzard (ranking 1 in 2012) (Newzoo, 2014), the company was not even included in the top 20 in 2009 (De Prato & al, 2010, p. 41). The development of the mobile Internet supported the expansion of the video game sector, which represents a significant share of sales on the Apple AppStore, e.g. half of all mobile phone's users play and the greatest video game players today are on Facebook and Apple. Finally, one can note that the Asian giants were born in the web video games (e.g. in Japan and China, South Korea to a lesser extent).

On-line and wireless video games have become the second largest product segments of the video games industry, reaching a 44% share (on-line games: 32%; mobile games: 12%) (Idate, 2012). Since 2004, the online and wireless market has grown with remarkable rapidity, driven by the increase in the number of broadband subscribers, the innovation in available games, the transition to handheld devices, and the newest generation consoles. This allowed indeed moving from console games/ packaged products and a handful of global publishers to a self-publishing model. It created opportunities for companies located in smaller markets such as Finland (Rovio), or New Zealand (Sidhe[52]), to reach a global market.

Social networks like Facebook contributed to popularize the games on these platforms (e.g. Zynga's "Farmville") offering simple games, however based on servers allowing the kind of interaction offered by Massive Multiplayer Online Games (MMOGs), fuelling audience and revenue growth in the online games segment.

New channels of distribution (on-line, mobile) have also been a key element of the strategy of the Finish company, the creator of "Angry Birds" so as to set up an innovative business model that frees the company from the overwhelming influence of publishers like EA. Indeed, the founders took the step of making and distributing their own video games on line, and to move later to free to play, another example of a winning strategy. The company is also going global. Table 1 offers a significant the view of the video games industry and its potential global development according to Rovio with the presence of the leading EU games clusters (Nordic clusters, London), and the locations in Asia (China, Japan, and South Korea). Indeed, Rovio opened its Chinese office at the beginning of 2012, and its first "Angry Birds" activity park in Shanghai at the end of that same year. The company invested in new business areas, such as animation and video distribution, ventured into new business models in games, and consolidated its market position in consumer products licensing, thereby "setting its foot strongly into the entertainment business" to quote the company (Rovio Entertainment Report, 2013, p. 1).

CONCLUSION: "TARZAN ECONOMICS"?

The ways to deal with the consumer involvement vary, from just a marketing angle based on smart treatment of data (Amazon, Netflix) to a more thorough implication in the production (video games). The right balance between amateurs (UGC) and professional or semi-professional contents is difficult to achieve, it fluctuate with media and distribution channels. Some pure players like "The Huffington Post" or Buongiorno found ways to design an appropriate business models, but business models remain an issue for most of the players, content providers and telecom operators alike.

The first part noted the shift from web to mobile in the backdrop of fast growing global mobile markets and sectors of the markets. Unfortunately, the revenue stream is far from following mechanically this impressive growth, nor is it easy to foresee this stream in a near future: as stressed by F. Filloud (2014) making users pay for mobile content has proven much more difficult than hoped for. The other element of the usual equation of the legacy business model, advertising, does not appear to be easier to define, so as to come up with modes of monetizing comparable to the web. Little is known of the behavior of consumers to advertising formats, they may trigger user rejection. Consumers may also react negatively to privacy or other regulatory issues that in turn may drive some regulatory intervention.

Besides, there is a tension between the logic of prototype firmly rooted in the media and content industry and the logic of demand that is gaining much momentum. Media consumption is gliding from push to pull as stressed. Again IT companies are better equipped to deal with this demand: we noted that Netflix's productions are highly successful in their distribution system. It means meeting the needs of consumers by creating more options for audiences to view content. IT companies exemplify original means to use technological resources, on a completcly different way from the incumbent media and content players. Instead of merely enhancing the legacy production process of contents, users 'interface or distribution channels, they managed to build their strategy around their activities (live-tweeting, fan video, fan arts, Tumblr ...), just as HBO does to involve the fans of its much noticed series "Game Of Thrones" (Quinton, 2014).

Although, content companies and especially more and more broadcasters are now eagerly embracing the new channels of distribution, most legacy players were initially reluctant to embrace the changes, a reaction often linked to the weight of their legacy business models, even with declining revenues (newspaper). This almost "catch 22" situation has been described as highly uncomfortable "Tarzan economics"[53]for incumbents. Besides, in the case on the book publishing industry, the cinema industry and the music industry, they often held conservative views stating that the Internet should be used to develop the existing business model rather than to radically change it, failing to understand the opportunities opening up. Even the "digital native", the video games industry went through some turbulences in the recent years due to changes in the business models (Le Diberder, 2012), the need for established players to adapt to the changes (Michaud, 2013), and some declining segments (consoles and PCs), the games market remains however in a healthy state with an exceptional growth dynamics forecasted.

By and large, the protective behavior of legacy content players, may have left too much room to the IT players especially in an area where scope matters more and more and where we are going through an upstream domination (Simon, 2012, 2014, Benghozi & al, 2014) under the leadership of huge companies with flabbergasting market caps.

This holds for mobile operators once protected by their "walled garden" model, they are also looking for the appropriate business models to accommodate the shift from paid voice and SMS, on the backdrop of declining revenues[54], toward more innovative and competitive pricing scheme. New data-centric tariff structures are appearing however often precipitating further declines in ARPU while at the same time reducing churn, increased net promoter scores, bringing more stable in-bundle revenue streams (Dewar, 2014).

The digital revolution of cultural industries is largely a revolution of the intermediation in its various components. New players emerge and offer unprecedented solutions for aggregating and distributing content, designing original terms of marketing and transaction adapted to this new framework (free subscriptions, micro-payments). Therefore, the weight of the technology is also the weight of intermediaries and economic actors supporting and supported by these technologies: Google, Amazon, Apple or other ISPs for example. Their importance is incommensurate with that of cultural actors and explains their ability to quickly build a place in the landscape of culture, using technological innovations to impose new economic models which strongly disturb the sectorial equilibria.

As a consequence, technological strategic partnerships cannot be merely apprehended through the traditional perspective of relations with some identified industrial partners, but they must additionally be conceived through the arrangement of a digital business ecosystem. The mobile ecosystem is still looking for workable and sustainable mode of cooperation for the production, distribution, and consumption of contents (edited or self-produced).

The new value networks are not as linear as the legacy value network. The value chain has been drastically altered and getting more complex. This leads the consultancy PwC to claim: "It's not a value chain anymore. It's an ecosystem of networked collaborators who commit to solve a specific part of the industry problem" (quoted by Avid White Paper, 2014, p. 4). Building such a network of collaborators is not an easy task either.

In this global ecosystem IT companies, US and Asian companies are taking the lead. As a final note, the global mobile market and its geography are changing and old and new companies are aiming for the Asian market as a way of counterbalancing more mature markets. Companies like Buongiorno, Rovio, but also Riot Games[55], a US games companies are already acting that way.

REFERENCES

Akamai (2014). *The State of the Internet*. Q1 2014 Report, Vol. 7(1). Retrieved from http://www.akamai.com/stateoftheinternet/

Appnation (2013). State of the App Economy Report. Retrieved from http://appnationconference.com/main/research/

Atelier Paribas. (2013). *Big data, big culture? The Growing Power of the Data and its Outlook for the Economy of Culture*. Retrieved from http://www.forum-avignon.org/sites/default/files/editeur/EtudeATELIER_FA_2013.pdf

Avid (2014). *Avid Everywhere, White Paper, A Vision for the Future of the Media Industry*. Retrieved from www.avid.com/static/resources/common/documents/whitepapers/Avid_Everywhere.pdf

Bacache-Beauvallet, M., Bourreau, M., Moreaux, F., (2011). *Portrait des musiciens à l'heure du numérique*. Paris: Editions rue d'Ulm.

Bain. (2013a). *Big Data: The Organization Challenge*. Retrieved from www.bain.com

Bain. (2013b). *The age of curation: From abundance to discovery*. Report for the Forum of Avignon. Retrieved from www.bain.com

Bakhshi, H.T. (2010). *Culture of Innovation? An economic analysis of innovation in arts and cultural organisations*. Nesta.

Benghozi, P.-J., Salvador, E., & Simon, J.-P. (2015). *R&D and innovation in the cultural-creative sectors: the case of cinema*. Retrieved from http://is.jrc.ec.europa.eu/pages/ISG/EURIPIDIS/documents/JRC95536.pdf

Boston Consulting Group. (2012). *The Value of Our Digital Identity*. Liberty Global Policy Series. Retrieved from https://www.bcgperspectives.com/content/articles/digital_economy_consumer_insight_value_of_our_digital_identity/

Buongiorno Annual report (2011). Retrieved from http://www.buongiorno.com/financial-information

Buongiorno company website. (n. d.). Retrieved from http://www.buongiorno.com

Busson, A., & Landau, O. (2011). La transformation de la chaîne de la valeur de l'audiovisuel. In La richesse des réseaux numériques - Actes du colloque Médias Proceedings of Aix Marseille University conference.

Byrne, D. (2012). *How music works*. Edinburgh, London: Canongate.

Carr, D. (2013). Giving Viewers What They Want. *The New York Times*. Retrieved from http://www.nytimes.com/2013/02/25/business/media/for-house-of-cards-using-big-data-to-guarantee-its-popularity.html?pagewanted=all&_r=0

China Internet Network Information Center (CCNIC). (2014). Statistical Report on Internet Development in China. Retrieved from http://www1.cnnic.cn/AU/MediaC/rdxw/hotnews/201401/t20140117_43849.htm

Cisco Visual Networking Index. (2013, February). *Global Mobile Data Traffic Forecast Update, 2012–2017*. Retrieved from http://www.cisco.com/en/US/solutions/collateral/ns341/ns525/ns537/ns705/ns827/white_paper_c11-520862.pdf

Cisco Visual Networking Index. (2014, February). *Global Mobile Data Traffic Forecast Update, 2013–2018*. Retrieved from http://www.cisco.com/c/en/us/solutions/collateral/service-provider/visual-networking-index-vni/white_paper_c11-520862.html

ComScore. (2013a). *Mobile in Future 2013*. Retrieved from http://fr.slideshare.net/Briancrotty/comscore-mobile-futureinfocusreport2013

ComScore. (2014). *The U.S. Mobile App Report*. Retrieved from https://www.comscore.com/fre/Insights/Presentations-and-Whitepapers/2014/The-US-Mobile-App-Report?&elqCampaignId=631

ComScore Device Essentials. (2013b). Retrieved from http://direct.comscore.com/Default.aspx

Costello, S. (2014a). IDC lowers tablet forecast on phablet growth. Retrieved from http://www.mobileworldlive.com/idc-lowers-tablet-forecast-phablet-growth?utm_campaign=MWL-D-20140604&utm_medium=email&utm_source=Eloqua&elq=0261f07a8f5044f4a5229868728f0b3a&elqCampaignId=1830

Costello, S. (2014b). Asia leading the way for apps. Retrieved from http://www.mobileworldlive.com/asia-leading-way-apps

Costello, S. (2014c). Take 2: LTE Broadcast learns from mobile TV missteps. Retrieved from http://www.mobileworldlive.com/take-2-lte-broadcast-learns-mobile-tv-missteps?utm_campaign=MWL_20140724&utm_medium=email&utm_source=Eloqua&elq=f275168051b04f6d85e33d64c63b79fc&elqCampaignId=2043

Davenport, T. H., & Dyché, J. (2013). *Big Data in Big Companies*. International Institute for Analytics (iianalytics.com). Retrieved from http://www.sas.com/resources/asset/Big-Data-in-Big-Companies.pdf

De Prato, G. (2014). The video games industry. In G. De Prato, E. Sanz, & J. P. Simon (Eds.). *Digital Media Worlds; The new media economy*. Oxford: Palgrave. doi:10.1057/9781137344250.0019

De Prato, G., Feijoo, C., & Simon, J. P. (2014). Innovations in the Video Games Industry: Changing global markets. *Communications & Strategies*, 94, 17–38.

De Prato, G., & Simon, J. P. (2015). "The next wave: Big Data? *Communications & Strategies*, 97, 15–39.

Dewar, C. (2014). Rebalancing the value from voice and SMS to data. Retrieved from http://www.mobileworldlive.com/rebalancing-value-voice-sms-data?utm_campaign=MWL_20140902&utm_medium=email&utm_source=Eloqua&elq=698c8f2d564f402380d40b1fec23acb1&elqCampaignId=2215

Erickson, K., Kretschmer, M., & Dinusha Mendis, D. (2013, January). *Copyright and the Economic Effects of Parody: An Empirical Study of Music Videos on the YouTube Platform and an Assessment of the Regulatory Options Parody and Pastiche*. Study III. Retrieved from http://zenodo.org/record/8374/files/CREATe-Working-Paper-2013-04.pdf

Ericsson Mobility Report. (2013). *On the Pulse of the Networked Society*. Retrieved from www.ericsson.com

Ericsson Mobility Report. (2014). Retrieved from www.ericsson.com/mobility-report

Feijoo, C., Gelabert, J., Lindmark, S., Matía, B., Tarín, C., & Villar, J. P. (2013). Public and Commercial Models of Access in the Digital Era. *European Parliament*. Retrieved from http://www.europarl.europa.eu/studies

Feijóo, C., Maghiros, I., Abadie, F., & Gomez-Barroso, J. L. (2009). Exploring a heterogeneous and fragmented digital ecosystem: Mobile content. *Telematics and Informatics*, *26*(3). 282–292. doi:10.1016/j.tele.2008.11.009

Ferguson, T. (2014). Phablets carving out their own app niche. Retrieved from http://www.mobileworldlive.com/phablets-carving-app-niche-flurry?utm_campaign=MWL-A-20140911&utm_medium=email&utm_source=Eloqua&elq=9ec8057650a74b4099530ab35c7ff091&elqCampaignId=2286

Filloux, F. (2014). Mobile Facts to Keep in Mind. Part 1. *Mondaynote*. Retrieved from http://www.mondaynote.com/2014/07/06/mobile-facts-to-keep-in-mind-part-1/

Flichy, P. (2010). *Le sacre de l'amateur*. Paris, Le Seuil: La République des idées.

Fowler, C., & Voci, P. (2011). *Brief Encounters: Theorizing Screen Attachments Outside the Movie Theatre*. Retrieved from http://www.screeningthepast.com/2011/11/brief-encounters-theorizing-screen-attachments-outside-the-movie-theatre/

Gillet, J. (2014). Measuring mobile penetration. *GSMA Intelligence*. Retrieved from https://gsmaintelligence.com/analysis/2014/5/measuring-mobile-penetration/430/

Gillet, J. (2013). Top-20 Global Mobile Operator Groups by Connections and Revenue. *GSMA Intelligence*. Retrieved from http://www.mobileworldlive.com/top-20-global-mobile-operator-groups-by-connections-and-revenue-q1-2013

Google. (2012, August). The new multiscreen world. Understanding cross-platform consumer behavior. Retrieved from http://www.google.com/think/research-studies/the-new-multi-screen-world-study.html#

Gubbins, M. (2014). *Audience in the mind*, Sampomedia. Report commissioned by Cine regio. Retrieved from www.SampoMedia.com

Handford, R. (2014b). ESPN, Hulu hail mobile video traffic but questions remain over advertising, network capacity. Retrieved from http://www.mobileworldlive.com/espn-hulu-hail-video-first-world-questions-remain-advertising-network-capacity?utm_campaign=MWL_20140910&utm_medium=email&utm_source=Eloqua&elq=1de7cc88861646dbab7a1eaca0392bc5&elqCampaignId=2274

Hanford, R. (2014a). China Mobile highlight radical changes with 5G. Retrieved from http://www.mobileworldlive.com/china-mobile-top-scientist-offers-string-5g-pearls?utm_campaign=MWL_20140923&utm_medium=email&utm_source=Eloqua&elq=c79149105ec54bb28ba1d429fefa4568&elqCampaignId=2360

Henley, J. (2014). L'exemple de The Guardian. In J.,M. Charon, J. Papet, (Eds.). *Le journalisme en questions* (pp 169-171).

Industry associations, trade press, consultancies and research institutes

GSMA Intelligence. (2014a). *The Mobile Economy. Asia Pacific*. Retrieved from www.gsmaintelligence.com

GSMA Intelligence. (2014b). *The Mobile Economy 2014*. Retrieved from www.gsmaintelligence.com

GSMA Intelligence. (2014c). Half a billion Chinese citizens have subscribed to the mobile internet. Retrieved from www.gsmaintelligence.com

GSMA Intelligence. (2014d). Chinese carriers dominate global operator ranking as M&A deals shake up US market. Retrieved from http://www.mobileworldlive.com/chinese-carriers-dominate-global-operator-ranking-ma-deals-shake-us-market?utm_campaign=MWL_20140924&utm_medium=email&utm_source=Eloqua&elq=e31fd99c468d440995cbeebe6fee5f83&elqCampaignId=2366

George, D., Stryjak, J., Meloán, M., Castells, P. (2015, July). The Mobile Economy 2015. *GSMA Intelligence*. Retrieved from www.gsmaintelligence.com

ITU. (2013). ICT Facts and Figures, International telecommunication Union, Geneva, Switzerland. Retrieved from http://www.itu.int/ITU-D/ict/facts/material/ICTFactsFigures2013.pdf

KakaoTalk. (n. d.). Retrieved from www.kakao.com/

Kerr, A. (2014, September 3-5). The Occupational Implications of Digital Games as a Service [Presentation] Proceedings of the Cost Conference, *The Dynamics of Virtual Work: the Transformation of Labour in a digital Economy,* Hatfield.

Le Diberder, A. (2012). Le modèle économique des jeux vidéo, un colosse en péril. In J. P. Lafrance (Ed.). *Les jeux vidéo. Quand jouer c'est communiquer, Hermes (Vol. 62,* pp. 136–143). Paris: CNRS.

Levy, D. A. L., & Newman, N. (Eds.). (2014). *Reuters Institute Digital News Report. Tracking the Future of News*. Retrieved from http://reutersinstitute.politics.ox.ac.uk/publications/risj.html

Malik, P., Vignesvara Ilavarasan, P. (2013). Information and Communications Technology Sector in India. In N. De Prato, & S.P. Simon (Eds), Asia in the Global ICT Innovation Network. Dancing with the tigers (pp. 19-43). Chandos, Whitney Oxforshire.

Matsa, K. E., & Mitchell, A. (2014). 8 Key Takeaways about Social Media and News. *Pew State*. Retrieved from http://www.journalism.org/2014/03/26/8-key-takeaways-about-social-media-and-news/

McClelland, S. (2012). The big picture/The rise of Asia. *Intermedia*, *40*(4). 16–17.

McDonald, A. (2014). Good companions – the second screen and multiscreen monetisation. *Digital TV Europe net*. Retrieved from http://www.digitaltveurope.net/221941/good-companions-the-second-screen-and-multiscreen-monetisation/

McKinsey. (2011). *Big Data: The next frontier for innovation, competition and productivity*. Retrieved from http://www.mckinsey.com/~/media/McKinsey/dotcom/Insights%20and%20pubs/MGI/Research/Technology%20and%20Innovation/Big%20Data/MGI_big_data_full_report.ashx

Michaud, L. (2013, December 4). *Evolution of video game industry ecosystem(s). Business models and trends in the evolving market for video games*. Proceedings of the Evolution of video games industry's ecosystems, innovation and regulation chair/mosaic conference, Paris. Retrieved from http://innovation-regulation2.telecom-paristech.fr/wp-content/uploads/2013/12/2013_GameCo_Evolution-of-business-models_Light.pdf

Minges, M. (2013). *Key Trends in the Development of the Mobile Sector.* Part 2. Statistical Appendix. Pp.115-134. World Bank. Retrieved from http://siteresources.worldbank.org/EXTINFORMATIONANDCOMMUNICATIONANDTECHNOLOGIES/Resources/IC4D_Part2_.pdf?utm_source=Research&utm_campaign=69d2fb0ede-snapshot-2014-05-22&utm_medium=email&utm_term=0_1791988daa-69d2fb0ede-294293285

MobiThinking. (2014). Retrieved from http://mobithinking.com/mobile-marketing-tools/latest-mobile-stats/e#smartphoneactivities

Morgan Stanley. (2012). *Tablet Landscape Evolution. Window(s) of Opportunities.* Retrieved from http://www.morganstanley.com/views/perspectives/index.html

News-Video-on-the-Web. (2014). *Pew.* Retrieved from http://www.journalism.org/files/2014/03/News-Video-on-the-Web.pdf

Odin, R. (1995). *Le film de famille: usage privé, usage public.* Paris: Éditions Méridiens Klincksieck.

Odin, R. (1999). *Le cinéma en amateur », Communications, n° 68.* Seuil.

Ofcom (2014). *Ofcom's Communications Market Report 2013.* Retrieved from http://stakeholders.ofcom.org.uk/market-data-research/market-data/communications-market-reports/cmr13/uk/

Podesta,J., & al, (2014, May). Big Data: Seizing Opportunities, Preserving Values. *Executive Office of the President.* Retrieved from http://www.whitehouse.gov/blog/2014/05/01/findings-big-data-and-privacy-working-group-review

Quinton, F. (2014). Game of Thrones, le transmédia couronné de succès. *INA Global.* Retrieved from t: www.inaglobal.fr/television/article/game-thrones-le-transmedia-couronne-de-succes-7542?tq=8

Ranjini Mei Hua, S., & Melissa, E. (2012). *Investigating the potential of mobile phones for e-governance in Indonesia.* Paper presented at the CPRAfrica2012/CPR7 conference. Retrieved from www.cprsouth.org

Reports consumer 2012. (2013). *Nielsen.* Retrieved from http://www.nielsen.com/libs/cq/core/content/login.html?resource=%2Fcontent%2Fdam%2Fcorporate%2Fus%2Fen%2Freports-download%2F2013&$$login$$=%24%24login%24%24

Riot Games. (n. d.). Retrieved from http://www.riotgames.com/about

RNCOS. (2011). *Global Mobile TV Forecast to 2013.* Retrieved from http://www.rncos.com/Report/IM194.htm

Rovio Entertainment. (2014). *Annual Report 2013.* Retrieved from www.rovio.com

Salmon, K., (2012). The manufacturing of innovation Management & creation, perspectives for economic growth. *Forum d'Avignon.* Retrieved from avignon.org.

Salmon, K. (2013). Creators, producers, distributors, consumers, public authorities...Who really has overall control? *Forum d'Avignon.* Retrieved from avignon.org.

Samarajiva, R. (2014). The potential of "big data" for improving public-service delivery in developing countries. Retrieved from http://lirneasia.net/search/

2013 Big Data Survey Research Brief. (2013). *SAS*. Retrieved from http://www.sas.com/resources/whitepaper/wp_58466.pdf

Sidhe. (n. d.). Retrieved from www.sidhe.co.nz

Silver, J. (2013, November 8). Panel 2. Stake-holders' view of evidence for copyright policy. In M. Kretschmer, & R. Towse (Eds). Proceedings of Digital Conference, Bournemouth University (pp. 58-62). Retrieved from www.copyrightevidence.org/create/esrc_evidence_symposium

Simon, J. P. (2012). *The Dynamics of the Media and Content Industries: A Synthesis*. Retrieved from http://is.jrc.ec.europa.eu/pages/ISG/MCI.html

Simon, J. P. (2014). Media in the changing media-IT- telecom ecosystem. In G. De Prato, E. Sanz, & J. P. Simon (Eds.). *Digital Media Worlds; The new media economy*. Oxford: Palgrave. doi:10.1057/9781137344250.0011

Simon, J. P. (2015). Le Big Data: un enjeu pour les industries creatives. *INA Global*. Retrieved from http://www.inaglobal.fr/numerique/article/le-big-data-un-enjeu-pour-les-industries-creatives-8065

PCs in Decline: TV Screen Takes Over As Americans' First Choice For Internet TV and Movies. (2013). *Strategy Analytics blog*. Retrieved from http://blogs.strategyanalytics.com/CM/post/2013/03/22/PCs-In-Decline-TV-Screen-Takes-Over-As-Americans'-First-Choice-For-Internet-TV-and-Movies.aspx

Syndicat National du Jeu Vidéo (SNJV). (2012). *Les chiffres des marchés du jeu vidéo dans le monde et en France*. Retrieved from http://www.snjv.org/fr/industrie-francaise-jeu-video/

Tencent website. (n. d.). Retrieved from www.tencent.com/en-us/at/abouttencent.shtml

The State of the News Media 2013. (2013). *Pew*. Retrieved from //stateofthemedia.org/2013/

Toffler, A. (1980). *The Third Wave: Democratization in the Late Twentieth Century*. New York: Bantam Book.

Transparency Market Research. (2013). *Big Data Market - Global Scenario, Trends, Industry Analysis, Size, Share And Forecast 2012 – 2018*. Retrieved from http://www.slideshare.net/wayne31jan/big-data-market-38557173

Trends In U.S. Video Gaming – The Rise Of Cross-Platform. (2012). *Nielsen*. Retrieved from http://www.gamesindustryblog.com/2012/03/trends-in-u-s-video-gaming-the-rise-of-cross-platform/

UK Online Measurement Company (UKOM). Tablets now account for 8% of website traffic: UKOM and comScore announce early data release date. Retrieved from http://www.ukom.uk.net/news/tablets-now-account-for-8-of-website-traffic-ukom-and-comscore-announce-ear

UNCTAD. (2010). *Creative Economy Report 2010 – Creative Economy: A Feasible Development Option*. Geneva. Retrieved from http://unctad.org/en/Docs/ditctab20103_en.pdf

Virgin Mobile. (2014). Custom: make it yours. Retrieved from http://www.virginmobileusa.com/custom/#/

Vision Mobile. (2014a). *App Economy Forecasts 2013-2016*. Retrieved from http://www.visionmobile.com/product/app-economy-forecasts-2013-2016/

Vision Mobile. (2014b). *App Profits and Costs. The ingredients of a successful app business and how to replicate them*. Retrieved from http://www.visionmobile.com/product/app-economy-profits/

Voci, P. (2010). *China on video: Smaller-screen realities, Routledge Studies in Asia's Transformations*. London, New York: Routledge.

Waring, J. (2014). China to continue to drive iPhone sales – despite subsidy cuts. Retrieved from http://www.mobileworldlive.com/china-continue-drive-iphone-sales-despite-subsidy-cuts?utm_campaign=MWL_20140820&utm_medium=email&utm_source=Eloqua&elq=c995dcfc0f634901998b4f6b791c4791&elqCampaignId=2149

Warman, P. (2013b). Mobile Gaming. A Global View on Impact and Opportunities. Retrieved from http://fr.slideshare.net/Markies/newzoo-mobile-first4junev1

Wee, W. (2014a). KakaoTalk reveals it has 140 million users, made $203 million in revenue last year". *Technostorm*. Retrieved from https://sg.news.yahoo.com/kakaotalk-reveals-140-million-users-060150446.html

Wee, W. (2014b). How KakaoTalk Grew From a Startup to a Giant Mobile Platform. Retrieved from *Techniasia*. http://www.techinasia.com/kakao-founding-story/

West, J., & Mace, M. (2010). Browsing as the killer app: Explaining the rapid success of Apple's iPhone. Retrieved from www.joelwest.org/Papers/WestMace2010-WP.pdf

Wireless Intelligence. (2013a). BRIC markets generating almost a quarter of global mobile sales. Retrieved from https://wirelessintelligence.com/analysis/2013/02/bric-markets-generating-almost-a-quarter-of-global-mobile-sales/368/

Wireless Intelligence. (2013b). Global and regional mobile revenue trends. Retrieved from https://wirelessintelligence.com/analysis/2013/01/global-and-regional-mobile-revenue-trends/367/

KEY TERMS AND DEFINITIONS

2G: 2nd generation mobile networks (GSM, CDMA 1x).

3G: 3rd generation mobile networks (WCDMA/HSPA, LTE, TD-SCDMA, CDMA EV-DO, Mobile WiMax) UMTS Rel. 99.

3G+: HSDPA, HSUPA, HSPA+.

APAC: Asia Pacific.

ARPU: Average revenue per user.

Basic Phone: Non-smartphone.

CAGR: Compound Annual Growth Rate.

CDMA: Code Division Multiple Access.

DL: Downlink.

EB: ExaByte, 10^{18} bytes.

EDGE: Enhanced Data Rates for Global Evolution (GSM Environment).

EVDO: Evolution Data Optimized (same family as IS95 and cdma2000).

GB: GigaByte, 10^9 bytes.
GSA: Global Supplier Association.
GSM: Global System for Mobile Communications.
GSMA: Global System for Mobile communication Association.
HoM/MIMO: High-order modulations /Multiple-Input Multiple-Output.
HSDPA: High Speed Downlink Packet Access.
HSPA: High Speed Packet Access.
HSUPA: High Speed Uplink Packet Access.
IMEI-TAC: International Mobile Equipment Identity –Type Approval Code.
IMS: Internet Protocol Multimedia Subsystem.
IoT: Internet of Things.
IP: Internet Protocol.
LTE: Long-Term Evolution.
M2M: Machine-to-Machine.
MB: MegaByte, 10^6 bytes.
MBB: Mobile Broadband (Defined as CDMA2000 EV-DO, HSPA, LTE, Mobile WiMax and TD-SCDMA).
Mbps: Megabits per second.
MMS: Multimedia Messaging Service.
Mobile PC: Defined as laptop or desktop PC devices with built-in cellular modem or external USB dongle.
Mobile Router: A device with a cellular network connection to the internet and Wi-Fi or Ethernet connection to one or several clients (such as PCs or tablets).
NGMN: Next generation mobile Networks.
OS: Operating System.
P2P: Peer-to-Peer.
PetaByte: 10^{15} bytes.
Smartphone: Mobile phones with open OS, e.g. iPhones, Android OS phones, Windows phones but also Symbian and Blackberry OS.
SON: Self Optimized Network.
TD-SCDMA: Time Division-Synchronous Code Division Multiple Access.
UL: Uplink.
UMTS: Universal Mobile Telecommunications System.
VLR: Visitor Location Register.
VoIP: Voice over IP (Internet Protocol).
WCDMA: Wideband Code Division Multiple Access.
WiMax: Worldwide Interoperability for Microwave Access.
Zettabytes: 10^{21}bytes.

ENDNOTES

1 The chapter is based on desk research, a review of literature, comment from experts and industry players. It stems from three research projects the authors are currently involved in: Big Data, innovation in creative industries, and videogames.

2 The views expressed are purely those of the authors and may not in any circumstances be regarded as stating an official position of their institutions.

3 GSMA, the mobile trade association, is careful with the way to compute the data. They emphasise the importance of measuring unique mobile subscribers as opposed to SIM connections, and the difference between mobile phone owners and mobile users (Gilet, 2014).

4 7.155 as of September 2014, 1.13 trillion US $ of revenues. GSMA Intelligence 2014.

5 3G: HSPA, EV-DO.

6 4G: LTE, TD-LTE and WiMAX. Samsung's flagship Galaxy S3 and Apple's new iPhone 5 are LTE-enabled devices.

7 According to Ericsson (2013). in 2013, total mobile traffic generated by mobile phones already exceeded that from mobile PCs, tablets and mobile routers for the first time.

8 A zettabyte is one trillion gigabytes or, as Cisco explains, the equivalent of everyone on the planet downloading ultra-HD video for 24 hours.

9 Marketed as 4G LTE, LTE stands for Long-Term Evolution, a standard for wireless communication of high-speed data for mobile phones and data terminals.

10 Annualised revenue for the period to Q2 2014. 90.44 billion US$ in 2013 (Gillet, 2013).

11 Brazil, Russia, India, China.

12 Or other portmanteau words like: "produsers" (Bruns). "creative audience" (Castells)." interactive audience" (Jenkins). The word "prosumer" was coined by the futurologist Alvin Toffler in 1980 anticipating this emerging phenomenon.

13 Obviously one should be careful with the real speed, as the World Bank notes in its "Key Trends in the Development of the Mobile Sector" report: "Manufacturers and operators cite ever-increasing bandwidth, but the average speeds fall far short. According to Akamai's analysis of 96 mobile networks across 58 economies carried out in the third quarter of 2011, peak speeds were around 8.9 megabits per second (Mbit/s). But average speeds were 1.8 Mbit/s (Akamai 2012).". (Minges 2013: 121). In the first quarter of 2014, average mobile connection speeds (aggregated at a country level) ranged from a high of 14.7 Mbps in South Korea down to a low of 1.0 Mbps in Argentina (Akamai, 2014: 3).

14 A phablet, another portmanteau of the words phone and tablet, is a class of mobile device designed to combine or straddle the functions of a smartphone and tablet. The Samsung's Galaxy Note is a well-known example, largely credited with pioneering the worldwide phablet market when launched in 2011.

15 It probably includes games.

16 Vision Mobile (2014a) gives a higher number for North America: 40%.

17 LINE Corporation is a Japanese company, but a subsidiary of South Korean company Naver Corporation. Line reported $338 million in revenue in 2013. Line also boasts to have over 350 million registered users. *Source*: Wee (2014a).

18 WhatsApp Messenger, founded in 2009, is a proprietary, cross-platform instant messaging subscription service for smartphones, based in Mountain View, California. The company employs 55 people. As of August, 2014, WhatsApp has over 600 million active users.

19 Rakuten is a Japanese e-commerce firm, Viber, a Cyprus-based mobile messaging startup.

20 Another California-based (Mountain View) mobile messaging company.

21 Pew Research examined the homepages of a mix of 32 local TV websites.

22 Based on online survey of almost 19,000 people in ten countries: the US, UK, Finland France, Germany, Denmark, Italy, Spain, Brazil, and Japan.

23 Based on the revenues from the 200 most popular games in the iPad, iPhone/iPod App Store and Google PlayStore.

24 For a more comprehensive overview see De Prato & Simon (2015).

25 1015 bytes of digital information.

26 Such as Apache Hadoop, Apache Pig, Apache Hive. Apache Hadoop is 100% open source, and is supposedly pioneering a fundamentally new way of storing and processing data. See: http://hadoop.apache.org/
http://www.cloudera.com/content/cloudera/en/about/hadoop-and-big-data.html

27 The author gives precise examples with urban planning, transportation, disaster risk reduction and response, health, and education.

28 Based on a survey: 1611 participants.

29 What the report describes as "the reinvention of the 1950s living room".

30 A US entertainment, social networking service and news website.

31 No similar data available from Pew in the 2014 "State of the News" report.

32 Based on online survey of almost 19,000 people in ten countries: the US, UK, Finland France, Germany, Denmark, Italy, Spain, Brazil, and Japan.

33 Small phones (e.g., most Blackberries). 3.5" or under screens, medium phones (e.g., iPhone). between 3.5"- 4.9" screens, phablets (e.g., Galaxy Note). 5.0" - 6.9" screens, small tablets (e.g., Kindle Fire). 7.0" - 8.4" screens, full-size tablets (e.g., the iPad). 8.5" or greater screens.

34 4oD is the on-demand service of Channel Four.

35 This category appears to be less obvious to grasp, but it basically includes handheld devices.

36 With headlines like: "Mobile TV Spreading in Europe and to the U.S." (O'Brien, 2008). Such headlines almost disappeared after 2010.

37 In 2006, the Commission encouraged the setting up of the European Mobile Broadcasting Council. Commission communication on EU policy for mobile TV (July 2007). "*Strengthening the Internal Market for Mobile TV*".

38 In 2011, the consultancy RNCOS was still predicting some 795 million users worldwide by the end of 2014.

39 In 2005, South Korea became the first country in the world to have mobile TV when it started satellite DMB (S-DMB) and terrestrial DMB (T-DMB) service on May 1 and December 1, respectively.

40 To allow mobile and IPTV to converge a combination of technologies like DVB-H/ DVB-HS/ unicast was required. By contrast LTE uses multicast.

41 Developed by Qualcomm, it was dropped in 2010.

42 Wimax was considered at that time as a big challenger for LTE. Since most wireless carriers opted for LTE. It was seen as the natural evolution path for current GSM / UMTS operators; and therefore supported by GSMA.

43 In December 2008, the Commission issued its « *Legal Framework for Mobile TV Networks and Services: Best Practice for Authorisation – The EU Model*". The document still however considered three models for authorising the service: an extension of existing Digital Terrestrial Television (DTT) rules to new services (model adopted by Italy). a plain wholesale model with regulation focusing on the wholesale operator, and an integrated approach, under which all players in the mobile TV value chain, including content providers, have to find an agreement on the services before the granting of authorisation.

44 During 2012, in addition to 'T Sports', a mobile sports broadcast service, SK Telecom launched 'Btv Mobile', a mobile IPTV service which provides 80 real-time channels for KRW 9,900 (approximately $9.60) per month. As of Q2 2014, 1.9 million of SK Telecom's customers had subscribed to the Btv Mobile service (Dewar, 2014).

45 A special interest group for the administration of the right of artists and musicians in France.

46 From an article in Wired magazine 2007 quoted in Bacache & al (2011).

47 It took the shape of the Android 4.4-powered Galaxy K Zoom. The device, which follows the Galaxy S4 Zoom's lead by having a digital camera-like rear body, is said to "*blend advanced digital camera technology*". *Source*: www.mobileworldlive.com/samsung-unveils-latest-camera-smartphone?utm_campaign=MWL-D-20140430&utm_medium=email&utm_source=Eloqua&elq=0a75892c53c7481c8ded4be162040d3f&elqCampaignId=1633

48 See for instance http://azbigmedia.com/scottsdale-living-magazine/lists/top-5-movie-parodies-youtube

49 www.huffingtonpost.com/shira-lazar/the-best-oscar-movie-paro_b_4855996.html

50 The authors built a sample of 8,299 user-generated music video parodies was constructed relating to the top-100 charting music singles in the UK for the year 2011, posted on You Tube. The find on average, there are 24 user-generated parodies available for each original video of a charting single.

51 Founded in November 1998, Tencent has grown into one of China's largest and most used Internet service portals. On June 16, 2004, Tencent Holdings Limited went public on the main board of the Hong Kong Stock Exchange. In 2012, total revenues were US $6 983.3 million, an increase of 54% over the year ended December 31, 2011. *Source*: Tencent website.

52 The largest game production studio in New Zealand. Sidhe provides services to and has products published by a range of world class partners and clients such as Warner Bros Interactive Entertainment, Sony Online Entertainment, Activison, and Ubisoft.

53 Jeremy Silver (Silver, 2013: 60) describes the notion of Tarzan economics: 'My business model is hanging from this tree and I'm hanging onto it like mad but I've got to get through this jungle. In order to get through this jungle I'm going to have to move to another economic model, which apparently is one of those lianas over there that's hanging from one of those trees. I don't know which one it is so I'm going to hang onto the one I'm hanging onto for as long as possible until I can make sure that the one over there is actually not going to come away in my hand. If I reach out for the one that's over there and it comes away in my hand I'm going to fall to the jungle floor and get devoured by insects. If I don't hold onto that one and hold onto this one I'm not going to make any progress because guess what? This one's really fraying quite badly too".

54 And free voice apps like What'sApp tolling the bell for chunk of revenues.

55 A company that released another global hit: "League of Legends" (2009) the most played PC game with over 67 million play on average per month (company website).

Chapter 2
Mobile Communications and the Entrepreneurial Revolution

Sergio Ramos
National University of Distance Education, Spain

Alberto Arenal
Technical University of Madrid, Spain

Cristina Armuña
Technical University of Madrid, Spain

Jesús Ferrandis
Telefonica, Spain

ABSTRACT

It is commonly accepted that entrepreneurship is a key activity in the current economic context, and entrepreneurs create value in form of new businesses ventures, and generate new jobs by mobilizing economic resources, which in combination has a positive impact in terms of increasing productivity and economic growth, among other effects. In this chapter, the authors acknowledge the importance of innovation in the mobile industry driven by entrepreneurs and firms as a source of disruptive changes that have completely reshaped the world, by presenting the key elements of current mobile industry and the emerging areas with the highest potential to attract entrepreneurs in the upcoming years.

MOBILE COMMUNICATIONS AND ENTREPRENEURSHIP IN CONTEXT

It is commonly accepted that entrepreneurship is a key activity in the current economic context, and entrepreneurs create value in form of new businesses ventures, and generate new jobs by mobilizing economic resources, which in combination has a positive impact in terms of increasing productivity and economic growth, among other effects (IEG World Bank, 2013). To this respect, international institutions like the United Nations (United Nations University, 2011) or the Organization for Economic Cooperation and Development (OECD, 2014) agree on the importance of innovation and entrepreneurship in sustainable economic growth and material and non-material welfare.

This perception is even more relevant in the recent situation of economic deceleration. Since the global financial crisis broke up in 2007, access to financing, one of the most significant resources for the creation and consolidation of small and medium sized enterprises, has been significantly reduced. This fact in combination to a severe decrease in demand for goods and services and the longer payment delays, have contributed in setting barriers for entrepreneurial activity to grow (OECD, 2009).

DOI: 10.4018/978-1-4666-8838-4.ch002

According to the EC High Level Expert Group on Small and Medium Enterprises and Infrastructure Financing (European Commission, 2013), project finance volumes in the EU-27 experienced a disproportionately large drop of over 40% from 2008 to 2009, slightly recovered in 2010 and 2011, before dropping by further 40% in 2012 to reach their lowest level since 2004.

Despite this difficult context, more and more stakeholders, from governments, private firms, universities, investors to academics and independent practitioners, have agreed on identifying that entrepreneurs can contribute to recover the economic growth path (World Economic Forum, 2009) and, consequently, to stimulate job creation (COADEC, 2014) .

An additional commonly accepted statement is the direct connection between technology and entrepreneurial activity. In particular Information and Communications Technologies (ICT) are considered one of the main drivers to foster entrepreneurial activity according to the results of the GEM Report (GEM Consortium, 2014). Furthermore United Nations stated (United Nations Conference on Trade and Development, 2011) that ICT are a determinant factor of the overall investment willingness in entrepreneurial ventures.

Within the framework of ICT, in this Chapter the enabling role of mobile industry to boost entrepreneurial activity is acknowledged through the presentation of main recent milestones in its evolution that have turned mobile communications into a source of opportunities for innovators and entrepreneurs, as a global marketing, commercial and distribution channel, as well as a platform for experimentation in new services, applications and contents that impact people lives, among other benefits[1].

Along the following sections, a set of specific paradigm shifts in mobile industry has been highlighted as main contributors to reconfigure entrepreneurial activity, as we currently know it. As a result, these disruptive changes has set the proper framework conditions for entrepreneurs to develop their ideas, using mobile technology as the entry door to other relevant industries in which innovators identify the biggest opportunities. Then, the last sections of the Chapter present some of the most appealing areas of activity for those involved in the entrepreneurial scene.

MOBILE INDUSTRY EVOLUTION: RECENT DISRUPTIVE MILESTONES

Following the mentioned approach, the present section cover three main milestones in the recent industry evolution that have contributed to set the appropriate conditions to create an entrepreneurial ecosystem around mobile:

- Smartphones paradigm, thanks to the massive adoption of these handsets as personal devices able to provide Internet connectivity and communications facilities everywhere defining a new user experience in mobile communications.
- Industry platformization, creating a complex ecosystem of players that compete and cooperate in contrast to the traditional mobile industry vertical integration of the value chain leaded by telecom operators.
- App Economy, acknowledging the role of entrepreneurs and software developers as third party players that stimulate the creation of new services, applications and contents in the mobile domain.

Smartphones Paradigm

It was 9 hours and 41 minutes on January 9, 2007 in San Francisco, when Steve Jobs introduced the iPhone by saying: "today Apple is going to reinvent the phone"(Apple, 2007). Since that moment everything changed, not only in the mobile industry but also in the Internet.

The iPhone was the first real smartphone, with the permission of previous Nokia N series handsets, and came to transform mobile user experience. It introduced a radically new user interface, built around a big touch screen, and eliminated the physical keyboard, venerated by most manufacturers. And, if Apple was the pioneer, Google was the responsible of democratizing smartphones by developing Android and offering it for free.

iPhone and Android smartphones rediscovered how users could take the most out of the Internet, with an experience really adapted to mobile devices. While previous handsets had tried to replicate the Internet user experience from the PC, by just using mobile-adapted version of desktop browsers, smartphones' experience turned around a brand new concept: the application.

From that moment, smartphone industry has shifted towards designing the best piece of hardware and letting users customize their mobile experience thanks to the emergence of a new disrupting industry like mobile apps (presented in the following section) in opposition to the old fashioned mobile Internet promoted by mobile operators, mainly based on de-facto walled garden strategies that stopped the creation of an ecosystem for third-party developers contents, services or applications (Ramos, Feijóo, González, Rojo, & Gómez-Barroso, 2004). This has introduced major changes in Internet usage experience. As a matter of fact, mobile access to the Internet is currently dominated by the use of native apps at the expense of mobile browsers[2].

In this new paradigm, great part of the offered value by the smartphones comes from the applications. And it was at this point where Apple and Google made a critical decision, introducing a new actor in the ecosystem. It was not people inside the companies who created the applications, but they opened their platforms to third-party developers. This measure was crucial to foster innovation and boost entrepreneurship in mobile bringing the concept of mobile platforms for services, applications and content provision as new business model paradigm.

The Role of Platforms as Key Enablers

Smartphone industry has adopted the role of platforms as a natural example of a business ecosystem in which players compete and cooperate. Nowadays almost every device is built over a specific platform constituted by an operating system and a set of developers' tools for creating new contents that are marketed through the platform.

This model creates a virtuous circle that connects developers, services providers and manufacturers in the ecosystem to the increasing number of users of a specific platform, which stimulates the creation of applications to feed the ecosystem. As the number of applications increases, users have greater interest to acquire and use a specific device attached to the corresponding platform.

In this scenario, platforms owners work as the gatekeepers (Ballon, 2009) of the ecosystem by connecting users and application developers through three main key activities (Visionmobile, 2014):

- **Development Tools:** Platforms offer their capabilities through open APIs (Application Programming Interfaces) that are used for building software applications.
- **Distribution:** Platforms work as the main channel for distributing contents to mobile devices. They include a marketplace, an application store, from where users can download applications. Monetization tools are also offered for developers to support their strategy when distributing apps.
- **Discovery:** Platforms are main prescribers for applications. They are key to promote, place, search and recommend apps.

iOS was the first to appear in the platform scene, in connection to the launch of the iPhone, with the conviction that hardware and software integration is the key to offer the best user experience. For this reason, iOS runs exclusively on Apple's devices (i.e. iPhone, iPad, iPod) and is not licensed to other manufacturers. It is a key differentiator for the iPhone and lets Apple keep a sustainable premium in price versus competitors, in a moment where smartphone's hardware moves towards commoditization.

The first commercial version of Android was launched in late 2008 as an alternative to iOS. Google followed a different business approach based on its experience on monetizing through advertising. In fact, the main objective of Google when launching Android was to extend its leading position in advertising business as Internet users moved from fixed to mobile, being then crucial to keep control of user experience on mobile devices and ensure users could access to all Google services from any personal device.

Furthermore, Google offered Android for free to a lot of different manufacturers as an "open" platform to be installed on the widest range of handsets, from the lowest to the highest tier. Most of the manufacturers, with just few exceptions, adopted Android as their platform, even at the expense of their own solutions (e.g. Samsung), and it quickly reached leadership among smartphone operating systems.

Wide adoption of Android and iOS has led to a duopoly in mobile operating systems, as they run in more than 95% of smartphones shipped during 2014, according to data from market intelligence firm IDC[3]. This duopoly is only broken in China, where local players distribute large scale of smartphones that fork Android (i.e. with a proprietary operating system running on top of Android), taking advantage of regulatory restrictions faced by Google to distribute its services in the country.

Some other platforms tried to consolidate a global third alternative, with success still to come. Windows Phone, the result of years of work within Microsoft to extend its leadership in software from the PC to the mobile, has been adopted by several vendors and is supported by a large community of developers, but has not achieved a solid position and lags behind. BlackBerry is another example of how former market leaders were disrupted and ousted (in this case, almost ousted). The company has been reduced to a minimal market share and currently struggles to survive under a new management after several attempts to be sold.

In the last years, some new alternatives have arisen (e.g. FirefoxOS, Tizen, Cyanogen), but they are not widly adopted yet.

The Concept of App Economy

In this context, the concept of App Economy gained popularity following an article in Business Week in 2009 (MacMillan, Burrows, & Ante, 2009), which pointed at the economic dynamism of the entrepreneurial companies developing apps in the Facebook and iPhone ecosystems. According to this article, the App Economy should be viewed not only as "a better delivery channel for software" but "as a collection of interlocking innovative ecosystems consisting of a core company, which creates and maintains a platform and an app marketplace, plus small and large companies that produce apps and/or mobile devices for that platform".

Therefore, the App Economy can be considered as a consequence of the mobile industry platformization, in which entrepreneurship flourishes, directly impacting in the way people communicate, work, buy or even relate and leisure. As an example of the achieved impact so far, 94.4 billions of Apps were downloaded globally in 2013 (EC, 2014) and predictions of the experts suggest that the number will multiply by two (Juniper Research, 2015) or even three (Gartner, 2013) until 2017.

Apps are a new way of understanding mobile communications, overtaking regular uses and shifting how value is created. In the early 2000s, mobile communications were used basically to support voice service and they supposed a small share of the total spent of users. Current trends however confirm that value provision has shifted from pure voice to a users-centric approach that covers their everyday needs and preferences in mobility (in health, education, transportation, leisure…).

The change is not only quantitative but also qualitative, as the way this value is delivered and monetized has been modified completely. Before the app revolution, telecom operators and device manufacturers were those who obtained the main revenues from mobile. Now, ecosystems rule and there is a rough competition not only between products, but also between industries. These factors configure a new scheme where competitors fight for controlling several or even all activities of the value chain.

According to the European Commission, the benefits from the App Economy range from the revenues and jobs directly linked to developers industry and sales of the apps to productivity gains obtained by users due to the intensive usage of mobile apps.

In this context, global app spending reached €20.3 billion in 2013 and it is forecasted to amount €69.3 billion in 2018 (EC, 2014), which involves a global growth rate of around 28%, higher than almost any other sector of the economy. With regards to job creation, the number of app developers all over the world is significantly growing, with estimations of 2.9 million in 2014 (Vision Mobile, 2014).

Mobile platforms and app stores allows to reach easily every corner in the world and, then, reduce the user acquisition cost and facilitate the monetization. Besides, new developing tools include friendlier interfaces that permit the increasing number of potential developers and even allow apps creation hardly without programming any code. All of these disruptive business and technological changes lower dramatically the entry barriers for developers and configure an ideal field for entrepreneurs to start their own apps based businesses.

UPCOMING BUSINESS OPPORTUNITIES

As presented in previous sections, the evolution of mobile industry and its recent changes experienced so far have favoured the creation of an entrepreneurial dynamism around mobile industry that shows an inherent potential of disruption through imaginative uses of hardware and software technology in mobility where personal devices, connectivity, services and apps play a key role to address existing and new user's needs that continue shaping the future of our economy and society.

Data about the use of mobile devices reveal the big opportunity around the mobile world. For example 2014 global mobile data traffic was nearly 18 times the size of the entire global Internet in 2000. Regarding the role of devices, by the end of 2014 the number of mobile-connected terminals exceeded the number of people on earth, and by 2018 there will be over 10 billion mobile-connected devices -including machine-to-machine (M2M) modules—, which means 1.4 mobile devices per person (Cisco, 2015). On the apps side, nowadays the 57% of users who download apps use them daily and the 24% of these downloaders use their apps for more than 30 minutes a day.

According to the European Commission (European Commission, 2014) a number of behaviour- and market-specific factors from the demand side will drive continued growth in the App Economy and its entrepreneurial ecosystem, including:

- Growing downloads per user.
- Growth of in-app purchases and other new monetisation methods.
- Growth in number of app developers (including the transitioning of other ICT skills and new entrants).
- Growth of consumer adoption of connected devices in emerging markets.

From the offer side, predictions reported by different consulting firms and analysts identify key emerging areas[4] in which the upcoming disruptions would take place and business opportunities area bigger[5].

These predictions suggest that the App Economy will continue growing fast in the upcoming years, extending its presence to other relevant industries in which technological innovations based on mobile platforms would be massively adopted.

However, it is important to highlight that the future evolution of the app movement depends as well on the existence (and performance) of other players in the mobile ecosystem that provides mobile connectivity, personal handsets and new electronic devices that open new possibilities to app developers.

Apps have enormously increased data transmission requirements over telecommunication networks. According to (Cisco, 2015), global mobile data traffic grew 81% in 2014, reaching 1.5 Exabytes per month. This is nearly 18 times the size of the entire global Internet in 2000. The 53% of this traffic was video.

Although mobile networks were initially conceived to support only voice and simple text messaging, in 2014 figures suggest a complete paradigm change as a smart device generates 29 times more traffic than a non-smart device and predictions reveal an incessant increase with 4G technologies. Although 4G connections represent only 2,9% of mobile connections today, they already account for 30% of mobile data traffic and will reach 66% of mobile data traffic by 2018.

Another technological trend of great importance is the appearance of devices like wearables and sensors as representatives of the so-called Internet of Things. They provide a new way of interaction among users and their environment based on continuous monitoring, tracking and information exchange. In this sense, the upcoming years will probably confirm their potential with the commercialization of watches, bracelets, glasses, or clothes that can obtain and provide information constantly and in more and more natural way.

This trend leads as well to an additional element in the future context for entrepreneurial activity and the mobile industry. We refer to the increasing amount of data that every user (human or machine) generates and is continuously stored and retrieved to extract additional information and patterns. The change in paradigm has transformed the classic personal hard drive into remote "cloud" storage where almost all information is accessible, but neither in the premises nor in the device of the user.

The advantages of this kind of storage are clear: accessibility everywhere and easy options to share. But on the other side, the protection of this information is a key issue, which is as well an opportunity for app developers and entrepreneurs.

In addition the vast amount of data from users in combination to other sources of information constitute an extraordinary source of opportunities around the concept of Big Data, another technological trend to follow.

All these elements suggest that the upcoming entrepreneurial opportunities will come in form of disruptive combinations of services, apps and contents, the massive usage of additional devices and data that transform users perceptions and experiences. In the following some of the most appealing areas for entrepreneurial activity are briefly presented to illustrate this new wave of innovative solutions.

Payment Services

Nowadays permanent use of personal mobile devices and the apps revolution have configured an scenario where online payment services are perceived as a natural element in the context of Internet and mobility.

Based on the European Banking Authority (European Banking Authority, 2014), e-wallets are the largest share of alternative payment methods and they are expected to grow further from 35% to 42% by 2015. In the same period, real time bank transfers are expected to double their share, growing from 12% to 20%. These shares will grow at the expense of paper-based options, which are expected to decrease from 22% to 14%.

Mobile devices are being used, and the tendency will increase in the near future, for a variety of payment scenarios such as leisure activities (concerts, restaurants, music tracks...), information (digital content, newspaper subscriptions...), travelling (flight, bus and train tickets), hotel booking, tolls, parking fees, taxis, and bank transactions, among others.

M-Health and Wellness

The Global Observatory for eHealth (World Health Organization, 2011) defined mHealth or mobile health as medical and public health practice supported by mobile devices, such as mobile phones, patient monitoring devices, personal digital assistants (PDAs), and other wireless devices.

Mobile health apps are expected to experiment a great growth, and it has a primary importance in developing countries. GSMA estimates that the amount saved in 2017 from the annual healthcare bill in developed countries as a result of mobile healthcare solutions would reach $400 billion.

Apps for wellness and health care show a significant growth, with more than 100.000 apps listed in the mHealth sections of major mobile platforms (Research2Guidance, 2014), reaching $26 billion revenues from apps download and wearable devices sales.

Gaming

In 2013, the mobile games sector generated $12,3 billion, 17,4% of all global game revenues. With close to $6 billion, Asia Pacific is the biggest mobile games market worldwide. Different analysts coincide to point out five major elements that would impact in the sector and foster a further take up (Casual Games Association, 2013):

- Two screens: tablets and smartphones offer different consumer experience, and from the point of view of players, each one has its own right of existence. So an environment of multiscreen consumers is substituting the usual behaviour of playing on one primary screen, which allows that mobile games growth is driven by simultaneous play.

- Accessibility: mobile devices allow consumers to play anytime and anywhere, which increases the time spent on games.
- Global reach: the penetration over the world of smartphones and tablets opens great opportunities for developers.
- Free-to-play: free-to-play is the dominant business model for mobile taking over 90% of global revenues. Ultimately this has been the choice of consumers, although nowadays almost the 40% of players pay for the apps so mix business models are coming up.
- Hardware innovation: the vertiginous rhythm of innovation in devices and wearables offers multiple innovative options to developers, including the chance to catch up with immersive console or PC games.

Nowadays there are more than 960 million players over the world and almost the 40% of them pay for the apps. Industry analysts predict that revenues from tablets and smartphones games app will reach $23.9 billion by 2016, which implies a revenues growth of 400% since 2012.

Education

Mobile apps and connectivity in the context of education offer the opportunity to establish new opportunities for teachers, students and parents, as well as customized learning tools and programs adapted to every need. Additionally, due to the increasing penetration of mobile broadband worldwide, it allows to extend educational initiatives to remote areas or in developing countries.

In this sense, GSMA estimates (GSM Association, 2011) that the number of children in developing countries that will have the opportunity to enjoy education due to mobile solutions until 2017 will reach 180 million.

Accessing online content, combining text with augmented reality technologies or real time exchanging of information are some of the trending areas, and predictions estimate that worldwide m-education market will reach $37.8 billion in 2020.

Smart Cities

A Smart City can be defined as "A city connecting the physical infrastructure, the IT infrastructure, the social infrastructure and the business infrastructure to leverage the collective intelligence of the city" (Harrison, 2010).

In this context, people, sensors and devices are interconnected and the generated data are managed to provide real time information and ubiquitous services, which contribute to transform the inherent city dynamics.

Innovative apps and services for consumers and firms can be developed processing these data through data analytics techniques and business intelligence tools to provide information about congested traffic routes, parking spaces, public transport delays, available cycles for renting, energy consumption at home, weather forecast, capacity of trash cans, or recycling points.

Automotive

The percentage of vehicles with intelligent devices and embedded connectivity will be almost six times greater in 2024 than in 2014, growing from 15% to 89% (ReportLinker, 2014).

The possibilities derived from connected cars affect a wide range of areas from the way of driving to the security systems. As an example one of the main current interest is improve the emergency services response in case of crash using an 'eCall', an initiative by the European Commission that was backed by the European Parliament in July 2012, which will be required in 2015 for all cars with embedded connectivity (European Commission, 2015).

According to Vision Mobile (Vision Mobile, 2014) apps future development for vehicles will follow four main strategies: in-vehicle dashboard (apps running in the built-in communication system), smartphone link (apps running in the mobile device linked to the vehicle), remote API (remote access to monitor and control the car), or Bluetooth dongle (using the on-board diagnostic port).

Tourism

It is widely acknowledged that tourism industry have been clearly impacted by mobile technologies, with 85% of leisure travellers using their smartphone while abroad, 30% using mobile apps to find hotel deals, and 15% downloading mobile apps to get information at local destinations (Euromonitor International, 2014).

According to the World Travel Market Association, due to the mobile nature of travelling, smartphones and tablets are expected to become the key customer service tool in the industry within the next five years, with a special focus on all aspects that ease travellers to enjoy personalized services and information in every destination.

In this sense the combination of mobile apps, enriched content including videos, photos, 3D models, audio and even augmented reality technologies, offer new opportunities to improve the tourist experience, for example in cultural destinations.

CLOSING REMARKS

As presented along the previous sections, entrepreneurship has become an activity of paramount importance for individuals and institutions as a major driver for innovation, economic growth and employment. Although entrepreneurial activity has been historically considered as the process in which an individual, small team or organization develops a business idea of any kind that implies a high level of risk given the inherent uncertainty and difficulties of its path for potential growth, the ICT revolution of the last decades, and in particular the disruptive capacity of mobile communications industry has multiplied the relevance of entrepreneurial activity.

Therefore, the massive deployment of mobile networks and the subsequent mobile service adoption (starting from voice to data), the evolution of personal devices from mere-voice oriented to smartphones with capabilities similar to a personal computer, the popularization of mobile Internet (mobile data connectivity) thanks to the combination of the previous factors and the increasing number of services and content providers adapted to mobile users, the platformization of the value chain that fully ease the creation of a complex ecosystem in which third party developers compete and cooperate to transform other industries, can be considered necessary factors, among others, that have boosted the entrepreneurial scene.

Mobile industry as it is currently known have modified the playing field in which entrepreneurs develop their ideas. Nowadays, entrepreneurs use mobile technologies within a platform to continuously experiment with new concepts than can be easily marketed given the accessibility and affordability of this process.

In this sense entrepreneurs have taken advantage of ICT and mobile technologies to change the way entrepreneurial activity is conceived. Currently the norm is to prototype and test[6] with real users using the minimum necessary resources instead of long periods of designing and manufacturing that require intensive investment before testing the market.

The App Economy as a new economic sector based on the previous factors is just an example of the impact of mobile industry disruptions in combination to a continuously evolving entrepreneurial mindset. It is expected that the upcoming years will show new business ventures in which mobile technologies and the next generation of personal devices and sensors will be applied to solve existing and future needs in different sectors, as presented before, leaded by the entrepreneurial movement.

REFERENCES

Apple Reinvents the Phone with iPhone. (2007). *Apple Press Info.*

GSM Association. (2011). *Mobile Education Landscape Report.*

Ballon, P. (2009). The Platformisation of the European Mobile Industry. *Communications & Stratégies*, (75): 15–33.

Casual Games Association. (2013). *Smartphone & Tablet Gaming 2013: Games Market Sector Report.* Newzoo.

Cisco. (2015). *Visual Networking Index: Global Mobile Data Traffic Forecast Update 2014–2019* [White Paper].

COADEC. (2014, September). The Startup Manifesto.

Consortium, G. E. M. (2014). *Global entrepreneurship monitor 2013 global report.* Recovered on February.

Euromonitor International. (2014). *WTM Global trends report.*

European Banking Authority. (2014). *Eba Consumer Trends Report 2014.*

European Commission. (2013). *Finance for Growth. Report of the high level expert group on SME and infrastructure financing.* doi:10.3917/reco.596.1075

European Commission. (2014). *Sizing the EU app economy.* Luxembourg; doi:10.2759/43252

European Commission. (2015). eCall: Time saved = lives saved. *Digital Agenda for Europe.*

Flurry. (2014). Apps Solidify Leadership Six Years into the Mobile... *Flurry Insights Blog.*

Gartner. (2013). Gartner Says Mobile App Stores Will See Annual Downloads Reach 102 Billion in 2013.

IEG World Bank. (2013). *World Bank Group Support for Innovation and Entrepreneurship. An Independent evaluation.*

Juniper Research. (2015). Apps & App Stores Mobile Market Research - Market Trends & Forecasts.

MacMillan, D., Burrows, P., & Ante, S. E. (2009). Inside the App Economy. *Business Week*, *4153*, 347–357.

OECD. (2009). *The Impact of the Global Crisis on SME and Entrepreneurship Financing and Policy Responses. Policy.*

OECD. (2014). *Financing SMEs and Entrepreneurs 2014*. Paris: OECD Publishing; doi:10.1787/fin_sme_ent-2014-en

Ramos, S., & Feijóo, C., González, a, Rojo, D., & Gómez-Barroso, J. L. (2004). Barriers to widespread use of mobile Internet in Europe. An overview of the new regulatory framework market competition analysis. *The Journal of the Communications Networks*, *3*(3), 76–83.

Research2Guidance. (2014). *mHealth App Developer Economics 2014. The State of the Art of mHealth App Publishing.*

Ries, E. (2011). *The Lean Startup: How Today's Entrepreneurs Use Continuous Innovation to Create Radically Successful Businesses*. New York: Crown Publishing Group.

United Nations Conference on Trade and Development. (2011). *Information Economy Report 2011. ICTs as an Enabler for Private Sector Development.* New York and Geneva: United Nations publication.

United Nations University. (2011). *Entrepreneurship and Economic Development* (W. Naudé, Ed.). London: Palgrave Macmillan.

Visionmobile. (2014). *Mobile Megatrends 2014.*

World Economic Forum. (2009). *Educating the Next Wave of Entrepreneurs. Unlocking entrepreneurial capabilities to meet the global challenges of the 21st Century.*

World Health Organization. (2011). mHealth: New horizons for health through mobile technologies. Global Observatory for eHealth series (Vol. 3).0

KEY TERMS AND DEFINITIONS

App Economy: The economic flows generated by entrepreneurs and software developers that stimulate the creation of new services, applications and contents in the mobile domain.

Big Data: The vast amount of data that is being generated from users and machines. Their analysis and treatment constitute an extraordinary source of business opportunities.

Cloud Storage: The remote "virtual" storage in which is called "the cloud". Instead of having all the information in hard drivers in the premises or in the devices, the information is stored and accessed in an online way with clear advantages such as accessibility everywhere and easy options to share.

Entrepreneurial Ecosystem: The different areas and stakeholders whose development, evolution and interactions stimulate or discourage the entrepreneurial activity.

Internet of Things: The network of "smart-things" (sensors, smart-meters, smartphones...etc) that access, send and exchange information through Internet.

Platform: The concept of platform is presented as a combination of hardware (personal device) and software (operating system) solutions that enables third party players' integration.

Wearables: Normal use devices such as watches, bracelets, glasses or clothes that include sensors to provide a new way of interaction among users and their environment based on continuous monitoring, tracking and information exchange.

ENDNOTES

1 A global survey conducted in 2012 by Time Magazine in cooperation with Qualcomm concluded that 93% of the people consider mobile communications crucial for developing any kind of entrepreneurship activity and managing small business.

2 According to data from mobile analytics company Flurry, in March 2014 apps commanded 86% of time consumers spent on mobile devices, which accounted for 2 hours and 42 minutes per day: "Mobile revolution" (Flurry, 2014).

3 IDC Worldwide Quarterly Mobile Phone Tracker, February 2015.

4 According to Portio Research, there are six types of application prevalent in the app market: Games, Productivity/Enterprise, Multimedia, Communication, Travel and Utilities.
Vision Mobile, in the report Business and Productivity Apps, identify a number of areas that currently attract considerable attention among businesses and where they see future value being unleashed in the business & productivity market: Vertical apps, Productivity/BYO, Mobile SaaS, Custom apps/services and MDM/MAM.

5 The industry predicts that by 2017, the revenues from apps markets will reach 24,5 billion dollars, which means duplicate the current revenues in two years.

6 This paradigm in entrepreneurial management is known as the Lean Startup (Ries, 2011) and has been widely adopted by almost every entrepreneur worldwide.

Chapter 3
Behavioural Targeting in the Mobile Ecosystem

José Luis Gómez-Barroso
National University of Distance Education (UNED), Spain

Juan Ángel Ruiz
National University of Distance Education (UNED), Spain

ABSTRACT

Personal information is a key intangible asset for companies. In particular, for those companies that collect personal data to attract advertisers through being a channel to reach their target markets, i.e., through offering targeted advertising. A universe of always located and always connected potential consumers enhance the potential of behavioural targeting, which is seen as the business model on which the success of new mobile content, applications and services relies. This chapter aims to explain the implications of the use of behavioural targeting within the mobile environment. It provides an overview of the exploitation of personal data, a comprehensive description of the behavioural targeting ecosystem, and a view of the promises and pitfalls associated to the rise of mobility.

INTRODUCTION

As the Internet became more and more accessible, and the number of eyeballs connected 24 hours a day around the globe continuously grew, it became clear that it is not only the most effective means of communication but also the largest breeding ground for business opportunities. However, finding how to turn opportunities into profitable ventures has proven to be a difficult matter. Over the years, many Internet business models failed to deliver on their promises. When looking at strategies of Internet giants, it becomes apparent that advertising is one of the few business models, if not the only, able to generate consistent and significant long-term profits.

The turning point was the adoption by search engines of a new way of financing the services they provided: around the start of the millennium, they began selling advertisements associated with search keywords. Since then, targeted advertisements allow those companies, as well as social networks and other online players, to provide "free" services to users while collecting data on who they are and what

DOI: 10.4018/978-1-4666-8838-4.ch003

they do and like. The use of personal information makes the difference on the effectiveness (but as well on the usefulness for users) of advertisements. This is not only true for firms producing digital products and services. Marketers see how social change is eroding the value of traditional advertisements (in print media, radio, and TV) and have turned their attention towards online behavioural-targeted marketing.

The "mobile revolution" accelerates and amplifies the underlying trends that were already at work. Exploiting the huge potential offered by a universe of always located and always connected potential consumers is the new big challenge. A challenge full again of opportunities but full as well of potentially dangerous implications for privacy.

Against this background, this chapter aims to explain the implications of the use of behavioural targeting within the mobile environment, both for companies and users. It is structured as follows: Section 2 provides an overview of the exploitation of personal data for marketing, and particularly for advertising purposes, where the firm's analytics capacities play an intensive role on understanding customer's behaviour; Section 3 offers a comprehensive description of the behavioural targeting ecosystem, identifying the different stakeholders and the relations amongst them; Section 4 characterizes mobility as a personal data capturing enabler, and underscore associated promises and pitfalls; the conclusions close the chapter.

THE VALUE OF PERSONAL INFORMATION AS AN INTANGIBLE ASSET

Customers and the data referred to them are, no doubt, a key intangible asset for companies. Hence it follows why they expend huge efforts in gathering data. Most firms manage and analyse this data with the intention of creating value for own new products and services through tailored marketing campaigns. A second situation is that of a particular type of company that collects data to attract advertisers through being a channel to reach their target markets.

Personal Information as an Asset for Own Use

Conceptually, marketing principles have not changed, even with the advent of the Internet: after discovering the likes of the customers and their purchasing habits, the companies try to aim better the messages to their (potential) customers proposing the "right" advertising campaigns, or actions in a broad sense.

However, in practical terms, profound changes have occurred in two directions. On the one hand, information and communication technologies allow firms to build up new relationships with their customers. For already quite a long time, many companies do not deal with consumers as a mass or as segments, but try to create individual relationships, managing markets of one, addressing each in terms of its stage of development (Blattberg & Deighton, 1991). On the other hand, companies are experiencing how digital customers behave differently compared with traditional customers (among many others, Belk, 2013; Berman and Kesterson-Townes, 2012; Nicholas et al., 2003). Therefore, they need to adapt their go-to-markets driven by the forces shaped by digital consumers, who can easily jump from one "shop" to a completely different one in seconds, changing their preferences as fast as a few clicks away.

These two forces have a direct impact in the companies who have to rethink strategies, production processes, human capital engagement and investments, from a holistic point of view. At the core of these transformations is, it should be underlined again, the customer, which becomes the centre of the companies' agenda. The more accurate the company's customers landscape is depicted, the more efficient will be the business decisions taken in the digital transformation arena. Therefore, it is more important today than ever to have real customer data obtained from customer analytics, interviews, surveys and/or captured from the users' interaction with the company's electronic channels. The objective is to identify customer's opinions and sentiments, the way (s)he moves across the information, the buying decision process… and, the cornerstone, uniquely identify the customer when possible.

To address those challenges, companies are using big data techniques to adapt their analytic capacities to predict the demand and to gain a deeper understanding of the customer conduct. The ability to rent rather than buy data storage and processing has lowered the entry barriers for the use of big data architectures (Varian, 2014), that have consequently been embraced by all size companies in recent years. In doing that, companies face a range of issues. Companies need to control data volume, velocity and variety to cope with the exponential increase of the number of connected persons and "things" (the "Internet of everything"), and to secure an efficient data governance, which is a reflection of how organizations value their diverse and expanding portfolio of data assets (Tallon, 2013). In particular, companies have responsibilities directly related with data privacy, like designing and implementing corporate policies on the ethical usage of the data, and ensuring the compliance with external regulations.

Being customer data an input of the production function of the companies, it might be considered, and consequently treated, as an asset. Indeed, no one disputes the need of regarding customers as assets that should be managed (a literature review can be found in Chang et al., 2012). However, it is only in recent years that companies have started to assign economic value to the data stored in their databases. Behind the estimation of this value there is a practical debate on the right procedure to discern if patents, employee training, software, R&D expenditure, certifications, and databases containing customer data and personal information are current expenses or capital accumulation, even if there is a general agreement between the scientific community and the business managers on the fact that intangible assets provide a sustainable competitive advantage that positively influences the innovation capacity of the organizations. Thus, the question on whether it is better to capitalize intangible assets or account them as expenses is one of the most controversial issues that can be found in the recent literature on assets (Bontempi & Mairesse, 2008).

Finally, it has to be considered that there is still a long way to understand the personal information's profitability differences between the regions where companies operate, and further research may be conducted to clarify if the data of a European consumer may be valued the same as one from the US, Asia or any other location. The reasons behind these differences may be linked to the attitudes when disclosing personal data, regulatory frameworks for consumer protection and/or personal data protection, and macroeconomic factors like the Internet connectivity levels, purchase capabilities, or others (Feijóo et al., 2014).

Personal Information as the Basic Asset for Online Advertising

Since the dawn of marketing, advertisers have struggled to target (potential) customers as effectively as possible considering the state of the art of the techniques available at each moment, and as a consequence advertising models has evolved to maximize results and revenues (Greengard, 2012). Segmenting has explicitly implied a huge effort for a long time, and it can be paradigmatically represented by the slot along the day, or by the month of the year, in which an advertisement is broadcasted on TV (e.g. fast food companies at lunch time on Saturday, cars at night, perfumes at Christmas, etc.). Advertisements targeted broad demographic segments (through television, radio, newspapers, magazines) and advertisers learnt by doing, by crossing data from sales (generic data obtained from field sales teams) and the dates in which the publicity was inserted in a particular media. On its turn, media markets brought about differentiated product spaces that guarantee self-selection of consumers towards the efficient product, overcoming in this way the difficulties of offering customized products to the end users (Kremhelmer & Zenger, 2008).

The firms' market addressability capacities have been leveraged by online data gathering together with the analysis capabilities of data mining systems. Technology offers businesses and marketing specialists the ability to collect (and, later on, to process) immense amounts of private data about individuals' interests or characteristics (whims, opinions, tastes, purchase history) as they surf the Internet and input personal information. All this information constitutes the profile, that companies build assembling and reassembling the information. At the end, profiles allow them to treat each customer or prospect as a unique and identifiable target.

Go-through gateways to the Internet are in the best position to collect and process personal data on an unprecedented scale. Search engines and social networks have replaced portals in this function and exploit their customers' databases to target the audience to whom advertisers would like to sell their products to. They have de facto become the in-between platform who links advertisers and potential customers.

As refers practical operation, targeted advertising works as follows. Advertisers only pay when the user clicks on the advertisement that redirects to the advertiser's website. If there is more than one offer for a specific advertising space, algorithms used by platforms assign the space to top bidders. This is typically the case of search platforms. Advertisers enter the text of the advertisement, the specific word or phrases ("keywords") in which they would want the advertisement shown, and the bid amount (the maximum amount they are willing to pay each time someone clicks their advertisement) and total budget into the system. Then, when a user sends a query, the search platform finds the set of advertisements containing the keywords that match the query and establishes the advertisements to be shown and the position in the web page. The amount of money to be paid is therefore determined by the bids of other competing advertisers, struggling for getting the best positions in the results pages (see details in Varian, 2009). Goto.com was the first company, in 1998, to price and display advertisements based on search terms or keywords and to hold continuous automated auctions to price the advertising of these keywords. Google was the first large search engine to use this model, and Yahoo and MSN eventually followed (Goldfarb & Tucker, 2008).

THE BEHAVIOURAL ADVERTISING ECOSYSTEM

For understanding the development of the advertising business around personal information and how new business models can be established, it seems necessary to depict a business ecosystem that helps to locate the players and the flows of information, money and services among them. The ecosystem metaphor is useful to refer to a high number of players that interact within a given environment in which none of them is able to control it completely; thus, both collaboration and competition occur at the same time.

Figure 1 offers a comprehensive approach to the behavioural advertising scenario. Money, personal data and products and/or services represent the vectors of value exchanged in the ecosystem. Users, platforms, advertisers and content producers are the main actors, but the list of companies that also capture and/or manage personal data, as result of their activity, must include banks, insurance companies, and data analysis companies, as well as public stakeholders.

Figure 1. Comprehensive business ecosystem on personal data
Source: Own elaboration

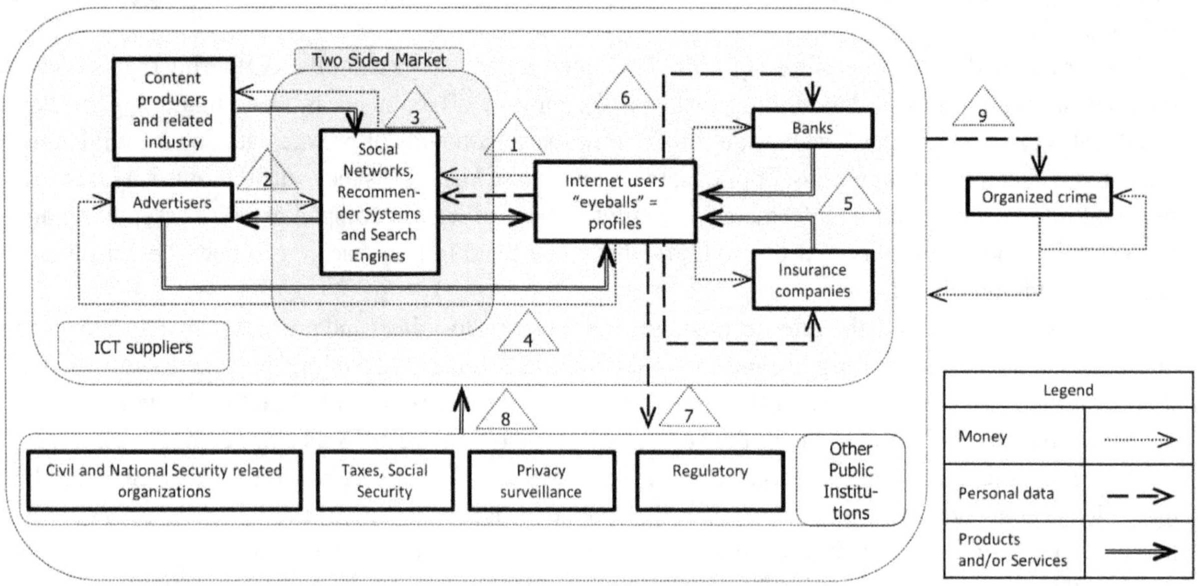

The centre of the diagram is occupied by a two-sided market "managed" by platforms. Platforms are usually search engines, social networks, or recommender systems. Other type of firms, though, can take this role of acting as an intermediary, provided they attract enough visitors to their websites or applications to generate revenue from advertising. To this extent, there is an open debate on whether mobile websites or dedicated applications ("apps") will capture higher attention quota from Internet users accessing contents (OECD, 2013).

On the one side of the market, platforms lure users (or "eyeballs" in the marketing terminology) that finally become profiles after federating from different information sources, for the most part the prints left by the Internet user in that particular platform. Triangle 1 (T1) represents this fact: personal data is traded off by users as long as they receive (usually) free services from the platforms. Platforms can deploy three basic strategies after collecting personal data (Christiansen, 2011):

1. Collect personal data, then anonymize and aggregate it to use it internally and/or to sell it to third parties. Without going into the details of the individuals, overall characteristics of a website's users are a valuable information for advertisers, indeed, the kind of data they were used to deal with. The platform itself could also use this information to strategize regarding content decisions and selling of advertisements.
2. Collect personal data, keeping personal data within the company but providing the opportunity for advertisers to specify a certain range of traits for target marketing. Advertisers pay a premium for advertisement placement directed to those users who meet the advertisers' specific criteria.
3. Collect personal data with the intention of selling the information, sometimes including specific profiles or names, to third parties. The third-party firms then compile that information into a comprehensive database for extensive and long-term data mining. While this is a growing industry, many platforms are reluctant to sell customers' information because doing so could limit their ability to sell targeted advertising.

On the other side of the market, advertisers pay money to the platforms, which place publicity on their webpages (T2). If the advertisement catch the attention of the targeted user and the operation is finally successful, manufactures or service providers get money from users in exchange for products or services (T4). Even if both play a role in the same side of the two-sided market depicted, content producers have been differentiated from advertisers. Content producers refers to application developers (e.g. sellers of games through the platform), or media creators. The scheme (T3) is similar to the bidirectional flow of money and services of T2, but sometimes, instead of the usual pay-per-click, it is agreed that the platform retains part of the money paid by users to content producers when buying an application (which is included also in T1).

Banks, data analysis and credit assessment companies, and insurance companies are also relevant actors as they hold private data. Indeed, these kind of firms have started to include in their annual reports information on how many customers they have, and how much they are worth (as an intangible asset). When an individual subscribes an insurance policy (or applies for a credit), his/her personal data is converted into a profile (T6) in exchange for the agreed service (T5). With regard to the public sector, it represents the largest personal data holder. Citizen information, since the person is born, is captured and managed (T7) by the public sector, who provides public services (T8) like health, defence, or education. More, privacy surveillance and regulatory institutions (including public or private non-governmental organizations) play a fundamental role when defending individuals' right to privacy. Finally, there is a lack of knowledge about the importance of organized crime in the trade of private information (T9) but it is clear that plays a non-negligible role, taking profit from continuous information leaks.

BEHAVIOURAL TARGETING IN A MOBILE CONTEXT

In the mobile domain, behavioural targeting models can be much enhanced by location information and context awareness. Thanks to this combination, it would be possible to go beyond behavioural tactics to create a feedback system with dynamic control. Indeed, "the advertiser can present an ad when one is near a target location, then track that person to determine whether the ad has had the desired response" (Wicker, 2012) and modify it if/as required. This is important because it helps companies to improve the user experience, preventing the effect of overwhelming him/her with contents that the user is not interested in. On the contrary, brands are capable of targeting users at a particular moment and/or in a particular place in which they have identified a potential need. Consequently, positioning information before the eyes of a customer becomes a triple challenge: reach consumers more effectively with the right product, where and when they are ready to make a buying decision.

This type of efficient push advertising strategies are a sort of programmatic advertising, where advertisements are automatically triggered by a particular event and deployed according to a specific set of rules. If not considered as intrusive by users (as seen later, this is the crux of the matter), it can help to create stronger connections between advertisers and their audiences. In addition, the long-tail of advertising niche markets (Barrows, 2010) derived from the variety of daily activities, local cultures and lifestyles of mobile users should be emphasized, under the assumption that aggregating huge numbers of small markets can be profitable for some firms. A decade ago, these were ignored because the price of identifying, tailoring products, and distributing to small niche markets exceeded the economic benefits. In contrast, today, the combination of big data techniques and affordable costs for streamlined package delivery process have fostered the attractiveness of understanding the tastes of a minority that could become a significant share of the market when massively reached.

Other relevant drivers for the development of mobile targeting strategies are the increasing availability of context-based metadata-enabled content (content enrichment or augmented reality) (Ynduráin et al., 2010), bio-parameters sensors and cognitive technologies, as well as the diffusion of mobile payment schemes that can encourage impulse purchases.

However, it is not all good news. After considering potential advantages that the mobile context offers, the drawbacks for those firms that rely on advertising to sell their products, or even for those companies that want to build a business model based on advertising, should be analysed. There are a number of them.

First, technical questions should be addressed. The adaptation of advertising formats to the mobile environment is not apparent (Nicolaou, 2013). Even worse, the type of terminal, desktop vs. handset, creates relevant differences. The effort can be even twofold: when it comes to advertising the media experience is drastically different from a tablet to a smartphone. Indeed, the large number of mobile devices with different specifications currently available affects consumers' ability to receive and view advertising messages properly (Chen & Hsieh, 2012). That could be accentuated in the future: projections of mobile traffic forecast four times more exabytes exchanged in 2017 (11.2) than in 2014 (2.8) while, apparently, many operating systems and devices will continue coexisting in the market. More, advanced mobile advertising requires further development of existing technologies; in particular mobile browser cookies do not operate in the same way as cookies in the desktop.

Second, information contained in mobile interaction has multiplied and it is neither apparent, as for now, understanding and managing it, and ultimately compiling the knowledge advertisers seek. Moreover, behaviour of users when using mobile handsets is different than the "fixed" behaviour that companies have learnt to understand. Mobile sessions are generally shorter, more casually triggered and full of distraction. This will be true, at least for as long as mobile broadband acts as a complement and not a substitute for fixed broadband. This is something, though, that is rapidly changing. Indeed, for many users, time spent on mobile devices has already surpassed time spent with desktop devices.

Third, a strong one-to-one connection with the audience in a very personal environment is an opportunity but also entails serious risks. The mobile device is the natural tool to bridge the physical world surrounding the user with the wealth of information on the net but it will be the user who is the one making decisions and using her/his preferences to create this link. The diffusion of bad practices could lead to a stricter regulation or even to undesirable situations such as a deep-seated lack of trust (Feijóo et al., 2010). Striking the right balance between privacy, personalization of services and sustainable business model remains then a foremost challenge (Gómez-Barroso et al., 2011).

In this regard, it is absolutely imperative to explore and understand the attitudes of users towards behavioural targeting, in particular in a mobile scenario. This is one of the areas in which research is being conducted in recent years. In general terms, the result from studies shows that users do not have highly positive attitude towards mobile advertising. Purcell et al. (2012) put figures to this rejection: 68% of participants in their survey were "not okay" with targeted advertising because they didn't like having their online behaviour tracked and analyzed. Some factors can help to gain acceptance of mobile advertising:

- Okazaki and Molina, (2012) find, no surprises, that perceived ubiquity (the ability to access information from any location at anytime) strengthens trust and positive attitude toward the mobile advertising. The survey was conducted with a sample of Japanese consumers and considers the perceived ubiquity as a two dimensions construct: time saving (Internet is an efficient way to manage my time, makes my life easier, Internet sites fits with my schedule) and spatial flexibility (information available at any time, overcoming spatial limitation, accession from any location).

- Amen (2010) investigates on the factors influencing consumer attitudes toward mobile advertising. The author concludes, again no surprises, that providing value for the consumers when using these kinds of mobile marketing services, through permissions and incentives, may drive to a higher user acceptance. In the same vein, Watson et al. (2013) conclude that acceptance can be enhanced by permission marketing, trust-building, creating a sense of being in control, and useful and entertaining website content.
- Message relevance improves response rates (Varnali & Yilmaz, 2012), as well as "infotainment" (information-based media content that also includes entertainment content) and credibility, as demonstrated by Liu et al. (2012). One of the hypotheses that this research manages is the influence of cultural aspects, being culture "the interactive aggregate of common characteristics that influence a group's response to its environment" (Liu et al. citing Hofstede, 1980). Indeed, culture is the value that most influences consumer behaviour and international marketing on the Internet (Shankar & Batra, 2009). For instance, the analysis of the reactions of different cultures may vary depending on the way information about individuals is shared, degree of transparency and frequency. All these cultural aspects may suggest links with the profitability differences of the personal information records from one geographical area to another.
- Moderate repetition and low time pressure also contributed to better mobile advertisement effectiveness (Rau et al., 2014). One of their findings is that mobile advertisement effectiveness is better under low time pressure context than under high time pressure context. Thus advertisers should preferentially choose time space when customers are relaxed, if they are able to access and use this information.
- Jayawardhena et al. (2009) talks of institutional trust (which refers to the wider trust that the consumer has on the institutional environment, including legal, cultural and political institutions, civil societal organizations, and the media) as the main factor affecting the consumers' decision to participate in mobile marketing.
- From the emotional perspective, Yang et al. (2013) researched on the links between utilitarian and hedonic considerations on the responses to mobile advertising, which is affected by the characteristics of the add communication and by users' voluntary choices of mobile technology. The study is conducted with Korean consumers and shows that a model that integrates technology evaluation and emotion based evaluation has a stronger predict capabilities on consumer's response to the mobile advertising. A close research can also be found in Lin et al. (2013) who highlight the fact that the enrichment of the ads with image features induces more favourable attitudes from the consumer.
- Finally, a number of papers analyze key drivers of users' attitude toward mobile advertising in the case of younger people. Martí Parreño et al. (2013) find that entertainment, irritation and usefulness are key drivers of teenagers' attitude toward mobile advertising. Sultan and Rohm (2008), in a study conducted in the United States and Pakistan, determine that the significant predictors of mobile marketing acceptance among young Americans are usage characteristics, privacy concerns, consumer innovativeness, attitudes toward mobile communications and the need for permission-based communications in the mobile space; of those factors, the only two that were backed by the Pakistani data were usage characteristics and permission-based involvement.

As a result of all the above, companies need to be open and willing to try different approaches. In this regard, as traditional advertising vehicles are more and more often rejected by consumers in the mobile environment, native advertising can be a solution. In native advertising, the advertiser attempts to gain attention of users by including references to the product in the context of the user's experience. Product placement (also known as embedded marketing) in movies, video games or music videos (Soba & Aydin, 2013) is the best example, even if the risks of lack of control on the way the brand is portrayed in the "scene", negative association with undesired values, and ethics on aggressive commercialization remain the same as in the past decades with the brands taking part of movies or music videos. New ways of interacting with products are becoming richer in terms of end user experience, like product linking (or "plinking"), that allows the viewer to click on objects and images and access to the website where the product or service is commercialized. A short history of product placement can be read in Glickman and Kim (2012), who also explain how technology drives new trends in product placement: virtual product placement (also known as "digital product placement", or "digital branded integration") in social media is a recent phenomenon and has enhanced a variety of approaches based on the way that networks connects their members. For instance, Twitter is used as meeting place for buyers and sellers of product placement spots. Moreover, brand companies are increasingly reaching agreements with celebrities for getting endorsement to their products in social media networks.

Tying in with the last point, the implications of social connections of users open too an enormous field for experimentation with new advertising approaches. Zhou and Dong (2014) analyse how network interaction in social media affects consumers' purchase intentions from four perspectives: the place, the content, the features, and the participants. The place where the interactivity is developed, which is represented by the community's goal and orientation, impacts in the success of the virtual community (in terms of posts, reactions and readers). It is also key the perception of the members about the consistency of the recommendations presented to the community, and the technical assurance (equivalent to the website's stability, design, privacy, security, reliability, ease of access, response time and flexibility). The second perspective is the content completeness of the product specifications information and the visual information provided in the comment provided to the community, which may reduce the risks of suspicions by the purchaser. The third aspect corresponds to the features, i.e. the openness of the message (providing positive sentiments and behaviour), the perceived independence, and the conversational capabilities that imply participating in a network that breaks the boundaries of time, space and own experience dissemination. The last point refers to the participants, and the perceived relationships established with them: purchase intention is positively impacted by the familiarity with the commenter, the degree of expertise assigned to that person, or the degree of intimacy.

In connection with recommenders, the understanding of their role is motivating some experimentation about "word-of-mouth" advertising strategies, known as viral marketing. The findings from Pescher et al. (2014) suggest that consumers who place high importance on the purposive value and entertainment value of a message are likely to enter the interest and referral stages. It is also worth not forgetting that some successful viral marketing examples have been very carefully planned. Long and Wong (2014) study how to minimize the number of targeted users in a social network ("seeds") that will spread the messages to their peers in the network (family, co-workers, friends…) who get "influenced" or become aware of that (new) product of service. This is done considering the use of incentives for these seeds, whose identification is the first bottleneck in this marketing technique design. In the age of big data, technologies ease large sample experimentations, and this is exactly what Facebook did during the first week of 2012, with a massive social engineering study in which the population was 689.003, and the final sample 155.000 participants selected because they "posted at least one status update during the experimental period" (Kramer et al., 2014). The study shows "that emotional states can be transferred to others via emotional contagion, leading people to experience the same emotions without their awareness (…) emotional contagion occurs without direct interaction between people (exposure to a friend expressing an emotion is sufficient), and in the complete absence of nonverbal cues".

CONCLUSION

The success of new mobile content, applications and services is linked with the appearance of affordable and scalable business models. As for today, advertising is the main model, although some impression remains that many firms bet on advertising possibly and simply because there was not a better idea on how else to make mobile applications profitable.

Behavioural targeting has become critical also for off-line businesses. Advertisers feel that they are not getting the impact and sustainability of their brands by just doing "business as usual" advertising. In particular, the progressing fragmentation of the audience structure calls for novel ways (ideally "real-time") to identify and approach the target audience. Moreover, since users are increasingly able to manage a number of media, a unique media advertising strategy (i.e., television) will miss a growing number of users or miss the time they devote to a different media usage. At the end, the advertising golden rule is still valid: to be where the potential impact is. One just has to look around to find easily out what the impact of mobile devices is.

Behavioural targeting is therefore seen as a main avenue for the development of a digital economy that every company (country) must adapt to. In such a scenario, it is easy to understand why personal data gathering has become a sort of obsession, and why big data is in the agenda of boards in almost every large company.

All this does not go without dangers. The risks of an excess of commercialisation of user profiles without the tools and institutional regimes in place to empower users deserves an additional reflection. Introducing mobile advertising in the personal communications space would require not just finding the relevant advertising that fits into the person's behaviour, but creating trust based on understanding, agreement and tangible truth, as well as some knowledge of the situation, physical and emotional, of the user.

Mobile behavioural targeting is still on its infancy. We are witnessing the development of tools and strategies that need time to ripen... or to die. On the medium term, its acceptance and success will be determined, no exception to any other rising social, cultural or economic trend, by its compatibility with the lifestyle of the moment, rather than by its technological capacity.

ACKNOWLEDGMENT

This research has been supported by the Spanish Ministry of Economy and Competitiveness through the National Programme for Research, Development and Innovation Oriented to Society Challenges (Project ECO2013-47055-R).

REFERENCES

Amen, U. (2010). Consumer attitude towards mobile advertising. *Interdisciplinary Journal of Contemporary Research in Business*, *2*(3), 75–104.

Barrows, J. (2010). The Jane Austen Industry and long tail marketing. *Yale Economic Review*, *6*(2), 36–38.

Belk, R. W. (2013). Extended self in a digital world. *The Journal of Consumer Research*, *40*(3), 477–500. doi:10.1086/671052

Berman, S. J., & Kesterson-Townes, L. (2012). Connecting with the digital customer of the future. *Strategy and Leadership*, *40*(6), 29–35. doi:10.1108/10878571211278868

Blattberg, R. C., & Deighton, J. (1991). Interactive marketing: Exploiting the age of addressability. *Sloan Management Review*, *33*(1), 5–14.

Bontempi, M. E., & Mairesse, J. (2008). *Intangible capital and productivity. An exploration on a panel of Italian manufacturing firms*. NBER (The National Bureau of Economic Research) Working Paper No. 14108. Cambridge (USA). Retrieved from http://www.nber.org/papers/w14108

Chang, W., Chang, C., & Li, Q. (2012). Customer lifetime value: A review. *Social Behavior and Personality*, *40*(7), 1057–1064. doi:10.2224/sbp.2012.40.7.1057

Chen, P. T., & Hsieh, H. P. (2012). Personalized mobile advertising: Its key attributes, trends, and social impact. *Technological Forecasting and Social Change*, *79*(3), 543–557. doi:10.1016/j.techfore.2011.08.011

Christiansen, L. (2011). Personal privacy and Internet marketing: An impossible conflict or a marriage made in heaven? *Business Horizons*, *54*(6), 509–514. doi:10.1016/j.bushor.2011.06.002

Feijóo, C., Gómez-Barroso, J. L., & Martínez Martínez, I. J. (2010). Nuevas vías para la comunicación empresarial: Publicidad en el móvil. *El Profesional de la Información*, *19*(2), 140–148. doi:10.3145/epi.2010.mar.04

Feijóo, C., Gómez-Barroso, J. L., & Voigt, P. (2014). Exploring the economic value of personal information from firms' financial statements. *International Journal of Information Management*, *34*(2), 248–256. doi:10.1016/j.ijinfomgt.2013.12.005

Glickman, L., & Kim, A. (2012). Product placement and technology: Developments, opportunities, and challenges. *The Entertainment and Sports Lawyer*, *30*(1), 30–41.

Goldfarb, A., & Tucker, C. (2008). Economic and business dimensions: Search engine advertising. *Communications of the ACM*, *51*(11), 22–24. doi:10.1145/1400214.1400222

Gómez-Barroso, J. L., Feijóo, C., & Compañó, R. (2011). Promising prospects in mobile search: Business as usual or techno-economic disruptions? *IEEE Signal Processing Magazine*, *28*(4), 131–135. doi:10.1109/MSP.2011.941135

Greengard, S. (2012). Advertising gets personal. *Communications of the ACM*, *55*(8), 18–20. doi:10.1145/2240236.2240243

Hofstede, G. (2001). *Culture's consequences: Comparing values, behaviors, institutions, and organizations across nations*. Thousand Oaks, USA: Sage Publications.

Jayawardhena, C., Kuckertz, A., Karjaluoto, H., & Kautonen, T. (2009). Antecedents to permission based mobile marketing: An initial examination. *European Journal of Marketing*, *43*(3/4), 473–499. doi:10.1108/03090560910935541

Kramer, A. D. I., Guillory, J. E., & Hancock, J. T. (2014). Experimental evidence of massive-scale emotional contagion through social networks. *Proceedings of the National Academy of Sciences of the United States of America*, *111*(24), 8788–8790. doi:10.1073/pnas.1320040111 PMID:24889601

Kremhelmer, S., & Zenger, H. (2008). Advertising and the screening role of mass media. *Information Economics and Policy*, *20*(2), 107–119. doi:10.1016/j.infoecopol.2007.11.003

Lin, H., Yeo, B., & Chen, Y. (2013). Mobilizing timely location-based advertising: A study of effectiveness on persuasion. *Australian Journal of Business and Management Research*, *3*(6), 15–29.

Liu, C. L., Sinkovics, R. R., Pezderka, N., & Haghirian, P. (2012). Determinants of consumer perceptions toward mobile advertising: A comparison between Japan and Austria. *Journal of Interactive Marketing*, *26*(1), 21–32. doi:10.1016/j.intmar.2011.07.002

Long, C., & Wong, R. C. W. (2014). Viral marketing for dedicated customers. *Information Systems*, *46*, 1–23. doi:10.1016/j.is.2014.05.003

Martí Parreño, J., Sanz-Blas, S., Ruiz-Mafé, C., & Aldás-Manzano, J. (2013). Key factors of teenagers' mobile advertising acceptance. *Industrial Management & Data Systems*, *113*(5), 732–749. doi:10.1108/02635571311324179

Nicholas, D., Dobrowolski, T., Withey, R., Russell, C., Huntington, P., & Williams, P. (2003). Digital information consumers, players and purchasers: Information seeking behaviour in the new digital interactive environment. *Aslib Proceedings*, *55*(1/2), 23–31. doi:10.1108/00012530310462689

Nicolaou, A. (2013). Best practices on the move: Building web apps for mobile devices. *Communications of the ACM*, *56*(8), 45–51. doi:10.1145/2492007.2492023

OECD. (2013). *The app economy. OECD Digital Economy Papers, no. 230.* Paris, France: OECD Publishing.

Okazaki, S., Molina, F. J., & Hirose, M. (2012). Mobile advertising avoidance: Exploring the role of ubiquity. *Electronic Markets*, *22*(3), 169–183. doi:10.1007/s12525-012-0087-1

Pescher, C., Reichhart, P., & Spann, M. (2014). Consumer decision-making processes in mobile viral marketing campaigns. *Journal of Interactive Marketing*, *28*(1), 43–54. doi:10.1016/j.intmar.2013.08.001

Purcell, K., Brenner, J., & Rainie, L. (2012). *Search engine use 2012*. Pew Internet & American Life Project Search. Retrieved from http://pewinternet.org/Reports/2012/Search-Engine-Use-2012.aspx

Rau, P. L. P., Zhou, J., Chen, D., & Lu, T. P. (2014). The influence of repetition and time pressure on effectiveness of mobile advertising messages. *Telematics and Informatics*, *31*(3), 63–476.

Shankar, V., & Batra, R. (2009). The growing influence of online marketing communications. *Journal of Interactive Marketing*, *23*(4), 285–287. doi:10.1016/j.intmar.2009.07.006

Soba, M., & Aydin, M. (2013). Product placement efficiency in marketing communication strategy. *International Journal of Business and Management*, *8*(12), 111–116. doi:10.5539/ijbm.v8n12p111

Sultan, F., & Rohm, A. J. (2008). How to market to generation M(obile). *MIT Sloan Management Review*, *49*(4), 34–41.

Tallon, P. P. (2013). Corporate governance of big data: Perspectives on value, risk, and cost. *Computer*, *46*(6), 32–38. doi:10.1109/MC.2013.155

Varian, H. (2009). Online ad auctions. *The American Economic Review*, *99*(2), 430–434. doi:10.1257/aer.99.2.430

Varian, H. (2014). Big data: New tricks for econometrics. *The Journal of Economic Perspectives*, *28*(2), 3–28. doi:10.1257/jep.28.2.3

Varnali, K., Yilmaz, C., & Toker, A. (2012). Predictors of attitudinal and behavioral outcomes in mobile advertising: A field experiment. *Electronic Commerce Research and Applications*, *11*(6), 570–581. doi:10.1016/j.elerap.2012.08.002

Watson, C., McCarthy, J., & Rowley, J. (2013). Consumer attitudes towards mobile marketing in the smart phone era. *International Journal of Information Management*, *33*(5), 840–849. doi:10.1016/j.ijinfomgt.2013.06.004

Wicker, S. B. (2012). The loss of location privacy in the cellular age. *Communications of the ACM*, *55*(8), 60–68. doi:10.1145/2240236.2240255

Yang, B., Kim, Y., & Yoo, C. (2013). The integrated mobile advertising model: The effects of technology- and emotion-based evaluations. *Journal of Business Research*, *66*(9), 1345–1352. doi:10.1016/j.jbusres.2012.02.035

Ynduráin, E., Feijóo, C., Ramos, S., & Campo, C. (2010). Context-aware mobile applications design: Implications and challenges for a new industry. *The Journal of the International Telecommunications Professional*, *4*(4), 16–28.

Zhou, X., & Dong, C. (2014). Study on the impact of network interactivity on purchase intention in the social media era. *International Journal of Business and Social Science*, *5*(6), 259–265.

KEY TERMS AND DEFINITIONS

Behavioural Targeting: Set of techniques that companies operating online use for delivering the (supposedly) right advertising message to the right audience on the right device and at the right time.

Big Data: Multidisciplinary concept that, overall, refers to the massive capture, management and data exploitation with the purpose of generating economic value.

Ecosystem: Applied to a communications mobile environment, the set of interconnected actors that interact exchanging data, information and economic value.

Intangible Assets: Those without physical or financial nature. They consist on computerised information (such as software and databases); innovative property (such as scientific and non-scientific R&D, copyrights, designs, trademarks); and economic competencies (including brand equity, firm-specific human capital, networks joining people and institutions, organizational know-how that increases enterprise efficiency, and aspects of advertising and marketing).

Mobile Behaviour: Individual's conduct in a communication mobile environment.

Personal Information: Equivalent to personal data, refers to any information related to an identified or potentially identifiable person.

Platform: In the context of this chapter, it refers to an intermediary (such as a search engine, a social network, or a recommender system) through which the other players in a two-sided market interact.

Search Engine: Internet web service that provides information on a structured information and human comprehensible design.

Chapter 4
A Reflection on Wearables and Innovation in the Mobile Ecosystem:
Two Possible Scenarios

Andreu Castellet
University of Murcia, Spain

ABSTRACT

This chapter explores the innovation landscape for the mobile content ecosystem building upon the hypothesis of the disruption potential of wearables. Google Glass is taken as the paradigm, as it offers direct access to network, becoming an eventual alternative to the smartphone. Equipment manufacturers lost traction as innovators right after the arrival of the iPhone in 2007, which brought a key innovation to the ecosystem. Observing the strategies of the different players operating in the supply side, two innovation scenarios are proposed for the debate: a sustaining-innovation based scenario, and a disruptive-innovation one. The main findings admit a clear disruption potential in wearables, mostly focused on the power of glasses as a driver to create a new wave of content, oriented towards augmented-reality features. However, it is still uncertain if these new devices can trigger a new era of innovation expanding its effects towards the whole ecosystem.

INTRODUCTION

The irruption of a new generation of mobile handsets, led by Apple's iPhone in 2007, paved the way for a whole set of events that transformed both the content and the telecommunications industries. The app store economy, the platform-ruled mobile media ecosystem, now a Google-Apple duopoly, was triggered in the first place by the eclosion of a new device. In the pre-iPhone era, mobile network operators had dictated the pace of the evolution and the innovation within the mobile content ecosystem. An innovative step from the terminal vendors sector changed the rules of the game forever.

DOI: 10.4018/978-1-4666-8838-4.ch004

However, little decisive innovation steps of the device makers have changed the picture in the mobile content business after that arrival of the iPhone, which was coincidental with Google's Android platform, and their competitors followed both incumbents.

Amid this ongoing discussion on the future of innovation within the mobile ecosystem, the aim of this chapter is to bring some ideas about the evolution of leading strategies in mobile content for the next future. Communication studies have often envisaged the mobile space as a mere channel, disregarding the fact that it involves a whole new experience that deserves more attention. However, over the years we can benefit of a remarkable corpus of research on user behavior, cognitive aspects of mobility, and the eclosion of mobile media. Innovation, though, is one of the research issues that need more attention efforts from the communication scholars' community.

We will look at the studies on innovation building in the first place upon Schumpeter's approach of "creative destruction", introduced in 1943, in which innovation necessarily involves the extinction of some incumbents in order to give way to new products and services. In that environment, price doesn't make the big difference; it's "the competition from the new commodity, the new technology, the new source of supply, the new type of organization (...) that commands a decisive cost or quality advantage and that strikes not at the margins of the profits and the outputs of the existing firms but at their foundations and their very lives" (Schumpeter, 2003: 84).

By the end of the Twentieth Century, another economist, Clayton Christensen (1996), went further on the topic establishing two types of innovation: "sustaining innovation", driven by the mere improvement of actual products; and "disruptive innovation", which aims to replace the actual product line, mostly acting on the "friction" issues (Akerman, 1998).

However, innovation impact is not an easy magnitude to evaluate. Before diving deeper into the matter, we are reminded by Hauser, Tellis and Griffin that: "a variety of disciplines address various aspects of innovation, including marketing, quality management, operations management, technology management, organizational behavior, product development, strategic management, and economics. Research on innovation has proceeded in many academic fields with incomplete links across those fields" (2006: 687).

We venture in more complex territories if we add the concept of disruption. Daneels (2004), reminds us that disruption can be only verified after a certain state of things, led by a number of incumbents, has been altered after the entry of a new product of service. According to that, we should not be able to fully recognize a disruption process unless it has reached its culmination.

That being said, we propose to focus our research on the innovation issue and its impact on the different players involved in the mobile ecosystem. In order to offer a pragmatic, not too perishable approach, we will deploy a two-fold study, based on two basic scenarios: in scenario number one, we will adopt the hypothesis of an ecosystem mostly driven by sustaining innovations, in which innovations in content, services and devices are mostly aimed to perform better the kind of tasks they have been conceived for so far; in scenario number two, we will start from a scenario of disruption, led by the impact on the different players of the new generation of wearable, prosthetic devices.

The kind of methods and materials we will use in order to fill both scenarios will be basically of the same kind. Secondary sources of different kinds are on the base of this chapter.

The pace of events that have shaped the mobile content ecosystem has been hectic enough to place the academic field well behind the newest innovations. Both supply and demand sides of the mobile media market have been better observed from the players point of view –only a fraction of which is usually publically available-, and from the set of expertise the mobile sector needs to set up their strategies, than from the strictly academic area. Thus, we will have been collecting data and insights from the original equipment manufacturers (OEM) sector; the network and communications equipment providers; the content aggregators and the platform leaders; also from the consulting business, which produces a regular flow of data and insights; also international bodies of various kinds, and scientific institutes devoted in several ways to connect with our field of study, have sourced this work.

Furthermore, the impact of mobile ecosystem news on media agenda is being recorded by news providers of all types, bloggers and a number of individuals that have helped us to build our prospective picture in terms of strategies and their effect on the media and on the telecoms businesses.

All the above written leads us to propose our research starting from the following questions:

1. Can mobile device producers retake leadership or influence in the mobile ecosystem innovation activities in a near future in a way comparable with what they triggered with the arrival of the iPhone?
2. In what ways has the creative destruction paradigm shown its impact on the mobile ecosystem, driven by the platform pattern of development?

BACKGROUND

This part has the only aim to offer the reader a certain perspective of the author's theory starting point on the matter, which can be also of some use to any newcomer to the topic of mobile media. However, it is not intended to summarize the history of global mobile media research, which is not the purpose of this paper.

Despite we are dealing with one of newest fields within the new media studies space, we don't ignore that our work starts with a thorough literature review on a wide topic, to the extent that the task of building a thorough corpus of knowledge on mobile media largely exceeds the most approaches made only from conventional communication studies. As Goggin and Hjorth point: "the process of theorizing mobile as media is still only in the early stages of being integrated into, taking advantage of, and enriching and complicating the heritage and traditions of studying media, So we still discern a 'disconnect' between mobile studies, on the one hand, and media studies, on the other hand, Just as much work in media studies is blissfully unaware (even still) of the state-of-the-art in, say, Internet studies, The same might be said of cultural studies, which, ironically perhaps given its remit and self-identities, has been relatively slow to engage with mobile media" (2014:1).

Since the very beginning of our short history, mobile media researchers have brought extremely valuable contributions to the communication studies. Many of those views came from non-native communication fields. The long-awaited arrival of UMTS or 3G networks, and their promise of a new range of content/service choices for European and US customers, was widely studied by Feijóo, et al., (2006), that emphasized on the potential for economic wealth of content delivery through mobile telephones, and start describing and connecting the different players within the ecosystem. Mobile network operators were then the main drivers in place, and the same authors were calling for an effort of terminal manufacturers in order to put at users' reach devices ready enough for consumption of the new data flows roaming through 2,5 G and 3G networks. As Dunnewijk and Hultén assessed, "the succession of newer generations' telecommunication techniques shifted the role of the network operators from being a network manager to a content organizer" (2007: 165). Jaokar and Fish (2006) were among the first to label those practices as "walled gardens" environments, in which telecom companies limited the users' range of options through a chain of alliances with OEMS and a small number of selected content and applications providers.

Figure 1. A layers' view of mobile ecosystem
Source: Adapted from Castellet and Feijóo, 2013.

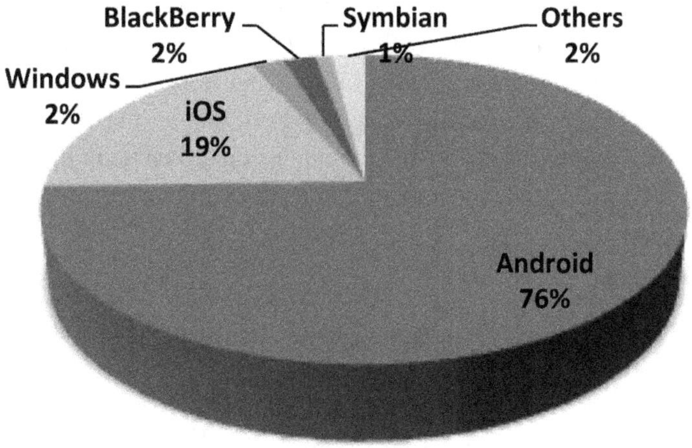

One of the most remarkable contributions of those views was the effort to identify and classify the different activities of the main players in the mobile content business as a whole. That kind of taxonomy is still valuable in order to analyze the ecosystem in terms of strategies (See Figure 1).

Also in the pre-iPhone era, Feldmann (2005) was exploring the different strategies –telecom companies, content and services providers, users- and its impact on the development of the ecosystem. Mobile content production and delivery were ordinary businesses in Japan since 1999 and NTT Docomo's i-mode, but that market formula failed everywhere else in the world, despite many attempts to replicate it in most European and North American markets.

By then, a number of views primarily related to business studies had noticed that the platform pattern that had made success in big sectors of technology business was shaping new relationships in the telecommunications environment. Some started speaking of "value networks" (Li & Whalley, 2002; Peppard & Rylander, 2006), as a concept that led towards a more complex view of firms interplay than the more classical Porter's concept of vertical integration (2008).

The apps and services eclosion triggered by the iPhone/Android arrival in 2007-2008 paved the way for approaches that added the platform component to the analysis of the ecosystem (Ballon, 2011; Gawer, & Cusumano, 2014; Goggin, Dwyer, & Martin, 2014). It was the assumption that the promise of a diverse, abundant, app economy had come to stay. Only a few had envisaged that potential before (Ahonen & Barrett, 2002; Feijóo et al., 2006). The app store/platform paradigm established itself as the prevalent environment in mobile, and also expanded many of its advantages to other consumer electronic businesses: connected TV, video games...

The Mobile Content Space

This chapter aims to integrate these, and other techno-economic views of mobile ecosystem, into the wide domain of communication studies, which has always been benefited by cross-discipline perspectives when looking at the communication experience, both in news and in entertainment content.

Scolari, Aguado, and Feijóo (2012) have worked comprehensively in taxonomies of the different kinds of mobile content, attending to a range of diverse criteria. For our current purpose, we will consider mostly their origin-oriented taxonomy, which will suit our needs, as we are mostly trying to focus on the supply side of the ecosystem. Those authors describe four main kinds of content, not mutually exclusive: a) adapted, for information intended to appear first in other media, which has been adjusted in order to be displayed and used on a mobile; b) repurposed, for creative content that have been reused and adapted to mobility; c) original or specific, for contents created having mobile in mind; and d) augmented, for contents conceived to take advantage of the specific features of mobile devices, in order to increase the value for users.

WEARABLES: A LONG-AWAITED PROMISE, COME TRUE WITH A HUGE SPACE FOR INNOVATION

Wearable devices have been ordinary news in the mobile scene for the last years, and their virtues have been praised by some researchers long before the emergence of mobile media: "The easiest way to improve intelligence is by augmenting the items we wear all the time—glasses, wrist-watches, clothes and shoes—with miniature computers, video displays, cameras and microphones" (Pentland, 1988: 91). If we go as far as Freud's work, we can find that idea of "prosthetic" devices attached to human body, tools that could made a man "a kind of prosthetic god". Prophetically, the Austrian thinker anticipated: "Future ages will bring with them new and probably unimaginably great advances in this field of civilization and will increase man's likeness to God still more" (1962: 39).

We cannot know if Sergey Brin, one of Google's founders, had Freud's vision in mind when thinking of Google Glass. In a more functional way, Brin explains that the main purpose of the new device is freeing human head of the need of looking down to be connected: "we ultimately questioned whether this is the ultimate future of how you want to connect to other people in your life, how you want to connect to information. Should it be by just walking around looking down?" (Brin, 2013).

Furthermore on the smartphone-related isolation issue, Brin makes his point by insisting on the need: "and also you want something that frees your eyes. That's why we put the display up high, out of your line of sight, so it wouldn't be where you're looking and it wouldn't be where you're making eye contact with people" (Brin, 2013). Brin's assessments, which bring light on the purpose behind Google Glass conception, allow a whole set of questions on the future of mobile media: can a non-immersive mobile device trigger a new evolution era in mobile media?

Anyway, the emergence of Google Glass, first as a prototype in 2011, and currently as an expected launch for the consumer market, has led the mobile wearables category of devices from futurology to news headlines. Opposite to the iPhone, whose appearance was carefully kept out of the public eye until it was unveiled to the media in 2007, a few months prior to its market entry, Google Glass ranks among the most widely announced pre-releases ever in the gadgets history (Google, 2012). Thanks to that flow of first-hand information, we know now that Google Glass will be a terminal fully connective with mobile networks, will be equipped at least with the same range of context-aware features, and will deploy a data processing power that compares to a smartphone's. In addition to that, its status of *prosthetic* device, attached to the eyes and the ears, resembling a pair of glasses, will benefit of an undetermined range of new features, related to augmented reality contents and services. ABI Research dares to forecast 2 million of Google Glass units sold in 2014 (Young, 2014).

At the same time, a diverse range of devices also labeled as *wearable* by vendors have entered the market of ancillary devices able to hold connection with smartphones and tablets. Some have been conceived to monitor a number of body functions, and fit into the vast category of m-health; others, not that dissimilar, perform some tasks as complements of sports and fitness activities, babysitting or even pets care; some can be used specifically for gaming. According to data compiled by Nielsen from Android users of Bluetooth-connected wearables, 60% of them are within the fitness category, followed by a 25% of smartwatch users; an 8% fit into the m-health segment, and the remaining 7% covers the rest of purposes. However, data about consumer interest on wearables are not that unanimous: a Fittkau and Maß survey (eMarketer, 2014) indicates that only a 1% of German customers were wearing a fitness band in 2014.

A particularly interesting category of consumer electronics products sold as wearables are *smartwatches*, a category under which coexist very different devices. Original equipment manufacturers (OEM) have shown a strong interest in occupying that niche of products, possibly trying to take advantage of their brand power; also, more smartwatches are in the pipeline of most of vendors who have not explored that field so far. The expectations of market potential for these devices are high: a Business Insider forecast (Danova, 2014), estimates that one in 20 smartphones will be connected to any kind of smartwatch in a near future, with sales surpassing globally 90 million units by 2018, with an average price of less than 100 $ each device. That study didn't get on time to cover Apple's Iwatch launch and its eventual influence in the market, specially its potential to increase the average price of terminals sold.

A query for "smart watches" in the amazon.co.uk shop under the "mobile phones & communications" category returned near to 1,000 results, from which we should exclude a number of secondary items, as screen protectors, etc. Anyway, the size of the market from the supply-side is relevant and is growing fast. Under the mentioned denomination coexists a diversity of objects: GPS receivers, NFC-or-Bluetooth-connected sensors. Some are merely able to give the time and warn on incoming messages; others can act as sports equipment, MP3 players and photo cameras... One of the very few things they all have in common is their primary wristwatch appearance. A watch with full-connectivity features is a rarity in 2014. One of the very few exceptions can be the Samsung Gear S watch (Lawler, 2014), scheduled to reach the market by the fourth quarter in 2014, equipped with a 3G modem, and ready to connect online and to make calls. According to ABI Research, the whole wearables market is reaching globally 90 million units sold in 2014 (Lawler, 2014).

In order to evaluate the disruptive potential of wearables seems reasonable establishing a more specific taxonomy among that vast category of objects. We will start by assuming, as a mere starting point, that "the purpose of wearable technology is to create constant, convenient, seamless, portable, and mostly hands-free access to electronics and computers" (Tehrani & Michael, 2014). Under such a view, any kind of the above mentioned items would be included.

Others, as Dudenhoeffer (2014), have created a taxonomy rather based on their appearance than on connectivity and data processing features. Such classification establishes the following categories: watches; glasses; clothing; smart bands; jewelry; and smart headgear.

But as we don't consider here appearance, but connectivity, as the ultimate potential for innovative disruption, we will focus on a taxonomy not based on appearance, vendors or primary use. For that purpose, we will start defining a wearable as a device that is worn on the body typically for long periods of time on a daily basis, which has processing and wireless capabilities. That definition is borrowed from a survey of Pyramid Research[1], a consulting firm, from which has been impossible to this author to get a proper attribution for the quotation.

Table 1. Changes in app developers' environment

	Telecoms World	Software World
Success factor	Installed base	Number of apps
Primary customer	consumer	developer
Speed of innovation	1 OS version every 2 years	5 OS versions/year
Time to market	1-2 years	1-2 weeks
Type of services	comms-centric	catering to entire needs portfolio
Risk-taking	predictability / de-risking	entrepreneurship / uncertainty
Access to innovation	100s of close partners	100,000s of developers
Business model	B2B licensing	B2C sales/ads/in-app sales
Channel to market	voice, text and web	smartphones
Discovery	On deck / on device	App store
First step	"we need to sign an NDA"	"we need to download the SDK"
Process	Waterfall: RFI, RFQ, deliver, QA	Agile: add feature, build, test, repeat
Attitude	"developers will come to us"	"we need to go to developers"

Source: Constantinou et al., 2011

Taking our taxonomy effort from that definition, for our current research purpose, we will not consider as wearable any device attached to the human body that cannot offer processing and *always on* connective features, unless not stated otherwise.

Professional Content Creators: App Developers and Legacy Media

App Developers

In our current mobile ecosystem, led by platforms, which act as necessary agents between users and content producers, the app has become the product unit. As Maghiros et al. (2007) clarified, apps, considered as part of the creative sector, include both creative content and processed information. They come often bundled, appearing as a single service to the public.

That comes as the result of a transformation story closely linked to the evolution of the mobile ecosystem, from the former "walled garden" order, to the current platform-dominated environment. Constantinou, Vakulenko and Kapetanakis (2011) describe the new set of rules under the idea of developers' activity change of environment, from "telecoms" to "software" paradigms, as displayed in table 1.

The promise of a "long tail economy", under which many could live on the success of blockbuster sales of a very few number of apps, has not delivered the expected wealth (Boudreau & Jeppesen, 2010). "Not many are ready to go through the trial and error process that involves market success: Angry Birds was creator Rovio's 52nd title", explains Vision Mobile in its survey of the developers' community (Wilcox & Voskoglou, 2014: 25). As Castellet and Feijóo point, "monetization and visibility are the two biggest challenges that the developer community is facing" (2013: 34) (See Table 2).

The income figures shown in table nr. 2 come from a world survey among mobile developers. The average developer that emerges from it employs three persons, and has to cover also a number of general expenses, as middleware tools, equipment maintenance and amortization, maybe premises, taxes, legal

Table 2. The four tiers of developers, according to Vision Mobile

Group	% of Developers	Revenue/Month per App ($)
We have nothings	47%	0-100
The Poverty Stricken	22%	100-1,000
The Strugglers	19%	1,000-10,000
The Haves	12%	10,000+

Source: Wilcox and Voskoglou, 2014

fees, etc. Once the net income figures come out, it is not hard to figure out how challenging the environment becomes to them, ruled by that "winner takes all" (Wilcox & Voskoglou, 2014) logic. Ahonen paints a still more depressing picture: "96% of all app developers do not break even" (2014e).

Another issue that Vision Mobile raises is the fact that 80% of app stores revenues come from games. Thus, monetization happens mostly in the games-making segment, but just a few ones can make a life on it: 57% of the games make less than 500 $ per month.

Scenario 1: Darwin Revisited

Given the figures above exposed, it's hard to imagine how the developers' community will keep growing at the same pace in the next years. Apart from the amateurs who make apps for fun or for hobby, and the thousands of outdated *zombie apps* buried in the app stores for years, an environment of sustaining innovation should push the developers community into a Darwinian process where only the best adapted will survive in the long term. There are not signs of change in users behavior that prompt to think that, either their loyalty or their apps budget, are going to change dramatically.

Developers have been mostly focused on the business of specific, repurposed and even adapted kinds of content, as we can figure out of the market weight of game apps. Just a very few have ventured into the promising fields of augmented content. A sustaining innovation environment, with increasing presence of ancillary wearable devices, can be a source of wealth to developers engaged with apps other than games, particularly the ones who work in the business of making apps for non-mobile companies, and also the enterprise/cloud-solutions developers (Wilcox & Voskoglou, 2014). The space for augmented contents, oriented towards augmented-reality of virtual-reality experiences should be much bigger than now.

Scenario 2: A Change of Course

With wearables leading the content demand, the developers' community can face a new kind of dilemmas. Since the birth of contemporary mobile platforms, content creators have been prompted to choose between two different starting scenarios: either developing for mobile and eventually porting the app to non-mobile environments was one way to make business; or porting to mobile, creating adapted, repurposed, original or augmented content.

In both cases, the result was a rather immersive experience of content consumption. But Google Glass can place the developers' dilemma in completely new terms: immersiveness or not immersiveness. In other words: as wireless glasses aim to take the mobile content-consuming experience beyond the absorbing environment involved in watching the handset screen, that "something that frees your eyes" claimed by Brin, the usual video games experience that has been successful so far, seems to directly conflict with the new value proposal, aimed to be compatible with, again in Brin's words, "making eye contact with people". That can be in coincidence by other efforts, particularly in automotive-related apps, to bring content to people minimizing the risk of distraction from their main task. Navdy, in the U.S., is working, for example on a system with a tiny projector able to bring data on the windscreen (Metz, 2014).

Thus, augmented reality developments seem to find the right environment for huge adoption, relegating virtual reality technologies to non-transparent screens, as smartphones, tablets, and similar handsets.

Legacy Media and the Mobile Ecosystem

Most of legacy media have regarded internet content enjoyment with a mixed-feelings approach: they see how users get engaged with new habits of media consumption and some brands are able to find new, abundant sources of wealth; on the other hand, they regard with mistrust the way the full digitalization of contents leads to a loss of their former control on distribution via copyright regulations.

That has brought a paradox: while most of persons want to enjoy, in mobility environments too, the content portfolio that legacy media own, those media have been in practice unable to fulfill users' wishes unless they accept the set of conditions imposed by outdated revenue models of copyright laws. Piracy, P2P sites, have filled the gap. At the same time, many content providers have felt threatened by the activity of aggregators, particularly search engines, in the belief that they are plundering a portion of their value, instead of assuming that these aggregators could be showing thousands of new users the way to their media sites.

The video games industry has been (see paragraphs above) the biggest exception to that, possibly because of its digital nature from its very beginning (Bogdanowicz et al., 2010).

The music industry has also found its way towards its integration in digital-consumption environments (Leurdijk & Nieuwenhuis, 2012), although it has taken more time to adapt to the new rules of the game. A major milestone in that process of adaptation comes from the success of Apple's iTunes Store, launched in 2003, which became the first multi-brand music store that was able to allow users to buy only the songs they wanted to purchase, at reasonable prices and with the support of the music industry. From then onwards, mobile has added to music consumption a social twist that has changed the way people enjoy music: from YouTube to Spotify, the smartphone has become the device of choice for music listening to millions of persons in all countries.

Apart from these two exceptions –video games and music- most of the incumbents in the legacy media set of businesses –news, TV/video, radio, book publishing- have not tried to create contents ready to take advantage of the specific features that the mobile experience brings to users. In terms of innovation the legacy media sector has not been involved with the mobile ecosystem in the same way the software developers have.

Scenario 1: A Gap between Two Worlds

The music industry and the video games sector have done efforts to adapt to the new reality of mobile. The rest of legacy media, however, have kept a cautious approach in a first phase, and later they have been trying to adapt/repurpose contents. Most of the news-related activities are still under the perplexity towards the new digital reality, sometimes more worried about the financial results than about understanding the new realities. As far as these attitudes keep prevailing, most of legacy media will be struggling to adapt or repurpose their catalogues of contents rather than creating original works conceived for mobile.

However, the movie and the TV industries, as hit-driven businesses (Wu, 2011), have tried to create specific, and eventually augmented contents, allowing creators to build true transmedia experiences, mobile-enjoyable, built upon blockbusters.

The gap between mobile-friendly and not mobile-friendly legacy media will determine their adaptation to the new environment.

Scenario 2: Legacy Media Go Augmented: Attention to News

In our scenario of disruptive innovation, the only chance for legacy media is to insert their vast portfolio of contents into the new reality. The same way a number of successful developers have survived acting as contractors for big media companies, aiming to repurpose or to help them into the mobile environment, big opportunities can arise for alliances between developers and legacy media. Thus, news organizations face the challenge to capitalize on the flows of information they work with, in order to create new forms of communication able to build a language suitable for new experiences based upon augmented-reality features.

Mobile Network Operators

Mobile network operators (MNO) do a wide set of activities within the internet world: they act often as internet service providers (ISP) in the mobile net, and take care of many of the tasks related to the physical layers of the system. In addition to that, they enjoy the privilege of personal relationship with the final users –either via direct billing or pay-as-you go models, and they sell the users most of their handsets – apparently *subsidized*-, and they do that in ways that can keep their customers locked-in.

If we give credit to all the assessments above made, it is hard to understand how MNOs are not in control of the content ecosystem. In addition to operating in an environment whose future course they cannot rule, they have reasons to feel that their typical revenue model is in serious danger (Tellabs, 2011), with declining average revenues per user (ARPU), and on the verge of falling into a "dumb pipe" model (De Wit, Mateus, et al., 2009; Thierer, 2004), constrained to pave the way, at their own expenses, for others –mostly aggregators and content/software producers-, who get the biggest part of the margins cake. MNOs' public complaints on the low reward they obtain for their commitment to develop networks, compared to the wealth they contribute to create for platform rulers, are commonplace (Thomas, 2014; Tomson & Orihuela, 2014).

In the first phase of the creation of the mobile media business, MNOs used to be closely involved with the activity of delivering content to users. In that early stage, "silo models" or "walled gardens" (Feijóo et al., 2008; Jaokar & Fish, 2006) happened to build an environment of restrictions to users' access to contents. Also, some agreements with handset vendors –mostly Nokia, the sector leader at the time-, allowed customers a convenient way of buying contents and applications through company m-stores, whose amounts were billed through the ordinary monthly bill of the MNO. It's not irrelevant to keep in mind that developers had to share profits with the MNOs on a 50/50 basis, once discounted the MNO commission.

However, the carrier-billing model has not vanished: it has grown a 50% globally in the 2012-2013 period, to reach 3 billion dollars (Heggestuen, 2014a), with Android's Google Play Store having reached agreements of this kind in 21 countries (Heggestuen, 2014b), according to BI Intelligence data.

With the content eclosion fueled by the iPhone launch and the new-style platforms and their app stores, the whole legacy distribution model collapsed. None of the MNO's attempts to restore their position in the platform era had succeeded by the end of 2014: neither the partnership with the Firefox Foundation and a number of telecom operators and device vendors from several continents has given its outcome, the HTML5-based Firefox OS, a minimal market visibility to grant a place for it among the ones who compete for the third position in the mobile ecosystem.

Nevertheless, another strategy seems to be more successful for MNOs to avoid complete commoditization: alliances, sometimes acquisitions, of content aggregators of any kind, particularly targeted towards the TV/online business. Telefonica's dominance over Spain's Sogecable pay-TV operator; Vivendi/Canal+ partnership; Spain's Vodafone acquisition of cable/TV operator ONO; and the long saga involving the US telecoms-cable TV-video producers/aggregators, with Netflix, AT&T, Time Warner Cable, Comcast, DirecTV and Verizon in the leading roles. "Distribution channels will continue to merge. Devices define our media diet — be it TV, computers, mobile, or/and whatever comes next", says Nielsen CEO (Barns, 2014).

Let's consider these facts and opinions under the light of some internet traffic figures (Cisco, 2014): internet mobile traffic grew 81% in 2013; in that year, video exceeded for the first time half of mobile data traffic; Cisco also estimates that by 2018, two thirds of mobile data traffic will come from video. That picture is pushing MNOs to establish the closest possible partnerships, when not acquisitions, with any kind of TV/video operators.

Scenario 1: Looking for the Right Place on the Freight Train

A scenario based on sustaining innovations will allow MNOs to deepen their knowledge of the strategies they have pursued so far. Despite consumers have abandoned the walled garden pattern, carrier-billing practices have not gone away, and now amount to a 12% of digital goods purchased via mobile (Heggestuen, 2014c), and can reach up to 13 billion dollars, counting on a BI Intelligence estimation (Heggestuen, 2014a). If mobile operators find a way to reduce what they are charging, that makes the transactions up to a 40% more expensive, they can hop on the big freight train the phenomenal growth of data consumption is. Without excluding any other strategies, direct-customer billing in a context of data traffic expansion looks as a winner bet.

Scenario 2: Between the Golden Fever and the Perfect Storm

In a context of disruptive innovation driven by fully connected wearables, carriers will have to face mobile traffic figures mounting at rates far beyond most of conventional forecasts. That could mean apparently more business for the industry. But it has a number of potential downsides that could become more damaging than benefits.

The need of increasing capital expenditure efforts in order to fulfill users' demand could become unsustainable, as ARPU evolution would be unable to offset its financial needs. Also, the close relationship between carriers and OEMs could have to face the dilemma of adding the wearable devices factor or not. It has to be said that subsidy strategies of MNOs have proved massive success with handsets, whereas in tablets has become a significant source of direct turnover. We have to keep in mind that, particularly in the Android segment, the proliferation of non-SIM card, only-wifi enabled tablets, has diminished the attractive of such offers.

In that effort to capture value from users, carriers will possibly exert mounting pressure on telecommunication regulators in order to force other ecosystem players –mostly the big platform rulers, well settled on the aggregation field- into sharing the whole investment in networks.

Content Aggregators: The Crossroads of Internet Content Business

The aggregation activity lies at the very core of the internet, at least since the net started to develop beyond the boundaries of a closed community restricted to academic purposes. Starting as mere listing pages, aggregators are now a necessary step for any web user, no matter the kind of access –fixed line or mobile- or the interest that leads him to browsing. As described by Kohlborn et al, "aggregation and composition are used to describe services that contain other services as sub-services (...). In a more general sense, (...) an aggregation would comprise multiple services and provide access to them in a single location" (2009: 199).

Internet aggregation is such a powerful layer of the system because of its ability to integrate services, tackling the "friction" (Akerman, 1998) issue that lies between our immediate content or transaction interests and the intricate complexities of the World Wide Web. Broder dove a bit into the cognitive interests of the search engine users in order to differentiate their basic purposes: "The intent behind a web search is often not informational - it might be navigational [show me the url of the site I want to reach] or transactional [show me sites where I can perform a certain transaction, e.g., shop, download a file, or find a map]" (2002:3). It's a list of motivations that we can extend to the whole of the aggregation space of activities.

Furthermore, filling the gap between users and content has fuelled the ambitions of some aggregators to the extent that they have become content producers themselves. As described by Christensen et al.: "it shouldn't be a surprise when new entrants like The Huffington Post and BuzzFeed, which began life as news aggregators, begin their march up the value network. They may have started by collecting cute pictures of cats but they are now expanding into politics, transforming from aggregators into generators of original content, and even, in the case of The Huffington Post, winning a Pulitzer Prize for its reporting. They are classic disruptors" (2012: 8).

We can notice the same trend in the video online business, with Netflix as a paradigm of the same trend: "Netflix's transition from delivering movies and TV shows through the U.S. Postal Service to beaming them over high-speed connections began in 2007 and is a well-known story. Less known is how Netflix found its way into the content business, a risky move that has embarrassed many tech firms" (Wu, 2013).

It's easy to perceive the global dominance of aggregators in the web just by looking at Alexa data. Alexa calculates the monthly sites audience through a combination of average daily visitors to the sites, and pageviews from users from that over the latter month. The site with the highest combination of visitors and pageviews is ranked #1 globally. Table 3 includes the 20 globally most popular web places in accordance with these data, along with a short description of their content.

All www sites ranking among the 20 most visited ones fill into the aggregators' category, being Wikipedia the only dubious exception. A look at the table also reveals the diversity of activities taking place under the same mediation function, sort of acting as a personal map or even a guide, catering for a universe of users with different interests.

Opposite to the content and applications developers, aggregators don't need to own the contents: they just put us on the way to get them. If we take into account their generic value proposition to the user, we can export Broder's non-exclusive taxonomy of users' needs to a different classification of activities

Table 3. Top 20 sites on the Web in September 2014

#	Name	Content and Details
1	Google.com	Search engine
2	Facebook.com	Social utility, Microblogging
3	Youtube.com	Social utility, video sharing
4	Yahoo.com	Contents portal, incl. search engine
5	Baidu.com	Search engine, Chinese
6	Wikipedia.org	Free encyclopedia, built collaboratively
7	Twitter.com	Social utility, Microblogging
8	Amazon.com	e-Commerce
9	Qq.com	Content portal, incl. Search engine, Chinese
10	Linkedin.com	Social utility, professional information sharing
11	Taobao.com	e-commerce, Chinese, linked to Alibaba Group
12	Live.com	Search engine, Microsoft owned
13	Google.co.in	Google's portal for India
14	Hao123.com	Baidu's online listings portal. Also Thai and Brazilian versions
15	Sina.com.cn	Chinese content portal, addressing the global ethnic Chinese community
16	Blogspot.com	Blog publishing service, owned by Google
17	Weibo.com	Chinese Social utility, Microblogging
18	Yahoo.co.jp	Yahoo's version for Japan
19	Tmall.com	Chinese luxury e-commerce site
20	Yandex.ru	Russian leading engine search. Also operates in Belarus, Kazakhstan, Turkey, and Ukraine

Source: www.alexa.com/websites.

that can cover also aggregators': search engines/assistants; content portals; e-commerce sites –including mobile- and app stores; and social network sites. Each player can do only one activity or bundle two or more into the same site.

Their apps are among the most used by smartphone owners, that have increased the time spent using "search, portals and social apps", to almost 11 hours per week, a 28% year-to-year surge, according to a 2014 US survey (Nielsen Newswire, 2014).

Search Engines and Personal Assistants

The search engine business is concentrated around the query/result-browse interaction triggered by the user initiative, while the assistant goes a step further: it assumes that what the user expects is not searching, but finding (Baeza-Yates, & Raghavan, 2010). In other words, from search to recommendation: "Long-term applications will flourish when search engines evolve into fully-fledged "recommendation engines", able to automatically update user preferences in real-time, social network patterns, and use patterns" (Gómez Barroso, et al., 2010:13).

When web search becomes mobile, it means a completely different experience from the fixed-line, computer-mediated search. Just to name its most relevant differences, we have to consider context, location, and network conditions –bandwidth, latency- (Church & Smith, 2009), and all the device interface differences, for instance the lack of a physical keyboard to write a query. In such an environment, the user expects a set of fewer, more accurate results, to choose from. Also, those results have to be more accurate, and all that has to be carried out through an input procedure able to limit keyboard use to a minimum, even considering the use of a voice-recognition input method.

Google leads the business overwhelmingly, with 88,15% of market share in the web search engines business (Statista, 2014) in July 2014, and 91,2% in mobile (Net Marketshare, 2014).

Content Portals

Content portals usually agglutinate a number of contents –news of any kind, online games...- and services –stock quotations, weather data...-, often a miscellaneous one, under a name. Starting from that they have developed some of the best known brands for delivering news and entertainment content, as Yahoo!, MSN, Terra, or Yandex, to name just a few. The development of the social web, added to the widespread use of mobile devices, have spurred the appearance of *social magazines.*

Well beyond the primary idea of content portals, but somehow related to them, social magazines are gaining increasing success among users, thanks to their ability to give the reader the power to build his own magazine, by integrating into a sort-of-magazine format a number of diverse sources in a flexible way: a new space of content convergence has emerged. As Benton describes: "the interconnection between mobile devices and social media has been a powerful one (...). New phone-first start-ups like Circa and Inside, products like Facebook Paper and Yahoo News Digest, and tablet emigrants like Flipboard are all presenting news in mobile-friendly ways. It's hard for the old guard to compete" (2014: 39).

These magazines often have closer relationship with content consumption habits through mobile apps –and devices- than the first portal sites, where web browsers are more popular as tools to access content.

E/m-Commerce Sites

E-commerce sites or m-commerce sites are shops than can address either the final consumer *–B-to-C* portals- or promote business among companies *–B-to-B* portals. The kind of items offered, the means of payment, even the transaction environment –fixed price or bid-, are traits that shape the different businesses grown under this category. Alibaba or Ebay are among the leaders of the segment. But the new era of mobile media trade started with the iPhone/Android arrival. Apple, the firm that disrupted the smartphone market with its iPhone device in 2007, was only able to establish itself as a platform when it opened itself to third-party developers setting up its own app store. It has been needed some time to establish that the step of creating an open app store to developers was not in the first Steve Job's roadmap: Apple's team of strategists had to convince him to cede a bit of his technology to developers in order to establish Apple as the second global mobile platform (Isaacson, 2011; West & Mace, 2010).

In the current state of things, the three leading mobile platforms that operate through an OS –Android, iOS and Amazon- need their own app store in order to get revenues and, each one in its own way, organize their relationships with their users. It's not Facebook's case (more extensively explained below), which has found its own strategy to take advantage of its users' base. Also, the rest of platforms struggling for any room in the mobile ecosystem, all OS-based –Blackberry, Windows, Firefox, Tizen-, need to have their own platform-ruled app store in order to play their cards in the game.

When we loook at the electronic commerce space and its influence on the mobile ecosystem development, we have to pay special attention at Amazon's case, that has taken advantage of its e-commerce starting point to place itself at the core of the fight for room in the war of platfoms (Schuermans, 2013; Vakulenko et al., 2011;Wilcox & Voskoglou, 2014), able to play in the league of the very big mobile platforms, in dispute with Google, Apple, and Facebook (Manjoo, 2011; Pulman, 2011).

However, m-commerce success keeps being one of the biggest challenges for e-commerce sites, even for the most successful ones. As Mace explains: "For online retailers, the single most frustrating thing about mobile technology, especially smartphones, is that it people using it don't buy a lot of stuff. (...) The industry rule of thumb is that a good commerce site on a personal computer will convert about 3% of shoppers to buyers (in other words, for every 100 online shoppers you make three sales). The conversion rate for smartphones is a third of that, about 1%" (Mace, 2014).

Unlike other activities in the mobile ecosystem, mobile commerce has still to deliver the wealth that other businesses have achieved through the fixed-line internet channel.

Social Media

Social networks or social media are concepts coined to describe a huge variety of sites that have only in common their function of hubs where people share contents: from conventional blogs, to micro-blogging sites as Twitter, video-sharing portals as YouTube or Vimeo, or spaces to share almost any kind of content, as Facebook, the leader of the segment with more than one billion accounts. Moreover, in the last years, some apps that started as person-to person direct-messaging services –WhatsApp, Line, WeChat, to name a few- and have become also spaces for sharing files, what possibly qualifies them for being considered as part of the social media space (Singh, 2014), although they first broke into the market as OTT alternatives to SMS and MMS phone messaging.

It's known that the success of smartphones and tablets has given the users ability to create and share contents in a way that has fostered the bounds between social media and mobile devices. Legacy media, consumer brands and businesses of all kinds deploy their visibility in the social media space, but what in the first place has attracted users' interest has been the abundance of user-generated content (UGC). Mobile devices have achieved so by "merging conventional content consumption [passive watching, listening or reading] with Internet and digital media interactive capacities, fostering thus an evolution from a watching-oriented logic of entertainment to a doing-oriented conception" (Aguado & Martínez, 2014: 188).

Facebook deserves a special attention in that respect, because has positioned itself among users as whole powerful ecosystem of its own, to the extent that some have labeled the social site as a new version of the old *walled garden* concept (Ingram, 2013). Making friends, playing games, watching celebrities' life and sharing content is just the tip of a vast iceberg of value Facebook has.

Facebook has built its power not only based on users' favor, but also leveraging on a very cleaver management of personal identity in the digital sphere. As the global leader in registered users, Facebook offers a system of *federated identity* to many massive sites, where users can spare themselves the chore of remembering usernames and keywords just by entering their Facebook IDs: that makes Facebook a privileged pipe of personal information from users of hundreds of portals that Facebook doesn't own, but of whose activity is aware. "Data mining about user profiles and behavior enacts a new source of value for digital content beyond monetizing audiences' attention through advertising. Both legacy media and users should be aware of that change in the value chain of digital entertainment in which mobile media are deeply involved" (Aguado & Martínez, 2014: 191).

Aggregators' business models doesn't typically rule out any of the typical revenue models: subscription o freemium revenues (Linkedin, app stores); carrier-billing models (some app stores, Spotify); mixed free and pay-per view strategies (YouTube); or in-app- purchases (app stores).

However, advertising-based revenue models tend to prevail in the non e-commerce area. It's easy to presume that, the lesser the user pays for a service/access to contents, the more advertising will be involved. But that assumption can become a thing of the past in the coming years, or at least can lose weight as a source of revenues to aggregators. The data flow generated by pervasive, context-aware devices, can become a source of wealth comparable to advertising: "The economics of personal information has much to do with the explosion of digital and mobile entertainment in the last decade. It is the relevance of mobile entertainment as a powerful source of personal behavioral information about users what explains the unnatural convergence of companies with different business models (like Apple, Google or Amazon) over the strategic centrality of digital entertainment content" (Aguado & Martínez, 2014: 190).

Scenario 1 for Aggregators: Surfing the Wave

In an environment where connectivity demands of users don't seem to stop, aggregators just can benefit from a sequence of sustaining innovations, including the positive impact on their activity of non-fully connected wearables, as headphones, m-health devices, fitness bands and glasses.

Search engines will have to face the challenge of improving their integration with voice and gesture recognition interfaces, in order to appear as valid tools for users that will be increasingly aware of any kind of context information around them. In this particular case, Google seems to be leading through Google Glass the efforts to trigger a second wave of disruption in search activities: "My vision (...) was that eventually you wouldn't have to have a search query at all. You'd just have information come to you as you needed it. And this is now, (...) that I think can deliver that vision when you're out and about on the street talking to people and so forth" (Brin: 2013).

Also, according to the users acceptation of augmented reality features embedded into their services, their market position will strengthen, particularly in convergence with the automotive and the home automation businesses connected to the Internet of Things (IoT) scenario.

Content portals and social magazines will keep improving services, possibly lagging behind the innovations brought by other ecosystem actors.

M-commerce, and particularly app stores, will have to face sooner or later the challenge of suiting the needs of users and the interests of content developers in a more satisfying way. According to some analysis, apps discovery –and therefore the need to introduce human-driven curation component-, can be a crucial issue in a near future (Gassée, 2014; Nash, 2014)

Scenario 2 for Aggregators: High Expectations, Still Higher Challenges

The disruptive innovation scenario puts before the aggregators a legion of demanding users that expect, at least, the same features in more enriched experience, apart from the challenges put by the new generation of devices. On the other hand, they will be in the middle of a gigantic new flow of more accurate context information on users' profile that can maximize their position in the big data business.

Search engines would have to undergo an added effort to process a whole new amount of the layered augmented reality data that will come necessarily along with the new devices. More functional visual and social features can be the next steps to suit the needs of their customers.

Content portals and social magazines seem placed to try their best to adapt their interfaces, possibly enhancing their interplay with the social web and the UGC.

In the case of social media, however, a huge evolution could be expected. Wearables can make of users pervasive producers of UGC of several kinds, and bring the sharing experience to unknown boundaries. The ability of players to offer new experiences through the creation of new services can be crucial in that respect.

M-commerce sites -app stores' case has already been treated more specifically in the contents producer section- can regard the new generation of devices as a two-edged sword. While wearables can reduce friction in m-transactions, for instance by enhancing visual searching, the conversions gap with fixed-line portals can become even wider, depending on the way users apply "showroming" or "reverse showrooming" practices (Adler, 2014) in high street stores. Those practices are closely linked to the way shoppers use their mobile terminals in brick-and-mortar retailers in order to compare prices of the same items.

Original Equipment Manufacturers (OEM)

Terminal vendors are the ultimate players responsible that allow physical access to the user. That should give them a privileged position within the ecosystem. However, that has not always allowed them to dictate the rules of the game. Let's take the smartphone segment as our starting point. From 2013 onwards, smartphone shipments have surpassed non-smartphone mobile handsets in sales. Assuming a global installed base of 1.9 billion units by mid 2014 (Ahonen, 2014c), roughly one in four human beings of the planet owns one. Far from reaching its maturity point, now the most accepted views forecast that in the next years, low-cost smartphones will replace most of current featurephones.

As Table 4 shows, the dramatic changes of leadership among device vendors have not happened as intensely in any other family of actors within the mobile ecosystem. The arrival of the iPhone in 2007 was a major event that disrupted the whole industry, and along with the Google/Android emergence, gave birth to a new version of platform to which Nokia, the dominant player with its Symbian OS ecosystem, was unable to adapt.

Table 4. Evolution of global smartphones market share

Vendor	2009	2013
Nokia	39	3.1
Blackberry	20	2
Apple	15	15.5
Samsung	4	31.5
Sony	3	3.9
HTC	5	2.7
Huawei	-	5.3
ZTE	-	4
Motorola	4	1.2
LG	3	4.8
Lenovo	-	4.7
Coolpad	-	3.6

Source: Ahonen, 2013; 2014

Nokia's crisis is not the only noticeable trend here. Apple remains the only big North American brand that has gained any strength in terms of market share: Blackberry is on the verge of becoming unnoticeable, and Motorola, first acquired by Google in 2012, was sold to China's Lenovo in 2014. Nokia's failure and the end of Sony-Ericsson partnership left European brands insignificant. Sony, the only Japanese relevant global player in the device business, has to fight for survival against two stronger competitors from Korea, Samsung –undisputed world leader in smartphones-, and LG. Furthermore, the awakening of the Chinese industry, with 385 million smartphones shipped in 2013 (CCS Insight, 2014), gives way to a number of entrants little known in some more developed markets, but very ambitious in their strategies: Coolpad, Huawei, ZTE, Lenovo, Xiaomi, to name just a few, which were unnoticed or little known a few years ago as mobile device makers, and now overcome HTC, another Chinese brand, but more focused on the most mature segments.

Although the iPhone has been so far the main turning point in the mobile devices, and perhaps in the whole mobile ecosystem, new kinds of devices have reached the stores and gained the interest of users. All that, along with the deployment of wifi-related access technologies and VoIP telephony, challenges some of the views on which mobile media theory has been built in its early stages.

Feldman said: "The parameters that are used to distinguish mobile devices are 'physical', 'electronic', 'IP-enabled', and 'telephony- enabled'" (2005: 7). The arrival of Apple's iPad in 2010, and with it the birth of the new category of tablet devices, powered by smartphone OSs and ready to extend mobile platforms scope using the same kind of apps, casts doubts on Feldman's definition. What should be understood by "telephony enabled"? A tablet connected through a SIM card? Would a wifi-connected tablet be also a mobile device? Should we change our notion of "telephony-enabled" device considering the popularity of OTT messaging and VoIP apps, that are IP-based services?

Many data about mobile content consumption tend to aggregate under the same field tablets and smartphones activity –to complicate things a bit further now we have to deal with the *phablet*, a smartphone-tablet hybrid, and the connected car has already entered some markets. Ahonen regards tablets as a light version of a laptop rather than as a mobile terminal: "a tablet is not a 'mobile' it is only an ultraportable. (...) Tablets immobilize people. Yes they are very VERY good replacements of laptops and in many cases can do what smartphones do - but they will not replace the smartphone. They are only ultraportable PCs and their market is a tiny slice in magnitude of what will be the global smartphone market" (2014d). There are also some research sustaining that there are different patterns for smartphones and tablet users, particularly in their attitude towards shopping (Adler, 2014).

Anyway, smartphones and other mobile handsets reach the final consumer through different market channels. With a very few exceptions, smartphone vendors sell most of their units through special agreements with MNOs, which subsidize a restricted choice of devices: in order to retain customers, these phones often keep the network customers locked-in through a special code that makes the phone only capable to connect to the designed network.

However, mobile operators have not repeated the smartphone-distribution pattern in the tablet segment. Most of tablets are being sold through, either ordinary consumer electronic retailers, or by OEM vendors themselves.

Scenario 1: Commoditization, or Strong Brands to Make a Difference

After the iPhone irruption in 2007, OEMs have lost their ultimate power of disruption as innovation drivers with ability to force the rest of the main ecosystem players to follow their lead. On the contrary, in an environment with a high degree of competition with new brands, particularly from Asia, they have been able to differentiate themselves only in minor features: screen size and definition, processing speed, camera features, battery, etc. This is how a CCS Insight report describes the current state of innovation in the handset industry: "Single-button, monobloc designs have become the norm in all sizes ranging from four-inch to seven-inch screens at all price points. In some ways this is a necessary phase that allows all types of user to buy with confidence, but it also means that innovation in hardware design is at a historic low" (2014: 8).

In a market where global standard bodies are in charge of the major specifications of the networks, and OS-based platforms determine the range of features users can access, OEMs have been regarded mostly as "contractors" (Castellet, 2012) of others' knowledge base. As Doyle summarizes: "The current pace at which new smartphone models are introduced annually is economically and ecologically unsustainable. (...) And mobile hardware is becoming far less important than the networks and services that support it" (2014).

Thus, in a scenario of sustaining innovation, the need to differentiate and produce new developments can bring ideas as the "modular handset" (Smith, 2014) promoted by Google, able to integrate a huge diversity of developers of different features. Others, as the Spanish firm Geeksphone, bets on a device able to play Firefox OS and Android simultaneously (Meyer, 2013).

Scenario 2: Back to the Innovation Leadership

A disruptive innovation scenario lead by fully connected wearables changes de rule of the game in many senses. It returns the innovation leadership to the OEMs players, forcing the rest of players –OS providers, mobile network operators, aggregators and content producers- to adapt their strategies to the new generation of devices.

Also, if we assume that Google is the frontrunner with its Google Glass product, the rest of OEMs will be forced to bring to users a range of devices able to compete with Google Glass, not mere hardware add-ons to smartphones. A wide gap will be created: between vendors able to bring high-added-value wearables to the public, and the ones who cannot and will have to go a new step further into the commoditization nightmare. And obviously, Google will have to be counted among the sales' leaders in the mobile devices segment.

Furthermore, under that scenario, mobile network operators could have to rethink their subsidizing strategy for handsets, as the new devices would be a huge driver of data traffic, and therefore a tool to increase business volumes.

Operating System Providers

The operating system (OS) is the genius behind our devices. Silverschatz and Galwin's definition for a computer OS applies also for the mobile handsets: "a program that acts as an intermediary between a user of a computer and the computer hardware" (1991: 3). "They mark out the boundaries of user experience. They are the standard that app creators, content providers and OEMs should bring to market devices able to operate under the provisions of the OS" (Castellet & Feijóo, 2013: 46).

In the mobile contents business, the OS has been the starting point to leverage a complex value network that lies in the origin of the two biggest incumbent platforms, Apple and Google (Ballon, 2009; Castellet et al., 2013; Telenor Group, 2011) (See Figure 2).

Figure 2. Market share on OS in smartphones installed base (2,11 M) as of 31 December 2014
Source: Ahonen, 2015.

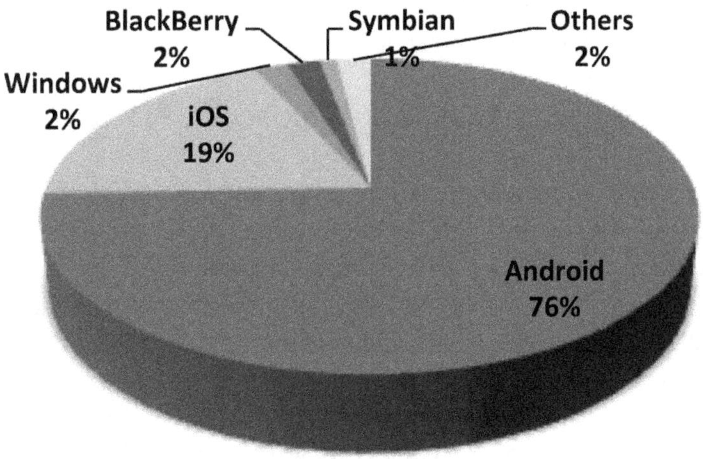

Google's Android OS prevails in the global market with Apple's iOS keeping a solid niche position as the iPhone inventor and owner of the high-end segment of the market, as figure nr. 1 shows. There is not a third solid player in the horizon, and anyway it would be eventually at a huge distance from the two incumbents.

All that said, it has to be observed that Android's figures show a dominance that, although true, has to be analyzed more closely. Within the Android category operate a number of players that have been taking advantage of the most basic features of Android OS, "the basic bones of a smartphone operating system" (Bright, 2014). That layer is the free-of charge layer of Android, also known as the Android Open Source Platform (AOSP) codebase. However, the Google added value services set as Maps, Search Engine, Google Play Store, Google Now, come in a second layer, that needs Google's certification, for which Google charges a small fee. The OEMs that include both Android layers–Samsung, LG, Lenovo, Sony, HTC, etc.– can also add a number of native features on their own.

But there is a number of OEMs that ignore Android's second layer, take it at no charge and add their own second layer with their own search widgets, map services, app stores, etc. Those OSs, considered Android *forks*, try to take advantage of the Android attractive without being tied to Google's revenue model. That has been the case of a number of Chinese companies –Alibaba, Baidu, Xiaomi-, and also of Amazon –with its Fire OS that power Kindle Fire and Phone Fire devices-, and even Nokia, that launched a line of cheaper smartphones under its Series X.

All those players can amount up to a third of Android OS global power (CCS, 2014). In the meantime, Google counterattacks with its Android One project (Subramaniam, 2014), aimed at add to its platform the lowest-price smartphone sector –up to 100$-, for developing markets, as India's and others.

Scenario 1: Innovation at Cruising Speed

All OS providers have been doing improving connectivity –NFC, Bluetooth, etc.- in order to suit the best users' growing interest in wearables. Thus, this sustaining innovation scenario doesn't seem to be damaging to the current state of leadership. Android seems to have overcome the perils of *internal* fragmentation inside its licensees' space: more than 85% of active Android certified devices belong to the 4.x. generation (Whitney, 2014); according to the same data, less than 1% of users are excluded of the last available features of Google Play Store. And when it comes to Apple, a 90% of their active users were updated to the iOS6 generation by the second half of 2014.

Scenario 2: First Advantage for Android but an Uncertain Picture

A disruptive scenario would increase the innovation effort of OS providers to comply with the new requests of a new generation of devices, attached to the human head and therefore more dependent of voice-and-gesture interfaces than the prevalent multi-touch screens. It's crucial in that respect to keep in mind that Google, the main Android's promoter, is also the creator and marketer of Google Glass, the first fully-connective wearable device scheduled to start massive sales by the end of 2014. Both Android and iOS are giving signs of being seriously engaged in updating their APIs and other system components in order to deliver an environment attractive enough for content/apps developers and for consumers. Anyway, there is not any warranty that backwards compatibility issues can eventually arise.

FUTURE RESEARCH DIRECTIONS

As only time can validate the feasibility of this two-folded analysis, the future development of events will determine the usefulness of this research. Anyway, there are two big research gaps than all the above written uncovers: a) the role in this innovation processes of big data/advertising activities, as they are closely linked to the evolution of media; and, b) an extensive study of users' attitudes and willingness to adopt the new generation of devices here studied, including the development of new kinds of UGC.

In the first case, this work cannot reach the big data/advertising and marketing sectors and their impact because of the lack of time and space to properly treat it, and that is a non-small limitation of our perspective.

In the second case, the users' perspective has been deliberately avoided, and only very little mentioned, because this chapter has been focused on the supply side of the ecosystem. If this perspective is attractive enough to other researchers, further papers, built upon sound and appropriate quantitative and qualitative tools, can add invaluable insights to the set of topics here covered.

CONCLUSION

The new generation of wearable terminals has the power to tilt the main innovation driving power from the operating systems' (OS)/aggregators conglomerate to the devices' area of the ecosystem, in a sense that resembles what happened with the arrival of the iPhone. The current OSs cannot cope with a new generation of devices without developing a new generation of software able to go far beyond the multi-touch revolution of the mature iOS/Android platforms. However, the two biggest incumbents –iOS-Apple and Android-Google platforms- doesn't seem threatened in their market dominance because of the lack of market penetration and innovation from the rest of the main players – Blackberry, Windows, and Firefox, mostly-. It's necessary to point, anyway, that even the iPhone irruption, with all its disruptive outcomes, didn't happen completely isolated from the aggregation activities –Apple had already its iTunes content store in connection with the new device, and Android phones appeared under the protecting sponsorship of Google-, and it's quite likely to guess that a close connection between the aggregation and the devices' makers will be necessary in any near-future scenario.

Furthermore, a scenario based on a disruption led by always-on, fully connected wearables, will have to cope with backwards compatibility, and OS fragmentation issues. The way these issues are addressed will be vital for the ecosystem incumbents in order to retain users' trust and developers' commitment to the platforms.

And although we have taken the fully connected wearable pair of glasses as the main driver of disruptive innovation, we should not rule out the chance of an equally powerful disruption led by glasses that work tethered to another device, a smartphone or a tablet, for instance. If we link the concept of innovative disruption to the suppression of friction, tethered glasses could be very convenient for users: as they are connected to the handset, the user doesn't have to learn a whole new interface; and these ancillary wearables are intended to cater to cheaper segments of the consumer market.

The impact on the content sphere of a widespread use of glasses cannot be assessed with the data available so far. However, what emerges from a close look at the features of the new devices is the fact that the user experience will change dramatically, in a way that will tend to favor augmented-reality-based content. This trend is also closely connected to the long-awaited promise of mobile content specifically tailored to the automotive sector, where many efforts are oriented towards a media-consuming experience that has to be compatible with the driving task.

The design of such a new kind of media becomes a challenge, not only to all kinds of content providers, but also to app stores, as the new services can adopt unsuspected ways of delivery to final users, eventually marginalizing the app store channel.

All the latter said, Schumpeter's "creative destruction" paradigm could be seen under a new light in an environment where products and services are created as an outcome of complex value networks, or platforms. Under these circumstances, players tend to perform different roles –sometimes main players, sometimes financial providers, sometimes competitors, for instance-: they are *creatively destroyed* in one role while they can become successful, even leaders, in another within the very same ecosystem. We can observe schumpeterian results of destruction if we look at the fate of some companies, that have been forced to shelter themselves in narrower niche markets –Nokia-Siemens or Ericsson, both retain still *mobile* business in the physical network activities, but very far from individual customers' reach-, while others have completely disappeared, being Palm one of the most prominent examples.

However, the different phases of the mobile content evolution, from the walled gardens era to the current platform-driven phase of business, make clear that in a climate of sustained innovation there have been shifts in the power to dictate the rules of the market game. Perhaps the "creative destruction" paradigm could allow some room for a "rely race" pattern, rather adapted to our industry ecosystems than to the less complex vertical integration model that very often doesn't reflect the complexities of our era.

REFERENCES

Adler, E. (2014a). 'Reverse showrooming': Bricks-and-Mortar retailers fight back. *Business Insider*. Retrieved from http://www.businessinsider.com/tablets-and-smartphones-are-used-completely-differently-for-shopping-and-buying-2014-4?utm_source=alerts&nr_email_referer=1

Adler, E. (2014b). Tablets and smartphones are used completely differently for shopping and buying. *Business Insider*. Retrieved from http://www.businessinsider.com/tablets-and-smartphones-are-used-completely-differently-for-shopping-and-buying-2014-4?utm_source=alerts&nr_email_referer=1

Aguado, J. M., & Martínez, I. (2014). Feeding digital omnivores: The impact of mobile media in digital entertainment. In G. Goggin & L. Hjorth (Eds.), *The Routledge Companion to Mobile Media* (pp. 181–194). New York, London: Routledge.

Ahonen, T. (2014a). The comprehensive app economics blog 2014. *Communities dominate brands*. Retrieved from http://communities-dominate.blogs.com/brands/2014/08/the-comprehensive-app-economics-blog-2014-yes-peak-app-is-apt-name-sheer-disaster-industry-with-only.html

Ahonen, T. (2014b). Enter the iflop, what will be seen as first apple failure after Steve Jobs. *Communities dominate brands*. Retrieved from http://communities-dominate.blogs.com/brands/2014/09/enter-the-iflop-what-will-be-seen-as-first-apple-failure-after-steve-jobs-but-the-first-edition-appl.html

Ahonen, T. (2014c). Seems like my work is done: As we shift from 'hunters' in mobile industry to 'farmers.' *Communities dominate brands*. Retrieved from http://communities-dominate.blogs.com/brands/2014/07/as-we-shift-from-hunters-in-mobile-industry-to-farmers-most-of-the-cool-stuff-is-done-and-dusted.html

Ahonen, T. (2014d). Smartphone top 10 for Q2 of 2014, same ole same ole. *Communities dominate brands*. Retrieved from http://communities-dominate.blogs.com/brands/2014/08/smartphone-top-10-for-q2-of-2014-same-ole-same-ole.html

Ahonen, T. (2015). Smartphone Year 2014 Final Stats: Brand top 10, OS top 4, installed base, regional split; and bloodbath Q4 results. *Communities dominate brands*. Retrieved from communities-dominate.blogs.com: http://communities-dominate.blogs.com/brands/2015/02/smarphone-year-2014-final-stats-brand-top-10-os-top-5-installed-base-regional-split-and-bloodbath-q4.html

Ahonen, T. T., & Barrett, J. (2002). *Services for UMTS: Creating killer applications in 3G*. Singapore: John Wiley & Sons. doi:10.1002/0470014172

Akerman, N. (1998). *The necessity of friction*. Heidelberg, Germany: Westwiew Press.

Baeza-Yates, R., & Raghavan, P. (2010). Next Generation Web Search. In S. Ceri & M. Brambilla (Eds.), *Search computing, LNCS 5950* (pp. 11–23). Berlin, Heidelberg: Springer-Verlag.

Ballon, P. (2009). Platform types and gatekeeper roles: The case of the mobile communications industry (pp. 17-19). CBS- Copenhagen Business School. Retrieved from http://www2.druid.dk/conferences/viewpaper.php?id=5952&cf=32

Ballon, P., & Van Heesvelde, E. (2011). ICT platforms and regulatory concerns in Europe. *Telecommunications Policy*, *35*(8), 702–714. doi:10.1016/j.telpol.2011.06.009

Barns, M. (2014). Expect a seismic shift in video consumption. *Re/code*. Retrieved from http://recode.net/2014/02/26/expect-a-seismic-shift-in-video-consumption/

Benton, J. (2014). The mobile majority. *Nieman Reports, 68*(2), 38-39. Retrieved from http://www.nieman.harvard.edu/assets/pdf/Nieman%20Reports/backissues/NRSpring2014.pdf

Bogdanowicz, M., de Prato, G., Feijóo, C., Nepelski, D., & Simon, J. P. (2010). *Born digital/grown digital. Assessing the future competitiveness of the EU video games software industry*. JRC Scientific and Technical Reports. Retrieved from http://ftp.jrc.es/EURdoc/JRC60711.pdf

Boudreau, K. J., & Jeppesen, L. B. (2014). Unpaid crowd complementors: The platform network effect mirage. *Available at SSRN 2475034.*

Bright, P. (2014). Neither Microsoft, Nokia, nor anyone else should fork Android. It's unforkable. *Ars Technica*. Retrieved from http://arstechnica.com/information-technology/2014/02/neither-microsoft-nokia-nor-anyone-else-should-fork-android-its-unforkable/

Brin, S. (2013). Why Google Glass? *TED Talks*. Retrieved from http://www.ted.com/talks/sergey_brin_why_google_glass

Broder, A. (2002). A Taxonomy of Web Search. *ACM Sigir Forum, 36*(2), 3-10. Retrieved from http://portal.acm.org/citation.cfm?id=792552

Castellet, A., Aguado, J. M., & Martínez, I. (2013). Los nuevos actores que cambian las reglas y el juego: Cómo Apple y Google han transformado la distribución de contenidos. In Zilles, Cuenca, & Rom (Eds.), *Breaking the media value chain* (pp. 333-42). Barcelona: Facultat de Comunicació Blanquerna, Universitat Ramon Llull. Retrieved from http://www.slideshare.net/acastellet/castellet-aguado-martinez-cicr2013

Castellet, A., & Feijóo, C. (2013). Los actores en el ecosistema móvil. In J. M. Aguado, C. Feijóo, & I. Martínez (Eds.), *La comunicación móvil. Hacia un nuevo ecosistema digital* (pp. 27–56). Barcelona: Gedisa.

Christensen, C., Skok, D., & Allworth, J. (2012). Be the disruptor. *Nieman Reports, 66*(3). Retrieved from http://www.nieman.harvard.edu/assets/pdf/Nieman%20Reports/backissues/Fall2012.pdf

Christensen, C. M. (1997). *The innovator's dillemma*. Cambridge, MA: Harvard Business School Press.

Church, K., & Smyth, B. (2009). Understanding the intent behind mobile information needs. *Proceedings of the 13th International Conference on Intelligent User Interfaces*, 247-256. Retrieved from https://wiki.ucop.edu/download/attachments/34668692/Understanding+the+Intent+Behind+Mobile+Information+Needs+p247-church.pdf

Cisco. (2014). *Cisco visual networking index: Global mobile data traffic forecast update, 2013–2018*. Retrieved from http://www.cisco.com/c/en/us/solutions/collateral/service-provider/visual-networking-index-vni/white_paper_c11-520862.html

Constantinou, A., Vakulenko, M., & Kapetanakis, M. (2011). *Mobile Megatrends 2011*. Vision Mobile. Retrieved from http://www.visionmobile.com/rsc/researchreports/mobile-megatrends-2011-(visionmobile).pdf

Danova, T. (2014). Smart watches will lead wearable computing to mass market status. *Business insider*. Retrieved from http://www.businessinsider.com/global-smartwatch-sales-set-to-explode-2014-3?utm_source=alerts&nr_email_referer=1

De Wit, M., Mateus, M., Moitry, J., & Acharys, A. (2009). Dumb pipe vs. Smart pipe? Better to design a valuable pipe. *Oliver Wyman*. Retrieved from http://www.oliverwyman.com/ow/pdf_files/CMT07-DPipeSPipe-final.pdf

Doyle, K. (2014). Smartphone obsolescence: How the personal cloud and IoT will disrupt the handset. *Gigaom research*. Retrieved from http://research.gigaom.com/report/smartphone-obsolescence-how-the-personal-cloud-and-iot-will-disrupt-the-handset/?utm_source=GeneralUsers&utm_campaign=15c55a60e5-v%3Awhatsnewd%3A08-11&utm_medium=email&utm_term=0_1dd83065c6-15c55a60e5-99640621

Dudenhoeffer, C. (2014). The existing wearable technology landscape. *Wearable world news*. Retrieved from http://wearableworldnews.com/2014/05/06/wearable-world-taxonomy/

Feijóo, C., Gómez Barroso, J. L., Marín-Andreu, Á., & Rojo-Alonso, D. (2006). An overview of contents exploitation scenarios in the new digital media (pp. 72-95). ENCIP. i2010: Comments and Contributions. Eurocpr 2006: Selected Papers.

Feijóo, C., Maghiros, I., & Gómez-Barroso, J. L. (2008). *Techno-Economic models for mobile content*. Proceedings of the 17th Biennial Conference of the International Telecommunications Society, Montréal, Canada.

Gassée, J. L. (2014). Tim Cook needs to hire human curators to fix the Apple App Store. *Quartz*. Retrieved from http://qz.com/247370/apple-tim-cook-desperately-needs-to-fix-the-app-store/

Gawer, A., & Cusumano, M. A. (2014). Industry platforms and ecosystem innovation. *Journal of Product Innovation Management*, *31*(3), 417–433. doi:10.1111/jpim.12105

Germany's consumers not sold on wearables. (2014). eMarketer. Retrieved from http://www.emarketer.com/Article/Germanys-Consumers-Not-Sold-on-Wearables/1011141/2

Goggin, G., Dwyer, T., & Martin, F. (2014). Moveable Media: Mobile internet and new policy modes. In M. Glowacki & L. Jackson (Eds.), *Creativity, Innovation and Interaction: Public Media Management Fit for the 21st Century*. New York: Routledge.

Gómez-Barroso, J. L., Compañó, R., Feijóo, C., Bacigalupo, M., Westlund, O., Ramos, S., & García-Jiménez, M. C. (2010). Prospects of mobile search. Seville: Institute for Prospective Technological Studies. Retrieved from http://ipts.jrc.ec.europa.eu/publications/pub.cfm?id=3159

Google. (2012). Project Glass: One day [YouTube video]. Retrieved from https://www.youtube.com/watch?v=9c6W4CCU9M4&feature=youtu.be

Heggestuen, J. (2014a). Carrier billing hasn't gone away. *Business Insider*. Retrieved from http://www.businessinsider.com/carrier-billing-hasnt-gone-away--here-is-why-it-has-got-developers-app-stores-and-mobile-carriers-so-excited-2014-2?utm_source=alerts&nr_email_referer=1

Heggestuen, J. (2014b). Carrier billing is another way to pay on phones. *Business Insider*. Retrieved from http://www.businessinsider.com/carrier-billing-is-another-way-to-pay-on-phones--and-its-creating-a-big-opportunity-to-monetize-mobile-users-2014-3?utm_source=alerts&nr_email_referer=1

Heggestuen, J. (2014c). Google is cutting deals with wireless carriers everywhere... *Business Insider*. Retrieved from http://www.businessinsider.com/google-play-and-carrier-billing-2014-4?utm_source=alerts&nr_email_referer=1

Ingram, M. (2013). Did Facebook miss a massive opportunity by building a walled garden instead of a truly open platform? *Giga OM*. Retrieved from http://gigaom.com/2013/07/24/did-facebook-miss-a-massive-opportunity-by-building-a-walled-garden-instead-of-a-truly-open-platform/

Insight, C. C. S. (2014). *Global smartphone market analysis and outlook: Disruption in a changing market*. Retrieved from http://www.lenovo.com/transactions/pdf/CCS-Insight-Smartphone-Market-Analysis-Full-Report-07-2014.pdf

Isaacson, W. (2011). *Steve Jobs*. New York: Simon & Schuster.

Jaokar, A., & Fish, T. (2006). *Mobile Web 2.0*. London: Futuretext.

Kohlborn, T., Korthaus, A., Riedl, C., & Krcmar, H. (2009). Service aggregators in business networks. In Enterprise Distributed Object Computing Conference Workshops, 2009. EDOCW 2009. 13th (pp. 195-202). doi:10.1109/EDOCW.2009.5331992

Leurdijk, A., & Nieuwenhuis, O. (2012). The Music Industry: Statistical Ecosystems and Competitiveness Analysis of the Media and Content Industries. *JRC Technical Reports*. Retrieved from ftp://ftp.jrc.es/pub/EURdoc/JRC69816.pdf

Li, F., & Whalley, J. (2002). Deconstruction of the telecommunications industry: From value chains to value networks. *Telecommunications Policy*, *26*(9/10), 451–472. doi:10.1016/S0308-5961(02)00056-3

Mace, M. (2014). The real meaning of the Fire Phone. *Mobile opportunity*. Retrieved from http://www.mobileopportunity.blogspot.com.es/2014/07/the-real-meaning-of-fire-phone.html

Maghiros, I., Abadie, F., & Pascu, C. (2007). *Introduction to the State of the Art of the Creative Content Sector*. Seville: Institute of Prospective Technological Studies.

Manjoo, F. (2011). The Great Tech War of 2012. *Fast company*. Retrieved from http://www.fastcompany.com/magazine/160/tech-wars-2012-amazon-apple-google-facebook

Marketshare, N. (2014). Search engine market share. Retrieved from http://www.netmarketshare.com/search-engine-market-share.aspx?qprid=4&qpcustomd=1&qptimeframe=Y

Metz, R. (2014). This Company Thinks Your Car Wants Google Glass. *MIT Technology Review*. Retrieved from http://www.technologyreview.com/news/529646/this-company-thinks-your-car-wants-google-glass/

Meyer, D. (2013). Want Android and Firefox OS in one phone? Geeksphone has something lined up for you. *Giga OM*. Retrieved from http://gigaom.com/2013/11/28/want-android-and-firefox-os-in-one-phone-geeksphone-has-something-lined-up-for-you/?utm_source=Triggermail&utm_medium=email&utm_term=Mobile%20Insights&utm_campaign=Post%20Blast%20%28sai%29%3A%20Netflix%20Is%20Deepening%20Its%20Personalization%20Research

Nash, T. (2014). Apple, developers think you are just one inconsistent loop. *Tapdaq Blog*. Retrieved from http://blog.tapdaq.com/insight/apple-developers-think-you-are-just-one-inconsistent-loop/

Netflix's war on Mass Culture. (2013). *New republic*. Retrieved from http://www.newrepublic.com/article/115687/netflixs-war-mass-culture

Newswire, N. (2014a). Smartphones: So many apps, so much time. Retrieved from http://www.nielsen.com/us/en/insights/news/2014/smartphones-so-many-apps--so-much-time.html

Newswire, N. (2014b). Wearable Bluetooth users, constantly connected consumers. Retrieved from http://www.nielsen.com/us/en/insights/news/2014/wearable-bluetooth-users-constantly-connected-consumers.html

Pentland, A. P. (1998). Wearable Intelligence. *Scientific American*, 90–95.

Porter, M. E. (2008). *Competitive advantage: Creating and sustaining superior performance* (Original work published 1985). New York: Simon and Schuster.

Pulman, S. (2011). Facebook as a mobile media platform. *Storytelling*. Retrieved from http://www.storytelling.es/facebook-as-a-mobile-media-platform/

Schuermans, S. (2013). The naked android. *VisionMobile*. Retrieved from http://www.visionmobile.com/blog/2013/11/the-naked-android/?utm_source=VisionMobile+subscriber+list&utm_campaign=2248ef3bba-2013_06+Newsletter&utm_medium=email&utm_term=0_efe0e34ffa-2248ef3bba-394601118

Silberschatz, A., & Galvin, P. (1991). *Operating system concepts*. New York: Addison-Wesley.

Singh, K. (2014). 10 Rising Social Networks you should explore. *Mashable*. Retrieved from http://mashable.com/2014/07/28/social-networks-on-the-rise/?utm_cid=mash-com-fb-socmed-link

Smith, C. (2014). *Project Ara development boards release date set for late July. BGR*. Retrieved from http://bgr.com/2014/07/15/project-ara-development-boards-release-date/

Statista. (2014). *Global market share of search engines 2010-2014*. Retrieved from http://www.statista.com/statistics/216573/worldwide-market-share-of-search-engines/

Subramaniam, N. (2012). Android One is Google's bet to sweep Indian smartphone market. *Tech2*. Retrieved from http://m.tech.firstpost.com/news-analysis/android-one-is-googles-bet-to-sweep-indian-smartphone-market-232006.html

Tehrani, K., & Michael, A. (2014). Wearable technology and wearable devices: Everything you need to know. *Wearable Devices Magazine*. Retrieved from http/:www.wearabledevices.com:what-is-a-wearable-device

Telenor Group. (2011). Customer loyalty lies with the operating system. Retrieved from http://telenor.com/en/news-and-media/articles/2011/customer-loyalty-lies-with-the-operating-system

Tellabs. (2011). *Tellabs "End of Profit" study executive summary*. Retrieved from http://www.tellabs.com/markets/tlab_end-of-profit_study.pdf

Thierer, A. (2004). Are "dumb pipe" mandates smart public policy? Vertical integration, net neutrality and the network layers model. *Journal on Telecommunications & High Technology Law*, *3*(2), 275.

Thomas, D. (2014). Telecoms groups blast EU net neutrality plans. *FT.Com*. Retrieved from http://www.ft.com/intl/cms/s/0/089476a4-9ca3-11e3-9360-00144feab7de.html?utm_content=buffer7eb3d&utm_medium=social&utm_source=twitter.com&utm_campaign=buffer&siteedition=intl

Thomson, A., & Orihuela, R. (2014). Phone carriers call for EU rules to hit facebook and google. *Bloomberg*. Retrieved from http://www.bloomberg.com/news/2014-08-31/vodafone-heads-to-spain-to-protest-eu-network-traffic-proposals.html

Vakulenko, M., Schuermans, S., Constantinou, A., & Kapetanakis, M. (2011). *Mobile platforms: The clash of ecosystems*. Vision Mobile. Retrieved from http://www.visionmobile.com/rsc/researchreports/VisionMobile-Clash-of-Ecosystems_v1.pdf

West, J., & Mace, M. (2010). Browsing as the killer app: Explaining the rapid success of apple's iphone. *Telecommunications Policy*, *34*(5-6), 270–286. doi:10.1016/j.telpol.2009.12.002

Whitney, L. (2014). KitKat chews up more than 20% of android devices. *CNET*. Retrieved from http://www.cnet.com/news/kitkat-chews-up-more-than-20-percent-of-android-devices/

Wilcox, M., & Voskoglou, C. (2014). *Developer economics. State of the developer nation Q3 2014*. Vision Mobile. Retrieved from http://visionmobile.com

Wit, M. d., Mateus, M., Moitry, J., & Acharys, A. (2009). *Dumb pipe vs. Smart pipe? Better to design a valuable pipe*. Oliver Wyman. Retrieved from http://www.oliverwyman.com/ow/pdf_files/CMT07-DPipeSPipe-final.pdf

Wu, T. (2011). *The Master Switch: The Rise and Fall of Information Empires*. London: Atlantic Books.

Young, E. (2014). 90 million wearables will be sold this 2014, analyst says. *TechnologyTell*. Retrieved from http://www.technologytell.com/gadgets/138712/90-million-wearables-will-be-sold-this-2014-analyst-says/

KEY TERMS AND DEFINITIONS

Context Sensors: In a mobile device, pieces providing information about the surrounding environment in a broad sense. GPS, cameras, pedometers and microphones can act as context sensors in a mobile device.

Ecosystem: A business/service changing environment made of different players –including final users- that interact one with another and are able to trigger changes and can need to change their behaviors in order to keep or extend their dominance in some particular activity.

Featurephone: Mobile phone with an operating system not obvious to its user, regardless of the scope of its features or its specs.

Friction: The set of obstacles or circumstances that lie between a particular consumption will and its full completion.

OS: Operating system. A set of software pieces, usually bundled with any mobile phone or terminal, that is in charge of the user interface, the primary handset functions and the context sensors. The OS determines ultimately the range of features of the device.

Platform: A way of business organization that needs the contribution of final users and business partners in order to develop a product of a service invented or promoted by a leadership –leadership can be shared by a number of different players. Platform rules differ depending on their foundations and their objectives.

Smartphone: Mobile phone equipped with an operating system that can be easily recognized by the user.

Wearables: Artifacts that can work as outfits, in a way similar to how a watch is attached to the forearm.

ENDNOTE

1 Pyramid Research: www.pyr.com.

Chapter 5
Mobile Music:
An Overview

Claudio Feijóo
Technical University of Madrid, Spain & Tongji University, China

Sergio Ramos
National University of Distance Education, Spain

Victoria Bleda
ContentWise.media, Spain

ABSTRACT

Music industry is arguably the foremost example of the advantages and disadvantages of having been exposed early on to digitalization. But not only has music been a pioneer in the area of digitalization, but it has also been one of the first sectors in the content industry to be massively "mobilized". It is therefore an area where lessons can be learnt for other mobile media and content market segments. From this framework, this chapter examines four main stages in the evolution of mobile music, investigates in some detail the emerging business models in the domain and concludes with the main challenges and opportunities for mobile music.

MUSIC IN THE HANDSET: PIONEERING DIGITAL MOBILE DEVELOPMENTS

It is beyond dispute that the music industry was the first to be significantly affected by the transformations caused by digitization. Analyzing the three basic stages that music content goes through from inception to consumption, namely, creation / production, distribution / access, and consumption / interaction (C Feijóo, Maghiros, Abadie, & Gomez-Barroso, 2009), it is rather clear that they have all been drastically altered by digitalization. Musical production, which is practically 100% digitized, has become considerably cheaper, with a variety of tools and techniques now available to creators that before were only available in the professional sector. Music distribution is increasingly conducted online in digital format; in fact, there are already a number of markets in which it represents more than 50% of total distribution (IFPI, 2013), in particular in three of the top 10 country music markets (IFPI, 2014). Similarly, music

DOI: 10.4018/978-1-4666-8838-4.ch005

consumption is more intense than ever and occurs through a wide variety of media and digital devices. In addition to its increasing ubiquity, the ease with which music – and the experience of listening to it – can be shared has made it a key element in social networks. In fact, to underline the relevance of music in those, it is enough to recall that in 2013, nine of the ten most followed people on Facebook – and seven out of ten on Twitter – were music artists and that nine of the ten most watched videos on YouTube were music videos (IFPI, 2013).

Of course, the role of the music industry as a digital pioneer has also brought with it a considerable number of new challenges that did not exist in the past. The most evident is the ease with which users can exchange and share digitized information – music. Independent of where the boundary is placed between the public and the private spheres, and going beyond avoiding economic hardship for legitimate content owners, it is certainly true that the traditional music industry has exhibited myopic behavior in terms of taking advantage of the opportunities of digitalization, in particular during the very long period of time in which it has been unable to provide an answer to the new needs and expectations of consumers. It has been necessary to wait until 2013 to see once again the modest growth of music of 15 years ago (IFPI, 2013). In any case, the effect of all of this has been the decline – at least in monetary terms – of the traditional music industry and the introduction of new digital players – providers – whose business models have only now started to compensate for the income generated before the shift to digital (Leurdijk & Nieuwenhuis, 2012). In 2012, digital music represented 5.6 billion dollars worldwide, representing 9% more than in 2011 and a third of the total income of the music industry (IFPI, 2013). Growth slowly continues with the latest figures available displaying that digital music at 5.9 billion dollars, this is, 39% of the total revenues of the industry (IFPI, 2014).

Mobile music, with its particularities, is a clear example of the advantages and disadvantages of having been exposed early on to digitalization, as will be explained in detail in the following sections. Therefore, it is a field from which other (mobile) sectors that are lagging in their digitalization process can learn many lessons. Not only has music been a pioneer in the area of digitalization, but it has also been one of the first sectors in the content industry to be massively "mobilized". Therefore, there is a considerable amount of valuable evidence that can be gathered from examining the (mobile) music industry.

Some general market data can help clarify the industry's relevance. According to market analysis firm Juniper, the total value of the global mobile music industry was 3.1 billion dollars in 2010, with a prediction that it will reach 5.1 billion in 2015.[1] Against this background, it is important to note that, of course, the most developed mobile music market has been the U.S., followed by the group consisting of Japan, Korea, and China, with Europe at a modest third place. In terms of the use of mobile music, according to data from the online market analysis firm Comscore, already in 2010, 24% of users in countries such as the United Kingdom, France, Germany, Italy, and Spain listened to music on their mobile phones, a 10% increase over the 2009 values. Available data from a survey administered by the firm NPD at the beginning of 2013 in the U.S. show that 56% of smartphone users regularly listen to music on their devices – 39% do it daily. The most popular services for listening to music – through "streaming" – were Pandora, Spotify, and Rhapsody. In addition, 60% of users had transferred part of their personal music collection to their devices. According to the last survey from ONTSI in Spain (ONTSI, 2012), the number of users with an integrated FM radio in their mobile handset is 68%, and of these, 39% use it. Mobile handsets with the capacity to play music in MP3 format amount to 57%, and this function is regularly used by 47% of those for whom it is available.

In the next section of this chapter the evolution of mobile music is framed into four main generations to arrive at current situation. From here the different possibilities for business models in mobile music are described. The final section blends the findings of previous section in the main trends that the authors consider will shape the future evolution of mobile music. The methodology used in the chapter has been simply desk research and review of the scarce literature available on this topic. Information obtained from these sources has been complemented with the own experience of authors for more than 20 years in different real-life projects and research endeavors in the music industry covering both traditional and online business models, as well as intellectual property regimes.

A BRIEF ACCOUNT OF MOBILE MUSIC: A TALE OF FOUR GENERATIONS

The links between music, mobility, and personalization – and the opportunities that accompany them – can be traced back to the battery-powered radio as the first portable, if not mobile, handset. However, perhaps the starting point of mobile music should be identified with the smashing commercial success of the Walkman by Sony, launched back in 1979. This revolutionary device, of which more than 200 million units were sold, was able to free consumers from having to be "tied down" to enjoy their favorite music. Undoubtedly, this device created the foundations of a user culture, which is the first trait that should be highlighted, linking music – the consumption of music – with freedom of usage, that is, consumption at any time and in any place.

What might be called the second generation of mobile music started with the digitalization of mobile communication systems – the mid-1990s – and largely extended to the launch of the first generation of smartphones on the market –approximately in 2003-4 in Europe. During this period, and also in the following one, as mentioned in the following, music had been the principal driver of the mobile content industry. The key element consisted of the massive adoption of ringtones by mobile consumers. From the perspective of the industry, ringtones represented a golden age: simple and extremely cheap to generate based on a musical theme, the income they produced was considerably higher than the expenses involved. Moreover, they mostly remained beyond the reach of phenomena such as piracy because of the closed – proprietary – character of the software and hardware in the mobile phones that used them. This is the second differentiating feature of mobile music: in one way or another, it has always been associated, from the point of view of the industry, with an environment that was "more controlled" than that of the general Internet.

Consumers could mainly obtain ringtones in three different ways. First, they could be pre-loaded on the handset and, therefore, would only require an agreement between the handset manufacturer – sometimes also the telecommunications operator was needed – and the music proprietor – typically the record companies. The role of the user was simply that of a passive consumer. The second, more advanced way, allowed users to download new ringtones from the operator's portal, typically through payment using an SMS premium. This allowed greater freedom of choice for the user and additional income for operators and the record companies. The third way was through a third-party website, a content aggregator. This method involved the effort of adapting the content to the myriad mobile handsets, each with specific software and hardware. This was known as the "off-deck" model, as opposed to the previously noted "on-deck" model with mobile operators. In any case, in the midst of this opportunity, a number of companies emerged that offered users all types of personalization possibilities for their handsets: not only ringtones but also wallpapers or other different merchandising. Unfortunately, some of these new companies abused the chance of charging users for services they did not fully understand or had not directly requested (EC, 2008), creating a considerably negative perception of this industry and, consequently, contributing to a notable delay in the introduction of interesting and reliable proposals for users and their subsequent adoption.

The third generation of mobile music included two distinct paths. From one perspective, the continuation of ringtones that had evolved to have a higher quality –"realtones" – or personalized ringtones for different functions – "ringbacks", for example – led to the gradual growth of the size of the mobile music market until the arrival and adoption of real smartphones in approximately 2008-9. During this time, even though it started with modest figures, the mobile music business consistently grew at a fast pace, completely in contrast to the rest of the music industry. The second element of this third generation was the arrival and success of digital music players based on solid-state memories. They were an improved and digitized version of the previously noted Walkman, with the iPod from Apple being a paradigmatic case. The huge storage capacity of these devices, their spectacular usability – especially compared with what was available in the market at the time – and the growing ease with which users could accumulate a personal collection or acquire new pieces were the key factors for its success. As a logical result, the integration of this device with the mobile handset was considered, given that they were both personal devices, which had considerable synergies.

While the business model in the case of mobile phones continued to be that of the previous generation, in the case of mobile players the situation was considerably different. To start with, users were able to go to the music store that had been set up by the device manufacturer to acquire and download music to their device. Again, the paradigmatic example is Apple's iTunes, which was launched on the market in 2003 and revamped for the launch of iPod. Alternatively or complementarily, they could directly download the digitized music they already had onto these devices, a process known as "side-loading". With this second procedure, mobile music began to no longer fall under the direct control of a particular agent in the ecosystem, which would have significant consequences. In this regard, it must be highlighted that some handsets already had the possibility of loading music ("full tracks") through either specific portals or side-loading, although their storage capacity and functionality were at that time still considerably limited compared with portable players.

The arrival of handsets with larger capacities – the real smartphones – and the transition to broadband mobile networks and affordable data plans led market analysts to posit that the success story of mobile music would continue indefinitely (Holden, 2008) through a new fourth generation of music in mobile devices. However, the reality has proven to be more complex and less linearly projected in the future.

In fact, what has happened is that the set of factors noted above deepened the gap between what users would want to do with their music and what music-related service providers and operators would allow them to do with it. Thus, still in 2015, consumers of mobile music spent a considerable amount of time using different software programs and connection media to transfer their audio libraries between the cloud, personal computers, diverse support systems, digital formats, and, finally and frequently painfully, their mobile devices so that their music could travel with them. Curiously, the key factor of this divergence has been the fact of convergence: a smartphone or tablet mobile handset combines the capacities and functionality of a mobile phone and a portable player, which forces users to "fill it" with music.

Of course, mobile music providers, which are part of what we may designate the new music industry (Preston & Rogers, 2011), have been getting closer to fulfilling the needs of users. However, they are still unable to cover the so-called "360 degrees" of user needs, which will most likely require a fifth generation of mobile music still to be fully developed. In any case, the business models and innovations generated by the fourth generation, as well as certain future perspectives, are discussed in the following sections.

To summarize this brief review of mobile music's evolution, we must note that although the relationship between mobile communications and music has, to date, been scarcely studied, practically all the steps in the evolution of data in the mobile ecosystem, which is unrelated to traditional voice communication, have had their precursors in mobile music consumption. Relevant examples are the convergence of functionalities in the same device – the mobile handset with radio and Walkman functionality; the first successful data download services in the form of ringtones; the use of the mobile phone as a second screen – second player – through the phenomenon of side-loading; the first incarnation of application stores as music stores – the case of iTunes; or, more recently, techniques such as streaming or cloud computing that new music services in the mobile device use by taking advantage of the growing penetration of smartphones and flat-rate data plans.

BUSINESS MODELS FOR MOBILE MUSIC

There are several different definitions of what is considered the music industry. In this chapter a broad definition is considered including the production and distribution of recorded music, the distribution and exploitation of associated intellectual property rights, and the organization of live events (Hesmondhalgh, 2007). All these activities are relevant in the case of music in the mobile environment. Within this industry, mobile music would refer to all those elements that make it possible to consume audio – which is different from conversational voice – on a mobile device. Thus, it entails also a broad definition that places the user's device at the center, taking into account all the technical media that have allowed music to reach the device, setting aside the fact that the process facilitator may have been a service provider or the user her/himself.

In fact, music content can reach the mobile device through two basic paths: through a data connection that uses mobile and/or wireless communication networks or through a procedure that allows the user to load music he or she is interested in onto the mobile device (the "side-loading" model noted above). These two models also exist for other types of mobile content, but in the case of music, it is fundamental to distinguish between the two to understand the opportunities and challenges with which this industry is confronted and the perceptions and experience of the user. The key issue is who controls the value chain and the experience of the user. In the first model, the service provider is in charge, while in the second, it is the manufacturer or provider of the device used by the consumer.

From the business perspective, the first model has a value chain that is similar to the traditional music industry, in which the provider of online services plays the role of distributor. The remaining stakeholders continue to be the same: authors/creators, producers/record companies, and related stakeholders (managers, editors, etc.). These providers' three basic types of business models are the typical ones for online services, namely, pay per download, subscription models (of very different types), or free models for the end user that are financed by ads. In the case of the second model, the objective is to facilitate the availability of all the music in which the user may be interested in the device. To that end, there are two basic strategies: either eliminating the distribution aspect so that there is no bottleneck or integrating the device with some distribution system – vertical integration – so that the user experience is as controlled as possible.

Based on these basic business models, offers – techno-economic models – can be established that are significantly differentiated, as explained below. As a reference, the section starts with the traditional model and its online equivalent. It must be also noted that, in general, mobile music providers use several of these models simultaneously, which means that the classifications are not mutually exclusive but instead simply indicative of the internal reasoning of each of the possible models.

Traditional Model: Physical Store

The traditional model has consisted of the sale of music through a physical medium, such as a vinyl record, cassette, CD, DVD, Blu-ray, USB, etc. The use of mobile devices, in some cases, makes it possible to complete the experience of the physical purchase and obtain additional information about the music – for example, using BIDI codes – or access exclusive launchings – for example, through localization or some type of game.

In 2013, the majority of the music that can be found in a user's playing device – be it a mobile phone or a portable device – is still acquired through this mechanism, this is, acquired from a physical media and then side-loaded to the device.

Quasi-Traditional Model: Online Store

This model is the immediate translation of the previous one to the digital ecosystem. The main online music stores are adapted in their interface to the mobile environment. In fact, some of them began as the evolution of the ringtones business model, as explained above.

The most interesting part of this model is the arrival of agents that adapt, store, and aggregate music so that it is available online and/or in a mobile medium. In addition, this model not only enables the direct purchase of a song or the record that contains it but also enables the user to acquire periodical awards, subscriptions or flat-rate plans for acquiring songs. Of course, online stores have all the advantages of digital storage, which translates into highly comprehensive libraries that can satisfy the needs of "long-tail" consumers. Downloads constituted 70% of the digital music business in 2012 (IFPI, 2013) although they are being rapidly substituted by streaming services, see below.

There are two different types of aggregators or license providers. First, there are "the majors" – EMI Music, Sony Music Entertainment, the Universal Music Group, and the Warner Music Group. Companies that wish to launch new online services on the market must reach an agreement with all or some of these companies if they wish to access a significant part of the audio-library of interest to users. The second type are usually called "independent" aggregators – that is, the aggregators do not belong to the major players – as in the case of companies such as ODA, IRIS, Finetunes, INgrooves, The Orchard, and Tunecore. Their rationale is simple: the digital environment evolved so rapidly that it is necessary for independent music to have an aggregator that is capable of reaching deals with a growing variety of services and digital platforms instead of the own independent label reaching agreements with providers.

The classical example of the online music store is Napster, a commercial name that was recycled into a legal store after the closing of the original P2P file sharing network. The store offered different possibilities for purchasing and subscribing to obtain songs for non-mobile and mobile devices, both as downloads and through streaming (see below). The subscription models were equipped with digital rights management technologies (DRM), while the songs that users purchase had no type of protection. Napster was launched in 2006 as a free model for music users through streaming, financed by ads, and was closed in March 2010 due to a lack of viability (see advertising model). It also reached an agreement with the computer manufacturer Dell to provide a one-year subscription to Napster for buyers of laptops (see bundle model).

Online Streaming

This model is similar to the previous one in terms of the relationship between stakeholders. However, it represents a radical paradigm shift, given that it represents a move away from considering music a product and toward considering it a personalized service. Thus, music is not stored in the user's device – at least, conceptually it is not supposed to be – and is consumed directly from the network during distribution.

In the case of the mobile environment, the start of these services has been somewhat delayed compared to the Internet version, despite the fact that they were the first ones in the mobile content environment. This delay has not just been the result of the cost mobile broadband access rates. It has mainly been an issue related to the rights and permits granted by "the majors" and to the limitations and competition that proprietary companies in mobile platforms have established, given that these platforms have also provided musical services "a la carte". Therefore, it is not an unexpected fact that some services that are free on the Internet are not free in the mobile environment.

From the perspective of 2015, this is the model that seems to have the greatest chance of success in the future and that several players are trying to replicate, with offers from companies such as Spotify, Deezer, WiMP and Tidal being the most paradigmatic. In fact, the main agents in the mobile environment, such as Apple and Google, prepared streaming services in 2013 especially geared toward their respective platforms. It certainly is a model that presents noteworthy advantages. In addition to the possibility of personalization, fresh new business models – pay for one day, for one week, during an event, pay for better quality, freemium models linked to user preferences, etc. – and the rights and licenses agreed upon with proprietors (authors, performers, record companies) can be used. Because they do not include downloading, these agreements are considerably less expensive, leaving a potential margin for profitability.

Somehow, the models based on these new types of subscriptions make it possible to move away from the extremes of paying either per song or monthly/annual flat rates, which introduces a great variety of possibilities that brings the value of music closer to the expectations of users. However, one must not forget that in 2012, subscription services and advertising only represented 30% of the digital music business (IFPI, 2013), so there is still a long way to go.

As noted above, Spotify is one of the prototypical examples of an online music streaming service. It was launched in 2008 and, in its basic version, allows for listening to music both "a la carte" and by theme. As is typical with new services in general, Spotify includes different functionalities to make the user experience easier (recommendations, history, song sharing, etc.). The business model includes advertising in a free version, a music cap by listening hours in the free version, "revenue sharing" with record companies to purchase music, and special payment features, such as a day pass, monthly subscriptions, listening to music without an internet connection and on a mobile device, etc. According to the company itself, it had three million subscribers in 2011, with the number growing to five million in 2012 and 20% of the total being active users. The same scheme used by Spotify is being followed by many other companies, all of them through specific mobile apps. Communication operators themselves, mobile or not, have launched their own online music "a la carte" services with varying levels of success. In any case, streaming services, especially those that include subscriptions, are frequently regional, given that they are supported by the cultural preferences of their users.

For all these services, one must note that after the failure of the majority of services based on DRM technologies and the transition of the main stores to "DRM-free" content for purchasing models, services based on streaming have emerged as facilitators of business models that offer a much more flexible framework for users and providers and relative independence from the type of device used to access such services.

Another type of online streaming service that deserves to be highlighted is that which detects the tastes and preferences of users automatically. Pandora, a service that searches for musical pieces similar to a given set as a function of a series of features that identify them (Ellis, 2006), is perhaps the most well-known case. It offers its services through the Internet and on several different mobile platforms, but it is limited to a specific number of countries due to the territorial licenses of music (a common problem for all the services considered in this section). It has both free versions, with advertising but limited in time, and paid versions. Similar to many other providers, it has gone through a series of financial up-and-downs, basically due to its licensing agreements with the record industry. In 2012, Pandora was responsible for 8% of all songs heard on the radio in the U.S. (IFPI, 2013).

Radio Model

The radio model is a variation of the prior model, but with a significant difference: the user only picks some features of the songs he or she wishes to listen to – style, artist, group of artists, random within a certain audio library, etc – and not the songs themselves. In this case, there are barely any differences between the general Internet model and the mobile model, and the business model is, of course, based on advertising.

Companies that offer streaming services also offer this type of model, although there are also specialized companies, with the most well-known case in the U.S. being Slacker. Particularly among the applications available in mobile stores, there are many, such as iHome, that allow for tuning into existing radio channels in any place around the world.

In addition, a growing number of mobile devices have an integrated chip that allows users to receive conventional radio emissions, eliminating, at least from the perspective of consumers, the boundaries between both media.

Models Based on Online Advertising

This model is related or complementary to the previous ones but is based on the basic premise of acquiring an Internet audience that is sufficiently segmented to sell these user profiles to advertisers – "targeted advertising" or "behavioral advertising", see Feijoo et al, (2010). As main instances, this is the model that Google has championed and Facebook broadly follows. In the case of mobile devices, data about user preferences are added to all those derived from the location and use of the mobile device (Gómez-Barroso, Feijóo, & Compañó, 2011). Therefore, the objective consists of gaining the greatest possible number of users, building a profile for them, and, following the theory of multilateral markets (Rochet & Tirole, 2006), simultaneously getting content developers and generators to join. Thus, there is a positive feedback loop between users, content, and advertisers. Many companies use this model: mobile search engines, mobile social networks, audiovisual channels, etc. In all these cases, there are variations in which music is either a fundamental element or the direct core of the business.

The social network that has come closest to music is MySpace, which, in fact, is partially owned by the major record companies. Its greatest appeal is the opportunity it provides to very easily create a music profile to which information, photographs, songs, and videos can be uploaded to share with others and to share music through streaming. From this standpoint, the business model, similar to online stores, is based on the creations of user profiles for advertisers – including diverse merchandising and concert tickets. In the specific case of mobile devices, music has been identified as one of the driving factors of mobile social networks (Scaglione & Hamoudia, 2010).

YouTube is perhaps the other prominent service that uses music and is geared toward personalized advertising. This service, owned by Google, is, of course, the most widely used video sharing website, but it is also an important tool for the music industry. Although it was initially a service only geared toward users who uploaded videos, it has evolved in a more professional vein, allowing artists and producers to create "official" channels for musical video clips, interviews, etc. and signing agreements with them to share the benefits they generate. An aspect of YouTube worth remembering from the perspective of music is that it is also the main place for finding information on musical training and related topics, which is a very important segment of music-related activities. It must be also noted that the difference between mobile licenses and Internet licenses described above makes the offer of musical videos different in one medium compared to another.

Of course, the major record companies have also launched their own attempts at capturing new value from the use of music. Among the most relevant cases is Vevo, an application that allows access to the library of musical videos of Sony and Universal. It integrates geolocation features – to ascertain what other users are watching in the same geographic area – and social network features for communication with the actual artists through Facebook and Twitter. It was also one of the three most-watched channels on YouTube in 2012 (IFPI, 2013).

Other video channels with predominantly musical themes, such as MTV, have also tried to use mobile screens to reach users. The success of tablets is creating new opportunities that more closely resemble the model of conventional television. This is also the case for all the new television "a la carte" services that use music – music programs and events – as part of what they offer.

The other interesting case that completes this section is Last.fm. It is an online music recommendation service (property of CBS) based on a collaborative social filter. That is, it makes recommendations as a function of the tastes of a specific user and the tastes of other similar users. Based on this, Last.fm has built one of the most widely used services in the field of online music, with information on artists, concerts, videos, photographs, etc. The recommendation service is free through its webpage or plug-ins for the main music players on computers and mobile devices, and Last.fm also offers an online streaming service for its music (which could be perfectly classified under this model), with a free version with advertising available in certain countries and a version without advertising available through a monthly subscription. One advantage over other tools that is has for new artists who want to get their work known is that they are paid for the music that they upload to the service and to which users listen.

Bundling with a Product or Service

This model is based on selling music jointly with another good or service that is necessary for its consumption. There are several variations of this model, depending on the good or service that is sold bundled together with the music and the conditions under which the music can be enjoyed. For example, it can be associated with a music player and a mobile phone, as well as with a USB device and a flat-rate data plan. In addition, the usage conditions can last for an unlimited time when the physical object is acquired or for a specified period of time or be subject to some type of subscription.

In many of these cases, the fundamental difference lies in the fact that there is a new stakeholder in the model who plays the role of intermediary between the distributor and consumer and who does the "bundling" of the music. This new agent uses music as an added value in its standard offer.

In the mobile environment, the most notable case was the "Comes with Music" service from Nokia. It consisted of the ambitious possibility of offering, with the purchase of a Nokia handset, a one-year subscription for unlimited music downloading. The service, which was launched at the end of 2008 and has been gradually closed, was part of the strategic shift by Nokia from being a mere manufacturer of devices to becoming an Internet service provider – a platform. However, the idea failed due to competition, due to the difficulties in acquiring reasonably priced licenses from the industry, and because the brand was unable to acquire a critical mass of users in a period in which smartphones were beginning to dominate the market and Nokia was losing share in this market segment.

A particular case of the previous model that deserves to be discussed in some detail is the relationship between music and video games. The use of music in video games has been common since their inception. However, over the years, the music in videogames has improved in quality, allowing for the introduction of real audio and the use of the entire library of the record industry. The advantage of this bundle is that users who play the game view the music as part of it and are willing to acquire it as part of the experience. Perhaps the most notorious case is the game Guitar Hero, which uses very well known songs and creates a bundle in which consumers pay for a game in which the music presents an essential value. Video games have also been used as promotional tools for artists who hold premieres or offer special promotions. As with the previous model, bundles create a double flow within the model. It is possible that the video game creator must pay to use the music, but it is also possible that the artists and record companies must pay to promote their productions through the video games. To a great extent, all of this resembles what occurs with music in the film industry.

In parallel with the actual games, game console manufacturers also want to use their communications channel with consumers to provide access to music online. In this regard, it must be recalled that portable game consoles are a noteworthy type of portable music reproduction system. The use of game consoles for online music is a strategy, for example, of Sony, which has also extended this model to all the devices it makes: televisions, video players, portable computers, etc.

Advertising Strategy

There are other models in which music is merely an advertising strategy that aims to improve or complete the user experience to sell some type of product or service. In this model, there is no musical transaction per se; instead, the intent is to simply benefit from the existing ecosystem of music distribution and consumption.

An example of this scheme appears in the sale of several different types of hardware or software for portable music playing. These models produce a "commoditization" of music, which does not require maintaining a direct relationship with the music industry, from which the manufacturer benefits. From the economic perspective, it would be a case of "free-riding", where investments are made by an existing industry and a third party benefits from them without any costs whatsoever, because, in general, there are no specific license requirements and only certain countries require compliance with some general rules derived from intellectual property rights.

An interesting case to highlight this model is Apple's iTunes, or, to be more precise, the platform created by iTunes and the different playing devices associated with it: the iPod, iPhone, and iPad. Currently, it is the leading music store in the world, ahead of all types of physical and virtual stores. In addition to direct purchases, iTunes is a type of software that allows users to manage their audio library, create lists, share recommendations, etc. Viewed as a whole, Apple's greatest source of income does not come from the sale of songs but rather from the sale of mobile handsets, players, and tablets. Added to this fact, it is noteworthy that the majority of the music in these handsets does not come from the iTunes store, the benefit obtained from the previous existence of digitized music seems clear. Simultaneously, the use of iTunes allows those who possess an Apple music player to become potential subscribers of an exclusive distribution network.

Information and Management of Music

Next model is based in information and management of music of interest for the consumer. This is a particularly important field in the mobile ecosystem, especially since the introduction of apps, which allow the consumer to benefit from the ubiquity of mobile connections.

In this area, there are considerably innovative proposals, such as Shazam, Midomi, and similar services. These are applications for mobile handsets that can identify songs, providing the name of the song and additional information about the song being played; they even work with a song that is hummed (Wang, 2006). They have the interesting trait of using a different interface from the usual keyboard – the microphone – to enable the input of information. The business model uses advertising and also "revenue sharing" from music purchases that are made based on or through the service. According to data from Shazam, 10% of the songs identified end up being purchased. There are also paid versions of the application with improved functionality. Regardless, similar to the previous case, these services benefit from pre-existing music.

Live Events

This model is based on taking advantage of all types of unique – exclusive – events in space and time. There are two main business models from the mobile perspective: the sale of tickets for the physical attendance of the event and the access to exclusive content related to the event. Surrounding these models, social networks can also be used, enabling users to share their experience in real time.

The most interesting feature in these models is the absence – disintermediation – of most of the traditional stakeholders in music. In theory, this would allow a direct relationship between artists and performers and the interested public. However, record companies have moved in to occupy this space, which had previously been occupied by managers and promoters. In the jargon of the music industry, these are called "360 degree contracts" between artists and companies; this designation quite clearly indicates that they cover all types of activities the artist may engage in that may generate profits. In addition, this is reminder that in the digital world, where a direct relationship between artists and consumers is possible, there is still the need for a marketing machinery of some type to generate the required notoriety, as will be discussed below. Perhaps this is the actual future role of the traditional record industry once the sale of physical support declines until it becomes a niche activity.

Branding, Merchandising and Market Place

The branding and merchandising model is very much related to the previous model because it is based on the "brand" value generated around artists to sell goods and services that are not directly related to their music. Typically, this activity is also part of the previously noted 360-degree contracts between artists and the music industry. To date these models are not characterized by a specific differentiated particularity in mobile technology other than being available for this platform.

THE FUTURE OF MOBILE MUSIC: CHALLENGES AND OPPORTUNITIES

The main challenges – and associated opportunities – faced by mobile music are described in the following subsections, starting with the perspective of users, followed by the structure of the industry and the effect of upcoming innovations, and concluding with the legal framework.

Music-as-a-Service

The first challenge has to do with the conversion of music from a product into a service. This transformation is part of a general process triggered by digitalization and the availability of handsets connected to broadband communication networks. Therefore, it affects not only music but also almost every type of content (audiovisual, games, etc.) and even many applications that are managed locally – a word processor, for example – and those that may be managed remotely – cloud computing. The main consequence for the industry and for users is that access to music, as opposed to owning music, becomes the greatest asset of interest. The main new digital agents – Apple, iTunes, Amazon, etc. – are strongly invested in this line, including the intensive use of the cloud to achieve device independence and facilitate the sharing of preferences and opinions. The remaining barrier in the mobile environment is the acquisition of a connection with sufficient capacity. However, with both Wi-Fi networks and the spread of fourth-generation mobile communications (LTE), it appears that this problem can be solved.

Transformation of Business Models and Industry Structure

Partly as a result of the previous trend, the business models on which music is based will be different from the traditional ones. As indicated in the previous section, there is a transition from a single model based on physical sales to a multiplicity of possible models that attempt to more closely approach the value and utility that consumer currently perceive in music and among which there is no clear winner.

Of course, not all models are compatible, and there are significant trade-offs between them. Being blind to this could lead to bubbles and senseless predictions in which everything grows; there are many examples. It is enough to consider that the success of streaming models means that models based on downloading will experience reductions to their income. If the user has the capacity to access music of interest in a way that is more or less free and that can be done at any time and place, why would she or he acquire it?

Of course, this transformation in business models has a direct impact on the music industry. As with all transitional phases, the onset of significant tensions, including the resulting "winners" and "losers" of the process, is entailed. Simultaneously, the transformation presents new and notorious opportunities for some agents. In fact, there has been an "empowering" of the agents located at the extremes of the ecosystem: the creators-artists, on the one hand, and the users-consumers, on the other. Setting aside the latter for a next section, the creators now have the most efficient tools, from an economic perspective, to bring their creativity to life and, at least in theory, to gain more direct contact with their potential audience. This translates into the traditional record industry losing control over the relationship with the audience while the channels that make their investments profitable have been diversified. This transformation has also caused this initial disintermediation to move toward a new re-intermediation process (Claudio Feijóo, 2013), in which the main players are new agents that are able to bring together creators and consumers in more useful and efficient ways.

The latter trend is not entirely new. It consists of the idea that the music business will continue to be based on marketing, even more so than before. Thus, promotion, "branding", and, in general, dissemination in the media will continue to be key elements of success. It is "only" a transformation of the particular media involved and the way in which this notoriety becomes a business.

The Importance of Users: Empowerment and the Personal Management of Music

The next trend has to do with the fact that there are still no appropriate tools for users to completely manage their digital audio libraries, irrespective of their device and adjusted to the usage experience of the environment in which they are located. This is why they still devote a considerable amount of time to "moving" songs between devices and locations.

Evidently, different providers attempt to take advantage of this need of users to try to turn it into an opportunity. The result is the launch of programs that envelop the user experience – iTunes – or storage systems in the cloud – Amazon and Google. However, barriers remain for the personal management of music under the conditions that users desire: portability among providers, uniformity of formats, less proprietary and closed schemes, quality of experience, etc.

The growing importance of users is also reflected in their new role as creators –or co-creators through their opinions and preferences – of musical content. The barriers to creating music have remarkably diminished, and this has caused the introduction of models that aim to simply reveal the creativity of these types of creators, not necessarily linked to subsequent commercialization. The music industry is still considering in what way it can participate in this new creative force.

In fact, the music industry, in particular the mobile industry, as discussed above, is experiencing a re-intermediation process with the introduction of new agents into the ecosystem that build their own silos – vertical integration models – in the value chain that are also replacing – or, at least, reducing – the action range of the traditional industry. These silos enable users to maintain control of the quality of their experience, but they also represent limitations in terms of the freedom of use of music.

Given that music is based on a context in which there is a substantial amount of pre-existing content, users demand a solution that is valid for the new providers and digital devices as well as for the older content that they already had. This tension between both types of approaches to the user has yet to be resolved in a thoroughly satisfactory way that takes into account certain regulatory aspects, as will be described below.

The State of Innovations

Precisely because of the existence of a huge quantity of content that users appreciate and because a large part of the business opportunities are based on users' having at their disposal a library of widely varied musical content, reaching agreements with the traditional industry to keep these models running continues to be the key. However, the industry is influenced by piracy and its own problems, and it has made the choice to allow the process of making music content available online to progress gradually in a controlled way. As a result, there have been numerous innovative proposals that, perhaps ahead of their time, have quickly failed. Among a growing number of examples, the most paradigmatic in the mobile environment is most likely the Comes with Music service that Nokia launched in 2008, which faced limitations due to music rights licenses and services by operators and other manufacturers and was unable to provide a sufficiently valued user experience, meaning that it was unable to acquire the critical mass of users that would have allowed it to take off.

Of course innovations in mobile music depend on a series of facilitating elements. Some of these factors are technical in nature, for example, the battery life on mobile devices, processing capacity, and resolution and interface usability. However, there are also other technological-economic in nature, for example, the capacity to transmit data through mobile networks or network coverage. In addition, there are pure economic factors, such as pricing strategies for mobile data connections and the music usage license conditions from the record industry. Finally, there is a set of elements related to the user experience, among which we can highlight the building of trust in users, privacy, interoperability, and the presence of environments that allow users to maintain control of how they use music.

In this scenario, the modest storage and processing capacity of mobile handsets has contributed to the exploration of cloud computing models, in this case, "music in the cloud" (Marquez, 2010). In fact, the only limitation to the success of these models lies in the maneuverability margin of flat-rate mobile data plans. Given the improvements in the conditions of these rates and also in Wi-Fi networks and other systems, users already use mobile devices to access content stored remotely, especially content related to music.

In fact, the industry has wagered much that cloud computing and streaming become the most important technological elements in offering attractive services to users. All the main stakeholders in this environment have acquired start-up companies to more quickly position themselves; observe the acquisitions of Lala by Apple, of Amie Street by Amazon, and of Simplify Media by Google. Furthermore, they have begun to offer services; for example, iCloud from Apple and Cloud Player from Amazon. The latter incorporates the interesting idea of scanning the user audio library and providing access to equivalent pieces in its own library to achieve device independence. Regardless, due to these innovations, there is – and will continue to be over the next few years – fierce competition to gain a stronghold in this market by way of a differentiated offer. Usability, an extensive library, and personal management capabilities for music are becoming the key elements in this differentiation.

One of the most interesting characteristics of new music providers, especially those based on streaming, is their ambition of becoming real platforms with music at their core, with the ability to attract the attention of developers, licensers, advertisers, and, of course, users. It is another way of applying the theory of multilateral markets discussed above. A couple of examples of interest are Spotify and Deezer. With the former, third-party applications such as Classify allow classical music fans to more easily search for what they are interested in and gain access to lists prepared by professional musicians. With the latter, third-party applications such as Songkick can be used to find concerts of interest that are near the user's location and that coincide, of course, with the user's favorite songs.

Another key aspect of mobile interfaces is their usability or, more precisely, their lack of usability in certain cases. For example, even with modern smartphones, it remains relatively expensive to key in the complete title of a song to search for related data. Therefore, it is not a surprise that there are a growing number of innovations that use other interfaces that are different from the keyboard to introduce and obtain information. It is precisely to satisfy this need to simplify and adapt the interface that music identification algorithms have been quickly and successfully adopted in mobile handsets (Ellis, 2006). Thus, a limitation of the mobile environment has been turned into an opportunity to offer innovative solutions, with the potential of being applied to other fields.

Geolocation is another fundamental example of a set of emerging mobile innovations. In the case of music, geolocation allows it to be linked to specific places and contexts: a mobile social network, a concert, a store, etc. A growing number of applications related to music use this functionality. From the perspective of augmented reality, geolocation makes it possible to add layers of musical information – or any type of audio – to the physical reality that surrounds us.

The mobile device also has the means to identify the environment in which it is located and the physical and emotional state of its user. This has led to the introduction of proposed solutions that adapt music to the environmental and personal situation of the user.

The Legal Framework

The last challenge has to do with the regulatory framework in which mobile music evolves, in particular legislation concerning intellectual property in the digital environment (GRETEL, 2005; Pérez, Feijóo, & García, 2002; Pérez & Feijóo, 2000). This legislation is perhaps the most developed of all content-related laws because it faced a number of challenges before any other type of content. Thus, once again, music is at the vanguard of the issues that are waiting to be resolved. It is, then, a field subject to considerable controversy and with potential for deep reforms.

The most relevant underlying issue concerns identifying the proper balance between the rights of creators – particularly remuneration rights – and the rights of consumers. In the digital world there is a growing consensus that, in effect, the creative industries must be supported with a favorable framework, but it continues to be unclear to the public – and to legislators – what this support should look like in practice or how it should be implemented. In retrospect, it is true that for a considerable period of time, there was a lack of music offers that made it possible "to keep users honest", to use the phrase that summarizes the actual philosophy of the industry. The latest available data (IFPI, 2013) seem to indicate that the trend of using legal music services is becoming entrenched, given that these services are, in effect, more closely approaching users' expectations of them.

In parallel with this discussion, there is starting to be a subtle confrontation between two antagonistic intellectual property models. On the one hand, there is the mostly Anglo-Saxon model of the pure market where intellectual property is subject to explicit contractual agreements that spell out who is the proprietor of the rights and, therefore, who shall hold the power to negotiate them. On the other hand, there is the so-called "continental model", in which intellectual property has strong social connotations of protection of the weakest side: basically, creators and artists. Here, business models are subordinate to this protective spirit. The result of this confrontation is not unimportant because it may tilt the scale toward the type of industry that will provide content to consumers and define the rules of the game in the triangle composed of creators, the industry, and users.

Finally, there is a territorial issue. Currently, legislation, especially the manner in which it is enforced, is completely fragmented in a multitude of rights, each of which has a limited geographical coverage. Even worse, they are not entirely compatible. Consequently, setting innovative models based on music in motion requires agreements to be painfully reached, one right at a time and one territory at a time. Evidently, a simplification of this framework is clearly a legitimate objective, at least for Europe.

REFERENCES

EC. (2008). EU crackdown of ringtone scams. Frequently asked questions. E. C. DG for Health and Consumers. Retrieved from http://europa.eu/rapid/pressReleasesAction.do?reference=MEMO/08/516&format=HTML&aged=0&language=EN&guiLanguage=en

Ellis, D. (2006). Extracting information from music audio. *Communications of the ACM*, *49*(8), 32–37. doi:10.1145/1145287.1145310

Feijóo, C. (2013). Soportes digitales y transformación de las industrias de contenidos. *El Profesional de La Informacion*, *22*(1), 5–9. doi:10.3145/epi.2013.ene.01

Feijoo, C., Gómez-Barroso, J. L., & Martínez, I. J. (2010). Nuevas vías para la comunicación empresarial: Publicidad en el móvil. *El Profesional de La Información*, *19*(2), 140–148. doi:10.3145/epi.2010.mar.04

Feijóo, C., Maghiros, I., Abadie, F., & Gomez-Barroso, J. L. (2009). Exploring a heterogeneous and fragmented digital ecosystem: Mobile content. *Telematics and Informatics*, *26*(3), 282–292. doi:10.1016/j.tele.2008.11.009

Gómez-Barroso, J. L., Feijóo, C., & Compañó, R. (2011). Opportunities in the mobile search market. *Computer*, *44*(11), 93–95. doi:10.1109/MC.2011.347

GRETEL. (2005). La explotación de los contenidos audiovisuales en la transición a los nuevos medios digitales. In C. Feijóo (Ed.), El sector audiovisual y su evolución. Madrid: Colegio Oficial de Ingenieros de Telecomunicación (COIT).

Hesmondhalgh, D. (2007). *The Cultural Industries*. London, U.K.: Sage Publications Ltd.

Holden, W. (2008). Making music with mobile. Basingstoke: Juniper Research. Retrieved from http://www.juniperresearch.com/shop/viewwhitepaper.php?id=98&whitepaper=56

IFPI. (2013). IFPI Digital Music Report 2013. Retrieved from http://www.promusicae.es/files/imagenes/file/DMR2013fullv9.pdf

IFPI. (2014). Digital music report. Retrieved from http://www.ifpi.org/downloads/Digital-Music-Report-2014.pdf

Leurdijk, A., & Nieuwenhuis, O. (2012). *Statistical, Ecosystems and Competitiveness Analysis of the Media and Content Industries: The Music Industry*. Luxembourg: Publication Office of the European Union. Retrieved from http://ipts.jrc.ec.europa.eu/publications/pub.cfm?id=5022

Marquez, I. V. (2010). Música en la nube. Experiencias musicales interactivas en la red. *Telos*, (83): 139–148.

ONTSI. (2012). Las TIC en los hogares españoles. Estudio de demanda y uso de Servicios de Telecomunicaciones y Sociedad de la Información. XXXVI Oleada (Abril – Junio 2012). Madrid. Retrieved from http://www.ontsi.red.es/ontsi/es/estudios-informes/xxxvi-oleada-del-panel-de-hogares

Pérez, J., & Feijóo, C. (2000). Convergencia, competencia y regulación en los mercados de las telecomunicaciones, el audiovisual e Internet. GRETEL. Madrid: Colegio Oficial de Ingenieros de Telecomunicación (COIT).

Pérez, J., Feijóo, C., & García, F. (2002). Nuevo diseño europeo de las telecomunicaciones, el audiovisual e Internet. GRETEL. Madrid: Colegio Oficial de Ingenieros de Telecomunicación (COIT).

Preston, P., & Rogers, J. (2011). Social networks, legal innovations and the "new" music industry. *EuroCPR*. Retrieved from http://www.eurocpr.org/data/2011/2_Preston_Rogers.pdf

Rochet, J.-C., & Tirole, J. (2006). Two-Sided Markets: A Progress Report. *The Rand Journal of Economics*, *37*(3), 645–667. http://www.jstor.org/stable/25046265 doi:10.1111/j.1756-2171.2006.tb00036.x

Scaglione, M., & Hamoudia, M. (2010). Online and mobile social networking: present and perspectives. *21st European Regional International Telecommunications Society Conference*. Copenhagen.

Wang, A. (2006). The Shazam music recognition service. *Communications of the ACM*, *49*(8), 44–48. doi:10.1145/1145287.1145312

KEY TERMS AND DEFINITIONS

Downloading: Process where music is consumed through a digital copy available at the user's device.

DRM (Digital Rights Management): System to assign specific rights of use to (audio) content.

Music Distribution: Process where music is made available to the general public by means of a physical package or a digital system.

Music Production: Process where audio content resulting from the creative interaction of authors, artists and producers is ready to be commercialized.

Ringtones: Simplified version of a music track that it is used as a audio tone for different functions in a handset, for instance when receiving a call.

Side-Loading: Process to load the audio library of a user in her/his mobile device without using third party's digital communications networks.

Streaming: Process where music is not stored in the user's device and is consumed directly from the network during distribution.

ENDNOTE

1 Of course, care should be taken with overoptimistic predictions. For example, in 2009, income derived from the music business was expected to grow by a factor of five in the next five years. However, the reality has proven to be substantially more modest, with a growth of 10%, including the consequences caused by the relatively expected fall in music ringtone sales.

Chapter 6
Examining Mobile Search Adoption
The Swedish Experience in the Uptake Phase

Andreu Castellet
University of Murcia, Spain

Oscar Westlund
University of Gothenburg, Sweden

ABSTRACT

The multimedia features brought by the rise of the mobile access to internet are tightly linked to the massive access of the public to mobile search. To date, there is limited knowledge on how distinct groups make use of mobile search in their everyday life. Attempting to fill that void, the purpose of this chapter is to describe and explain how mobile search behaviours have gained traction among distinct groups in Sweden from 2010 to 2012. The findings evidence a growing uptake among all groups over time. Smartphone ownership and age score the highest odds ratios, revealing smartphone users and 16–29 year olds as most inclined to use mobile search. In addition, a high education and use of a personal or company subscription show significant results.

INTRODUCTION

Amid the widespread development, diffusion and appropriation of mobile-specific information and communication technologies (ICTs), especially smartphones and tablets, the mediascape is changing in a multitude of ways. The pervasive nature of mobile devices enables timely and prompt mobile media and communication (Koivumaki et al., 2008), and the patterns of use formed are typically easy to maintain and routinise (Oulasvirta, Rattenbury, Ma, & Raita, 2011). Technological developments and convergence, alongside the rise of mobile ecosystems nowadays totalling more than a billion mobile applications for Android, Apple iOS, Microsoft Windows Mobile and so forth, have made mobile media an increasingly

DOI: 10.4018/978-1-4666-8838-4.ch006

important area to understand, and upon which to strategically act (Westlund, 2013). It has implications for the diverse agents of media firms trying to stay relevant in relation to their audiences through media innovation (Lewis & Westlund, 2015; Westlund & Lewis, 2014). Mobile search is becoming increasingly important to both citizens and diverse actors (e.g. companies and authorities).

Search engines, which have long since been the main gateway for most Web users, have turned towards facilitating mobile search experiences. Similar to the wired Internet, a prerequisite for the mobile web and its applications is supplying data, content and utility services in an efficient and user-friendly manner. Mobile content and mobile search have been marked as representing truly important areas of development (Feijóo et al., 2009), not least for the salient case of the European Union (Reding, 2008; Gómez-Barroso et al., 2010).

Much previous research in mobile media and communication has focused on "who" (types of users) or "how" (types of uses) (Wirth et al., 2008). Scholars have pointed out the need for deeper research on mobile media (Scolari et al., 2012) and its impact on users' behaviour, which includes mobile search. This chapter integrates these two perspectives by focusing on mobile search usage on a daily or weekly basis ("how"), depending on distinct socio-demographics ("who"). Importantly, the chapter does not focus on the adoption and use of mobile devices in general. Instead, it focuses on a sort of second stage of adoption (Westlund, 2008; Li, 2013), here involving the more specific use of mobile search functionality.

Various kinds of reports–both from academia and the industry – have witnessed the rapidly increasing appropriation of mobile media use around the globe (Campbell, 2007; Chan, 2015; Kitamura, 2013; Martin, 2015; Sasseen et al., 2013; van Damme et al., 2015; Verbrugge et al., 2013; Wei et al., 2014; Westlund, 2014; Westlund, 2015; Westlund & Färdigh, 2015). In addition, several studies have documented and compared mobile media use from a cross-cultural perspective involving the United States and Germany (Humphreys et al., 2013), Europe and the Americas (Newman & Levy, 2013) as well as Sweden and Japan (Westlund, 2010). A salient finding from many of the studies is that mobile media have become interwoven with the rhythms of everyday life, being used off and on over the course of the day and night when users are awake, while they are at both home or work/school, and while on the go (Dimmick et al., 2011; Ericsson, 2011; Taneja et al., 2012; van Damme et al., 2015; Westlund et al., 2011). Some empirical evidence finds that mobile Internet enhances rather than replaces Internet desktop use (Ongena et al., 2012). On the other hand, a number of more specific analyses on news access via mobile devices and computers suggest that some groups use both resources in a complementary manner, some have displaced computers in favour of mobile, whereas others have not adopted mobile news at all (Westlund & Färdigh, 2012; 2015).

As demonstrated from assessment by the World Economic Forum, for instance, Sweden has achieved comparatively significant advancements of digital media (Dutta & Bilbao-Osorio, 2012). It is a country with a broad diffusion of Internet and mobile media, in which the prominent position of legacy media such as newspapers and (public service) broadcasters have been accompanied by a plethora of agents offering media content and services (Westlund & Weibull, 2013; Westlund, 2010). The Swedish case is particularly illustrative of the uptake of the relationship between smartphone ownership and mobile search use, which comprises the 2010–2012 period. When looking at developing markets that are transitioning from feature phone to smartphone widespread ownership, it is feasible to sustain that prior experiences can lead from a better understanding of transitional processes to a higher use of ICT-related devices. The global dimension of the communication phenomena seems to support this argument (Campbell, 2007).

Sweden is considered a leading country in the adoption of IT technologies and services, as the United Nations International Telecommunications Union (ITU) indicates. In 2010 Sweden was listed by this organisation among the very few states, along with Norway, Ukraine and the United States, that had started to offer mobile communication services based on 3G standards (ITU, 2010). According to its ICT Development Index (IDI), Sweden ranks as the second highest country among the 157 states represented for the 2010–2012 period, only surpassed by Korea (ITU, 2012, 2014). Furthermore, smartphone penetration in Sweden was Europe's highest at the time, with a penetration rate close to 60 per cent by the end of 2012 (Ipsos Media CT, 2013).

STUDY RATIONALE

Extant literature on mobile search usage suggests there is limited robust empirical research tracing the uptake of mobile search, over time and among distinct groups. This chapter will thus attempt to fill that void. The purpose of the chapter is:

To describe and explain how mobile search behaviors have gained traction among distinct groups in Sweden over the course of 2010 to 2012.

The chapter will analyse similarities and differences depending on socio-demographics (age, gender and education) as well as device conditions (smartphone ownership and subscription model). The chapter analyses robust data from three annually conducted cross-sectional postal-based surveys, representative of the Swedish public (Swedes aged 16–85 years), offering a record for understanding the growing significance of mobile search. The analysis will explore usage patterns regarding two levels of frequency: weekly vis-à-vis daily use. Weekly use (inclusive of daily) provides a broader measure of the degree to which the population, and distinct groups, has started using mobile search. Daily use, on the other hand, provides a record for those who have embraced and domesticated mobile search into their everyday life. This measure corresponds to research promoting an understanding of the naturalisation of media use in everyday life (Bakardijeva, 2011), and also how people do not live their lives with media, but instead live media lives (Deuze, 2012).

The chapter draws on data from three nationally representative scientific omnibus survey project datasets. The SOM Institute at the University of Gothenburg in Sweden has conducted postal-based surveys annually since 1986 through systematic probability sampling of the Swedish population. Over the full period, the net response rate has varied from 60 to 70 percent, with the lower figures in more recent years. This chapter analyses data collected in 2010, 2011 and 2012, years for which the net response rates were 60, 57 and 57 per cent, respectively. For each of these years, several unique questionnaires were sent to representative samples of 3000 Swedish citizens. The questionnaires were sent by post in order to verify the respondents' identities, but recipients could also fill out an online questionnaire by using a unique identification. Some questionnaire items were used in several samples. However, the questions concerning mobile search were only included in one of these. The total number of responses applicable for the mobile search analysis was 1509 in 2010, 1524 in 2011 and 1548 in 2012.

Extant Literature on Mobile Search Usage

In the early years of the Internet, mobile search was largely overlooked in academic research and industry reports (Westlund et al., 2011). However, from the beginning of the 21st century onwards, a body of knowledge has been generated by academics (Bohlin et al., 2007; Bouwman et al., 2009; Church et al., 2008; Haddon & Vincent, 2008; Kamvar & Baluja, 2006; 2007) and industry observers alike (by Compete or Comscore, for instance). In their literature review, Westlund et al. (2011) discuss that most research on mobile search has analysed usage patterns by means of large-scale logs from search engine databases.

Some research was also conducted on the extent and frequency of mobile search usage among different groups. Some of these latter analyses allowed for cross-service examinations, such as comparing mobile search with other mobile media or mobile communication functionalities. Few scholars have conducted quantitative research on the uptake of mobile search. At the same time, qualitative analyses have distinguished pronounced differences in such uptake and use, with *connected* users having embraced mobile search, whereas *traditionalists* and *diversionists* have refrained from doing so (Westlund et al., 2011).

The set of data studied here, focused on smartphone and search engine use, appears as particularly suitable for an analytical approach under the lenses of the Technology Acceptance Model (TAM), and Innovation Diffusion Theory (IDT). The authors do not ignore some remarkable efforts to unify views on the adopting technology research process (Venkatesh et al., 2003). However, the corpus of historic data with which they work on this occasion, mostly linked to the ownership of a smartphone, leads them to consider the ease of use (Davis, 1989; Pagani, 2004), and also the perceived usefulness, above other factors which are part of the extensive field of technology acceptance. In their case, the perceived usefulness interprets the role of mobile search acceptance likened to carrying out job tasks, in the sense that "would enhance his or her job performance" (Davis, 1989: 320). Most of the works on mobile search have started from Broder's taxonomy (2002), which aimed to guess the intention behind the web search task. That led him to establish three search categories: *informational* (looking for an answer or a set of answers related to a question); *navigational* (looking for an URL); or *transactional* (looking for the way to perform a specific action in cyberspace).

MOBILE SEARCH: SERVICE AND CONTEXT

The idea of mobile search as an experience different to desktop web search (Agosti et al., 2012; Baeza-Yates et al., 2007; Baeza-Yates, & Raghavan, 2010; Broder et al., 2007; Church et al., 2008; Gómez-Barroso et al., 2010) has gained traction among researchers. That idea builds upon the assumptions that: a) the network/data plan to which the user has access when on the move determines dramatically the quality of the search experience; b) users' interests can differ when using mobile search compared to desktop search; c) mobile devices enhance the search experience when they add some unique features of their own, mostly related to the set of context-aware sensors – GPS features, camera, microphone, multi-touch interface, wearable sensors connectivity, accelerometer – which most smartphones and tablets currently offer; d) The combination of context factors and device interface peculiarities change the way the user actually interacts with the device, with the need to simplify the query-input method considered a major issue.

Mobile search develops, therefore, as a result of the convergence of mobile networks and the evolution of search engines. As Westlund et al. assessed: "Mobile search can intelligently be integrated with other mobile applications, such as is the case for iPhone, Android-based phones or any other of the various emerging platforms" (2011, p. 691). That convergence between mobile search genuine actors and other activities within the same media ecosystem is currently an expanding reality. Two examples from the most social-oriented area of the mobile business can validate it: Facebook is working hard to improve its message search services (Frier, 2014), while Pinterest claims its status as a general search tool for images (Madrigal, 2014). The picture-sharing-based social network aims to follow the path that brought YouTube to supremacy in video sharing/searching.

User context is also important to how web search develops. The way in which search engines have applied personalisation leads to what Pariser calls a "bubble", in which customers live: they stay in loneliness, with their interests narrowed; they do not perceive they are actually inside the bubble; and they have not been asked permission to be put into that environment. Pariser's assessments gain relevance when one connects them with Lessig's views. Lessig devoted a significant part of his research to connecting system architecture and regulation worlds, in a way that shows how architecture, or "code", becomes a primary regulator of activity on the Internet: "In real space, we recognize how laws regulate—through constitutions, statutes, and other legal codes. In cyberspace we must understand how a different "code" regulates (...). As William Mitchell states, this code is cyberspace's "law". Joel Reidenberg first termed it "Lex Informatica", or even better, "code is law" (2006, p. 5).

That can lead to exploring the boundaries between search and research activities, as users make use of the same resource for both activities. However, for the purpose of this chapter, they will be considered together as a whole. Whereas research has a more knowledge-driven aim, search looks at query results with a more business-driven goal. Pariser (2011) tells the story of two different users, who, from an "Egypt" query on Google, obtain very different results: one gets mainly information about the country–country statistics, Wikipedia, the Tahrir Square demonstrations, etc., whereas the other gets a page filled mostly with tourist information such as resorts, hotels, and flights. Google has actually treated the first one, regardless of the user's ultimate needs or intentions, as a researcher, whereas the second user query was processed as a "search".

"Google is the lens through which we view the world. Google refracts, more than reflects, what we think is true and important. It filters and focuses our queries and explorations through the world of digitized information", points out Vaidhyanathan (2011, p. 7). That is the result of a process of personalisation of search, through which search engines use navigational data from users in order to offer what the engine dealing with the query finds can cater to the user's interest in the best way. Pariser explains that Google uses 57 user's signals to customise results. Some authors, such as Liao and Fu (2013), have tried to further explore how selective exposure to information can interfere in the process through which users build their own points of view on current issues, noting that factual information tends to be more appreciated by users with a higher degree of personal attachment to the topic sought. The fact that Google is the web search leader does not mean it is the only one to implement personalisation tools. Yahoo, for instance, almost tripled U.S. users of its *Today box* feature in two years by changing its personalisation algorithm (Boyd, 2011). The success of the *Today box* case prompted Yahoo to extend its core technology to other search/aggregating services.

Data from surveys show that consumers have mixed feelings on web engines. According to the American Consumer Satisfaction Index, the survey on Internet portals and search engines indicates a growing confidence in the sector, particularly in the biggest and smallest players. "Overall search engine site performance is excellent and mobile compatibility is top-notch. Search engine users also rate the freshness of content to be far better than newspaper websites or social media sites. The amount of ads is still the worst aspect of the user search experience, but ads are better tolerated on search engines than on social media or news and opinion websites" (ACSI, 2014).

Table 1. American customer satisfaction index on search engines

Company	2013	2014	% Change
Google	77	83	8%
All Others	70	77	10%
Bing (Microsoft)	76	73	-4%
MSN (Microsoft)	74	73	-1%
Yahoo!	76	71	-7%
AOL	71	70	-1%
Sector Average	76	80	5%

Source: ACSI E-Business Report 2014.

Furthermore, search users report mostly positive experiences, according to a Pew Research Center survey among the U.S. population (Purcell, Brenner, & Rainie, 2012). However, the same survey indicates that 65 per cent of persons asked have negative feelings about a search engine collecting data about their searches and using it to rank their future search results. Those users see that practice as restrictive, because it limits the kind of information they can get online. Also, they consider their privacy to be in jeopardy because of the personal data collection that those user-tailored search results involve.

Google has achieved not only the best consideration, but also absolute leadership in mobile search. As shown in Table 1, beyond Google there is a slim market share of less than one-tenth, in which only Yahoo! and Bing are hardly noticeable by the public. Consequently, Google hegemony in mobile search extends much further than the share of the mobile OS market led by Android, which amounts for approximately 60 per cent of the installed base. However, there is a significant number of devices working on *forked* versions of Google's OS (CCS, 2014) – Baidu and Alibaba in China, Amazon's global Fire OS, as well as some Nokia handsets – that do not offer the Google Search widget embedded in their desktops (See Table 2).

Search engines are among the most visited sites. According to Alexa[2] global rankings, four search engines – Google, YouTube, Yahoo and Baidu – were listed among the "5 Top sites" of the web by September 2014. As this chapter focuses on the case of Sweden, it is worth noting that the same source indicates that Google leads the *Top five* rankings; google.se and google.com are the top two, followed by facebook.com, youtube.com and aftonbladet.se – a newspaper site – in that order. The task of adapting

Table 2. Global market share of mobile search as of July, 2014

Search Engine	Total Market Share
Google - Global	91.21%
Yahoo - Global	5.91%
Bing	2.26%
Baidu	0.13%
Ask - Global	0.12%
AOL - Global	0.02%

Source: Net Applications[1].

the search engine experience to mobility becomes even more challenging due to the contextual changes which users can experience all the time. That condition makes it difficult to find interface solutions that are versatile enough. As Norrie points out:

The most serious challenge for general search services is the fact that the settings are constantly changing as users are on the move. This means that while a system designed to use audio input and output may offer a solution that leaves users hands-free and may be ideal in some settings, it is problematic in noisy environments and useless in settings that require silence. Other solutions designed to work well in indoor settings, may not work so well in outdoor settings. Although the idea of context-aware applications is that they can adapt to context, we have yet to see solutions that really can adapt to the wide variety of contexts. (2011, p. 49)

Search services respond to a vast number of user queries. Users' trust in their search engine option plays a crucial role in the business: plenty of access to the whole–or at least, to the maximum amount of data available–of the information available in cyberspace, and, no less important, an implicit act of confidence in the whole way their query will be treated. This chapter will focus on this specific point in an attempt to study the way in which the query results are elaborated.

The mobile search activity tends to be studied as a relevant part of a wider sector within the mobile ecosystem and all of cyberspace, the "aggregators" (Feijóo et al., 2009) domain, which comprises, depending on the different sources, a number of different tasks (Castellet & Feijóo et al., 2013; Feijóo et al., 2009; Hu, 2008; Kohlborn, 2009), ranging from search engines to content portals; m-commerce sites and app stores; and the whole plethora of social media, to name the main ones. All those activities share many traits with the ones that have shaped web search from its very beginning (Broder, 2002). One is accustomed to noticing search features as a core element of their value proposition to users.

Furthermore, search plays a crucial role in a user's mobile experience to the extent that innovation in search has been assumed to be a crucial source of wealth for other players. As Gómez-Barroso et al. predicted, only a couple of years after the launch of the first iPhone, "no attractive services and applications were offered to users. Recognizing this fact, the mobile industry started to improve the end-user experience, acknowledging that the mobile internet is not simply about repurposing the internet, but creating a relevant service and experience 'made for the mobile'" (2010, p. 24).

Also, as the iPhone strengthened the user's attachment (Aguado & Martínez, 2014) to the device and the services it provides, the company that created that smartphone was the first to offer a personal assistant feature bundled with its OS. That tool, known as Siri, merged into a single service voice recognition, the personal assistant, and predictive technologies, which was a relevant innovation at the time.

The development of cheap, efficient voice-recognition technologies is at the core of the OS providers' ambitions, and they are merely scratching the surface of its possibilities in mobile services. Google has integrated voice recognition in several languages in its mobile search widget. However, the real challenge for search engines, particularly in mobile, is not being able to search, but rather to find. "There's a new generation of user-interface theory that says there should not be a user interface; the information should just be around you" (Manyika, 2013, p. 2), admitted Google's Eric Schmidt.

The everyday interaction of users with Siri has produced a long list of mistakes and funny anecdotes. All that is not sufficient to silence views that envisage the great disruptive potential in those personal assistant services: "Technology like this is increasingly going to disintermediate websites in the same way that websites disintermediated brick and mortar and print media" (O'Reilly, 2011). "Long-term applications will flourish" (2010, p. 13), Gómez-Barroso et al. predicted, too. Google has also worked on the search-find friction issue, with predictive assistants such as its Google Now app. Furthermore, projects such as San Francisco's Expect Labs involve predictive assistants, with financial support from Samsung, Intel and Telefonica, beyond platform environment restrictions. Expect Labs is developing the MindMeld API (Mind Meld, 2014), a tool able to predict what a user may need, by 'listening' to the content of his/her own phone conversations. That extends much further than Google Now's features, a predictive assistant from the family of Google-Android apps.

Those disintermediators act in many ways, although taking advantage of the sound recording and processing capabilities of mobile rank among the most developed so far. Shazam is one of the most popular ones, with more than 100 million active users per month. The service, created in 1999, was one of the first to appear in Apple's App Store in 2008. Shazam works as an engine able to identify a song from a radio station or a music player thanks to a complex recognition system, which also needs to be connected to a vast database of songs. Shazam has expanded its business in two directions: more social and more commercial at the same time. Shazam offers users the opportunity to buy the song through the iTunes Store, Amazon MP3 Store, or to listen to it on Spotify. Paid and ad-sponsored versions of the service co-exist. Shazam also offers a service of television programme recognition, which only works with the television shows that are part of the Shazam-TV station agreements. Once it has reached a certain stage of development of the recognition algorithm, it can be applied to several kinds of user's needs, given that the databases are huge enough. "We don't have better algorithms. We just have more data", admits Google's scientist Peter Norvig (O'Reilly, 2011).

Visual search is another promising field with vast disintermediation potential. According to Möller et al. (2012), visual search can be considered in three fields: one-dimensional codes – bar codes; two-dimensional codes – QR codes; and object recognition through pictures taken of an object. Visual search is becoming increasingly popular in brick-and-mortar shops, when customers do "showrooming" (Milchen & Wood, 2014), and compare prices for the same product from a store through one of the aforementioned methods of visual search. Apart from a large number of scanning applications for barcodes and QR codes, the Goggles app from Google is possibly the most well-known tool in that domain. Goggles can scan objects through the camera of a smartphone/tablet, and then should be able to recognise them. That step becomes the starting point for a user's query on the Google search engine.

DESCRIBING MOBILE SEARCH BEHAVIORS AMONG DISTINCT GROUPS

This section presents weekly and daily mobile search usage among Swedes from 2010 to 2012. The descriptive findings in FigureS 1 and 2 evidence that in 2010, there were 5 percent daily users and 13 percent weekly users of mobile search among the Swedish public aged 16 to 85 years. Corresponding figures for 2011 were 14 and 32 percent, and for 2012, these figures were 25 and 43 percent. In conclusion, the daily use quintupled in only two years, reaching one-quarter of the population. Moreover, the total number of frequent users tripled, currently accounting for more than four out of ten Swedes.

Turning to findings on usage among specific groups, the findings show a relatively equal uptake among men and women (men have adopted mobile search slightly more). There are, however, pronounced differences regarding patterns of mobile search usage depending on age during 2010 to 2012. As of 2012, a total of 56 percent of 16–29 year olds used mobile search daily, compared to 4 percent among 65–85 year olds. While usage has increased among all age groups over this time period, people aged 16–29

Figure 1. Daily use of mobile search among groups 2010-2012 (percent)
Source: The National Swedish SOM surveys from 2010, 2011 and 2012.
Note: The Swedish public here includes a representative sample of citizens aged 16–85 years. The four definitions of educational level are defined according to the highest level of education completed: low (elementary/comprehensive school), middle-low (high-school), middle-high (some higher studies but no degree), high (college or university degree).

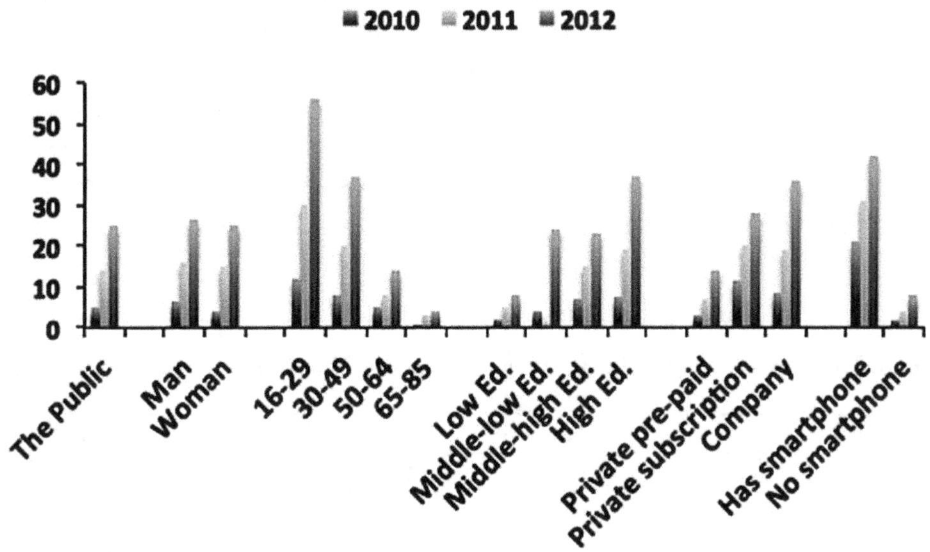

Figure 2. Weekly use of mobile search among groups 2010-2012 (percent)
Source: The National Swedish SOM surveys from 2010, 2011 and 2012.
Note: The Swedish public here includes a representative sample of citizens aged 16–85 years. The four definitions of educational level are defined according to the highest level of education completed: low (elementary/comprehensive school), middle-low (high-school), middle-high (some higher studies but no degree), high (college or university degree).

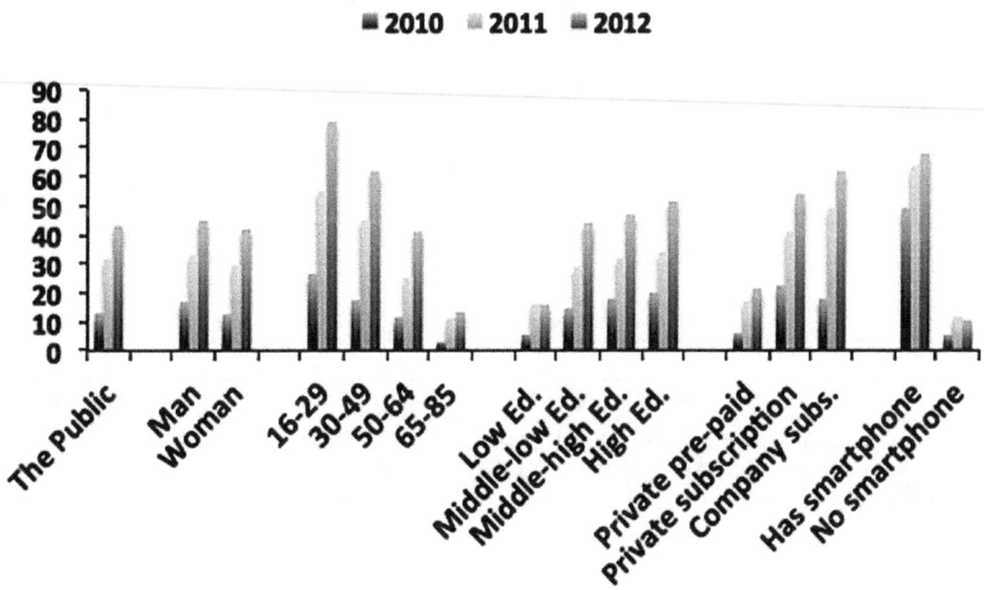

and 30–49 demonstrate much higher scores than the other age groups. Moreover, there is a consistent pattern depending on educational level between groups and over time: the diffusion of mobile search increased with a higher level of education. For instance, the 2012 figures for highly educated vis-à-vis low educated were 37 and 8 percent. In other words, the educated prove to be more likely to use their mobile device to search for information. Moreover, a similarly consistent pattern emerges regarding payment model: mobile search is least common among people using a private pre-paid card, whereas it is much more common with holders of a subscription, especially in the case of company subscriptions. Finally, patterns of use vary most significantly when comparing owners of smartphones with those using feature phones. The weekly use of mobile search among smartphone owners has increased from 46 percent in 2010 to 71 percent in 2012. Corresponding figures for non-owners of smartphones were 4 and 8 percent, respectively.

Table 3. Explaining daily use of mobile search among groups 2010-2012 (correlations and odds ratios)

	Pearson's r			Odds Ratio EXP [ß]		
	2010	2011	2012	2010	2011	2012
Sex						
Man	.074**	0.014	0.015	1.967	1.083	1.071
Woman	-0.74**	-0.014	-0.015	0.508	0.923	0.934
Age						
16-29	.111**	.213**	.302**	2.766	3.674	5.298
30-49	0.049	.118**	.180**	1.554	1.975	2.351
50-64	-0.031	-.109**	-.139**	0.723	0.438	0.445
65-85	-.121**	-.194**	-.289**	0.041	0.129	0.094
Education						
Low	-.091**	-.129**	-.197**	0.195	0.278	0.197
Middle-low	-0.038	0.008	-0.012	0.69	1.05	0.942
Middle-high	0.049	0.024	0.026	1.624	1.17	1.157
High	.081**	.085**	.163**	2.074	1.668	2.212
Payment Model						
Private pre-paid	-.123**	-.163**	-.194**	0.196	0.297	0.304
Private subscription	.110**	.154**	.158**	2.616	2.6	2.212
Company subscription	.075**	.074**	.182**	2.042	1.578	2.419
Owns Smartphone						
Yes	.335**	.389**	.434**	16.983	12.581	19.495
No	-.335**	-.389**	-.434**	0.059	0.079	0.051

Source: The National Swedish SOM surveys from 2010, 2011 and 2012.

Note: Anti-logarithmic regression coefficients (odds ratio). For each of the variables used, the reference category is all of the other variables in that category (i.e. all other age groups) where $\beta = 1.000$. **. Significant at the 0.01 level (two-tailed). *. Significant at the 0.05 level (two-tailed).

Explaining Mobile Search Behaviors among Distinct Groups

Logistic regression has been used for the explanatory analysis of nonlinear relations. Table 3 and 4 present the odds ratios (anti-logarithmic regression coefficients), which is a statistical measure for probability. For each group, a reference category is used for which β = 1.00. The groups scoring higher than 1 mean that the likelihood of them engaging in such use is higher than the reference category, whereas a score lower than 1 means they are less likely. For each of the analyses presented here, the reference group consists of all of the other groups combined. For instance, the score for 16–29 determines whether they are more or less likely in mobile search than all three other age groups together. Table 1a and 1b also report Pearson's correlations, which here function as a complementary measure to the odds ratios; they essentially reveal the interrelationship between specific groups and mobile search. It is worth noting that the levels of significance reported are the same for the correlations and the odds ratios.

Table 4. Explaining weekly use of mobile search among groups, 2010-2012 (correlations and odds ratios)

	Pearson's r			Odds Ratio EXP [ß]		
	2010	**2011**	**2012**	**2010**	**2011**	**2012**
Sex						
Man	.092**	.055*	0.027	1.742	1.266	1.116
Woman	-.092**	-.055*	-0.027	0.574	0.79	0.896
Age						
16-29	.178**	.212**	.305**	3.097	3.145	6.534
30-49	.082**	.193**	.266**	1.642	2.381	3.245
50-64	-.075**	-.101**	-.133**	0.581	0.597	0.537
65-85	-.169**	-.281**	-.387**	0.128	0.161	0.107
Education						
Low	-.141**	-.189**	-.287**	0.207	0.283	0.162
Middle-low	-0.02	0.02	0.007	0.885	1.097	1.03
Middle-high	.064*	.058*	.058*	1.548	1.33	1.33
High	.094**	.093**	.194**	1.802	1.567	2.409
Payment Model						
Private pre-paid	-.177**	-.239**	-.289**	0.241	0.289	0.236
Private subscription	.149**	.186**	.225**	2.474	2.297	2.612
Company subscription	.107**	.187**	.245**	1.952	2.438	3.072
Owns Smartphone						
Yes	.507**	.587**	.634**	19.16	17.43	30.13
No	-.507**	-.587**	-.634**	0.052	0.057	0.033

Source: The National Swedish SOM surveys 2010, 2011 and 2012.

Note: Anti-logarithmic regression coefficients (odds ratio). For each of the variables used, the reference category is all the other variables in that category (i.e. all other age groups) where β = 1.000. **. Significant at the 0.01 level (two-tailed). *. Significant at the 0.05 level (two-tailed).

Turning to the statistical and explanatory data for daily use of mobile search, the results reveal that in 2010 men were 1,967 times more likely than women to use their mobile device for search. The equivalent score for weekly use was 1,742. The analysis did not produce any significant differences between men and women during 2011 and 2012 for daily or weekly usage. In conclusion, the findings suggest that over time men and women have become equally likely to turn to their mobile device for information search.

Regarding age, there are significant and salient differences, with the youngest being keenest on turning to mobile search, and the oldest seemingly refraining from such use. In 2010, the 16–29 year olds were 2,766 more likely than others to turn to mobile search on a daily basis, and 3,097 times more likely on a weekly basis. These figures rose slightly in 2011, and then took a massive leap for 2012. That year the equivalent figures for 16–29 year olds were 5,298 and 6,534, respectively. Commanding closer attention to the three other age groups, one finds that the 50–64 and 65–85 year olds score far below 1, signalling that they are less likely than their counterparts to engage in mobile search. The 30-49 year olds, however, are also more likely than the other age groups to use mobile search. Moreover, their likelihood has increased over time, from 1,642 in 2010 to 3,245 in 2012 (daily use).

The analysis of influence relating to educational level reveals a clear pattern: people with limited education are least likely to engage in mobile search, whereas those with high educational levels are most likely to do so. The two middle groups do not distinguish themselves in a significant way. More specifically, the results reveal that in 2010 the highly educated were 2,074 times more likely than others to use mobile search daily, and 1,802 times for weekly use. The corresponding figures for 2012 (2,419 and 2,409) reveal fairly stable patterns, albeit a slight increase concerning weekly usage. In comparison, the odds ratios for daily and weekly use among low educated in 2010 were 0,195 and 0,207, and 0,197 and 0,162 for 2012. As expected, the patterns are perfectly reversed.

Turning towards a similar scrutiny of odds ratios depending on payment model, it appears salient that users with pre-paid cards are least likely to engage in mobile search, whereas the subscribers are most likely to do so (both private and company subscriptions). In 2010, the likelihood of daily use among private subscribers (2,616) and company subscribers (2,042) far exceeded that of pre-paid users (0,196). The corresponding figures for 2012 reveal a slightly smaller difference between pre-paid users (0,304) and private subscribers (2,212). Company subscribers, on the other hand, turned out to be slightly more likely to engage in mobile search on a daily basis. The figures for weekly use are relatively similar and stable concerning pre-paid users and private subscribers, whereas the scores for company subscribers reveal a more pronounced leap in likelihood (from 1,802 in 2010 to 2,409 in 2012). Evidence seems to indicate a coincidence in behavior which connects with the theories focused on social influence vectors, mostly-job related, that can lead users to feel a subtle pressure to step into using new IT devices (Karahanna, Straub, and Chervany, 1999).

At last this chapter turns to smartphone ownership, which represents the category responsible for the most significant differences in odds ratio. In 2010, the likelihood for a smartphone owner to use mobile search (compared to those without) was 16,983 regarding daily use and 19,16 for weekly use. These figures had risen significantly by 2012, especially the likelihood among weekly users (30,13), but also concerning daily users (19,495). One should elaborate on these findings in relation to the continuous uptake and diffusion of smartphone devices. It seems plausible to argue that the diffusion of smartphone devices has resulted in those items becoming more mainstream, involving both early adopters and the early majority (Rogers, 2003). The differentiating factor has consequently increased. That is, people who desire to use search and other networked functionalities with their mobile device most likely acquire such a handset, whereas those who feel less desire do not (See Table 4).

CONCLUSION

This chapter has presented unique findings describing and explaining the uptake of mobile search in the salient case of Sweden. The most significant descriptive findings have shown significant differences in users' behaviour. Three main gaps regarding frequency of use can be noted in the period analysed. Firstly, *the age gap* evidences that the young are engaging with mobile search six times more often than the elderly. Secondly, *the education gap* witnesses that the number of mobile search users has quadrupled more among the well-educated compared to the least educated. Thirdly, *the mobile device gap* reveals that ownership of a smartphone constitutes an exceptionally strong dividing line between users and non-users of mobile search.

The explanatory analysis has revealed important similarities and differences among distinct groups, largely in support of the descriptive findings. Some of these differences have increased over time, and some have decreased. The logistical regressions show that men were initially more likely than women to use mobile search. However, by the end of the study period, these differences had disappeared. The 16–29 and 30–49 year olds have always been more likely to turn to mobile search than the older age groups. Moreover, their relative likelihood of doing so has increased substantially over the period of study. The analysis of educational level shows stable patterns over time: the highly educated are most likely to turn to mobile search, the opposite applies to the least educated, whereas there are no significant figures for the two groups in between. Turning to payment model, those with a subscription score the highest likelihood throughout the period, whereas users with pre-paid cards are significantly less likely to turn to mobile search. Finally, smartphone ownership stands out as the most significant differentiating factor. This finding suggests that smartphone owners are thirty times more likely than their counterparts to utilise mobile search on a weekly basis.

FUTURE RESEARCH DIRECTIONS

On the basis of the findings presented, future research should acknowledge and examine more specifically the growing significance of mobile devices for how people search and retrieve different sorts of information (transportation, maps, shopping, tourist information, etc.), in different situational contexts (depending on space and time) and in different cultural contexts. Moreover, mobile devices in most cases are personal and application providers engage intensively in tracing, analysing and capitalising on personal data. Consequently, many opportunities arise for more specific studies and inquiries. These include but are not limited to research on how personalisation data is being matched with context-sensitive layers, relating to positioning services and the ubiquitous affordances of the mobile device (Nyre et al., 2012; Goggin et al., 2014). Such can be used for the purposes of commercial enterprises as well as authorities trying to establish functioning systems for crisis communications, for example.

This study, which tackles user behaviour from a broad perspective, can also lead the way to further research based on two main criteria of group discrimination: a) attending to the kind of queries via web browser/search widget-app; b) attending to users' behaviour when performing queries in smartphones versus tablets. Both new fields of research can also eventually incorporate age-discriminating criteria. However, that research will have to be based on the use of specific metrics tools of user activity which is not covered in this chapter. It would be particularly interesting for future research studies to determine to what extent the incorporation of simpler ways of input for search queries can change the rate of engagement of older ones, and to what extent those new search technologies have helped to fill the age gap in frequency of use. As the smartphone ownership is one of the main traits which has determined in the past the frequency of use, one can presume that, considering that Sweden currently ranks among the countries with the highest rate of smartphone use, the smartphone question will no longer be an issue.

REFERENCES

Accessing the news in an age of mobile media: tracing displacing and complementary effects of mobile news on newspapers and online news. (2015). *Mobile Media & Communication, 3*(1), 53-74. doi:10.1177/2050157914549039

ACSI. (2014). *ACSI e-Business Report 2014*. Retrieved from http://www.theacsi.org

Agosti, M., Crivellari, F., & Di Nunzio, G. M. (2012). Web Log Analysis: A Review Of A Decade Of Studies About Information Acquisition, Inspection And Interpretation Of User Interaction. *Data Mining and Knowledge Discovery*, *24*(3), 663–696. doi:10.1007/s10618-011-0228-8

Aguado, J. M., & Martínez, I. (2014). Feeding Digital Omnivores: The Impact Of Mobile Media In Digital Entertainment. In G. Goggin & L. Hjorth (Eds.), *The Routledge Companion to Mobile Media* (pp. 181–194). New York, London: Routledge.

Baeza-Yates, R., Dupret, G., & Velasco, J. (2007). A Study of Mobile Search Queries in Japan. *Proceedings of the international World Wide Web Conference*. Retrieved from http://lahuen.dcc.uchile.cl/~gdupret/pubs/banff_japan.pdf

Baeza-Yates, R., & Raghavan, P. (2010). Next Generation Web Search. In *Search Computing* (pp. 11–23). Springer. doi:10.1007/978-3-642-12310-8_2

Bakardjieva, M. (2011). The internet in everyday life: Exploring the tenets and contributions of diverse approaches. In M. Consalvo & C. Ess (Eds.), The handbook of internet studies (pp. 59–82). Oxford: Wiley-Blackwell.

Bohlin, E., Olsson, T., & Westlund, O. (2007). Mobile Barometer 2007 – An Analysis of the Swedish Mobile Society. Tokyo: Mobile Society Research Institute Report, NttDoCoMo. Retrieved from: http://www.moba-ken.jp/wp-content/pdf/research07-01.pdf

Bouwman, H., et al. (2009). Reconsidering the actual and future use of mobile services. *Information Systems and e-Business Management*, 7, 301–317.

Boyd, E. (2011). Brains And Bots Deep Inside Yahoo's CORE Grab A Billion Clicks. *Fast Company*. Retrieved from http://www.fastcompany.com/1770673/brains-and-bots-deep-inside-yahoos-core-grab-billion-clicks

Broder, A. (2002). A Taxonomy Of Web Search. *ACM Sigir Forum, 36*(2), 3-10. Retrieved from http://portal.acm.org/citation.cfm?id=792552

Broder, A., Fontoura, M., Josifoovski, V., & Riedel, L. (2007). A semantic approach to contextual advertising. *Mobile Data Engineering Laboratory, Department of Computer Science and Engineering.* Korea University. Retrieved from http://websense.korea.ac.kr/projects/ads/a_semantic_approach_to_contextual_advertising.pdf

Campbell, S. W. (2007). A Cross-Cultural Comparison of Perceptions And Uses Of Mobile Telephony. *New Media & Society*, *9*(2), 343–363. doi:10.1177/1461444807075016

Castellet, A., & Feijóo, C. (2013). Los Actores En El Ecosistema Móvil. In *La Comunicación Móvil. Hacia Un Nuevo Ecosistema Digital.* Barcelona: Gedisa.

Chan, M. (2015). *Examining The Influences Of News Use Patterns, Motivations, And Age Cohort On Mobile News Use: The case of Hong Kong, Mobile Media & Communication, 3(2).* OnlineFirst.

Church, K., Smyth, B., Bradley, K., & Cotter, P. (2008). A Large Scale Study of European Mobile Search Behaviour. *Proceedings of the 10th international conference on human-computer interaction with mobile devices and services.* ACM Press. doi:10.1145/1409240.1409243

Deciphering Trends in Mobile Search. (2007.). *Computer*, *40*(8), 58–62. doi:10.1109/MC.2007.270

Deuze, M. (2012). *Media Life.* Cambridge: Polity Press.

Dimmick, J. W., Feaster, J. C., & Hoplamazian, G. J. (2011). News in the interstices: the niches of mobile media in space and time. *New Media & Society*, *13*(1), 23–39. doi:10.1177/1461444810363452

Dutta, S., & Bilbao-Osorio, B. (2012). The Global Information Technology Report 2012. Living. In *A Hyperconnected World, Insight Report.* Geneva: RO-Kundig.

Feijóo, C., Maghiros, I., Abadie, F., & Gómez-Barroso, J. L. (2009). Exploring A Heterogeneous And Fragmented Digital Ecosystem: Mobile Content. *Telematics and Informatics*, *26*(3), 282–292. doi:10.1016/j.tele.2008.11.009

Feldmann, V. (2005). *Leveraging mobile media: Cross-media strategy and innovation policy for mobile media communication.* Heidelberg: Physica-Verlag.

Frier, S. (2014). Facebook test lets users search old posts by keyword. *Bloomberg* [Web page]. Retrieved from: http://www.bloomberg.com/news/2014-08-28/facebook-test-lets-users-search-old-posts-by-keyword.html

Goggin, G., Martin, F., & Dwyer, T. (2014). Locative News. *Journalism Studies.* doi: 10.1080/1461670X.2014.890329

Gómez-Barroso, J. L., Compañó, R., Feijóo, C., Bacigalupo, M., Westlund, O., Ramos, S., et al. (2010). *Prospects of Mobile Search*, European Commission, JRC, Institute for Prospective Technological Studies (IPTS), EUR 24148 EN 2010. (http://ipts.jrc.ec.europa.eu/publications/pub.cfm?id=3159)

Haddon, L., & Vincent, J. (2008). Children's broadening use of mobile phones. In G. Goggin & L. Hjorth (Eds.), *Mobile technologies: From telecommunications to media* (pp. 37–49). Abingdon: Routledge.

Humphreys, L., Von Pape, T., & Karnowski, V. (2013). Evolving Mobile Media: Uses And Conceptualizations Of The Mobile Internet. *Journal of Computer-Mediated Communication*, *18*(4), 491–507. doi:10.1111/jcc4.12019

Insight, C. C. S. (2014). *Global smartphone market analysis and outlook: Disruption in a changing market.* Retrieved from http://www.lenovo.com/transactions/pdf/CCS-Insight-Smartphone-Market-Analysis-Full-Report-07-2014.pdf

Ipsos Media, C. T. (2013). *Our Mobile Planet: Sweden.* Retrieved from http://services.google.com/fh/files/misc/omp-2013-sw-en.pdf

ITU. (2010). *The World in 2010. ICT Facts and Figures.* Retrieved from http://www.itu.int/en/ITU-D/Statistics/Documents/facts/ICTFactsFigures2010.pdf

Kamvar, M., & Baluja, S. (2006 April 22–27). A Large Scale Study of Wireless Search Behavior: Google Mobile Search. Proceedings of the SIGCHI conference on human factors in computing systems, Montreal (pp. 701–709). Retrieved from http://www.esprockets.com/papers/kamvar-baluja.chi06.pdf

Karahanna, E., Straub, D. W., & Chervany, N. L. (1999). Information Technology Adoption Across Time: A Cross-Sectional Comparison Of Pre-Adoption And Post-Adoption Beliefs. *Management Information Systems Quarterly*, *23*(2), 183–213. doi:10.2307/249751

Kitamura, S. (2013). The Relationship between use of the internet and traditional information sources: an empirical study in Japan, *SAGE Open,* April-June, 1–9.

Kohlborn, T., Korthaus, A., Riedl, C., & Krcmar, H. (2009). Service Aggregators in Business Networks. In Enterprise distributed object computing conference workshops, 2009. EDOCW 2009. 13th (pp. 195–202). doi:10.1109/EDOCW.2009.5331992

Labs, E. (2013). Intel capital, Samsung venture investment corporation, and Telefónica Digital make strategic investment. Retrieved from http://www.prnewswire.com/news-releases/intel-capital-samsung-venture-investment-corporation-and-telefonica-digital-make-strategic-investment-in-expect-labs-205361051.html

Lessig, L. (2006). Code: Version 2.0. New York: Basic Books. Retrieved from http://codev2.cc/download+remix/Lessig-Codev2.pdf

Lewis, S. C., & Westlund, O. (2015). Actors, Actants, Audiences, and Activities in Cross-Media News Work. *Digital Journalism*, *3*(1), 19–37. doi:10.1080/21670811.2014.927986

Li, X. (2013). Innovativeness, Personal Initiative, News Affinity And News Utility As Predictors Of The Use Of Mobile Phones As News Devices. *Chinese Journal of Communication*, *6*(3), 350–373. doi:10.1080/17544750.2013.789429

Madrigal, A. C. (2014). What is Pinterest? A Database Of Intentions. *The Atlantic* [Web . Retrieved from: http://www.theatlantic.com/technology/archive/2014/07/what-is-pinterest-a-database-of-intentions/375365/?single_page=true

Manyika, J. (2013). The impact of disruptive technology: a conversation with Eric Schmidt. *Mckinsey & Company.* Retrieved from http://www.mckinsey.com/insights/high_tech_telecoms_internet/disruptive_technologies

Martin, J. (2015). *Mobile News Use and Participation in Elections: A Bridge for the Democratic Divide? Mobile Media & Communication, 3(2).* OnlineFirst.

Measuring the Information Society 2012. (2012). Retrieved from http://www.itu.int/en/ITU-D/Statistics/Documents/publications/mis2012/MIS2012_without_Annex_4.pdf

Measuring the Information Society 2013. (2013). Retrieved from http://www.itu.int/en/ITU-D/Statistics/Documents/publications/mis2013/MIS2013_without_Annex_4.pdf

Milchen, J., & Woods, C. (2014). Seven ways businesses and communities can counter "showrooming". *AMIBA*. Retrieved from http://www.amiba.net/resources/showrooming

MindMeld. (2014). The MindMeld API: Introducing multilingual support. Retrieved from https://www.youtube.com/watch?v=14xQK2h9yGY&utm_source=Expect+Labs+Mailing+List&utm_campaign=d32c75f15a-Multilingual_Video&utm_medium=email&utm_term=0_bcc914a475-d32c75f15a-99209801

Mobile news: A review and model of journalism in an age of mobile media. (2013). *Digital Journalism, 1,* 6–26.

Möller, A., Diewald, S., Roalter, L., & Kranz, M. (2012). MobiMed: Comparing object identification techniques on smartphones. *Proceedings of the 7th Nordic Conference on Human-Computer Interaction: Making sense through design* (pp. 31–40). doi:10.1145/2399016.2399022

Newman, N., & Levy, D. A. L. (2013). *Reuters Institute Digital News Report 2013: Tracking the future of news*. Reuters Institute for the Study of Journalism, University of Oxford.

News Consumption in an Age of Mobile Media: Patterns, People, Place and Participation. (2015). *Mobile Media & Communication, 3(2).* doi: 10.1177/2050157914563369

New(S) Functions for the Mobile. (2010). *New Media & Society, 12,* 91–108. doi:10.1177/1461444809355116

Nyre, L., Bjornestad, S., Tessem, B., & Oie, K. V. (2012). Locative Journalism: Designing A Location-Dependent News Medium For Smartphones. *Convergence (London), 18*(3), 297–314. doi:10.1177/1354856512441151

O'Reilly. Tim. (2011). Some context for thinking about the future of media. Madrid: Ficod. Retrieved from http://www.slideshare.net/timoreilly/ficod-2011-pdf-with-notes/download

Ongena, G., Bouwman, H., & Gillebaard, H. (2012). Displacement And Supplemental Effects Of The Mobile Internet On Fixed Internet Use. *International Journal on Media Management, 14*(4), 279–299. doi:10.1080/14241277.2012.684192

Oulasvirta, A., Rattenbury, T., Ma, L., & Raita, E. (2012). Habits Make Smartphone Use More Pervasive. *Personal and Ubiquitous Computing, 16*(1), 105–114. doi:10.1007/s00779-011-0412-2

Pagani, M. (2004). Determinants Of Adoption Of Third Generation Mobile Multimedia Services. *Journal of Interactive Marketing, 18*(3), 46–59. doi:10.1002/dir.20011

Pariser, E. (2011). The Filter Bubble: What The Internet Is Hiding From You. Penguin UK.

Purcell, K., Brenner, J., & Rainie, L. (2012). *Search Engine Use 2012*. Pew Research Center's Internet & American Life Project. Retrieved from http://pewinternet.org/Reports/2012/Search-Engine-Use-2012.aspx

Reding, V. (2008, November 13). Digital Europe: The Internet mega-trends that will shape tomorrow's Europe. Speech at the European Internet Foundation Special Event. A view of the Digital World in 2025, Brussels. Retrieved from http://europa.eu/rapid/press-release_SPEECH-08-616_en.htm

Regalado, A. (2014). Why did Google pay $400 million for DeepMind? *MIT Technology Review*. Retrieved from www.technologyreview.com

Rogers, E. M. (2003). *Diffusion of innovations*. New York: Free Press.

Rogers, E. M. (2003). *Diffusion Of Innovations. New York*. New York: Free Press.

Sasseen, J., Olmstead, K., & Mitchell, A. (2013). Digital: As mobile grows rapidly, the pressures on news intensify. The state of the news media 2013: An annual report on American journalism. Pew Research Center's Project for Excellence in Journalism, Washington, D.C., 2012.

Scolari, C. A., Aguado, J. M., & Feijoo, C. (2012). Mobile media: Towards a definition and taxonomy of contents and applications. *International Journal of Interactive Mobile Technologies*, *6*, 29–38.

Taneja, H., Webster, J. G., Malthouse, E. C., & Ksiazek, T. B. (2012). Media Consumption Across Platforms: Identifying User-Defined Repertoires. *New Media & Society*, *14*(6), 951–968. doi:10.1177/1461444811436146

2014. The production and consumption of mobile news. InGoggin, G., & Hjorth, L. (Eds.), *The Routledge companion to mobile media* (pp. 135–145). New York: Routledge.

Traffic and market data report. (2011, November). *Ericsson*.

Vaidhyanathan, S. (2011). *The Googlization Of Everything: And Why We Should Worry*. Berkeley & Los Angeles, CA: University of California Press.

Van Damme, K., Courtois, C., Verbrugge, K., & de Marez, L. (2015). *What's happening to News? A Mixed-Method Audience Centred Study on Mobile News Consumption. Mobile Media & Communication, 3(2)*. OnlineFirst.

Venkatesh, V., Morris, M. G., Davis, G. B., & Davis, F. D. (2003). User Acceptance Of Information Technology: Toward A Unified View. *Management Information Systems Quarterly*, *27*(3), 425–478.

Verbrugge, K., Stevens, I., De Marez, L., Veeckman, C., Coppens, P., Van Hamme, L., & Lievens, B. (2013). Mobimeter 2013: A Report On The Adoption And Use Of Mobile Media Technologies In Flanders. Gent, Belgium.

Wei, R., Lo, V., Xu, X., Chen, Y. K., & Zhang, G. (2014). Predicting mobile news use among college students: the role of press freedom in four Asian cities. *New Media & Society*, *14*(4), 637–654. doi:10.1177/1461444813487963

Westlund, O. (2008). From mobile to phone to mobile device: news consumption on the go. *Canadian Journal of Communication*, *33*, 443–463.

Westlund, O., & Färdigh, M. A. (2012). Conceptualizing Media Generations: The Print-, Online- And Individualized Generations. *Observatorio Journal*, *6*, 181–213.

Westlund, O., Gómez-Barroso, J.-L., Compañó, R., & Feijóo, C. (2011). Exploring the logic of mobile search. *Behaviour & Information Technology*, *30*(5), 691–703. doi:10.1080/0144929X.2010.516020

Westlund, O., & Lewis, S. (2014). Agents of media innovation: actors, actants, and audiences. *The Journal of Media Innovations*, *1*(2). doi:10.5617/jmi.v1i2.856

Westlund, O., & Weibull, L. (2013). Generation, life course and news media use in Sweden 1986–2011. *Northern Lights*, *11*(1), 147–173. doi:10.1386/nl.11.1.147_1

Wirth, W., Von Pape, T., & Karnowski, V. (2008). An integrative model of mobile phone appropriation. *Journal of Computer-Mediated Communication*, *13*(3), 593–617. doi:10.1111/j.1083-6101.2008.00412.x

KEY TERMS AND DEFINITIONS

Aggregators: In the Internet, aggregators integrate different services under the umbrella of a single site. They can live on their own content, on others' content, or both ways.

Bubble: Environment in which the user is surrounded by a number of restricting circumstances that appear transparent or unnoticed to him.

Context-Aware Applications: In this chapter, applications intended to integrate user queries with the data provided by, for instance, by GPS, cameras, pedometers and microphones that are inside the mobile device.

Innovation Diffusion Theory: Theory that seeks to explain the pace at which different cultures integrate new ideas and technology. It originated by the sixties in the 20th Century.

OS: Operating system. A set of software pieces, usually bundled with any mobile phone or terminal, that is in charge of the user interface, the primary handset functions and the context sensors. The OS determines ultimately the range of features of the device.

Smartphone: Mobile phone equipped with an operating system that can be easily recognized by the user.

Technology Acceptance Model: Theory that aims to explain how users accept and use a technology. It appeared in its current formulation by the beginning of the 21st Century.

ENDNOTES

1 Net Applications monthly data can be browsed from www.netapplications.com.

2 Monthly global and country data can be browsed from www.alexa.com.

Chapter 7
From Personal to Mobile Healthcare:
Challenges and Opportunities

Elena Villalba-Mora
Technical University of Madrid, Spain

Ignacio Peinado
Getafe University Hospital, Spain

Leocadio Rodriguez-Mañas
Getafe University Hospital, Spain

ABSTRACT

mHealth is an emerging field that comprises the delivery of health through mobile technologies, being part of a broader concept: eHealth. With the rapid advances and adoption of mobile technologies and applications, mHealth opens new opportunities to deliver healthcare and wellness promotion through personal mobiles. This will transform the current delivery of services linked to particular places to a delocalization and personalization of services delivery. Although the adoption and integration of mobile technologies is high across the globe, the particular adoption of mHealth remains low and scattered due to some remaining challenges. This chapter presents the current picture of mHealth and it later discusses the opportunities both at market level and with regards to healthcare delivery. Then, it presents future research directions as big data; and it closes with conclusions on this challenging emerging field.

INTRODUCTION

eHealth lies in the *intersection of medical informatics, public health and business; referring to health services and information delivered or enhanced through the Internet and related technologies* (Eysenbach, 2001). The term exceeds a mere technical development, being also a way of thinking, an attitude, and a commitment for networked, global thinking, to improve health care locally, regionally, and worldwide by using information and communication technology (ICT).

DOI: 10.4018/978-1-4666-8838-4.ch007

Mobile Health (hereafter mHealth) is a component of eHealth. No standardized definition of mHealth has been established yet. One of the most cited definitions of mHealth is the one provided by Istepanian et al. (2006), which defined mHealth as *emerging mobile communications and network technologies for healthcare*. In 2011, The World Health Organisation (WHO) proposed a broader definition that included also sensors as part of the mHealth paradigm, as *medical and public health practice supported by mobile devices, such as mobile phones, patient monitoring devices, personal digital assistants (PDAs), and other wireless devices* (WHO, 2011).

Different regions across the World are facing different challenges that in their context push mHealth deployment and adoption. The 2011 WHO Survey showed that the imperative cut to healthcare costs is driving mHealth in higher-income countries, whereas in lower-income countries the driver is the need for access, mainly to primary care.

In higher-income countries, healthcare systems must offer and assure high-quality and affordable care to all citizens. Thus, these systems must tackle the demographic change, the mobility of patients and the associated rising healthcare costs. As an example, the proportion of European population aged 65 years and over is predicted to rise from 17.1% in 2008 to 30.0% in 2060, and those aged 80 and over rising from 4.4% to 12.1% over the same period (Eurostat, 2008). One of the direct consequences will be the increased prevalence of chronic diseases, which will be leading to a higher number of people with compromised health and quality of life, higher demand for healthcare services and loss of productivity due to reduced capacity of the workforce. Chronic diseases significantly strains healthcare services causing an increase in the healthcare expenditure. This expenditure in European countries, measured as a percentage of Gross Domestic Product (GDP), is expected to rise from 9% in 2007 to around 16% by 2020 (PWC, 2005). Chronic diseases are now responsible for the consumption of the vast majority of healthcare resources (more than 70% in higher-income countries).

In contrast, in emerging countries the need to access care where there are not services in place have pushed the adoption of mHealth as an alternative. According to the 2011 WHO survey, in Africa and Asia, most mHealth services focus on improving the efficiency of the healthcare workforce and systems; and in the prevention of infectious diseases (e.g. in India, South Africa and Kenya).

ICT and mobile technologies may be used in many aspects related to the health sector, healthcare delivery, public health, research and innovation, education and training of professionals and citizens and patients' empowerment and lifestyle. For instance, mHealth may delocalize the care delivery, which is nowadays mainly performed in hospitals when an acute event already has occurred, which represents a reactive model. Through the use of mHealth, healthcare shifts from reactive, hospital-based care to a proactive, preventive care, including self-care (Bonfiglio, 2012). The benefit of this delocalization is twofold: on the one hand, healthcare can be provided in remote areas like rural zones assuring quality of health delivery for all. On the other hand, healthcare systems can be used at home by patients and informal caregivers, whereas they are remotely followed by professionals. Professionals can remotely assess patients' health status periodically and early detect an event. This way the outcome of the patient improves by reducing co-morbidities and even mortality. Moreover, to have information between face to face encounters may help the professional to assess the patient's evolution and to better understand the diseases and treatments' effectiveness.

This chapter aims to present and discuss the challenges and opportunities of the mobile health. Following this introduction of the topic, the next sections describe the background of personal healthcare applications and the current picture of mHealth around the globe. Later, a section discusses the challenges and opportunities of mHealth. Finally, a conclusion closes the chapter.

BACKGROUND

The integration of ICT in its modern form in the healthcare process began as early as in the mid-1960s, mainly driven by the military and aerospace sectors. First applications of ICT in the healthcare domain were basically telemedicine solutions that allowed physicians to communicate amongst them and with patients. From the 1980s, the widespread use of mainframes and personal computers created new possibilities for the delivery of health care. In a first phase, integration efforts were twofold: first, there was a need to digitalize all the information about the patient and to ensure it could be accessed in a secure way by different members throughout all healthcare tiers; second, technologies such as the telephone or the emerging Internet could be used to provide patients with healthcare outside the clinical environment.

During the 90s of the 20th century public funding European programs focused at ensuring the link between all different care tiers, which represents the earliest technological solutions, also known as telemedicine. The second phase started in the early 00s of the present century, with the objective of connecting individuals to the Health Information networks, i.e telehealth and telecare. This second phase aimed to support the emergence of Personal Health Systems (PHS).

The concept of PHS emerged in the late 1990s, representing a paradigm shift within the domain of healthcare by putting the individual citizen at the centre of the healthcare delivery process. In its presentation "From e-Health to m- and p-Health" (EHRA Summit 2010, 22-23 March 2010 at Sophia, Antipolis), the European Commission defined PHSs as "*a new generation of disruptive eHealth tools*" with the following characteristics: 1) they place the individual person in the centre of the healthcare delivery process (i.e. person-centred care); 2) they aimed for high quality, personalised care at the point of need; and 3) they are expected to make a better use of the available healthcare resources.

PHS include sensors, persons' interaction, management of measured data including context data (e.g. health status data, mood, environmental data, date and time or place) and they close the loop with a professional (e.g. physicians, general practitioners or nurses). Services and products based on ICT and mobile technologies may include remote monitoring using sensors and local analysis of the current patient's status and evolution through time. Sensors can be on-body sensors like wearable sensors to assess the heart status (e.g. heart rate sensors, respiration rate, skin temperature, blood pressure, etc.), portable sensors to assess diabetes (e.g. weight, glucose meter, etc.) or ambient sensors to study person's behaviour and detect possible problems (e.g. presence sensors, movement sensors, light detection, humidity, etc.)

As defined within the PHS2020 initiative (Codagnone, 2009), Personal Health Systems (PHS) assist in the provision of continuous, quality controlled, and personalised health services to empowered individuals regardless of location, consisting of:

- Ambient and/or body (wearable, portable or implantable) devices, which acquire, monitor and communicate physiological parameters and other health related context of an individual.
- Intelligent processing of the acquired information and coupling of it with expert biomedical knowledge to derive important new insights about individual's health status.
- Active feedback based on such new insights, either from health professionals or directly from the devices to the individuals, assisting in diagnosis, treatment and rehabilitation as well as in disease prevention and lifestyle management.

During the late 1990s and early 2000s, several European projects implemented PHS that were, in most cases, tested in small scale pilots. Very few of these initiatives have been finally integrated into the routine practice. Moreover, different economic studies estimated that, by the end of the 2000s, PHS penetration was really small and still a niche market (Codagnone, 2009). Nonetheless, the 2011 WHO Survey showed that in Europe these services represents one of the more recent drivers of mHealth, This was not anticipated by other studies(Frost & Sullivan, 2008), which stated that the European mHealth market was only worth over €1 million at that year.

Some of the reasons that have prevented PHSs to be integrated into the routine practice include:

- Need to provide users with dedicated systems or ICT set-up (i.e. internet or broadband connection). Moreover, desktop web access is much more common in developed countries, while mobile access is becoming pervasive in developing countries.
- Inability of the sensors to acquire environment and subjective data.
- Most sensors were too illness-specific, and some of them might be considered as too invasive by patients (i.e. wearables).
- Inability to process big and multimodal data.
- Lack of integration with working Health Information systems.

According to the roadmap developed by the European Commission within their PHS2020 initiative, in order to succeed the future truly personalised and efficient PHS will need to have the following characteristics:

- Ability to capture the very peculiar characteristics of individuals (vital and physiological signs, but also their generic outlook, as well as their clinical history, and their socio-demographic and socio-economic condition)
- Ability to ensure awareness of very punctual contextual conditions (location, activity being performed, emotional status, physical and chemical conditions in the environment, etc.)
- Ability to process such information in an intelligent way in order to support traditional action.

Those characteristics may be fulfilled though mobile technologies which are personal devices that may incorporate sensors, process information and communicate no matter the location. This will be elaborated in further detail in section "Towards mobile healthcare".

MHEALTH: CURRENT PICTURE

The rapid advances in mobile technologies and applications, the integration of mobile health into existing eHealth services, and the continued growth in coverage of mobile cellular networks have driven the uptake of mHealth. According to the International Telecommunication Union (ITU), there are now over 5 billion wireless subscribers; over 70% of them reside in low- and middle-income countries. The GSM Association reports commercial wireless signals cover over 85% of the world's population, extending far beyond the reach of the electrical grid (WHO, 2011).

According to the latter reference (WHO, 2011), the four most frequently reported mHealth initiatives were: health call centres (59%), emergency toll-free telephone services (55%), managing emergencies and disasters (54%), and mobile telemedicine (49%). With the exception of health call centres, emergency toll-free telephone services, and managing emergencies and disasters, around two thirds of mHealth services are in the pilot or informal stage.

Higher-income countries show more mHealth activity than do lower-income countries. Countries in the European Region are currently the most active and those in the African Region the least active. mHealth is currently more integrated into processes and services which have historically used voice communication through conventional telephone networks, as health call centres, toll-free numbers and emergency services using mobile communications. The least frequently use of mHealth is in relation to surveillance, public health, and decision support systems. These require enhanced capabilities and infrastructure to be implemented, and therefore may not be a health priority in Member States with financial constraints.

Within the field of content / application development, it is worthy to mention the noticeable growth of available mHealth applications available in the market. By November 2012, the GSMA mHealth deployment tracker (Mobile tracker, 2015), a customised tool that collates mobile health products around the globe, accounted for 300 mHealth deployments all around the world. By November 2014 the same tool reported a total number of 1125 mHealth solutions.

PatientView published in October 2012 a report with a review of mobile health apps in Europe. It also launched an online version: www.myhealthapps.net in November 2013. Both compile selected and rated apps which take into account the views of patients, carers, families, and the healthy public in tandem with other quality approval processes. They all nominate their favourite apps to myhealthapps.net. They provide reviews and, where possible, some assessment of the level of usage of the app amongst the people they represent. A concerted attempt is made to include apps that cover every aspect of life that encourages better health (PatientView, 2012). Up to 200 apps can be found though their name or depending on their classified target. For instance: accident and emergency, cancer, children' health, Alzheimer disease, communication disability, arthritis, coeliac disease, diabetes, diet, exercise, HIV/AIDS, medication reminder, mental health amongst others. The majority of the apps are in English, although some of them also cover different languages as French, Finnish or Dutch.

TOWARDS MOBILE HEALTHCARE

While eHealth encompasses all aspects related to the application of ICT into the healthcare provision, mHealth focuses on the application of a specific set of technologies, such as wireless communications, voice transmission, third and fourth generation mobile communications, global position systems (GPS), short-range data transmission technologies such as Bluetooth, Zigbee or Near-Field Communications (NFC).

The widespread adoption of mobile technologies can help overcome some of the limitations faced by eHealth and PHS solutions, basically related to reach target population and set-up costs. Moreover, besides their storage, communication and processing capabilities, modern smartphones can include up to 18 sensors, including accelerometers and gyroscopes.

Opportunities for Healthcare

The use of mobile technologies within the scope of healthcare provision opens big opportunities to deploy, in a cost-effective way, systems and services. Depending on their target, services provided by m-health solutions include services used by patients and those aimed at improving the healthcare systems per se (Vishwanath, 2012).

First, the solutions that provide services to patients implement the personalized health (p-health) paradigm, putting the patient at the centre of the healthcare process and empowering them to take control of their own health condition.

mHealth permits the shift in patients' role from the passive one they usually have to a more proactive and participating one. For that, patients must be educated and empowered. That is to say, they are to understand their health status, their symptoms and diseases' course; and coproduce their health and well-being treatment and management. Patient-centred solutions can be divided into the following categories:

- **Wellness:** Wellness and health promotion solutions use messages, multimedia and/or direct communication in order to empower patients to keep a healthier lifestyle, by improving their nutrition, fitness levels and lifestyle habits.
- **Prevention:** Mobile prevention solutions allow healthcare producers to launch prevention campaigns using mobile technologies such as SMS or MMS in order to spread awareness about unhealthy behaviours or to control disease outbreaks.
- **Treatment Compliance:** Treatment compliance refers to the use of mobile communication technologies to assess the compliance of patient to medical treatments.
- **Monitoring:** Mobile monitoring systems use a mobile device as a hub to acquire data from different sources, perform some lightweight processing and eventually send all data to a dedicated repository. Physiological data of the patient can be acquired using embedded sensors or external medical sensors, such as blood pressure cuffs, glucometers, etc.
- **Diagnosis Systems:** These provide patients with mobile-based telemedicine solutions in order to remotely communicate with the professionals.

Second, the solutions aiming at strengthening healthcare systems include mobile technologies in order to enhance the training and performance of health professionals as well as to optimize healthcare resources, comprising:

- **Emergency Response Systems:** Emergency response systems make use of mobile technologies in order to better coordinate all tiers of care in case of an emergency or catastrophe.
- **Healthcare Practitioner Support:** Including the use of mobiles to provide access to medical information and training to health professionals or the use of Decision Support Systems (DSSs) that provide health professionals with advice on clinical diagnoses using complex software algorithms.
- **Healthcare Surveillance:** Healthcare surveillance refers to the use of mobile devices to input and transmit data that will be used to track diseases for surveillance programmes.
- **Healthcare Administration Tools:** These aims at optimizing certain processes in the management of care delivery at the healthcare centres.

Market Opportunities

According to Internet Live Stats, by July 2014 almost 3 billion people in the world had an Internet connection out of 7.243.784.121 world inhabitants, hence reaching a penetration rate of over 40% (Internet Live Stats, 2015). According to the data provided by the ITU, Internet penetration rates range from over 75% in Europe to a 32% in the Asia-Pacific area and 16% in Africa (ITU, 2013). On the other hand, GSMA estimates that by Nov 2014 there were more than 7 billion mobile connections (including M2M) in the world. In developed countries, the penetration rate of mobile-cellular was over 100% (meaning that there were more functioning mobile devices than people), while in developing countries the penetration rate was approximately 89% (GSMA Intelligence, 2015). These data show that the use of mobile technologies for delivering healthcare can create a greater impact by reaching a larger population. For instance, in Africa the difference between the penetration rates of mobile devices and internet is roughly 47% of the population.

In July 2013, PwC and GSMA released a study (Vishwanath et al, 2012) that estimated that the mobile market revenue will reach above US$ 23 billion across all stakeholders by 2017. According to this study, Europe and Asia Pacific will have most of the market share, accounting for 30% market share each, followed by developed North America (US and Canada) with a 28% market share and with estimations for Latin America and Africa for approximately 7% and 5% market share, respectively.

According to (Vishwanath et al, 2012), depending on the category of the service, patient-centred solutions are estimated to account for approximately 94% of the overall mobile health revenues (US$ 21.62 billion). From this estimated revenue, almost US$ 15 billion (65% of the total) will be generated by the deployment of Monitoring services. Monitoring services comprise chronic disease management services, that are expected to generate revenues for US$ 10,7 billion in 2017, and support for independent living services, that are estimated to earn US$ 4,3 billion. Still within the scope of patient-centred solutions, diagnostic services are foreseen to account for 15% of the overall mHealth revenues. On the other hand, healthcare strengthening solutions are estimated to account for approximately 6% of the overall market in 2017, mostly from Healthcare practitioner support.

Besides, it is foreseen that by 2017 3.4 billion people will own a smartphone and half of them will be using mHealth apps. According to recent estimations, there are currently 97,000 mHealth apps available across multiple platforms on the global market, of which approximately 70% target the consumer wellness and fitness segments; and the other 30% target health professionals, facilitating the access to patient data, consultation and monitoring, etc. (Research2Guidance, 2013)

The impact of each the different categories of mHealth applications will strongly vary depending on the development level and the prevalence of various diseases of the region. Higher-income countries will benefit the most from lifestyle changing systems, post-acute monitoring applications and chronic disease management solutions. For instance, monitoring of patients with metabolic diseases such as diabetes or obesity are expected to comprise around 40% of the chronic disease management in the US, 23% in Germany or 20% in Brazil, while remote monitoring of cardiovascular conditions can reach up to 47% of the chronic disease management market share in the US. In higher-income countries where patients are usually well covered by the public health system, there will be more resilient to changes in the care model. Therefore, other stakeholders such as healthcare providers or call centres will have to adopt the role of driving force in the deployment of mHealth-based care solutions. On the other hand, in developing countries there is much more limited access to health care, in terms of number of physicians, number of hospital beds or available technologies. Within these countries, diagnosis, treatment and basic prevention solutions will offer a possibility to overcome some of the limitations of their health systems.

The stakeholders that will benefit from the growth of the mHealth market share comprise mobile operators, device vendors, healthcare providers and content/application players. Mobile operators will benefit the most for the growth of the mHealth market share, with an estimated 50% share of the overall mHealth market in 2017. This revenue will arise mostly from the deployment of monitoring services. According to the estimations provided in the PwC and GSMA study (Vishwanath et al, 2012), the rest of the pie would be as follows: device vendors would get almost 30% of the overall market share (US$ 6.6 billion), while content / application players and health care providers would get approximately 10% of the overall market share (US$ 2.6 billion and US$ 2.4 billion respectively).

Up to now, most health and medical applications are being developed by small and medium sized enterprises around the globe, with 30% of mobile app developer companies are individuals and 34.3% small companies (i.e. 2-9 employees) (IDC, 2011). Recently, mainstream industry companies have started the integration of mHealth solutions within their platforms. In June 2014 Apple presented its app 'Health' and the 'Apple Healthkit' platform. The 'Health' app aims at becoming the central point where users store and check all data regarding their health condition. Moreover, the 'Apple HealthKit' platform provides a series of APIs that allow third-party developers and/or medical sensor manufacturers (i.e. Fitbit, Jawbone UP, Misfit, iHealth) to integrate their data with the 'Health' app.

In the same context, Google has recently presented Google Fit, a platform that also gathers data from third-party applications and wearable devices in a unique repository. Moreover, Samsung has developed their own health platform, Samsung SAMI (Samsung Architecture for Multimodal Interactions), a vendor-agnostic, open platform that takes data from multiple sources via open API and run complex analytics on it for deeper insights into the user's health. All of these platforms provide application developers and device manufacturers with tools to integrate their data in a general-purpose platform, hence fostering the integration with other data captured by other third-party applications. Nevertheless, no national-level, public similar initiatives have already been developed. The failure of similar initiatives such as Google Health or Microsoft HealthVault can lead to the conclusion that patients have some concerns of providing private companies with their health data. In order to succeed, a platform that aggregates health data about patients should get the confidence of citizens, with clear terms and conditions and a conscientious respect to national and international legislation about the protection of sensitive data.

Challenges

The process of diffusion of new technologies starts with their technical development, and it culminates in full adoption when they are integrated and normalised into the health care processes (Fonkych and Roger, 2005). Studies (e.g. Mair, 2007) that analyse the integration of health information related technologies into medical practice state that they need to be embedded in the health care system and that the physicians play a key role in the process through the adoption of the technology. In addition, there is a need to adapt the adoption models specifically to the healthcare context (Holden 2010), considering its particular drivers and barriers. That is, not only there is a need of user adoption, but also clinical or medical adoption. So far, the use of mHealth within the healthcare context is still limited and many challenges remain (Villalba, 2013).

With regards to the adoption curve for mobile healthcare, a 2011 study reveals (Boehm, 2011) that only 23% of mobile consumers have used any mHealth solutions, whereas 77% did never used them and 67% stated they did not foresee to use their mobile for health related activities.

In countries were the healthcare services are managed by public funds, there is a need of policy commitment which must shift their current focus towards investments on innovative solutions that include mHealth. Policy makers claim that evidence of positive impact of mHealth on patient's outcome and healthcare resources management is needed. Although several studies demonstrate that PHS and related technologies may contribute to improve the healthcare delivery (Barlow, 2007) (Stroetmann, 2010) (Steventon, 2012), their deployment and adoption have so far been slow and fragmented, as shown by (May et al., 2001) in the case of the United Kingdom and by (Villalba et al., 2013) in eight countries across Europe. Thus, the remaining main barrier seems to be the reluctance from the public decision makers to integrated mHealth.

Other barriers include financial aspects which also need a change in the financing legislation when including public funds. In relation to the economic model there are two different scenarios: countries were healthcare services remain a public universal service as in Europe (e.g. France, Spain) or Canada; and those with a private model (e.g. EEUU) or co-payments in mixed models (e.g. Israel). In the first group, the economic model relies on the policy commitment to such innovative services. Whereas, in the latter relies on the particular willingness of healthcare providers to improve their care provision.

Data protection and security is another crucial challenge which may act as a barrier if not adequately addressed. Data collected through apps or services by individuals, app developers, health professionals, advertising companies or public authorities must be protected. Large quantities of information (e.g. data from sensors, data introduced by the user, location) are currently referred to as Big Data. The concept and implications of Big Data will be discussed in future research directions section. This information in most cases contains personal data that are directly or indirectly identified or identifiable as a natural person. Besides, health data are particularly sensitive requiring special protection. Thus, mHealth solutions must contain specific and suitable security and safety, such as encryption of patient data and appropriate patient authentication mechanisms to ensure security and to mitigate associated risks.

In Europe, personal data protection is a fundamental right (Article 8 of the Charter of Fundamental Rights of the European Union and Article 16(1) of the Treaty on the Functioning of the European Union (TFEU)). Therefore, mHealth services must comply with personal data protection rules, with information on the data subject, data security, and the lawful processing of personal data, including health and medical data. Moreover, the current Personal Data Protection Directive is under revision to respond to new challenges that arose after the fast spread of new technologies while ensuring the individuals' effective control over their personal data (Directive 95/46/EC, 1995). The new Directive will includes the principles of "data minimisation", "data protection by design", and "data protection by default" to make sure that data protection safeguards are taken into account at the planning stage of procedures and systems (European Commission, 2014).

Furthermore, before mHealth is completely integrated into our lives, the patients, their caregivers, and all citizens in general must be able to use mobile applications regardless their functional capabilities. That is to say, mHealth must be usable and accessible.

According to the ISO 9241, usability stands for the effectiveness, efficiency and satisfaction with which specified users achieve specified goals in particular environments: 1) effectiveness: the accuracy and completeness with which specified users can achieve specified goals in particular environments; 2) efficiency: the resources expended in relation to the accuracy and completeness of goals achieved; and 3) satisfaction: the comfort and acceptability of the work system to its users and other people affected by its use. Usability is key to enable the widespread of mHealth turning the users into empowered citizens that coproduce their health, especially those with multiple chronic conditions that use massively the health resources; and in preventing healthy citizens to become ill.

Accessibility refers to the design of products, devices, services or environments for people with disabilities (Henry et al, 2014). Several guidelines (such as W3C's Web Accessibility Guidelines – WCAG - 2.0) and legislation (such as Section 508 of the US Rehabilitation Act in the United States or Mandate 376 – European Accessibility Requirement for Public Procurement of Products and Services in the ICT domain in the European Union) have worked in order to ensure the public procurement of accessible information and communication technologies (ICT) products and services. Most of the efforts performed so far have been dedicated to ensure the accessibility of the web and the web content. Nevertheless, no standardized guidelines or legislation have been deployed in order to ensure the accessibility of mobile terminals and applications. The most relevant mobile operating systems (i.e. iOS and Android) comprise a minimum set of Assistive Technologies (ATs) that provide users with functional impairments with solutions that provide alternative interaction methods that suits their needs. However, there is still a long way in order to ensure fully accessible mobile devices.

FUTURE RESEARCH DIRECTIONS

In the last decade mHealth research has focused mainly on health promotion, self-management and communication of specific diseases as diabetes, obesity or mental health. These conditions account for more than half of the examples reported in the scientific literature (Fiordelli, 2013). However, the evidence for the health benefits of mHealth has been built not through traditional, large-scale and time-intensive randomized controlled trials, but by a small pilot projects. Rapid technological innovation has led to a proliferation of untested methods, which are small-scale and, for the most part, have little policy coordination (Heerden, 2012). The latter authors stated seven key recommendations to define the research agenda of mHealth reducing resource wastage and programme duplication. These recommendations include the need to generate evidence based solutions and validate in large scale trials. Besides, mHealth should be interoperable with existing eHealth initiatives; and it should adopt and implement the same standards. The implementation of mHealth needs to be participatory and needs to involve all relevant stakeholders (users, clinical professionals, policy makers, mobile operators, etc.). mHealth programmes also need to contribute to sustainability of healthcare and equity in health. And last, mHealth needs to focus on health rather than on the technology.

Nonetheless, technological research is on the spot itself, including the miniaturization of sensors, new materials, new interaction technologies and methods, better communication features and standards; and in data mining, known as big data.

Big Data refers to data sets with sizes beyond the ability of commonly used software tools to capture, curate, manage, and process data within a tolerable elapsed time (Snijders, 2012). Most definitions of Big Data focus on the size of data in storage. The volume of data may be the most relevant characteristic of Big Data, there are other important characteristics of Big Data, namely data variety and data velocity. The three Vs of Big Data (volume, variety, and velocity) constitute a comprehensive definition (Laney, 2001) (Russom, 2011).

mHealth can facilitate the mining of large amounts of health data (e.g. of measurements, medical images, symptom descriptions) which can be stored in large databases with the potential to boost healthcare research and innovation (European Commission, 2014). As an example, according to Pentland et al. (Pentland, 2009), personal sensor data are expected to grow from 10% of all stored information to approximately 90% within the next decade.

It is expected that Big Data will contribute to epidemiological research. This will allow researchers and scientists to improve patient treatment by looking for patterns on a larger scale or by achieving new conclusions, for example, through the relation between the development of a medical condition and environmental factors. Big data can also contribute to a reduction of trial periods for medication or to the development of more advanced mechanisms for early detection and prevention of diseases (European Commission, 2014).

CONCLUSION

mHealth represents the particularisation of eHealth through the use of mobile devices and technologies. It is one of the most important emerging markets across the globe, with different opportunities, challenges and barriers depending on the region. In higher-income countries, the necessity to reduce the healthcare costs, mainly associated to the demographic change, and the increase prevalence of chronic diseases, has driven the investment on mHealth. There is much evidence that through mHealth healthcare provision may be more efficient and costs could be reduced. Therefore, many states are starting to invest in mHealth. In Europe, the European Commission is allocating effort and funds in mHealth, and even it is starting to work in the implementation of the needed legislation. On the other hand, in lower-income countries the need to provide care to the citizens where there are not available infrastructures in place has been the main driver. There are also differences in the type of services deployed. In higher-income countries there are services focused on chronic patients themselves, whereas in lower-income countries most services aims to improve the efficiency of the healthcare workforce and to prevent infectious diseases.

So far, the advances on the technology have been the main driver to push the deployment of eHealth as this happened in other sectors as banking or leisure. However, the demand in this case did not adopt the technology so easily, due for instance to legal or financial barriers. Mobile technologies in the recent years have sake the technological arena with so intensity that now the final users, the citizens themselves are pulling the development and integration of new applications; forcing the healthcare actors and managers, and policy makers to take into account mHealth in their agendas.

For many years PHS did not take off until the mobile devices, especially the smart phones were in almost every pocket in the world. With a mobile, the user may fulfil the requirements that were not solved by traditional PHS: the personalisation, the intimacy and the mobility. These characteristics are key to adopt definitely mHealth as a reality to provide healthcare, allowing the active participation on the persons, making the shift to a more proactive care.

Today, the dominant form of mHealth is characterized by small-scale pilot projects that address single issues in information sharing and access. There are only limited larger mHealth implementations (primarily supported by public-private partnerships). However, some challenges still remain. For instance, the adoption by citizens in higher-income countries is growing in the last years, but older population are in most cases chronic patients who may experience problems to use mHealth. There, usability and accessibility are crucial factors in order to allow them benefit from new services based on mobile technologies.

Data protection is another issue at stake that may hinder the deployment of mHealth solutions. mHealth solutions may collect big quantities of information that is personal data which have to be protected according to law. For this, all stakeholders including legal actors have to assure security principles.

Regarding the future, the research community in the world is seeking for new applications of mobile Big Data integrating the information that can be gathered from the mobile devices and their sensors. Also, more and more researchers are validating mHealth at larger scale including clinical trials which will open definitely the spread of mHealth both at commercial and at healthcare level.

REFERENCES

2008: Population projections 2008–2060 in EU27. (2008). Eurostat.

Barlow, J., Singh, D., Bayer, S., & Curry, R. (2007). A systematic review of the benefits of home telecare for frail elderly people and those with long-term conditions. *Journal of Telemedicine and Telecare*, *13*(4), 172–179. doi:10.1258/135763307780908058 PMID:17565772

Boehm, E. (2011). *Mobile Healthcare's Slow Adoption Curve*. Forrester Research, Inc.

Bonfiglio, S. (2012). The Role of ICT in a Healthcare Moving from "Clinical-Centric" to "Patient-Centric". In Impact Analysis of Solutions for Chronic Disease Prevention and Management (pp. 250-253). Springer Berlin Heidelberg.

Cocosila, M., Coursaris, C., & Yuan, Y. (2004). M-healthcare for patient self-management: A case for diabetics. *International Journal of Electronic Healthcare*, *1*(2), 221–241. doi:10.1504/IJEH.2004.005869 PMID:18048222

Codagnone, C. (2009). Reconstructing the whole: present and future of Personal Health Systems.

Directive 95/46/EC. (1995, October 24). European Parliament.

European Commission. (2014). Green Paper on mobile health ("mHealth").

Eysenbach, G. (2001). What is e-health? *Journal of Medical Internet Research*, *3*(2), e20. doi:10.2196/jmir.3.2.e20 PMID:11720962

Fiordelli, M., Diviani, N., & Schulz, P. J. (2013). Mapping mHealth research: A decade of evolution. *Journal of Medical Internet Research*, *15*(5), e95. doi:10.2196/jmir.2430 PMID:23697600

Fonkych, K., & Taylor, R. (2005). *The state and pattern of health information technology adoption*. Rand Corporation.

Frost & Sullivan. (2008). *Wireless Healthcare Technologies in Europe*. Mobile.

Heerden, A. V., Tomlinson, M., & Swartz, L. (2012). Point of care in your pocket: A research agenda for the field of m-health. *Bulletin of the World Health Organization*, *90*(5), 393–394. doi:10.2471/BLT.11.099788 PMID:22589575

Henry, S. L., Abou-Zahra, S., & Brewer, J. (2014, April). The role of accessibility in a universal web. *Proceedings of the 11th Web for All Conference* (p. 17). ACM. doi:10.1145/2596695.2596719

Holden, R. J., & Karsh, B. T. (2010). The technology acceptance model: Its past and its future in health care. *Journal of Biomedical Informatics*, *43*(1), 159–172. doi:10.1016/j.jbi.2009.07.002 PMID:19615467

IDC (2011). *Worldwide and U.S. Mobile Applications, Storefronts, Developer, and In-App Advertising 2011-2015 Forecast: Emergence of Postdownload Business Models*. International Data Corporation - IDC Corporate USA.

GSMA Intelligence. (2015). Definitive data and analysis for the mobile industry. Retrieved from https://gsmaintelligence.com/

Internet live statistics. (2015). Retrieved from http://www.internetlivestats.com/internet-users/

Istepanian, R., Laxminarayan, S., & Pattichis, C. S. (2006). *M-health*. Springer Science+ Business Media, Incorporated.

ITU. (2013). ICT Facts and Figures. Retrieved from http://www.itu.int/en/ITU-D/Statistics/Documents/facts/ICTFactsFigures2013-e.pdf

Laney, D. (2001). 3D data management: Controlling data volume, velocity and variety. *META Group Research Note, 6*.

Mair, F. S., May, C., Finch, T., Murray, E., Anderson, G., & Sullivan, F. et al. (2007). Understanding the implementation and integration of e-health services. *Journal of Telemedicine and Telecare*, *13*(5suppl 1), 36–37. doi:10.1258/135763307781645112

May, C., Mort, M., Mair, F., & Williams, T. (2001). Factors affecting the adoption of telehealthcare in the United Kingdom: The policy context and the problem of evidence. *Health Informatics Journal*, *7*(3-4), 131–134. doi:10.1177/146045820100700304

Mobile tracker 2015: http://www.mobileworldlive.com/mhealth-tracker. Retrieved on January 2015.

PatientView. (2012). A review by patient groups and empowered consumers. Retrieved from http://myhealthapps.net/

Pentland, A., Lazer, D., Brewer, D., & Heibeck, T. (2009). Improving Public Health and Medicine by use of Reality Mining. *Studies in Health Technology and Informatics*, *149*, 93–102. PMID:19745474

PWC. (2005). *HealthCast 2020: Creating a sustainable future*. Price Waterhouse Coopers.

PWC. (2012). *Emerging mHealth: Paths for growth*. Price Waterhouse Coopers.

Research2Guidance (2013). *The mobile health global market report 2013-2017: the commercialisation of mHealth apps* (Vol. 3).

Russom, P. (2011). *Big data analytics*. TDWI Best Practices Report, Fourth Quarter.

Snijders, C., Matzat, U., & Reips, U.-D. (2012). Big Data: Big gaps of knowledge in the field of Internet. *International Journal of Internet Science*, *7*, 1–5.

Socio-economic impact of mHealth. *Touching lives through mobile health - Assessment of the global market opportunity*. (2013).

Steventon, A., Bardsley, M., Billings, J., Dixon, J., Doll, H., Hirani, S., ... & Newman, S. (2012). Effect of telehealth on use of secondary care and mortality: findings from the Whole System Demonstrator cluster randomised trial. *BMJ: British Medical Journal, 344*.

Stroetmann, K. A., Kubitschke, L., Robinson, S., et al. (2010). How can telehealth help in the provision of integrated care? World Health Organization. Regional Office for Europe, 2010.

Villalba, E., Casas, I., Abadie, F., & Lluch, M. (2013). Integrated personal health and care services deployment: Experiences in eight European countries. *International Journal of Medical Informatics*, *82*(7), 626–635. doi:10.1016/j.ijmedinf.2013.03.002 PMID:23587432

Villalba, E., Salvi, D., Ottaviano, M., Peinado, I., Arredondo, M. T., & Akay, A. (2009). Wearable and mobile system to manage remotely heart failure. Information Technology in Biomedicine. *IEEE Transactions on*, *13*(6), 990–996.

Vishwanath, S., Vaidya, K., Nawal, R., Kumar, A., Parthasarathy, S., & Verma, S. (2012). *Touching lives through mobile health-Assessment of the global market opportunity*. Pricewaterhouse Coopers.

World Health Organisation. (2011). mHealth (Vol. 3). Geneva, Switzerland: WHO Press.

KEY TERMS AND DEFINITIONS

Big Data: Large collection of data sets from different sources that traditional data processing tools are not able to capture, curate, manage, and process data within a tolerable elapsed time.

Digital Health: It is a newer term similar to eHealth or Health It, including more wireless sensors, genomics, imaging, and health information systems.

eHealth: The most commonly term used referring to any usage of information and communication system or technology related to health and healthcare.

Empowerment: To give responsibility to the patients and citizens to cooperate with their own care.

mHealth: The particularisation of eHealth when this is supported by mobile technologies and devices.

Personal Health Systems: Health systems that place the individual at the centre, creating product or services that are personalised and produce empower patients or citizens.

Telecare: The particularisation of eHealth to provide social care. This term is no longer in use.

Telehealth: A broader concept which includes also prevention and patients.

Telemedicine: The particularisation of eHealth to those services mainly focused on interchange of information amongst professionals. This term is no longer in use.

Usability: A characteristic that picture to which extent a product or service or app can be used by users with effectiveness, efficiency, and satisfaction being satisfied.

Chapter 8
Mobile Big Data:
A New Frontier of Innovation

Shivom Aggarwal
IE Business School, Spain

Abhishek Nayak
IE Business School, Spain

ABSTRACT

Mobile technologies have given rise to tremendous amounts of data in real-time, which can be unstructured and uncertain. This growth can be attributed as Mobile Big Data and provides new challenges and opportunities for innovation. This chapter attempts to define the concept of Mobile Big Data, provide description of various sources of Mobile Big Data and discuss SWAI (Sources Warehousing Analytics Insights) model of Big Data processing. To understand this complex concept, it is important to visualize the Big Data ecosystem, respective players. Moreover, mobile computing, Internet of things, and other associated technologies have been discussed in light of marketing and communications based applications. The current trends in Mobile Big Data and associated value chain help us understand where the next frontiers of innovation are and how one can create value. This is linked to the future aspects of the Mobile Big Data and evolution of technologies from now onwards.

1. INTRODUCTION: WHAT IS MOBILE BIG DATA?

Mobile technology has exploded in the past decades from a research tool available in advanced economies to handheld devices in developing nations to access internet. The growth has been fueled by reduction in hardware costs, accessibility of services, increase in computation power (in range of 200 billion times since 1980s) and increase in usage for day-to-day activities. On the other hand, people from different age groups, especially teenagers and millennia generation have learnt to use these mobile technologies in creative, collaborative and collective manner. Moreover, the mobile manufacturers have built in a series of sensors to facilitate the usage of phones in a diverse ways, as well as, network externalities worked in the favor of developers creating millions of applications for the same. Some of the examples of such sensors are GPS, infrared, NFC, accelerometer, gyroscope, microphone, high resolution camera and

DOI: 10.4018/978-1-4666-8838-4.ch008

Bluetooth. This created a fertile ground for data to be generated, captured and analyzed in a systematic manner. And companies, NGOs, governmental entities, inter-governmental agencies (such as United Nations) are finding this new challenge of massive, real time, unstructured data quite intriguing and highly valuable at the same time. This forms, what we can assert as "Mobile Big Data", but to understand this complex concept, we first have to fathom the notion of Big Data itself and then, we can visualize Mobile Big Data as a special case of the same.

This chapter attempts to define the concept of Mobile Big Data, then provides detailed description of accompanied sources and respective applications. To understand this complex concept, it is important to visualize the Big Data ecosystem, described in the next section. Then, we discuss the Mobile Big Data value chain and associated disruptive technologies. This brings us to the final section comprising of a holistic perspective and put forth insights on Big Data influence in policy making and business decisions (with a specific case of Mobile operators). We take the case of mobile operators who have ready access to huge volumes of subscriber data, their data and network usage, consumer behavior, spending patterns and other activities. Moreover, this access to individual level data is also enhanced by their role as last mile connection provided where the new information is gathered continuously and accurately. In sum, we conclude and discuss the future aspects of Mobile Big Data.

1.1 The Concept of Big Data

Big Data is a complex and elusive concept. From epistemological perspective, Big Data is often referred to as a term with definition based on the size of dataset, although there is no formal or informal demarcation above which a dataset shall be considered "Big", but as of 2014, may be larger than some terabytes. Big Data can be seen in the light of not only size but also heterogeneity, incompatibility, unstructured formats, real-time generation & analysis and other accepted technical possibilities. This has fuelled the popularity and pertinence of the term in public policy, corporate plans, and research agendas in scientific or philosophical manner. A more abstract definition could be "data whose size forces us to look beyond the tried-and true methods that are prevalent at that time" (Jacobs, 2009). In other words, "Big Data refers to datasets whose size is beyond the ability of typical database software tools to capture, store, manage, and analyze" (Manyika, et al., 2011). But according to these definitions, it ought to be a moving target which depends on the current stage of technology development and associated parameters will continuously change over the course of time. One of the most important parameter that must be used to define "Big Data" is based on the structural construct of the dataset. Traditional relational database managements systems have inbuilt capabilities to store, manage and analyze quite large enough datasets, subjected to one attribute, the datasets should be structured enough. But, nowadays, the data created via social media, different device interfaces in context of smart cities, different public agencies, etc. is highly unstructured and entails a completely different set of technologies to store, manage and analyze it. A more eloquent definition of "Big Data" can be stated as "a large dataset with/without a combination of structured and unstructured data, which cannot be managed/analyzed by traditional database management methodologies."

On the other hand, (Feijoo, Gómez-Barroso, & Aggarwal, 2015)provided an economic-based perspective to the understanding of the Big Data concept. Viewing from the lens of value, it has been declared as a new asset class (World Economic Forum, 2012)[1] and the basis of "a drift toward data-driven discovery and decision-making" (Lohr, 2012). Moreover, a new influx of data has been encountered in the form of "Open Data" which adds another layer of complexity to the existent discord. "Open Data" can be defined as "a dataset available/produced in a machine readable format with interoperability features such that it can be used by any entity or person without a need to further process it and it follows and/or adheres to the open standards of data storage/movement over the technological mediums." With the Open data movement, there are many governments all around world, who are providing open data through different agencies such as Open-data.europa.eu (European Union Open Data Portal, 2011)[2], Data.gov[3] (US Government GSA, 2013) and Data.gov.uk[4] (UK Government, 2013). The Podesta review report[5] concludes that government data should be viewed as national resource and responsibly made available for deriving social value from it. The added level of complexity is generally overlooked in the academic literature, due to the different types, formats, structure, file systems, etc. of the data provided by different public agencies, government departments, NGOs, non-profit organizations, corporations of disparate forms. Thus, this open data along with varied level of complexity takes a form of "Open Big Data" (Marton, Avital, & Blegind Jensen, 2013). The "Open Big Data" can be measured on two different planes of parameters, the first (Marton, Avital, & Blegind Jensen, 2013) level of distinction can be based on dimensions of "order" and "relationality", and second can be the underlying structure of the overall dataset. Nevertheless, we can comprehend the Open data subsumed within the over-arching term of Big Data itself.

As (Feijoo, Gómez-Barroso, & Aggarwal, 2015) summarize the Big Data concept in the form of ensuing dimensions[6], namely:

- Volume, a measure of the scale of the data such that above a consensus based limit, a given dataset can be considered "Big".
- Variety, a measure of the forms or underlying structural dimension which encompasses the technological aspects of compatibility. The data could be structured, semi-structured, unstructured or combinations of either.
- Velocity, a measure of the speed and direction of the incoming or outgoing data-streams, especially in real-time.
- Veracity, a measure of uncertainty & validity associated with a given dataset as most of the datasets are collected for very specific purposes, while we use them for different domains (far from the original one). This brings measurement errors, inaccuracy, validity and quality into perspective.

Another distinction that need to be made apparent is with "data science", which could be viewed as an umbrella label encompassing all aspects of "Data" in general using a scientific enquiry, including Big Data itself. Data science is not just about statistics and analytical methodologies available to generate insights from the data, as the data in itself has changed its characteristics over time (Dhar, 2013). In sum, the key issue revolves around what to do with the data that can be useful rather than just acquiring large amounts of data.

From a pragmatic perspective, the "Big Data" concept can be encountered with use-case approach, especially domain specificity. One such (and very crucial) specific domain is Mobile technology. Aforementioned Mobile Big Data can now be defined (in the light of above discussion) as a large dataset generated by mobile technology (devices, applications, computing, etc.) with/without a combination of structured, semi-structured and unstructured data, generated/analyzed in real-time or not, which cannot be managed/analyzed by traditional database management methodologies. Mobile Big Data will become more important going ahead as the usage of mobile technology is on the rise (at a tremendous pace) and this will bring more challenges than answers. And also allied definitions & concepts will mutate accordingly, although, we can fairly assume the core notions won't change.

1.2 Key Sources of Big Data

One of the key issues with Big Data is to figure out where it is generated and based on that, we bring the analytical expertise to build suitable solutions. At first we can delineate the sources of Big Data based on the position of the source:

- **Internal Sources:** This consists of data sources that reside behind the firewalls of the organizations (could be public or private).
- **External Sources:** This are the data sources outside the environment of any given organization and also includes the Open Data provided by government/ non-government entities.

This simple classification of the sources don't do justification to the wider scope of Big Data. Thus, we need a more detailed distinction on the basis of real-life sources, in order to understand it better. Such an assortment can be built with bottom-up approach and by identifying distinctive sources (or potential sources), such as[7]:

- **Archives:** Contain the historical system of records of organizations, customers, outside as well as internal environments. These may include scanned documents, statements, forms, medical records, customer correspondence, financial statements, past product descriptions, etc.
- **Machine Logs:** Contain log data created by various mechanisms, such as, event logs, server logs, application logs, business process logs, call detail records, mobile locations, mobile usage logs, etc.
- **Sensors:** Generate new sources of data, collected for various purposes by corporations or governments (and are often real-time). Some of the examples are smart electric meters, GPS, mobile accelerometer, mobile gyroscope, microphones, road cameras, car sensors, household appliances, cable boxes, assembly lines, traffic recorders, etc.
- **Social Media:** This brings tremendous amount of most unstructured data such as posts, micro-messages, online feeds, pictures, videos, etc. Some of key examples are Facebook, Twitter, Instagram, LinkedIn, Google+, Pinterest, Chatter, Yammer, etc.
- **Software and Applications:** Consist of enterprise-level software such as ERP (Enterprise Resource Planning), CRM (Customer Relationship Management), project management, HR (Human Resource) management, revenue management, inventory management, etc. as well as personal applications such as documents processors (Google docs, MS Word), web browsers, games (mobile, console, desktop), music apps, wen conferencing, etc.
- **Public Sources:** Include statistics, census, and other holistic datasets, such as data on weather, healthcare, financial markets, public finances, population indexes, etc.

Upon identifying the key sources of Big Data, it becomes more evident what kind of data we are dealing with and helps in creating viable analytical strategies. As we can see, many of the data sources are predominantly based on (or related to) mobile technology and thus, come under the umbrella of Mobile Big Data. Thus, a new frontier of innovation emerges in the form of Mobile Big Data and brings its own challenges and diaspora of opportunities.

2. BIG DATA ECOSYSTEM

Evolution of Big Data ecosystem was not linear, but rather random, exponential and complex, as the technologies required at different stages of Big Data processing described later in Section 2.1, were developed not in tandem with each but quite distinctively. This gave rise to multiple options and possibilities at every stage. Moreover, this lead to the inception of disparate types of players in the landscape such as large multi-nationals and innovation based tech startups. Understanding the Big Data ecosystem is essential as the vast proliferation of technologies in this competitive market has left corporations and decision makers in quandary of how to build a Big Data strategy and what solutions (or combination of solutions) to adopt at each stage of processing.

2.1 Stages of Big Data Processing

Big Data processing entails several stages to tackle the inherent complexity due to four dimensions of Volume, Variety, Velocity and Veracity (as discussed above). We need a simplified model to fathom these stages. SWAI (Sources Warehousing Analytics Insights) model (as shown in Figure 1) is a sequential model with four stages such that prior stage has to be completed before moving onto the other. For the purpose of clarification, we are not saying that none of the stages can co-exist, which may happen in a large organization where different tasks are performed simultaneously. Rather, SWAI model assumes that there is one massive dataset fulfilling the criteria mentioned above (to define Big Data) and in order to process this given dataset, following stages are required.

Figure 1. SWAI model of big data processing

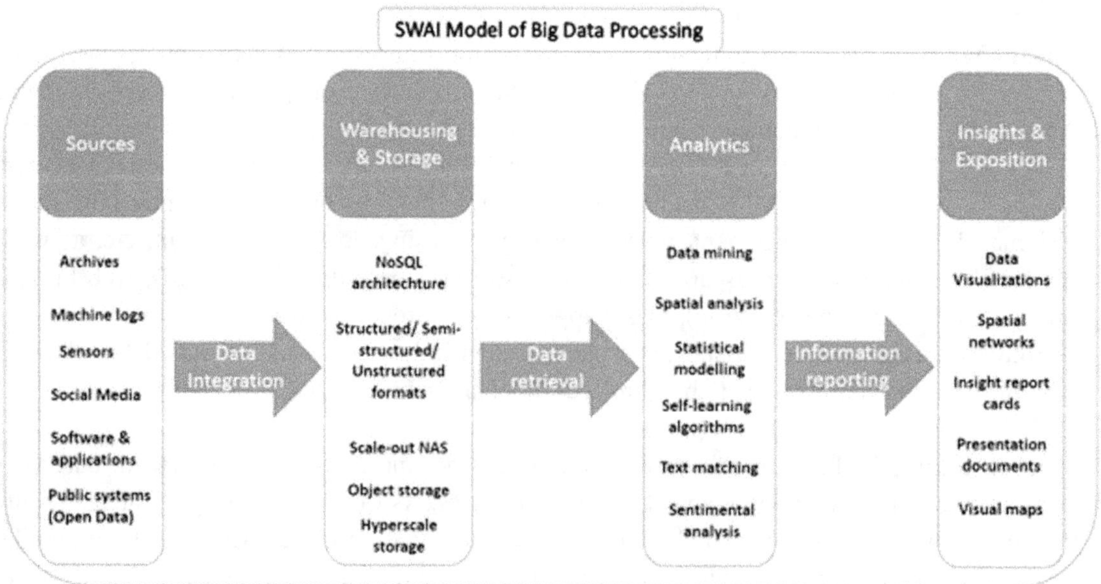

- **Sources:** Data can be sourced from disparate sources and an exhaustive classification has been provided in Section 1.2. It can be extracted or fed from different sources (namely, Archives, Machine logs, Sensors, Social media, Software & applications and Public systems) simultaneously/sequentially. The key here is to first identify the respective sources and then, based on specific business goal/objective, select some or all the sources. Upon selection of sources, the data may be of the form of Big Data (satisfying the above discussed definition). Next, this data is integrated into a DataStream or Data-blocks and is stored.
- **Warehousing (and Storage):** This stage is crucial in terms of storing the data (as it may determine the efficiency of all the following stages). Data warehousing is used to separate analytical workload from transaction workload and further enables to consolidate data from several sources. In many cases, ETL (Extraction Transformation Loading) step is included within the warehouse environment. Preprocessing of data in order to establish validity and reliability is performed here, which can be viewed as a part of transformation of data, in accordance with predefined rules (follows the logic of business goals as specified in previous step). Big Data is then loaded into a form of data marts that can be physically located anywhere and follow principles of distributed computing. Now, the storage is very different from traditional (relational databases), especially due to the underlying structure of the Big Data. It is generally stored in NoSQL (No Structured Query Language) architecture in order to provide a pertinent environment for unstructured data. Some of the examples of possible Big Data storage are Scale-out NAS (Network Attached Storage). It represents a data storage structure (at file-level) connected to a computer network handling multiple heterogeneous clients simultaneously. Another example is that of Object storage – a horizontal data structure which is in contrast with hierarchical file system, such that here files are located by unique identifiers. Next is Hyperscale storage – a hyper-dimensional storage structure with numerous commodity hardware-based nodes of compute. Then, data is retrieved, although not necessarily through structured queries, for the purpose of analysis.
- **Analytics:** This is the core stage of SWAI model. This stage consists of various techniques such as statistical modeling, data learning, spatial analysis, sentimental analysis, self-learning algorithms, text matching, etc. But the choice of technique (or a combination of techniques) depends on two key parameters, namely, characteristics of the data under exploration and business (or research) question to be investigated. This layer may also include analysis of behavioral patterns and theories established in various scientific disciplines such as complexity theory, game theory, network theory, etc.
- **Insights (and Exposition):** This is the final and context based stage. Different forms of outputs for insights generated from analytics stage are built or coded here. These insights could be in the form of data visualizations (to help non-technical actors to understand the insights), spatial networks (when underlying insights are inherently networked), standard or graphic report cards, presentations and visual maps. This stage acts as a bridge that connects plain data with rich information that can be used to make better decisions.

Although the stages of Big Data processing are well defined, but the evolution of players have been not been linear and brings another layer of complexity. Thus, in order to polish our understanding of Mobile Big Data, we need to explore various players and their roles in the Mobile Big Data landscape.

Figure 2. Big Data players[8]

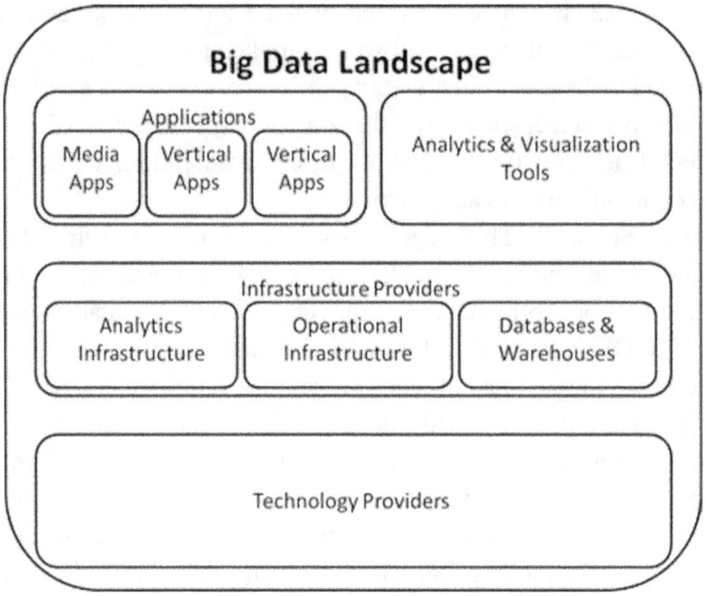

2.2 Mobile Big Data Landscape: Big Data Players and Their Roles

(Feijoo, Gómez-Barroso, & Aggarwal, 2015) provide a comprehensive typology of players in Big Data landscape and their respective roles. We will restate the typology provided and analyze the players to understand what kind of companies are there, what they are doing, how their solutions complement, supplement or alternate with others.

We have already discussed the SWAI model that describes different stages of Big Data processing. In hindsight, we have observed that the evolution of companies capturing different stages is not linear, rather very complex. The evolution didn't happen sequentially but rather simultaneously. For instance, it wasn't the case in which first all the players, targeted Warehousing stage and then new players or same players developed solutions in Analytics stage, but rather both stages began together. This simultaneous development of both aspects and constant player switching between the two formed the current landscape. The different analytical capabilities have been used to understand the new form of data management structures that inherit big data landscape are designed to meet the velocity, volume, variety and veracity requirements of the underlying data. To handle such disparate datasets, a new wave of offerings has culminated into a diaspora of vendors (Figure 2), which can be classified broadly into four categories as defined by (Feijoo, Gómez-Barroso, & Aggarwal, 2015):

- **Technology Providers (TP):** Companies found that will have to first build core technologies to fuel and support the Big Data infrastructures. This gave rise to the domain of Technology Providers. This entails a convergence in coherence between base providers and users/providers who built on top of these technologies (Lind & Zmud, 1991). Due to lack of capabilities of RDBMS (Relational Database Management Systems) to handle unstructured, the solutions offered by these players use nodal structures, massive parallel processing, NoSQL queries, etc. From a perspective of Mobile

Big Data, these players are very crucial as they tend to determine the pace as well as direction of technological evolution in nascent field of Mobile Big Data. Some of the potential players which offer such solutions are Hadoop, MapReduce, Cassandra, etc.

- **Infrastructure Providers (IP):** In case of Mobile Big Data, we need players that provide back-end architecture. Such architecture can be leveraged for storage, analyses and retrieving of data (which can be structured, semi-structured, unstructured or some combination of three types). Infrastructure providers evolved in parallel with advancements in technology providers' category. Some of the players offer standalone solutions with massive data warehousing capabilities and others offer combo-solutions for data storage and handling. They can be further sub-classified as Analytics Infrastructure (AI), Operational Infrastructure (OI) and Databases & Warehouses (D&W) (Feinleib, 2012). Due to the high velocity aspect of most of the Mobile Big Data, infrastructure providers face many challenges with respect to working with archival datasets. Some of the players which offer Mobile Big Data solutions are: ORACLE, SYBASE, MYSQL, Cloudera, TeraData, InfoChimps, etc.
- **Analytics and Visualization Tools (A&VT):** These players form the core of the Big Data value chain and offer variety of solutions. With the objective of supporting human decision making with automated algorithms, sophisticated analytics has the potential to substantially improve decision making, minimize risks, and unearth valuable insights that would otherwise remain hidden (Manyika, et al., 2011). Key offerings provide methodologies based on different theoretical paradigms with objective of creating Business Intelligence. These tools tend to have Graphical User Interface (GUI) and client-server architectures. Mobile Big Data requires tremendous amount of computation power and sophisticated analytical techniques, thus, many of the players in this category are large corporations such as: SAP, Hyperion, Microsoft, DataSpora, TIBCO, Ayata, Platfora, COGNOS, HP (Autonomy), etc.
- **Big Data Applications (BDA):** This is the most divided category and is continuously evolving at a very high pace. Most of the players are new innovative startups and have developed targeted solutions. Some of the solutions as described by (Feijoo, Gómez-Barroso, & Aggarwal, 2015): are Social Media applications for real-time consumer targeting, studying consumer trends, Vertical Applications for interpreting consumer demand and deliver relevant user experiences, Log Data Applications for log management services or even Data-as-a-service through cloud infrastructure. Along these lines some identifiable niche players are: Loggly, Media Science, Bloomreach, INRIX, GNIP, etc.

Since digital data can be transferred, processed and replicated, it presents tremendous reuse opportunities, speeding up the scientific investigations and better management decisions (Lynch, 2008). This Big Data Landscape is continuously evolving and brings new challenges every day, thus attracting new forms of players to counter these challenges. There are too few data scientists to go around, the technologies are new and in some cases exotic, but both the trends in technology and in the business payoff are unmistakable. There is strong evidence bolstering the fact that data-driven decision makers are better off (McAfee, Brynjolfsson, Davenport, Patil, & Barton, 2012). On the other hand, Mobile Big Data is bringing equal amount of opportunities and challenges, and we need to identify and build on them. In the next sections, we will discuss the concept of mobile computing and how it fits with big data value chain, some of the key trends in mobile usage and tremendous growth in mobile applications, increasing smart phone penetration and mobile becoming the primary data source in future for private as well as public sector. We will introduce the concept of internet of things briefly and how it is related to Mobile Big Data and how such data can be used for real-life applications.

3. MOBILE COMPUTING AND BIG DATA VALUE CHAIN

Mobile computing as a concept can be defined as "process of computation on a mobile device such as a smart phone, tablet, hand-held devices and other small wearable devices through which a set of distributed computing systems or service provider servers participate, connect, and synchronize through mobile communication protocols"[9] (Koudounas & Iqbal, 1999). In other words, it is a generic term used to refer to a variety of devices that allow people to access data and information from where ever they are[10,11]

More recently, the concept of mobile computing has been linked with the concept of ubiquitous computing that essentially focuses on the concepts of 'portability' or 'mobility' that allows user to use a computing device anytime and communicate with anyone, even when they are on move. Rapid development in the field of mobile communication and technological devices has increased the importance of mobile computing. Moreover, data generated by numerous transactions happening on these devices and the potential of harnessing valuable insights from this data have led to the creation of an entirely new field of research.

This is where a bridge has been created between mobile computing and the data that it produces. Researchers and practitioners alike have started to look at this combination from a Big Data perspective where they are looking at this large-scale mobile data as input to find valuable insights about human communication, interaction and individual traits. These insights can be eventually applied to determine and predict consumer purchase behavior and future intentions. The figure 3 describes how mobile computing seamlessly integrated into the Big Data value chain. It also provide in insight on the process through which huge volumes of data are converted into valuable insights.

3.1 Mobile Big Data Value Chain

The value chain of Mobile Big Data is very similar to that of Big Data in general. The key objective is to 'extract value from the volume of data' (Figure 3).

Mobile Big Data Production

At this stage, data is in a raw unprocessed form, i.e., unstructured or semi-structured. Moreover, format of data from different sources is not standardized. As described earlier in Section 1.2 & 1.3, sources are the key and have to be identified and harmonized in order to proceed to other Big Data processing stages. Various sources of data within mobile can be combined to get valuable insights about the individual and his/her interaction and consumption patterns. GPS or maps that provide information about the current location of the individual and moreover his past activities, numerous mobile apps, social media usage and various web applications are key sources of data as far as mobile devices are concerned. The key stakeholders or data handlers at this stage are mainly mobile operators and application developers.

Figure 3. Mobile big data value chain

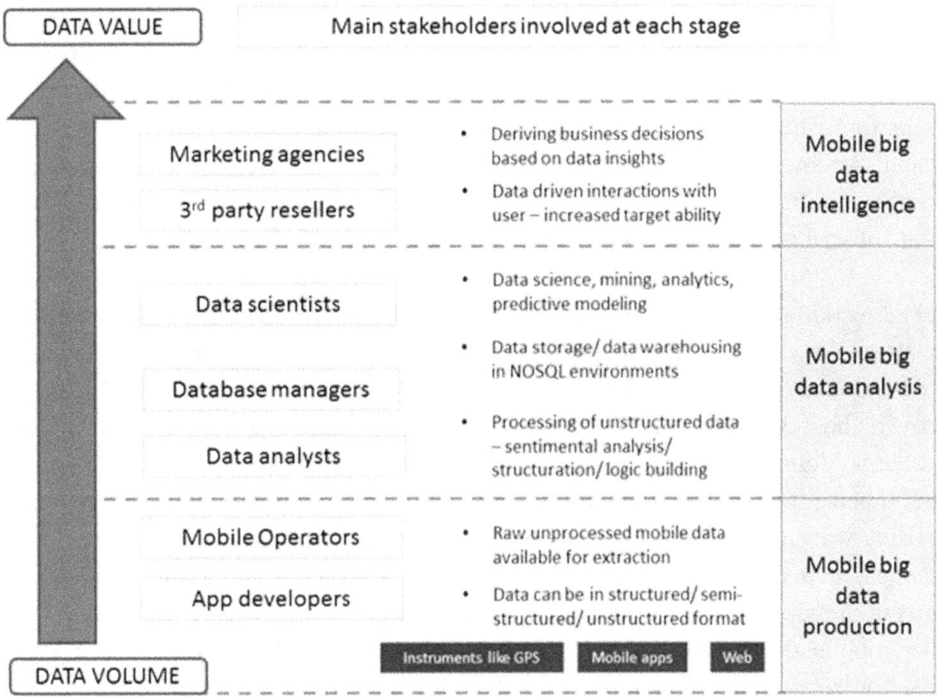

Mobile Big Data Analysis

As we go up the value chain, data is processed and stored in the appropriate formats. In this stage, unstructured data obtained from various sources is converted into a form which may have some underlying structure (depending on the business objective). The main stakeholders at this stage are mainly mobile operators and data centers.

Aggregated data is available for analysis to data scientists. They could be both a part of mobile operators and/or any third party analytics company. At this stage, data scientists look for general patterns in data. Once some relationship is identified between variables, they use predictive models to test the strength of the correlation and the robustness of the predictive ability of the model.

Mobile Big Data Intelligence

Once the insights are available from the previous stage, they could be either used by mobile operators for their own value added offers or they could be sold to marketing agencies or third party resellers. Key idea of this stage is to derive business level information from the variable level information. This information can be used by various parties to create better interactions, value offerings and enhanced targeted customer service.

3.2 Key Trend in the Mobile Usage and Applications

Mobile usage has increased rapidly and some of the trends are path breaking and have important implications for Mobile Big Data extraction. Adoption of smart phones by the consumers has been increasing at a rapid rate. It is estimated that number of mobile connected devices would reach 10 billion by 2020 (Chen, Chiang, & Storey, 2012). Majority of the mobile traffic now belongs to internet data usage and hence a deep insight into how consumers are using the data becomes a necessary requirement. Latest reports about subscriber usage behavior provide following valuable insights[12]:

- Social networking is the most used activity on mobile phones and accounts for 50 percent of data usage alone. About 40 percent people who have smart phones use it for social networking and this number is bound to increase in the future.
- Moreover, about 37 percent of the people use data for information and navigation, which provide valuable opportunities in the domain of location based services.
- About 50 percent of smartphone users are served ads. This again shows the potential for targeted advertising when combined with insights from Mobile Big Data.
- Another useful insight into the mobile usage shows the growing importance of video as a medium. 52 percent of mobile data traffic is video content.
- Further, mobile usage trends related to various mobile applications shows a lot of potential for 3rd parties. For instance, 50 percent of the total health apps are related to fitness.
- Finally, another key trend shows how improved technology is further enhancing mobile data usage. About 50 percent of webpages load in 5 seconds or less in 2014. The same number was 8 seconds or less in 2013.
- Moreover, about 33 percent of the people leave the website if it takes 5 seconds to load, describing the changing behavior patterns.

Apart from the usage behavior, one also needs to look at the actions taken by the users. A report by Microsoft suggests that almost 70% of the smart phone users conduct search and take action within an hour of starting[13]. This also necessitates real time processing (in line with the velocity dimension as discussed in Section 1.1) of Mobile Big Data so that insights can be obtained immediately and future actions of the user can be predicted.

3.3 Mobile Big Data and Value for Mobile Operators

Given the huge volume and potential richness of everyday mobile data, one question that lingers on the mind of every mobile operator (i.e., the carrier of data) is what to do with this data and how to extract the rich potential that everyone is talking about. Mobile operator's advantage is further enhanced by their significant reach to all different demographics of customers essentially tapping the huge variance in subscription, data and application usage patterns. Integration services also allow the operators to assemble most complementary services together compared to other providers with single service offerings. Over the years, mobile operators have also developed significant analytics based capabilities that they are already using to provide enhanced services to their users. All these factors could play a huge role in elevating the status of mobile operators as dominant big data players and help them in delivering the best value proposition for their subscribers.

What Mobile Big Data offers today in the form of transactions such as text messages, instant messaging, location based services, video feeds and a plethora of mobile apps is definitely unique. What makes Mobile Big Data unique is the nature of transactions that happen in real time, are location based and generated at a higher pace than never seen before. So given this unique nature and extreme potential of Mobile Big Data there are mainly three questions that the mobile operators are asking to themselves[14]:

- How to tap into this data and what to do with it?
- How this data should be analyzed, cleaned and made worthy of use?
- How to monetize this data?

Apart from these main questions the mobile operators also need to figure out the structure of potential partnerships with third party providers in terms of analyzing data and finding ways to monetize it.

The New Four V's of Mobile Big Data

Before understanding the key opportunities in the Mobile Big Data segment, let's try to understand some factors that are behind the creation of these opportunities. We discussed four dimensions of the Big Data in Section 1.1 (i.e., Volume Variety Velocity and Veracity). But some industry experts feel that the fourth dimension when we are investigating real-time data generated from multiple devices, should be Value created from it. Thus, we discuss the dimensions of Big Data in the specific case of Mobile Big Data. Using the 4 V's framework we can analyze the factors through four themes – Volume, Variety, Velocity and Value[15]

- **Volume:** Just the sheer size of mobile transactions is one of the biggest opportunities in this segment. One example of this humongous growth in the size of mobile transactions is that the mobile payment transaction volume is stipulated to grow to $721.4 billion in 2017 from $163.1 billion in 2012, a whopping 342% growth in just 5 years[16]. This volume of data could be viewed as an opportunity for various parties working in consumer analytics, mobile applications and others.
- **Variety:** The next factor which makes Mobile Big Data more relevant today is the variety of information that one can collect from a mobile. Today, a mobile generates different kinds of data that mainly can be grouped in three types – Structured data such as a payment transaction, semi-structured data or unstructured data from sources such as location based services or use of social media. Companies can combine data from various sources and derive meaningful insights from this large and growing data.
- **Velocity:** Another important factor that makes big data valuable is the speed with it is being created. Due to multi-tasking facilities (using multiple applications such as maps, chat, videos at the same time) available on a mobile phone, users simultaneously generate a huge variety and amount of data At the same time availability of sophisticated tools have also enabled real time analysis of this data. One of the best examples could be - by using real time location based services a marketer can reach a consumer with a coffee deal or coupon on his mobile, assuming he is near a coffee-shop. Consumer can instantly use the coupon to drink a coffee if he is interested. The entire transaction is done in real time with-in minutes and data is available to the marketer for analysis, instantly.

- **Value:** Any type of data could have huge volume and growth, high variety and high speed of collection and analysis, but does it automatically make it valuable? Probably not. This is where Mobile Big Data is different as this data when analyzed could have trend changing implications in the field of marketing and sales, better consumer experience and market changing innovations.

The above factors play a huge role in bringing out the true potential of Mobile Big Data and provide a list of opportunities for various parties mainly mobile operators.

Opportunities in Mobile Big Data for Mobile Operators

Looking back at the value chain of Mobile Big Data, mobile operators can have opportunities to monetize data at each stage of processing. They can start by selling raw data to third parties as they have access to a tremendous amount of user information such as demographics, data usage, bill amount and others. Verizon with around 98 million retail customers has already taken up an initiative to monetize the user demographics. As a part of this initiative, they are targeting big events such as super bowl games in the US. Recently, Verizon showed that fans from Baltimore outnumbered fans from San Francisco by three to one inside the Super Bowl stadium. This type of data, otherwise very difficult to obtain has tremendous value to the marketers. Telefonica, the biggest mobile operator in Spain has also started a similar initiative (Leber, 2013).

In the next stage, of data analysis, mobile operators can choose to invest or not in data processing as predominantly they are data providers. They can either set up their own data analytics centers or collaborate with third party analytics companies to process and analyze the data. As we move up the value chain of Mobile Big Data, operators can definitely gain a lot from the insights available from the previous stage. They can use these insights to create their own value added offerings or sell these insights to marketing agencies or third party resellers. By analyzing the customer activity across the mobile network, operators can improve their infrastructure management and decrease operational costs. Further, they can create cross category value offerings for their own customers.

Telefonica's initiative called 'Dynamic Insights', aggregates data from various sources and generates further understanding on the collective behavior of people in terms of different segments of population (Leber, 2013). This data can be further used for better advertisement targeting, predicting consumer purchase behavior, forwarding coupons and deals at the right moment and creating platforms where consumers can easily transact. Thus, we see that mobile operators have plenty of room where they can pick and choose their involvement in the entire Mobile Big Data value chain. In order to determine their approach they need to undertake a careful assessment of the various challenges that might act as roadblocks in harnessing the full potential of Mobile Big Data. Main challenges are discussed in the next section.

3.4 Challenges in Harnessing Mobile Big Data Potential for Mobile Operators

Mobile operators have a role to play at the source of mobile big data value chain, given their access to huge volume of raw unprocessed data. Further, factors such as integrated services, variety of data sources, high penetration, and infrastructure for data analytics give them a significant advantage as big data players as compared to single source providers. However, these same factors could be deemed as challenges until they realize the expertise and techniques to leverage on their advantage. Some of the main challenges are discussed below.

Variance in the Data

A primary challenge in deriving value out of Mobile Big Data is the variance in the data. Data is collected from different carriers and platforms, different types of devices. The complexity increases due many reasons and sheer number of handsets around. Moreover, different type of applications produce different data and one needs a common platform to analyze such data. It is estimated that the number of mobile applications is expected to reach 44 billion by 2016 leading to different types of datasets (Chen, Chiang, & Storey, 2012).

In another industry report[17] Mobile Big Data has been characterized as vast amount of data that is available to mobile operators into the following types:

- Customer demographic data – Gender, age, contact information.
- Usage data – voice call duration, text message, instant messaging, data downloads and browsing history.
- Transaction data – Purchase history, tariff, bill amount, contract type.
- Marketing data – Campaigns successfully accepted, offers, contact history.

This data can be further segregated to answer 4 questions about consumers:

- Who you are? Demographic.
- What you like? Psychographic.
- What are the characteristics of your network? Socio-graphic.
- What you do? (places you go and activities you pursue) Behavioral.

Potential for Privacy Abuse

Mobile Big Data can be pretty intrusive considering the amount of personal demographic information and behavioral information that is available. There is a potential threat of invasive tracking of individuals as every mobile device can be traced back to the individual using Mobile Subscriber Integrated Services Digital Network Number (MSISDN) or International Mobile Subscriber Identity (IMSI) numbers contained on the phone's SIM card[18].

From the consumers' perspective, response to privacy breaches have been very critical. In a survey by a market research company, Gfk found that 63% find it creepy the way that some companies use information about them, while 90% prefer to do business with companies that respect their policy[19]. However, if companies act responsibly they can find ways to collect this data. Consumers are more willing to share data if a real value is offered in return. Consumers must be able to excise their right through appropriate permission and preference management systems.

Thus, it is a prerequisite that proper privacy controls regulations are in place. This way useful demographic and behavioral information can be obtained at a segment or individual level but without revealing personal information about subscribers.

Integration with Fixed Internet Data Could Be Critical

There is debate on whether to treat mobile audience differently from the web audience in terms of their behavior. However, in absence of any evidence for treating mobile audience differently from web audience, it is safer to assume that both audiences are similar. This however, presents few challenges on its own. Integration of mobile data with web data would be difficult because of the difference in the data structures. Further, it would be difficult to undertake real time analytics of mobile data as compared to fixed internet data because very few companies have the ability to leverage this aspect of Mobile Big Data.

Provide Relevance and Not Redundancy

One of the key challenges is to how to use the insights available from the Mobile Big Data. Companies should strive to provide relevance rather than redundancy. For instance, pushing discount coupons using location based service to take advantage of the consumers when they are near to a certain store is definitely one of the applications of Mobile Big Data. The challenge is that the consumer might not be interested in consuming that product in that particular moment. A high frequency of such instances might give a feeling of 'spam' and the consumer might block the application. This is where Mobile Big Data can go a step further than just providing discount coupons based on location but figuring out whether a consumer is interested in pursuing the purchase at that current moment or not. This can be done by understanding consumer usage patterns and also by letting the consumer choose when he/she wants to be exposed to such information. Thus, giving relevant information at the right time rather than redundant information that can be perceived as 'spam' is one of the biggest challenges of Mobile Big Data application.

3.5 Internet of Things and Mobile Big Data

Internet of things (IoT) can be defined as internet of objects and their virtual representations in an internet like structure (Ashton, 2009). It is based on the unique addressability of a large number of objects which may be RFID-based tags, sensors, actuators, or other embedded devices, which can collect and transmit data in an automated way (Aggarwal, Ashish, & Sheth, 2012). IoT consists of many aspects of day to day life of an individual and tracks individual behavior in an environment of connected homes and cities. IoT can be considered as an evolving 'network of networks' that encompasses domains such as telephone services, security, lighting, energy consumption, transportation and many others[20]. IoT is simply the point in time when more "things or objects" were connected to the Internet than people (definition by Cisco Internet Business Solutions Group (IBSG)).

IoT can be considered as a giant leap in the internet evolution as it has added sensory capabilities to internet (ability to measure temperature, pressure, light sensitivity, vibration, touch, moisture, stress). This way one can already monitor individual behavior on these parameters and help us become more aware of our own usage.

Given the complex nature and the volume of this sensory data, big data tools are required for processing and analysis of this data as the current systems and database are not equipped to deal with such data. Moreover, data analysis gets more complicated with IoT in terms of security and data privacy and interaction as we are dealing with a lot of personal usage and transaction data. Add to this a large number of connected devices (that are increasing in number) and producing different streams of data and we reach a point where one requires distributed, ubiquitous and real time processing (Aggarwal, Ashish, & Sheth, 2012).

Thus, IoT presents another important application where the computing and processing power of Mobile Big Data can be harnessed.

4. CONCLUSION

Big Data is an innovative notion and various conceptualizations provided by scholars render it complicated to understand. But, the allied opportunities are tremendous as we are living in an age where the data will keep growing and the pace of growth will increase as we move into the future. It is wise to conjecture that the progress and advancements in Big Data will fuel not only business growth, but also scientific endeavors such as particle physics, space exploration, biological evolution, genetic structuration, etc. "Big data can be used for great social good, it can also be used in ways that perpetrate social harms or render outcomes that have inequitable outcomes", Podesta report (2014)[21]. This brings the policy making issues at the heart of Big Data usage and value creation discourse. The policy making should consider Big Data as national resource and present effective plans for managing the related issues of privacy, security and education. The increasing use of mobile technologies in day-to-day lives has brought the Mobile Big Data at the crossroads of innovation and we ought to understand it today, because as we go into next decades we will observe the omnipresence of its concepts in everyday business operations. The key dimensions of Big Data – Volume, Variety, Velocity and Veracity – provide a better understanding, which paves way for exploring different sources of the same. Once sources are identified, the appropriate processing is entailed, which is captured in SWAI model discussed above. Based on the processing stages, many players have been identified with respect to their roles in Mobile Big Data landscape.

The current trends in Mobile Big Data and associated value chain help us understand where the next frontiers of innovation are and how one can create value. This is linked to the future aspects of the Mobile Big Data and propositions regarding the evolution of technologies from now onwards. We have also discussed some of the challenges that are faced when we work with complex technologies like Big Data. In sum, the technological advancements will bring new dimensions to the concept, modify trends, change the landscape and new dynamics will emerge. In our opinion, it seems inevitable that Mobile Big Data will become part of everyday decision making from professional to personal level, as the reduction in cost of hardware as well software will make it affordable. The other emerging technologies such as Internet of Things and Artificial Intelligence will act as catalysts in the fast paced evolution of Big Data technologies. This chapter sheds some light on the increasing importance of Big Data and mobile content, in the context of emerging technologies and how it fuels growth of new business models.

4.1 Proposed Strategies for Future

We have already talked about both the internal and external opportunities that mobile operators can pursue leveraging Mobile Big Data. However it is pertinent that they have a full-fledged strategy in place. Extracting value out of Mobile Big Data is not only about setting appropriate IT solutions and analytics systems in place but considering the insights as part of the overall business strategy. Here we have presented some of the key guidelines to create a successful Mobile Big Data strategy.

Identify Key Objectives

The first essential step in developing a successful Mobile Big Data strategy is to clarify the objectives. These objectives should be related to the overall business objectives of the company. Companies can follow a top-down approach by first establishing their objectives and look to find insights from the Mobile Big Data that would help them in achieving their objectives. On the other hand, the company can follow a bottom-up approach where they can just look at the data first and use the insights obtained in setting up objectives.

Establish Proper Technological Infrastructure

Sheer volume and the velocity of the Mobile Big Data have created a situation where the mobile operators need to have an advanced infrastructure in place. Further, they need to constantly upgrade and improve the infrastructure. Companies like Verzion, AT & T, British Telecom and China Mobile are trying to manage massive data on systems that were designed to handle data volume and velocity of five to ten years ago[22]. The limitations of current and legacy infrastructures are quite visible as there is always a trade-off between speed of data loading and its analysis. For instance legacy systems such as Oracle is good at processing but they lag behind in providing very large scale data analytics. Thus, mobile operators need an infrastructure that can handle both volume and velocity without compromising on either.

Collaboration Is the Way to Go

Given the massive volume and variety of data, key partnerships are a necessity in order to link all the pieces of the puzzle. Various stages of data production, cleansing and analysis would require a network of mobile operators, data centers, data scientists, third party analytics and marketers in order to extract maximum value from the Mobile Big Data .

Moreover data applications are not limited to one domain and require extensive breadth of knowledge to find their optimum use. This is where mobile operators can combine with other parties such as banks for instance to combine transaction data and demographics with telephone usage data to make it more valuable to third party analytics companies.

Operational excellence can be achieved as the partnerships can help mobile operators scale up their operations very quickly. The entire value chain from mobile data production to Mobile Big Data intelligence can be trans-passed faster and the mobile operators can move improved applications and products faster to the market.

Enforce Proper Controls and Policies on Data Usage

Mobile operators need to create an environment where customers trust the intentions of the company in terms of the way they use customer's data. There are mainly two aspects of the mobile data usage policy that mobile operators need to understand – Implement proper Security and authorization controls to protect customer's identity and privacy; secondly implement rules and norms about level of data collection about an individual. By ensuring proper permission management systems, consumers can control the amount of data they want to share. In such a trusted environment, consumers would be more open to share data. About the second aspect, mobile operators need to take a call on how they will use the data internally and what should be their privacy protocols on how much personal detail of an individual is available to them.

Use Insights to Improve Business Intelligence

After acquiring proper technology and infrastructure and by ensuring proper controls and policies mobile operators would be able to find insights about the behavior of consumers. Depending upon the objective, whether it is customer acquisition or sales growth or both mobile operators can further take business decisions using the insights. Mobile operators can use these insights to improve their internal offerings or monetize them by selling them to third party analytics companies. Of course it is easier said than done, as real time analysis of Mobile Big Data is very difficult. Further, data is continuously going in volume and variety, hence even the insights are evolving. For instance, as more data is available for analysis about individuals or customer segments, it is quite possible that previous learning is no longer applicable. Consumer usage behavior is also affected by trends and fads in the market leading to multiple changes in the way they use their mobile applications. Hence, obtaining insights and using them to improve business intelligence is a highly interactive and interactive process. Mobile operators need to work upon processing continuously reforming data in real time rather than just analyzing historical trends.

4.2 Potential Applications of Mobile Big Data

Some of the primary applications of the Mobile Big Data are discussed below:

Customer Engagement

Data obtained from customer's activity on mobile can produce tremendous wealth. By tapping into this data companies can understand consumer behavior in a better way. This can result in an interactive relationship between the company and the consumers and lead to a stronger customer engagement. US based retailer Target collects valuable information about customer purchase behavior through its mobile app. This mobile app facilitates easier product search, store search and comparison of prices with other stores. This customer usage information can help Target to identify best suitable timings for pushing coupons and deals. Also, it can help them in improving the overall shopping experience and customer loyalty by providing mobile accessible shopping lists and purchase suggestions that are customized based on customer's profile.

Predicting Future Purchase Intentions

As data about a customer segment grows in volume, more and more information is available for analysis. With better machine learning algorithms the results are getting more detailed and valuable. This iterative process of constant learning can eventually learn in real time about the customer reaction. This can be used in applications such as predictive modeling and adaptive contextual marketing. Some of the queries raised about the behavior of consumers could be:

- Is the collective purchase spending showing an increasing or decreasing trend?
- What is the demographic profile of the customers who show increasing trend in spending?
- What is demographic profile of the customers depending upon the content they search or download?
- Which customers have the highest lifetime value and what is their demographic/ psychographic profile?
- Using the customer profile can we predict what customers are likely to be of higher value?

Targeted Advertising

Advertisements are one of the key driving forces behind mobile marketing. Reach of advertising has already increased to more than half of the smartphone subscribers. Mobile ads are much simpler than the traditional online ads due to the smaller size and lower resolution. However, mobile ads could be more effective in catching attention as the mobile ads cover more space as a percentage of screen size as compared to that of traditional online ads. Moreover, the amount of data volume generated by advertisements is also increasing. Another feature of mobile ads is that they can leverage different type of content. For instance, video ads are in a ratio of 1: 20 to that of banner ads[23] but this will definitely improve in the future as the media grows richer in visual and interactive content.

Advertising when combine with valuable insights about consumer behavior obtained from Mobile Big Data can help in addressing the needs of customers in real time. This is where individual behavioral profiling and location based services can collaborate to push advertisements at the most appropriate time.

REFERENCES

Aggarwal, C. C., Ashish, N., & Sheth, A. (2012). The Internet of Things: A Survey from the Data-Centric Perspective. In C. C. Aggarwal (Ed.), *Managing and Mining Sensor Data* (pp. 383–428). New York City: Springer.

Ashton, K. (2009). That 'internet of things' thing. *RFiD Journal*, *22*, 97–114.

Chen, H., Chiang, R. H., & Storey, V. C. (2012). Business Intelligence and Analytics: From Big Data to Big Impact. *Management Information Systems Quarterly*, *36*(4), 1165–1188.

Dhar, V. (2013). Data science and prediction. *Communications of the ACM*, *56*(12), 64–73. doi:10.1145/2500499

Feijoo, C., Gómez-Barroso, J. L., & Aggarwal, S. (2015). Economics of Big Data. In B. &. Latzer, Internet Economics (p. 27). Cheltenham Glos: Edward Elgar Publishing Limited.

Feinleib, D. (2012). Big-Data-Trends. *Forbes*.

Jacobs, A. (2009). The pathologies of big data. *Communications of the ACM*, *52*(8), 36–44. doi:10.1145/1536616.1536632

Koudounas, V., & Iqbal, O. (1999). *Mobile Computing: Past, present and future*. London: Department of Computing, Imperial College London.

Leber, J. (2013, April 12). How Wireless Carriers Are Monetizing Your Movements. *MIT Technology Review*.

Lind, M. R., & Zmud, R. W. (1991). The influence of a convergence in understanding between technology providers and users on information technology innovativeness. *Organization Science*, *2*(2), 195–217. doi:10.1287/orsc.2.2.195

Lohr, S. (2012). The age of big data. *New York Times*.

Lynch, C. (2008). Big data: How do your data grow? *Nature*, *455*(7209), 28–29. doi:10.1038/455028a PMID:18769419

Manyika, J., Chui, M., Brown, B., Bughin, J., Dobbs, R., Roxburgh, C., & Byers, A. (2011). *Big data: The next frontier for innovation, competition, and productivity*. McKinsey Global Institute.

Marton, A., Avital, M., & Blegind Jensen, T. (2013). Reframing Open Big Data. *ECIS 2013 Completed Research*, 146.

McAfee, A., Brynjolfsson, E., Davenport, T. H., Patil, D., & Barton, D. (2012). Big Data: The management revolution. *Harvard Business Review*, *90*(10), 61–67. PMID:23074865

KEY TERMS AND DEFINITIONS

Artificial Intelligence: Reflects the concept of intelligent machines (or programs) which are able to make independent decisions using data after training on a small dataset.

Data Warehousing: It signifies a mechanism of storing data in massive central repository such that the stored data can be used for analysis later.

Mobile Big Data: A special use-case of Big Data in which the focal devices for collecting and generating the Big Data are mobility-oriented devices.

Mobile Computing: A process of computation on a mobile device such as a smart phone, tablet, hand-held devices and other small wearable devices through which a set of distributed computing systems or service provider servers participate, connect, and synchronize through mobile communication protocols.

Mobile Big Data Intelligence: Gathering insights using Mobile Big Data for a pre-specified purpose.

Internet of Things (IoT): An internet of objects and their virtual representations in an internet like structure.

SWAI (Sources Warehousing Analytics Insights): It is a model represents a sequential model of Big Data processing with four stages such that prior stage has to be completed before moving onto the other.

Targeted Advertising: A form of advertising using customer intelligence generated suing techniques such as Mobile Big Data analytics, such that the adverts are highly customized for a particular customer.

ENDNOTES

1 Details can be accessed at http://www3.weforum.org/docs/WEF_ITTC_PersonalDataNewAsset_Report_2011.pdf

2 See http://open-data.europa.eu/

3 See http://www.data.gov/

4 See http://data.gov.uk/

5 https://www.whitehouse.gov/sites/default/files/docs/big_data_privacy_report_may_1_2014.pdf

6 IBM (2012) put forth initially three Vs of Big Data, namely, Volume, Velocity & Variety, and in 2013, added the fourth dimension of Veracity at an abstract level.

7 Rijmenam (2014) identified varied sources of big data and can be viewed in the form of an infographic at https://datafloq.com/read/understanding-sources-big-data-infographic/338.

8 Feijoo, Gómez-Barroso & Aggarwal (2014) provided detailed explanation of the Big Data landscape and associated model on p5.

9 Technopedia. (2014). Mobile Computing Device (MCD). Retrieved 15 November, 2014, from http://www.techopedia.com/definition/8270/mobile-computing-device-mcd.

10 Bucki, J. (2012). Definition of Mobile Computing. Retrieved 24 November, 2014, from http://operationstech.about.com/od/glossary/g/Definition-Of-Mobile-Computing.htm.

11 Wattenhofer, R. (2004). MOBILE COMPUTING Retrieved 25 November, 2014, fromhttp://www.disco.ethz.ch/lectures/ss04/mobicomp/lecture/1/Chapter1Introduction4Slides.pdf.

12 See the report at https://www.citrix.com/content/dam/citrix/en_us/documents/products-solutions/citrix-mobile-analytics-report-september-2014.pdf.

13 See the report on http://www.mmaglobal.com/files/documents/the-mma-primer-on-mobile-analytics.pdf.

14 See the report on https://www.accenture.com/nl-en/insight-data-monetization-big-data-mobile-operators-summary.aspx.

15 See the report http://www.intel.es/content/dam/www/public/us/en/documents/reports/intel-corp-big-data-policy-position-paper.pdf.

16 See the report http://www.statista.com/statistics/226530/mobile-payment-transaction-volume-forecast/.

17 See the report http://piranpartners.com/wp-content/uploads/2014/12/Mobile_Big_Data_and_Analytics_April_014.pdf.

18 See the report http://lists.w3.org/Archives/Public/public-mw4d/2009Jun/att-0025/the-truth-about-mobile-analytics.pdf.

19 See the report http://piranpartners.com/wp-content/uploads/2014/12/Mobile_Big_Data_and_Analytics_April_014.pdf.

20 See the report http://postscapes.com/cisco-internet-of-things-white-paper-how-the-next-evolution-of-the-internet-is-changing-everything%20.

21 https://www.whitehouse.gov/sites/default/files/docs/big_data_privacy_report_may_1_2014.pdf.

22 See the report https://www.infobright.com/index.php/big-data-internet-things-part-one/#.VazP-Pmqqko.

23 See the report https://www.citrix.com/content/dam/citrix/en_us/documents/products-solutions/citrix-mobile-analytics-report-september-2014.pdf.

Section 2

On the Semantics and Pragmatics of the Mobile Content Evolution

Chapter 9
Driving Media Transformations: Mobile Content and Personal Information

Juan Miguel Aguado
University of Murcia, Spain

Inmaculada J. Martinez
University of Murcia, Spain

ABSTRACT

The consolidation of the mobile ecosystem is deeply influencing the adaptation of traditional media industries to the digital transformations of business, consumption and audiences. Legacy media digitization comes along with a perfect storm where different kinds of crisis converge. This is also the ground where an eventual response to uncertainties about future may appear in the form of new opportunities and possibilities. Simultaneously, mobile content evolution brings forth a new habitus of consumption, an increasingly complex set of social rituals, content format syntax and technical requirements that constitute new consumption scenarios. This chapter aims at discussing a conceptual framework on how the techno-economic drift of the mobile ecosystem matches the evolution of content industries dynamics and content consumption scenarios in terms of dysfunctions and challenges. The theoretical ground is built upon the works from a three-year research project on the evolution of mobile content and the economic and sociocultural relevance of personal information management.

INTRODUCTION

The consolidation of mobile communication technologies has given rise to an entire new ecosystem within digital media landscape. Mobile ecosystem involves a complex and changing net of players (including telecomm operators, handset suppliers, software developers, content aggregators, content producers, new advertising players and users) that are becoming increasingly influential in the sphere of media and content industries (Feijóo, Pasu, Misuraca & Lusoli, 2009).

Legacy media digitization comes along with a perfect storm where different kinds of crisis (of financial resources, of business and management models, of audiences, of distribution systems…) converge. The dramatic drop of advertising expenditure (Perry, 2012) and of pay-per-access income (Pew Research

DOI: 10.4018/978-1-4666-8838-4.ch009

Center, 2013), however, appears not to be exclusively due to global financial crisis, to the increasing fragmentation of audiences or to the fragmentation of users' time and attention in a growing number of devices and media practices (Purcell et alt. 2010). It is also rooted in deeper transformations that concern the whole of the cultural content value chain and which involve a reorganisation of players and influences among them (De Prato, Sanz & Simon, 2014). The impact of the mobile ecosystem in digital content industries is transferred to traditional media industries in the form of innovation vectors or trends that may give answer to some of the pending questions on the future of media: content formats, distribution systems, the role of personal information and new value networks in cultural industries.

The transformations triggered by the process of digitization are as well the ground where an eventual response to uncertainties about future may appear in the form of new opportunities and possibilities. On addition, media digitization is no only a matter of economic or technological challenges. These are intrinsically linked to social transformations. The social and cultural consequences of the mobile revolution have been –and still are- extensively dealt with by many scholars and researchers. But these changes also concerns to the social and symbolic side of mobile content. In this context, mobile content evolution brings forth a new habitus of consumption, an increasingly complex set of social rituals, content format syntax and technical requirements that constitute what we identify -following Goffman (1959)- as consumption scenarios.

In this chapter we intend to outline a conceptual framework for the study of the interrelations between mobile ecosystem players and media in order to understand how the techno-economic drift mobile technologies are driving matches the evolution of content industries in terms of dysfunctions and challenges. Our interest, however, is not solely focused on the economic landscape, but as well in the socio-cultural transformations that accompany the mobile side of the digital revolution.

In broad terms, the resulting picture strengthens the traditional media industry's perception of a collision with software and digital industries, to which cultural content has become a central issue. The conflict points delimited in this chapter (that is, the critical innovation vectors where traditional media show to be more dependent) are basically three: the influence of new distribution channels, the characterisation of content consumption in terms of social interactions and the increasing relevance of personal information in the redefinition of cultural content business models. These aspects constitute as well a relevant argument for consolidating mobile content environment as a specific field in the current development of media studies.

MOBILE PLATFORMS AND THE RE-INTERMEDIATION OF CULTURAL INDUSTRIES

Mobile ecosystem has evolved in the last six years from a network centred structure, in which telecomm operators had a privileged position, to a platform centred structure, in which software developers (especially OS suppliers) influence the relations amongst the different mobile players. In this process of platformisation (Ballon, 2009) the main players tend to group together, in a loose or tight scheme, all of the required roles for a common set of hardware, software and other techno-economic specifications (Feijóo, Gómez-Barroso, Aguado & Barrero, 2012). Mobile platforms enact complex multi-sided markets with multiple interdependencies amongst involved players (DePrato, Sanz & Simon, 2014). Apple's iOS and Google's Android are examples of dominant mobile platforms that tend to create a more or less closed environment in which content becomes a relevant asset and a distinctive product at the same time.

Figure 1. Structure of mobile platforms
Source: Elaborated by the authors

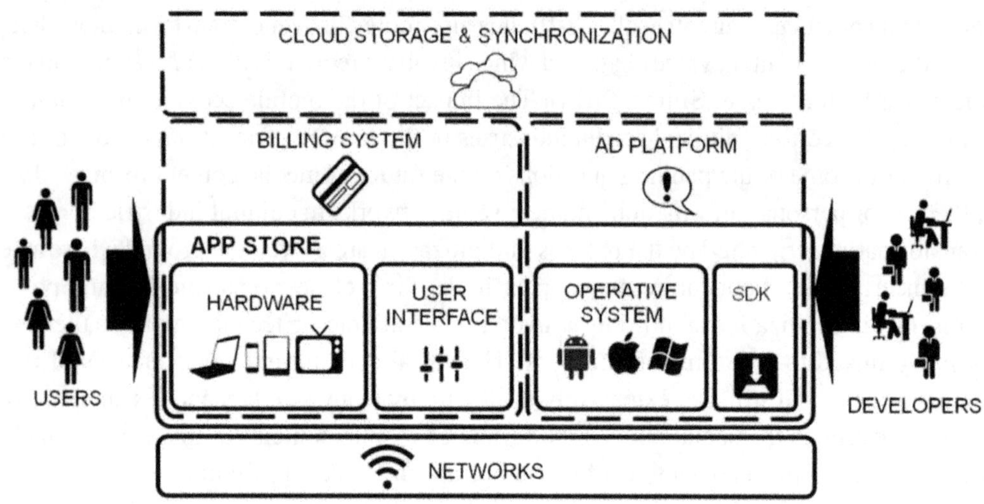

These dominant mobile platforms integrate in an OS-based structure different roles, services and products that contribute to enable access to the mobile information environment: hardware (device specifications), access software (user interface), networks (mobile telecomm operators), distribution (application and content stores), billing systems and third party environments (like specific SDKs for application developers and service providers, and application publishing systems). Application and content stores play a key role in articulating mobile platforms around a viable distribution channel. Mobile platforms also include different added value services, be it addressed to third party players or to end users, such as cloud storage (Apple's iCloud or Google Drive, for instance), multi-device cloud-based synchronisation or mobile specific advertising platforms (like Apple's iAd or Google's AdMob) (see figure 1).

The process of platformisation is not limited to the mobile environment, however. The relevance of mobile application and content stores (like Apple's iTunes/Appstore or Google's Google Play) as an evolution of online software distribution and the consolidation of mobile apps as a prevalent user interface for accessing content and services become a key issue in the reconfiguration of the content industries landscape. Due to its success, the model of application stores -the front of mobile platforms- is being exported to other media environments, like connected TV, desktop PCs, game consoles and photo cameras. As a part of that cross-device scenario, cloud content storage and synchronization services have become a strategic move for digital players. These services make possible to expand the structure of mobile platforms to a broader and increasingly mobile-influenced digital environment (Feijóo, Lindmark et al., 2013). But they also open the possibility for media and content industries to take advantage of cross-media ubiquitous offer. For instance, the cross-device strategy of The New York Times (so called 'NYT Everywhere') involves an integration of different products, access channels and income models.

The importance of mobile platforms is not accidental in this respect. Mobile technology introduces three key aspects that facilitate the drift of the digital ecosystem towards a cross-media landscape: personalization, integration and ubiquity. Mobile platforms perform a decisive role in integrating information environments into ubiquitous personalized access, fostering multi-device and multi-purpose coordination of actions in different contexts (searching, buying, having entertainment, self-expressing, socializing, accomplishing specific tasks, etc.) (Google & Ypsos, 2012). The search for diverse wearable devices –like Google Glass or Apple's iWatch- seems to be part of that same process.

The fact that controlling distribution channels is a crucial strategic aspect of media business needs no further explanation. Indeed, despite the importance traditionally given to content producers, it is upon distribution models that legacy cultural industries have always shaped their business and their social influence (DePrato, Esteve & Simon, 2014). And since they are a relevant part of ICT hardware or data based business models, taking over digital distribution channels has also become a decisive factor for the mobile ecosystem players (Manjoo, 2012).

Consequently, platformisation of the mobile and the ICT environment is about how the different players relate to control the value chain of ubiquitous information and content. In its actual shape, platformisation involves a set of latent conflicts, the most relevant of which concern telecomm operators, on one side, and content producers, on the other side. The first ones try to resist being trapped in the same 'dumb pipe' scheme that they already experienced in the Internet in late 90's (Aguado, Feijóo & Martínez, 2011; De Prato, Sanz & Simon, 2014). The latter (including legacy media), expect a neutral mobile ecosystem over neutral mobile networks that operate as a powerful ground for new forms of access to content and related services (Feijóo, Maghiros, Abadie & Gómez Barroso, 2009).

The relevance of OS developer-based mobile platforms lies in their success in implementing distribution channels and controlling users' access to services and content, which used to be the main asset of legacy media in the pre-Internet era. Application stores are the window for content providers and app developers to access users. They provide also the tools for content and service discovery. Consequently, application stores play a crucial role not only in shaping the market (Feijóo, Gómez-Barroso, Aguado & Barroso, 2012), but also in determining users' perception and access routines to mobile content (Bloem, van Doorm, Duivestein & Sjöström, 2013). As a consequence of that, mobile applications have become the prevalent user interface in the mobile Internet (Khalaf, 2013).

Mobile platforms' capacity to personalize and to provide direct access to end user has also fostered the disintermediation process of cultural consumption that already started with the Social Web in the first decade of the 21st Century (De Prato, Sanz & Simon, 2014). The very same channels and tools that allow content producers (such as legacy media) reaching mobile consumers favour also the transformation of content production into an object of cultural consumption. iBook Store or iTunes, not to mention Amazon Store or Google Play, become that way not only repositories for the products and services of legacy or new digital content industries, but also a direct platform through which professional and amateur content producers (writers, musicians, photographers, software developers, etc.) can directly access their audiences without depending on traditional, over-sized, cost-increasing cultural intermediaries.

This disintermediation process facilitated by mobile platforms is part of a re-intermediation strategy through which mobile platforms take the place of traditional cultural industries as content consumption intermediaries (Aguado, 2013; Feijóo, Lindmark et alt., 2013; De Prato, Sanz & Simon, 2014). Their success in controlling digital distribution channels and in shaping content discovery, access and management in an increasingly cross-device environment is well illustrated by one simple fact: in late 2012, the yearly income of Apple's iTunes and App stores (iPhone's and iPad's) exceeded in USD 300 million the combined turnover of The New York Times; Simon & Schuster (the publisher of the best-selling Steve Jobs' biography); Warner Bros. film studios (with popular film franchises like Batman or Harry Potter); and Time Inc. (the largest magazine publisher in the U.S.) (Lee, 2012).

Despite being relevant, typically considered factors such as ubiquitous availability of content, quality standards or cost-utility balance do not provide a sufficient explanation in understanding the privileged position of application stores as a dominant content distribution channel. The failure of mobile television constitutes a good example in that respect (Aguado, Feijóo, Martínez & Roel, 2012). The relational and context-attached nature of mobile devices underlies new forms of understanding digital content and its insertion into everyday consumption practices. In a sense, users' perception and attitudes towards content in mobile (and, by extension, in cross-device) environment have changed more than content producers'.

UNDERSTANDING CONTENT CONSUMPTION IN THE MOBILE ENVIRONMENT

Both in the popular media discourses and in the mind of users mobile devices are basically personal portable devices characterised by their attachment to situations and social interactions (Aguado & Martinez, 2010). Using mobile devices then involves different forms of availability and access to others' communicative presence. Digital content is obviously not alien to that premise. In fact, users' perception on the value of mobile content appears to be highly determined by two factors: the capacity of the user to act upon it and the ease of integrating it into social interactions. Experiences with users clearly emphasize a difference between traditional media (which are to watch, listen to, read…) and mobile media (García Jiménez, 2012). In the mobile context, users increasingly expect not only to access content, but to do things with content (and with others around content). In this very sense, mobile devices contribute to enhance the digital merge of cultural content and social communications. As a result of the evolution of mobile content, users' attitude and content consumption rituals have increasingly evolved towards an intervention paradigm (Scolari, Aguado & Feijóo, 2012).

The consolidation of mobile applications as the prevailing interfaces for mobile media (Khalaf, 2013) and as the prevalent distribution model (De Prato, Sanz & Simon, 2014) should be explained also as a part of that shift. Mobile applications are small pieces of software developed as specific task oriented tools. As any other tool, they do not solely involve the functional purpose they are made for, but also the symbolic frame and the operative instructions under which they are intended to work. It is in this sense that mobile applications involve also an interface: they enact meaning-action frames that presuppose the user (intentions, attitudes and competences) and an object of action (possible results) in a coherent, connected way (Scolari, 2009). As a tool, an app is something that allows (and is intended for) doing things. Since mobile content is increasingly accessed through native (platform based) or web applications (Khalaf, 2013), mobile environment emphasizes the specific connection between mobile content and user actions. Unlike media oriented computer software, mobile applications bring forth a tight semantic link among object (content), user (presupposed intentions, attitudes and competences) and specific tasks (Scolari, Aguado & Feijóo, 2012). Especially for users under age 45, their relation to his or her apps is clearly different from that to his or her desktop computer software. They feel mobile apps and software linked to their everyday life and to the emotional aspects of it. What they have installed in their smartphones (and even which apps they frequently use or what makes them to eventually uninstall an app) has to do with their, with their personal preferences and with their momentary impulses. The two ideas –"content is something to act upon" and "content and apps speak about who I am"- ideally merge in the perception and uses of mobile content applications.

Figure 2. A conceptual frame for the classification of mobile content applications (Scolari, Aguado & Feijóo, 2013)

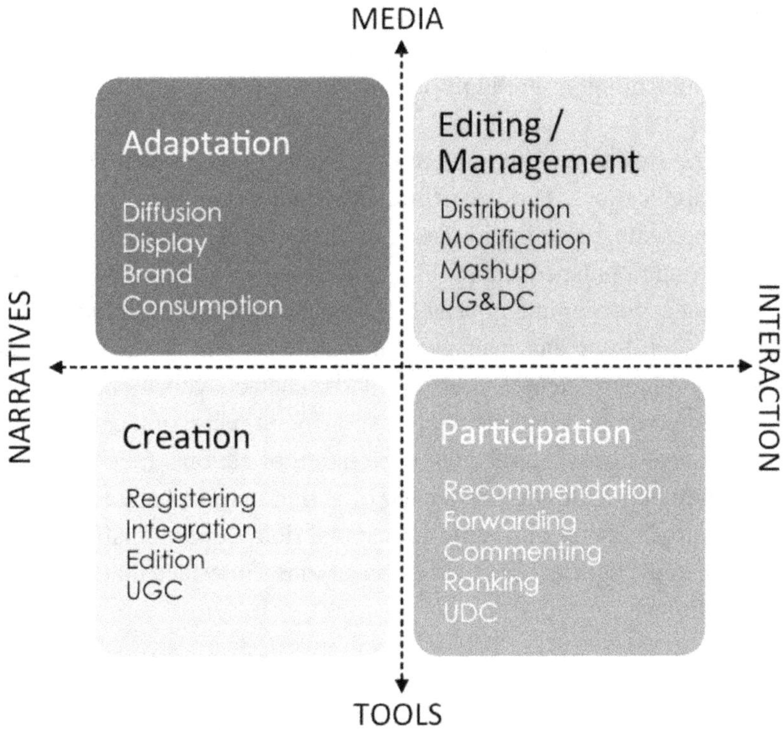

The functional evolution of mobile content apps in the last five years supports these assumptions. Drawing on a semiotic analysis of a sample of 40 content addressed mobile apps, we have developed a conceptual model defined along two axes (see figure 2): media / tool orientation (considering a scale from the mere content display to high intervention capacity) and narrative / interaction involvement (from narrative focused apps –in displaying or instrumental terms-, to communication focused apps).

That scheme allowed us developing a general taxonomy of mobile content applications by differentiating four clear fields (Scolari, Aguado & Feijóo, 2012):

1. Adapted or repurposed mobile content apps, addressed to conventional media diffusion and consumption
2. Creative mobile content apps, addressed to facilitate registering, producing and editing user generated content (photos, videos, music, audio, etc.)
3. Content management mobile applications, focusing in organizing and transforming conventional media content (libraries, playlist managers, etc.), including some aspects of user generated and distributed content
4. Participation mobile apps, addressed to integrate mobile media and user generated content into social conversations (including, from the perspective of media, the so called 'user distributed content').

These broad categories were characterised by the prevalence of a given frame of presupposed actions and tasks:

1. Contemplation, which involves passive consumption of conventional content (reading, watching or listening)
2. Intervention, which involves acting upon content to transform it (editing, managing, merging…)
3. Conversation, which involves using content as a source of meaning in social interactions (linking, recommending, commenting, resending…).

Figure 3 illustrates the structure of the involved prevailing actions that characterises the different kinds of mobile content applications. The size of circles in the graph represents relevance, while position (inclusion/exclusion) represents subordination. Accordingly, content apps within the category of adaptation (typically including traditional media apps, like mobile magazines or mobile news apps) presuppose contemplation as the main action guiding interface design and user's attitude, with intervention capacity having a subordinated role. Editing and management mobile content applications (such as content aggregators) perform contemplation oriented interfaces and related conversational functions, however with a more relevant role of intervention capacity. Applications within creation category (typically photo or video editors and music tools) invert the former correlation of actions, focusing interface in intervention and subjecting contemplation to the prevalence of acting upon –characteristically- user generated content. Finally, mobile applications within the conceptual field of participation (such as mobile social networks and messaging apps) perform a conversation-oriented interface that subordinates intervention and contemplation capacities.

Figure 3. Structure of presupposed actions in mobile content applications
Source: Scolari, Aguado & Feijóo, 2013.

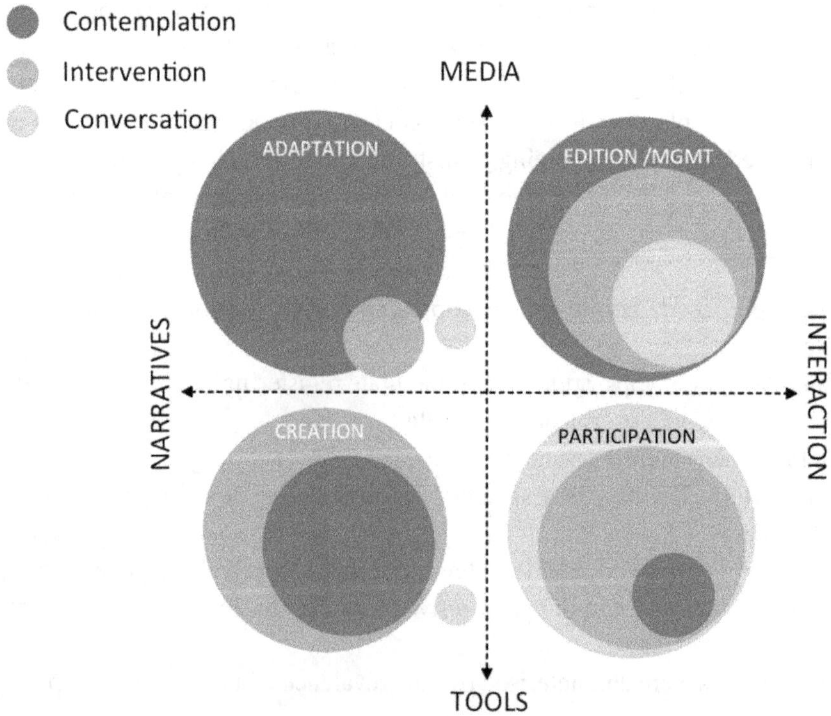

One of the interesting points in this taxonomy is that categories are not closed either in conceptual nor in evolutionary terms. Characteristically, for example, networked mobile video games pose a singular merge of both content/tool and narratives/interaction values, involving tight interdependence among contemplation, intervention and conversation, hence they are located close to the middle point of the conceptual space. Furthermore, the model allows us locating a given app in a concrete place –according to the relevance of its features and the prevailing actions involved- and observing how it evolved through the conceptual space as long as its features changed in the consecutive software actualizations.

The model makes possible to visualize a drift of mobile applications towards the centre of the conceptual space: those applications from the creative and adapted areas progressively include more and more social interaction related features and instrumental functionalities. On the opposite side, those applications from the participation area progressively include more content related features. As a matter of example, the chart in figure 4 depicts the score of the functional attributes[1] for the actualizations of Twitter app (in iOS) along the 13 versions updated from March 2013 to March 2015. It depicts a clear drift towards a functional relevance of content display and content management within an interaction driven app.

Some of the evolving consumption scenarios and the content formats emerging attached to them support these arguments. The growing importance of multiscreening (Google & Ypsos, 2012) and its semantic translation into transmedia narratives (Scolari, 2009) seem to be directly related to users' activity with content. The so called 'second screen apps' constitute a kind of software implementation for a consumption scenario that involves synchronising different content and social conversation related activities. Users have incorporated conventional media consumption (like watching TV series, sports

Figure 4. Functional evolution of Twitter app during 2013-2015, from version 5.4 to version 6.25.1
Source: Elaborated by the authors

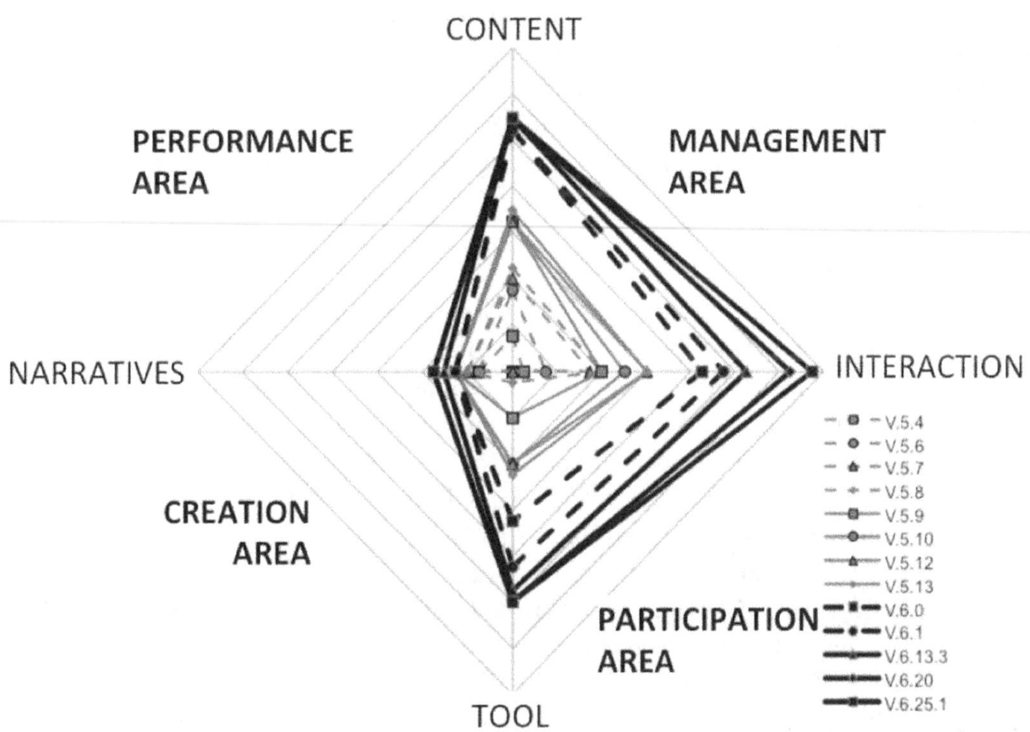

or movies) to their mobile mediated social conversations in live time: for example, tweeting to friends about an episode of a favourite TV series in virtually co-present watching (Pew Research Centre, 2012). Content industries have turned their interest to mobile social networks and mobile applications as a new source for audience metrics (Noguera et alt., 2013) but also as a vehicle for promoting and expanding transmedia content (Scolari, 2009). Second screen apps contribute to systematize that integration between content and social interactions, be it connecting social conversation to a given content (like *The Walking Dead - Walkers Kill Count*, a social network game-like application about guessing the number of kills in an episode of the popular zombies' TV series-, providing extra content and information –like most of Disney's second screen apps-, promoting content related participation –like in the case of apps complementing media coverage of the last USA Presidential Election - or expanding stories in participatory terms –like 'twittersodes' that expand the story of TV series in the form of characters' conversations in which users can participate -.

Media users, in sum, increasingly conceive content as a part of their social interactions and, consequently as a matter to act upon, be it re-creating (transforming mainstream media content into forms of parody, irony, protest or support), co-creating (merging it into user generated content) or channelling (recommending, forwarding, commenting, etc.). This process that shifts content conception from the logic of seeing to the logic of doing poses a deeper connection between user identity and content consumption with two relevant consequences in both media and digital environments: it enters content consumption into the content distribution value chain and constitutes social relations as a specific object of cultural consumption.

PERSONAL INFORMATION AS A NEW VALUE DRIVER

In the previous sections we have outlined a conceptual background for the influence of mobile ecosystem's distribution structures and content formats in transforming conceptions about digital content consumption. But how this affects legacy media and digital industries from a business perspective? Describing mobile platforms as relevant players in a process of re-intermediation of cultural industries demands understanding why and how digital industries have embraced content business as an important part of their strategy. What has content business to offer digital players and what should legacy media learn form digital businesses?

Despite having substantially different business models, all of the big Internet companies –especially the so called "Internet Fantastic Four": Apple, Google, Amazon and Facebook- have rapidly become mobile centred companies giving strategic relevance to content aggregation and distribution (Manjoo, 2012). The central role content and media play in an e-commerce giant like Amazon needs no explanation. However, the convergence of the other Internet giants, with such different businesses like hardware (Apple), search engines (Google) or social networks (Facebook), may deserve a deeper look.

One relevant point these companies manifestly share is their effort to become the centre of a multi-screen platform, with mobile devices (smartphones and tablets) in its centre, but including in different degrees connected TV, desktop computers and game consoles, with cloud synchronisation and storage services (like iCloud, Amazon Cloud Drive or Google Drive) and OTT services (like Apple TV) above them. Amazon, for example, has developed a threefold strategy consisting of providing low-price hardware (Kindle Fire), an integrated distribution channel that complements Internet ecommerce portal (Amazon Appstore), and multi-screen content access (Amazon Instant Video, Amazon Game Studio), building a platform-like structure upon a tailored version of Android OS. Facebook, in turn, has adopted the shape of a platform-nested environment both in desktop and mobile devices, providing platform-like added services such as search, application store or content access (movies, games, etc.) under the umbrella of users' social ties.

One of the reasons for this strategic convergence over cultural content lies in the increasing importance of personal information in cross-device platforms. Data mining about users has become in the last five years a key aspect of Internet service companies (Acquisti, 2010): processing personal information and creating user profiles facilitates enhancing segmentation and adapting both advertising (like in Google and Facebook cases) and content or hardware supply (Amazon and Apple) to user profile and context (Punie, Lusoli, Centeno, Misuraca, & Broster, 2009; Manjoo, 2012). Information about user profile and user behaviour is not only the base for adaptive services in a context-addressed ubiquitous information environment. It is, at the same time, the main asset to support large scale selling or advertising based business models (Acquisti, 2010; Laurila et alt., 2013), including content distribution. For instance, in 2013 Google acquired Behavio, a company specialized in monitoring mobile devices activity (location, movement, proximity, app activity, etc.) and forecasting behavioural patterns. Facebook's strategy –integrating apps, search, browsing, contacts and social networks- is also addressed to settle an access ground to user behaviour without depending on the filters of other platforms, thus adding value to both its metrics and channels (Filloux, 2013).

Cultural content (entertainment, news, advertising, etc.) is a core element in everyday mobile activity: searching, creating, buying, sharing, commenting or simply consuming content (reading, watching or listening to) takes a big part of what we usually do with our smartphones or tablets (AOL-BBDO, 2012). Furthermore, cultural content plays an increasingly relevant role in our communications (Google & Ypsos, 2012; Bloem, van Doorm, Duivestein & Sjöström, 2012). Interviewed users –especially those aged under 45- agreed their attitude and expectations towards mobile device highly depended on content availability, even in the case of communicating to others. Natural behaviour around content becomes that way a valuable source of information not only about users and their world, but –precisely because of that- also about user attitudes towards devices, content, services and brands.

The personal and pervasive condition of mobile devices makes them extremely suitable as information gathering tools. Unlike desktop Internet user profiles, mobile user profiles are consistent (they are about a given user behaviour) (Aguado, 2012) and complex (they include context and location aware parameters). Pervasiveness and ubiquity considerably broaden the range of contextual scenarios in which information is both used and gathered. They also make possible to weave continuity and coherence among these different scenarios. The singular relation between mobile device and user, different from other digital devices, brings forth a specific horizon for both the operative and the economic value of personal information.

According to Acquisty (2010) and Feijóo & Gómez Barroso (2013), personal information involves data about users (user identification, like name, address, economic data derived from purchasing or banking, health related data, etc.) but also about what users do (behavioural information related to browsing, search, use of software and services, etc.) and information about how and when they use services (frequency, stickiness, etc.). Besides facilitating user profile coherence derived from processing usage activity, mobile devices provide additional relevant data concerning location and context related information, allowing to situate the 'how' and the 'when' in relation to behavioural and identity information.

By controlling software access to services, third party developers innovation (via SDKs and OS specifications), content distribution channels (via content and application stores) and hardware specifications (including built-in sensors), mobile platforms have a privileged position in accessing and managing not only mobile personal information. As long as the barriers between conventional Internet and mobile Internet dissolve, mobile platforms get a significant advantage in a cross-device ubiquitous information environment (Feijóo & Gómez Barroso, 2013). That concentration of the capacity to access and manage personal information in the hands of mobile-centred Internet players disrupts digital business models and raises concerns about privacy protection and transparency (Xu, Rossom, Gupta & Carroll, 2012; Luchetta, 2013; Feijóo & Gómez Barroso, 2013).

EXPANDING DIGITAL CONTENT VALUE CHAIN

In the case of traditional media industries, involved as they are in their very digital transformation, all the above mentioned involves a radical change in the value chain of cultural content. Legacy media business model was characteristically content-centric media dependent. In short terms, content used to be the core and the end of the value creation process. Through controlling production and distribution, legacy media ensured the attention of wholesale audiences, which was in turn monetized in pay-per-access schemes or, more often, via advertising. The whole process ended up in content consumption: there was no value after content was read, watched or listened to. In order to maximize control over production and distribution, the whole productive structure was designed according to media requirements (printed paper, radio or screens), including expensive, complex –and often slow- distribution infrastructures. The confluence in the last decade of digital media development and financial crisis puts that model in serious troubles: the dramatic drop of advertising expenditure (Perry, 2012), the rigidity of over-sized media dependent production structures (Westlund, 2012) and the problems attached to licensing and intellectual property management (Punie, Lusoli, Centeno, Misuraca & Broster, 2009) seriously challenge the viability of conventional media industries.

Legacy media confront thus an urgent need of adaptation to deep changes in production, distribution and consumption. Because of the different structures, players and consumption scenarios that characterize digital environment, old media business models show important dysfunctions in that adaptation: digital advertising is less effective as a revenue source than it used to be traditional advertising, and pay-per-access formulas resist to work in a content saturated environment with hundreds or thousands of free access alternatives (Wolff, 2013). Digital environment, in addition, raises two disruptive conditions that alter the structure of legacy media business: de-materialization of content and its integration into users' social dynamics.

According to the general drift towards dematerialization that characterizes digitization (De Prato, Sanz & Simon, 2014), cultural content is becoming increasingly less constrained by the requirements of the médium and consequently more adaptable to a diversity of consumption scenarios. Dematerialized contents can be accessed through, and deployed in, different media (mobile devices, paper, big screens, etc.), whether in complementary or in alternative forms. New cross-media distribution models take advantage of this, and mobile platforms are becoming the basis for multidevice, malleable content circulation where controlling distribution channels becomes even more crucial. Cross-media distribution fosters (and is equally favoredby) cross-media consumption. The growing interest in transmedia storytelling (Scolari 2009) is connected with that transformation of content and its impact in distribution and consumption (Google & Ypsos 2012)

The second disruption factor coming along digitization and mobility is socialization. Mobile technology poses an increasing merge of media and communication that facilitates the integration of media into social dynamics. It is true that the internalization of media activities as a part of the social global conversation is not strictly speaking a new phenomenon: almost any media effects theory considers in one way or another how media activity gets into everyday life. Furthermore, commenting latest news while having a coffee with colleagues or recommending a movie to friends are natural ways in which media become part of our social interactions. The difference now is that media related social interactions are not just *about* media, but rather *within* media, and they are taking place in the same digital environment (and often in the same consumption scenario) in which content is accessed and consumed. In other words, social interactions are attached to and share space with media consumption. From the perspective of media industries, this opens a brave new horizon of possibilities, expanding the value chain of cultural content to include what comes after (or around) content consumption. In this respect ubiquity, personalization, integration and pervasiveness are key characteristics that ensure a natural merge of media consumption and the social situations they take place in.

Social networks (and especially mobile social networks) contribute to create new added value for media. The merge of media use and social networks contributes to settle user information management as a relevant asset for digital cross-device oriented media (Purcell et alt, 2010; Noguera, Villy, Nyiro, De Blasio & Bourdaa, 2013). In that context, emerging media players tend to provide the highest number and diversity of possible interactions in order to gather intensive data about user behaviour. The effective management of user information allows tailoring services and products to user's profile and context and (b) it adds value to profiled advertising lowering costs and increasing efficiency through programmatic buying and through escalating inventories and buyers' networks (Feijóo, Gómez Barroso and Martínez, 2010).

However, social networks do not only constitute a new vein of personal information (that enhances content as a source of user data): as argued before, they are changing the focus of media consumption from passive reception to media related social interactions. This change in focus is not about media including (and getting advantage of) social interactions. It is rather the opposite: social interactions including (and getting advantage of) media. In other words, it is the social relationship that constitutes a valuable object of consumption (Noguera et alt., 2013). Media become interesting to the extent they contribute (as a language, as a topic, as an object of parody or emotional positioning, as a matter of re-creation and transformation) to the ritual of social interaction.

As a result of these disruption factors, the digital content value chain is being transformed as well. It is no longer a stable linear process of value creation with its core in distribution-focused legacy media companies. On the contrary, it is rapidly becoming a complex network of heterogeneous players coming from different ecosystems (telecom players, software industries, traditional media, users, new regulators…). The privileged position of some of these players (the so called Internet giants) in accessing end users through content distribution centred platforms drives a re-intermediation process in which legacy media looses its traditional control over audiences. In exchange, as content consumption and social interaction merge, new possibilities for value creation emerge.

Content consumption related social interactions contribute to expand content value chain at least in two ways. In first place, significantly amplifying the role of content consumption as personal information driver. Social conversation around content –and even the use of actual content narratives as a social language (Aguado & Martínez, 2010)- efficiently merges three relevant spheres of information about a person: identity presentation, personal contacts and socio-cultural situation (who you are, to whom you relate and what/where are you doing). As stated in section 3, enhancing the capacity to do things with content is a prerequisite to socially integrate content consumption in a ubiquitous communication environment. In second place, by merging content to social interactions users become a relevant distribution channel. In fact, content curation, rating and sharing has become a relevant activity for social networking sites and apps. Some of them, like Facebook and Linkedin, are including content recommendation and curation functionalities. Similarly, some content curation oriented aggregators, such as Flipboard, are clearly shifting towards content centred social networking functionalities.

CONCLUDING REMARKS

The impact of the development of a platform centred mobile ecosystem has contributed in a relevant way to shape the horizon of problems and challenges that legacy media and traditional content industries face in adapting to the digital environment (Aguado, Feijóo & Martínez, 2012). The kind of transformations induced by the mobile ecosystem is not limited to diffusion related aspects commonly considered, like mobile device penetration or content access normalisation. Mobile ecosystem evolution is actively contributing to change the nature of players involved in the digital content value chain and the structure of strategic relationships among them (Feijóo, Lindmark et alt., 2013). These changes result as well from a mobile-grounded transformation in users' conception about digital content consumption (Aguado, 2013). This transformation is marked by aspects like intervention capacity and content integration into social interactions, which support the consolidation of content mediated social interactions as an emerging form of cultural consumption. Three main innovation vectors condense these changes:

Firstly, the implementation of new distribution systems, characteristically shaped by the pattern of mobile application stores, which constitute the core of the mobile content ecosystem, disrupts traditional content intermediation structures. The so called 'platformisation' (Ballon, 2009) over application oriented environments favours a re-intermediation processes by means of which legacy media are displaced from value attached relevant positions and raises a category of cloud-based, multi-device accessible and medium independent content that challenges the viability of advertising or copy-centred pay-per-access based traditional business models.

In a complementary way to that, new distribution models and new software based content formats facilitate the redefinition of users' perception about valuable cultural content, driving new content formats that take advantage of the mobile/digital ecosystem's possibilities (integrating user's activity with content, adapting content to user's situation and context, etc.). The consolidation of mobile applications as a prevalent interface involves in this respect a shift from a 'seeing-oriented' logic to a 'doing-oriented logic' that particularly fits mobile's characteristic relational and context-attached nature and constitutes a user-centric environment –instead of the traditional producer-centric one (Westlund, 2012). The application of taxonomy oriented conceptual model for the functional analysis of mobile content applications (Scolari, Aguado & Feijóo, 2013) supports the observation of a shift in mobile content consumption from a contemplation paradigm to an intervention paradigm. This shift allows expanding traditional content value chain by including user activity with content (sharing, transforming, commenting, etc.) into monetizable digital processes (Noguera et alt., 2013).

Finally, the integration of what users do with content into cultural content value chain is implemented upon the availability of personal information and the capacity of digital players to process this information in the forms of profiles that enhance the identity and situation driven adaptation of advertising, products and services (Feijóo & Gómez-Barroso, 2013).

REFERENCES

Acquisti, A. (2010). *The Economics of Personal Data and the Economics of Privacy. 30 Years after the OECD Privacy Guidelines*. Brussels: OECD.

Aguado, J. M. (2013). La industria de contenido en la era post-PC: Horizontes, amenazas y oportunidades. In Notícias e Mobilidade: Jornalismo e Dispositivos Móveis (pp. 5-28). Covilha, Portugal: LabCom, Universidade de Beira Interior.

Aguado, J. M., Feijóo, C., & Martínez, I. J. (2011). Contenidos digitales y comunicación móvil: Hacia la transformación de Internet. *Derecho a Comunicar*, *2*(3), 1–15.

Aguado, J. M., Feijóo, C., Martínez, I. J., & Roel, M. (2012). Mobile Television, A Paradigmatic Case of the Uncertainties and Opportunities of the New Media Ecosystem. In A. Abruzzese, N. Barile, J. Gebhardt, J. Vincent, & L. Fortunati (Eds.), *The New Television Ecosystem* (pp. 141–158). Frankfurt am Main: Peter Lang.

Aguado, J. M., & Martínez, I. J. (2010). Liquid digital selves: Mobile Media Implicit Cultures, Social Relations and Identity Management. *Encyclopaideia: Journal of Phenomenology and Education*, *15*(1), 63–88.

AOL-BBDO. (2012). Seven Shades of Mobile. The Hidden Motivations of Smartphone Users. Research. Retrieved from http://advertising.aol.com/sites/default/files/ content/download/ attachment/research/ aol-bbdo-7-shades-mobile-abstract-final.pdf

Ballon, P. (2009). The Platformisation of the European Mobile Industry. *Communications & Stratégies*, *75*(3), 15–34.

Bloem, J., van Doorm, M., Duivestein, S., & Sjöström, A. (2013). *The App Effect*. Amsterdam: VINT Bariet.

Castellet, A., Aguado, J. M., & Martínez, I. J. (2012, December 7-8). Towards a medium-less ecosystem: The impact of app stores in the digital content environment, *I Technological Innovation in a Mobile World Conference*, Murcia.

De Prato, G., Sanz, E., & Simon, J. P. (2014). *Digital Media Worlds. The New Economy of Media*. New York: Palgrave-Macmillan. doi:10.1057/9781137344250

Feijóo, C., & Gómez Barroso, J. L. (2013). Hacia una economía de la información personal. In J. M. Aguado (Ed.), *Feijóo, C. & Martínez, I. (2013), La Comunicación Móvil: Hacia un Nuevo Ecosistema Digital* (pp. 305–322). Barcelona: Gedisa.

Feijóo, C., Gómez Barroso, J. L., Aguado, J. M., & Barrero, S. (2012). Mobile gaming: Industry challenges and policy implications. *Telecommunications Policy*, *36*(3), 212–221. doi:10.1016/j.telpol.2011.12.004

Feijóo, C., Gómez-Barroso, J.L. & Martínez, I. J. (2010). Nuevas vías para la comunicación empresarial: publicidad en el móvil. *El profesional de la información, 19*(2), 140-148.

Feijóo, C., Lindmark, S., Villar, J. P., Tarín, C., Gelabert, J., & Matía, B. (2013). *Public and commercial modes of access to content and information in the digital era. Culture and Education. IP/B/CULT/IC/2012-018*. Brussels: European Parliament.

Feijóo, C., Maghiros, I., Abadie, F., & Gomez-Barroso, J. (2009). Exploring a heterogeneous and fragmented digital ecosystem: Mobile content. *Telematics and Informatics*, *26*(3), 282–292. doi:10.1016/j.tele.2008.11.009

Feijóo, C., Pascu, C., Misuraca, G., & Lusoli, W. (2009). The next paradigm shift in the mobile ecosystem: Mobile social computing and the increasing relevance of users. *Communications & Stratégies*, *75*(3), 57–77.

Feijóo, C., & Ramos, S. (2013). La música en el móvil o el dispositivo como soporte. In J. M. Aguado (Ed.), *Feijóo, C. & Martínez, I. (2013), La Comunicación Móvil: Hacia un Nuevo Ecosistema Digital* (pp. 157–186). Barcelona: Gedisa.

Filloux, F. (2013) Facebook Home: Another Android Lock Pick. *Monday Note*. Retrieved from http://www.mondaynote.com/2013/04/07/facebook-home-another-android-lock-pick/

García Jiménez, L. (2012, December 7-8). The use, consumption and perception of smartphones by digital natives. Proceedings of *I Technological Innovation in a Mobile World Conference*, Murcia.

Goffman, E. (1959). *The presentation of self in everyday life*. New York, NY: Doubleday.

Google & Ypsos. (2012). The new multi-screen world. Retrieved from http://www.thinkwithgoogle.com/insights/library/studies/the-new-multi-screen-world-study/

Khalaf, S. (2013). Flurry five-year report: it's an app world. The web just lives in it. *Flurry Analytics*. Retrieved from http://blog.flurry.com/bid/95723/Flurry-Five-Year-Report-It-s-an-App-World-The-Just-Web-Lives-in-It

Laurila, J., Gatica-Pérez, D., Aad, I., Blom, J., Bonet, O., Do, T., et al. (2013). The mobile data challenge: Big data for mobile computing research. *Nokia Research*. Retrieved from http://research.nokia.com/files/public/MDC2012_Overview_ LaurilaGaticaPerezEtAl.pdf

Lee, E. (2012). Apple's ITunes Would Be One of World's Biggest Media Companies, Retrieved from http://go.bloomberg.com/tech-blog/2012-12-03-apple%E2%80%99s-itunes-would-be-one-of-world%E2%80%99s-biggest-media-companies/

Luchetta, G. (2013). The Law and Economics of Intermediaries of Personal Information. Centre for European Policy Studies. Retrieved from http://ssrn.com/abstract=2228358

Manjoo, F. (2011, November). The Great Tech War of 2012. *Fast Company*. Retrieved from http://www.fastcompany.com/magazine/160/tech-wars-2012-amazon-apple-google-facebook

Noguera, J. M., Villi, M., Nyiro, N., De Blasio, E., & Bourdaa, M. (2013). The role of the media industry when participation is a product. In N. Carpentier, K. Schroder, & L. Hallett (Eds.), *Audience Transformations. Shifting Audience Positions in Late Modernity* (pp. 174–192). London: Routledge.

Perry, M. (2012). Newspaper Ad Revenues Fall to 60-Yr. Low in 2011. Retrieved form http://mjperry.blogspot.com.es/2012/02/newspaper-ad-revenues-fall-to-50year.html

Pew Research Centre. (2012). One in ten 'dual-screened' the Presidential Debate. Retrieved from http://www.people-press.org/2012/10/11/one-in-ten-dual-screened-the-presidential-debate/1/

Pew Research Centre. (2013). *The State of the News Media 2013*. Retrieved from http://stateofthemedia.org/2013/overview-5/

Punie, Y., Lusoli, W., Centeno, C., Misuraca, G., & Broster, D. (2009). *The Impact of Social Computing in the EU Information Society and Economy*. JCR Scientifical and Technical Reports. European Commission. Retrieved from ftp://ftp.jrc.es/pub/EURdoc/JRC54327.pdf

Purcell, K., Rainie, L., Mitchell, A., Rosentiel, T., & Olmstead, K. (2010). Understanding the participatory news consumer. How internet and cellphone users have turned news into a social experience. *Pew Research Centre. Project for Excellence in Journalism*. Retrieved from http://www.pewinternet.org/~/media//Files/Reports/2010/PIP_Understanding_the_ParticipatoryNews_Consumer.pdf

Scolari, C., Aguado, J. M., & Feijóo, C. (2013). Una ecología del medio móvil: Contenidos y aplicaciones. In J. M. Aguado, C. Feijóo, & I. Martínez (Eds.), *La Comunicación Móvil: Hacia un Nuevo Ecosistema Digital* (pp. 79–106). Barcelona: Gedisa.

Scolari, C. A. (2009). Transmedia Storytelling, Narrative Worlds, and Branding in Contemporary Media Production. *International Journal of Communication*, *3*, 586–606.

Scolari, C. A., Aguado, J. M., & Feijóo, C. (2012). Mobile Media: Towards a Definition and Taxonomy of Contents and Applications. *International Journal of Interactive Mobile Technologies*, *6*(2).

Westlund, O. (2012). Producer-centric vs. Participation-centric: On the sahping of mobile media. *Northern Lights: Film & Media Studies Yearbook*, *10*(1), 107–121. doi:10.1386/nl.10.1.107_1

Wolff, M. (2013, January 4). This tipping-point for paywalls does not fix newspapers' larger crisis. *The Guardian*. Retrieved from http://m.guardian.co.uk/commentisfree/2013/apr/01/tipping-point-paywalls-newspapers-crisis

Xu, H., Rossom, M. B., Gupta, S., & Carroll, J. M. (2012). Measuring Mobile User's Concern for Information Privacy. Proceedings of the *33rd International Conference on Information Systems*. Orlando, FLA. Retrieved from http://faculty.ist.psu.edu/xu/papers/Xu_etal_ICIS_2012a.pdf

KEY TERMS AND DEFINITIONS

Consumption Scenarios: A consumption scenarios involves the physical, social and symbolic aspects of a situated content consumption by a user. In its physical aspect, it includes hardware, software, and the usual spaces (public or private) in which it is supposed to take place. The social dimension of a consumption scenario includes presupposed rituals and roles, whereas the symbolic dimension involves meaning frameworks and values.

Ecosystem: Referred to techno-economic environments, like the mobile communications one, an ecosystem is here understood as a complex net of interdependent players and environmental factors that (through co-operation and competition) internally generates its own order. The ICT ecosystem, for instace, would include four characteristic layers that group along diverse functions different groups of players: Equipment providers, network operators, content and applications providers, and users.

Interface: Any kind of meaning-action frames that presuppose and connect users (intentions, attitudes and competences), actions and objects of action (tokens) in a coherent and meaningful way (including pragmatic recipes and expectable results).

Mobile Content Applications: Any mobile application whose identifying functions are address to operate about content in terms of performance, creation, management or participation.

Personal Information: Personal information or personal data refer to all those data derived from the digital identity, digital records and online or off-line activities of a given user. Mobile personal information includes also data obtained from the mobile devices' built in sensors (thus converting off-line activities, like walking or running, location, etc., in digital information). According to different policy and regulation traditions, personal data can be understood in either restricted or expanded conception. The restricted conception refers to that information strictly related to digital identity and personal records (like financial parameters, health records, etc.). The expanded conception of personal data includes data derived from what users do (buying, searching, playing, navigating…), how they do and where or when they do, regardless if this is connected or not to digital identity. Digital profiles are coherent sets of personal data.

Platformisation: It refers to the process in which ICT players tend to group together, in a loose or tight scheme, all of the required roles for a common set of hardware, software and other techno-economic specifications. In the mobile environment, since 2007, platforms have been gaining increasing relevance around OS and hardware providers. Apple's iOS and Google's Android are examples of software based mobile platforms.

Value Chain: A value chain or a value network identifies the process, players and relations involved in the creation of value around a given product or service, and how this value is distributed amongst the different players.

ENDNOTE

1 The scoring model is an adaptation of functional content analysis in which the different actualization items are attributed binary values (0,5 to 1). Versions and actualization items not directly dealing with content (such as "adaptation to retina display" or "minor bug solving") were not considered.

Chapter 10
Mobile Content Adaptation:
An Analysis of Techniques and Frameworks

Christian Sonnenberg
Florida Institute of Technology, USA

ABSTRACT

The mobile environment presents numerous challenges to users and developers alike. Ideally, mobile content should be designed to enhance a user's experience rather than simply serve up the same content originally designed for other platforms. This chapter presents a study of techniques for "content adaptation", the process of reformatting content and displaying it optimized for a mobile device. Adaptation techniques consist of a varied number of mechanisms by which content is adapted into a format more "friendly" to mobile devices. The idea of "friendly" can range from simple accessibility requirements to a dynamic redesign. Included in this discussion is a summary of the advantages and disadvantages of current techniques. Examples of different techniques are presented and illustrated along with the system framework necessary to implement them.

INTRODUCTION

The advent of the mobile environment, much like the early days of the PC and Internet-capable consumer electronics, has generated a diverse range of devices and formats by which users interact with traditional content. This freedom and flexibility has benefited users by providing cheap, wireless access to content. However, the inherent nature of smaller devices coupled with the relative youth of mobile formats has created an environment that largely ignores the context of the mobile user. The traditional methods of content creation and delivery that apply to a standard desktop simply do not apply in the mobile world. While these concerns affect cross-platform content distribution mechanisms, namely the Mobile Web, they also influence the development lifecycles of platform specific content such as mobile applications.

DOI: 10.4018/978-1-4666-8838-4.ch010

BACKGROUND

Great strides have been made in the field of device independence, the ability for content to reach a user regardless of which type of digital device or machine they choose to view it on. Tools and systems have been developed to reduce the need to generate multiple versions of content or apps and reach as many devices as possible. Some of these mechanisms will be discussed in detail in later sections. However, while device independence is a goal that has largely been achieved, the usability of such content has not been adequately addressed. Digital content that gets displayed on a screen is designed for human consumption. Therefore it is subject to important sociological and cognitive processes that can aid or hinder the success of such content adaptation.

In order for content to be deemed successful there must be a match between the content creator's objectives, the user's goals when accessing the content, and the content design. Furthermore, content access is distinctly affected by the vehicle by which it is deployed to. A mobile device carries certain contextual attributes that can affect cognition, create distraction, inhibit usability, and decrease performance. Research shows that distractions in the environment can have negative impacts on both usability and performance (Deegan, 2014). However, it is important to note that certain distractions can decrease performance, while others can decrease usability. The distinction is one of technical vs. sociological effects. Performance defines the speed of completion and the resource cost of content adaptation from a technical perspective. Usability involves the satisfaction and success of a user's interaction with content from a sociological/cognitive perspective.

This chapter presents an analysis of content adaptation techniques from the perspective of performance and usability. Nielsen defines usability as a combination of quantitative and subjective values including learnability, efficiency, memorability, error rate, and satisfaction (Budiu & Nielsen, 2012). Broadly stated, it is a combination of the user's perception of how useful and an interface is and the measurable efficiency gains in using it. In addition to these five attributes, Nielsen defines utility as the ability of a system to meet the needs of the user. This is generally considered a separate precondition towards usability. When a product fails to provide utility then it does not offer the basic needs to perform a usability analysis.

Digital content takes on a vastly different form in the mobile environment. It is at the same time simplified and complicated. An attribute such as less screen space means that there is less content to interact with, thus a simpler interface. However, using a touch screen and walking at the same time is a more complicated task than using a mouse while sitting. Some complexity can be desirable. When an interface is considered too simple, it can be seen as dull and uneventful. It has been noted that people prefer a middle level of complexity: too simple and users are bored, too complex and users are confused (Norman, 2010). Therefore, it is crucial to examine and understand the factors, both technical and sociological, that affect complexity and ultimately usability. Without a proper understanding of the various factors, content creators and providers put users at a disadvantage.

DEVICE FACTORS

There are wide number of attributes and restrictions that desktop users rarely have to contend with: issues such as screen size, memory restraints, input mechanisms, and bandwidth constraints. Currently, many content providers simply ignore these factors, instead delivering the exact same content to mobile and desktop users regardless of what type of device they use. A number of studies show that desktop sites exhibit poor usability metrics when accessed on mobile devices (Gallant, Boone, & LaRoche, 2014). A mobile user's needs are often quite different from a desktop user. For example, a mobile user accessing flight information while navigating an airport is likely only interested in a specific portion of an airline's website, such as flight delays, while a desktop user can afford more time and screen real estate to explore more advanced options like frequent flyer miles and package deals. The following sections will detail some of the factors that come into play and how drastically they affect the mobile content experience.

Screen Size

Screen size is one of the most understandable and prominent features of mobile devices that affects how content is delivered. The simple fact that mobile device screens must be smaller than laptop or desktop to aid in enhanced mobility means that there is less screen real estate to work with. Only a limited amount of content can be displayed to a user at any given time, therefore this choice in content may be a critical determining factor in how successful the content creator is.

The importance of screen size stems from the strengths and weaknesses of human cognition. A smaller screen will limit the total viewable space, limiting contextual reference for the user. The less a user can immediately see means there is more for them to recall later. Short term memory is unfortunately weak for humans, therefore the greater the amount of content the user has to retain in order to navigate or understand a mobile site's context, the worse their overall comprehension will be. Due to limited screen size, mobile content may require extensive scrolling, multiple pages, menus, or interfaces, which can introduce serious usability concerns. The more elements that a user has contend with in order to filter out the content they want from the content they don't need adds additional levels of cognitive workload.

Input Mechanics

Another important factor that affects mobile usability is the method of user input. Mobile devices have less physical space to provide input, without the luxury of having a mouse and keyboard like a desktop user might have. Most mainstream mobile devices rely on a combination of touch screen manipulation and physical or simulated buttons to interact with. Both of these approaches have distinctive drawbacks; touch screens can lead to mistaken clicks or "fat fingering", while buttons reduce flexibility and take more time to accomplish tasks than traditional mouse manipulation (Ham, et al., 2006).

While manipulation may be difficult on a mobile device, inputting text can be especially challenging, requiring more precision than simply navigating a page. Data entry on a mobile device can be tedious and error prone; the common practice of providing a scaled down version of the QWERTY keyboard on a touch screen leads to mistaken clicks as the human finger is limited in precision when buttons go below a certain size. Keys are usually small and too close to each other, while users are often in motion when using the device, further increasing the error rate (Ham, et al., 2006). Studies at MIT report that an ideal size of 45-57 pixels is optimal for users of mobile buttons (Dandekar, et al., 2003). This is unfortunately much larger than typical interfaces allow. Uniform Resource Locators (URL's) are a particularly difficult data entry to type, containing special characters and syntax that must be input precisely otherwise the user will not be able to visit that web site (Mobile Web Best Practices 1.0, W3C Recommendation). Forms inputs may also require additional precision, as data must be entered in an expected order while the entry elements may be too small for quick navigation.

Modern mobile devices have begun to alleviate some of these issues, providing such features as back buttons, autocorrect options, bookmarks, speech recognition and auto-fill for forms. Many of these features are ones that have been passed down from existing desktop implementations. These features manage to give users some flexibility and efficiency, especially for more experienced mobile users, but they don't necessarily address the root issue. Speech recognition is notable for being a mobile-driven technology that is beginning to define and shape the unique mobile experience rather than an extension of traditional computing techniques. This is worthy of its own discussion as a means of content access in later sections and chapters.

Operating Systems and Browsers

The software that manages and runs a mobile device, specifically the operating system, plays a large part in way content is handled. In the past decade the operating system market for mobile devices has been a constantly shifting landscape, with a number of providers and technologies falling by the wayside. As of this writing, Android holds the lead for market share in the mobile environment, but Apple's iOS and Windows Mobile have enough of a stake to influence current trends. In fact, many companies that are no longer entrenched as operating systems still influence the environment through patents and partnerships with larger tech firms.

While a definitive analysis of each and every system is perhaps beyond the scope of this chapter, it is worth mentioning some of the current matchups if only to emphasize the difference between platforms. This disparity leads to the need to standardize cross-platform content delivery. Android, as the market leader, emphasizes free apps and open source distribution, as it itself is free and open source to the device makers. This means Android devices are typically cheaper at the expense of less control over operating system versions. The fragmentation of Android versions is a key contributing factor towards the reluctance of developers to embrace the platform, while the scale and reach of the device is the main draw. Apple's iOS platform in contrast emphasizes hardware integration, tightly coupling the version of the OS across all devices. This enforces a level of consistency across their devices and garners much more positive user and developer experience ratings.

Historically, mobile operating systems have employed specific protocols to allow access to Web content. The Wireless Application Protocol (WAP) was designed to provide a means of viewing Internet content on mobiles, but required content developers to make entirely new WAP-friendly sites. However, modern mobile browsers are now fully HTML compatible and no longer require the WAP conversion. While this has alleviated some device dependency issues, the mobile web browser has not evolved to keep pace with the technology advancements of devices in order to create a better experience to the user. These browsers are built on top of the same visualization techniques used by the desktop oriented web browsers, delivering the same layout regardless of platform (Dharmasiri, et al., 2013). This common practice is well suited to eliminate device dependence, but falls short on usability metrics.

NETWORKS

As mobile Web usage has overtaken traditional desktop usage, so too does the increasing demands on the wireless networks that support it. Unfortunately, the wireless and mobile telecommunications infrastructure is not growing at the same rate; companies that supply access will begin throttling and limiting usage. Much of this stems from the very nature of wireless networks. A wired network can simply add more cables or higher throughput cables like fiber optic to handle increasing demands. Wireless networks rely on a bounded range of frequencies on the electromagnetic spectrum, therefore there is only so much communication that can take place on these frequencies before they need to be placed on a wired line. As wireless carriers gradually move users to granular, high-priced models, the amount of data a particular mobile Web page sends will be of much higher concern than one in a desktop environment.

Typically, mobile networks are slower than fixed, wired data connections and will have a measurably higher latency (Guirguis & Hassan, 2010). The result is longer retrieval times for web pages, amplified by the length and size of content and the frequency a user retrieves new pages. Elements that were considered satisfactory for fixed broadband connections, such as large images or streaming video, may not be quickly accessed on a mobile environment. Mobile Web pages, just like any web page, can contain content that the user has not specifically requested, such as advertising media, which further increases bandwidth demands. User's attention spans will wane the longer a page takes to load until they reach a threshold of patience and abandon the task. Much like the early days of the Web, a web page's total size plays a critical role in how usable and accessible a piece of content is in the mobile environment.

CONTENT FACTORS

Mobile Web vs. Mobile App

It is important to note the distinction between content delivery mechanisms as that of Mobile Web content and Mobile App content. Both formats were created to provide essentially the same service, exhibiting content and reaching audiences on a mobile device. However, each avenue has particular traits that must be considered as a content developer. Much of what is discussed throughout this chapter is with regards to Mobile Web content, as that avenue requires much more thought towards adaptation. The reason is because Mobile Web is a lens by which traditional content is often seen. Mobile Apps are applications that are always specifically built with mobile devices in mind. This has the benefit of considering more of the factors and taking advantage of mobile-specific features like accelerometers and GPS, at the expense of added development time and cost. Mobile Web content in contrast is considerably cheaper to develop and deploy, but is dependent upon a delivery mechanism, namely the browser, to view and interact with. Therefore, capabilities and usability of Mobile Web content suffers more so than Mobile App content. Mobile App users are also more invested in terms of time and focus when accessing content. This will be discussed in more detail in the section on user context.

Format

Content viewed on a mobile device is generally run through several mechanisms to add an additional presentation layer that makes it palatable in that platform. This is a process typically known as "transcoding". This is not always an ideal process, as content can experience a number of issues that desktop web sites do not. One particular difference is the lack of multiple windows, specifically the ability to open more than one browser window at a time. On the Mobile Web, only one page can be displayed at one time, which means previous pages can only be accessed in sequential order. In a desktop site, a user could copy and paste information from one form to another very easily between windows. If a mobile user wishes to transmit information found in one page to another site, they would have to directly memorize this information instead or constantly click back and forth between the two pages.

Mobile devices are also not very supportive of client-side scripting, a widely used feature in most Web sites to accomplish such features as saving user preferences, viewing videos, popup windows, or loading plug-ins. In addition, some mobile devices do not support common Web protocols and technologies, such as Flash animations, which may lead to completely inoperable pages. While the mobile user may not necessarily need these advanced functionalities, if no alternatives are provided it could cause a loss in comprehension of the content.

Size and Latency

As mentioned previously, mobile networks have certain inherent limitations. Thus, mobile web browsers are considered slower compared to desktop web browsers when dealing with large file sizes and computation-intensive operations. In addition, even if a page cannot load some advanced feature, like a streaming video, it still often downloads the entire page's contents regardless. This can be an inefficient use of precious bandwidth and degrade the user experience. In order to improve the performance of web browsers, companies and researchers have worked on a number of different techniques, including parallel optimization, web page pre-fetching, and speculative loading. These techniques are designed to increase the download speed by parceling out data retrieval or filtering out useless or broken content. The specifics of some of these techniques are discussed in more detail in later sections with regards to how they actively adapt pages to reduce page size.

Sociological and User Factors

Digital content is put in a tricky position. It must be considered usable for human users but at the same time it should be usable by devices. For example, the inclusion of certain key phrases on a mobile Web page may be inconsequential from a human's perspective, but the search engines and adaptation mechanisms that parse content may view these words with extreme care. However, while technical factors are important, the user needs to be a primary driver in understanding how content should be deployed and adapted.

In many ways the success and satisfaction that users have with content is derived from a perception of fulfillment along many factors. These factors can be broadly categorized as privacy, trustworthiness, and navigational complexity (Sohaib, 2012). Privacy involves the level of information a company utilizes upon agreement to exchange digital content with a user. Many users do not have an adequate understanding of the extent of information they share during such transactions, but are generally willing to part with a nominal amount of it, such as emails and browsing history. Privacy concerns scale in accordance with the level of personal concern. Financial information and unique user credentials are a much higher tradeoff for content transaction costs.

Trustworthiness is the willingness for a user to be vulnerable to the actions of the content provider based on an expectation that the provider will perform beneficial and not harmful actions towards the user. In a general sense this is the extent by which users are willing to sacrifice resources, such as time or money, to access content. This is another perception value that is modified by many factors. Cultural, age, gender, socioeconomic status, and many other factors can shape a user's perception of what can be considered trustworthy. It is therefore important to scale content presentation accordingly in order to reach more users. This includes such features as authorization/authentication mechanisms, data integrity systems, and confidentiality agreements.

Navigation pertains to the level of difficulty a user perceives in interacting with content. Research has demonstrated that cognitive overload is an important aspect of usability (Adams, 2007). Mobile devices are particularly sensitive to the effects of cognitive overload, due to their deployment in multiple task settings and limitations of size. Users rely heavily on a foraging paradigm for browsing and accessing content. This means that every option is performed with a certain level of risk/reward analysis. Users perceive what is most relevant to them and ignore the rest. The difference in the mobile environment stems from the amount of information that is shown to forage immediately from. The cognitive process is shifting from users to adaptation mechanisms that choose what content they feel the users want to see.

Usage Context

The manner in which users operate and the environment within which they use their device is defined as the usage context. Users may be in a number of different states when using a device. They could be moving or sitting, actively engaged or using the device as a peripheral task. It is this operational context that puts different priorities on Mobile content. Mobile content consumption is at times sporadic and seen as a secondary task, the cognitive expectations therefore are often far less than they might normally be if using a desktop or laptop. Mobile users can't be expected to process large amounts of content when their concentration and attention isn't fully devoted to that particular task.

Several surveys have been done to analyze just how mobile users interact with their devices. Interestingly, the most popular usage of mobile devices is "killing time", usage that is devoted specifically to filling intermittent gaps in primary tasks with a form of entertainment or communication (Global mobile statistics, 2012). Mobile devices excel at this task because they are readily and quickly available when users are waiting around for something to happen. Favorite time-wasters include activities like social media, games, and videos. Users that are tangentially interacting with their device may often prefer ambient interaction, such as voice recognition, to access content, while their hands are free for other tasks. However, this is not to say that mobile users specifically prefer this type of content. Some users, especially those using advanced Mobile Apps, will devote afford the same levels of attention they would on a desktop application.

According to information foraging theory, users seek to maximize their cost/benefit ratio when seeking content (Fu & Pirolli, 2007). Specifically, people want more entertainment and less interaction overhead. Interaction costs, the steps taken to retrieve content, are inherently greater in mobile devices. Consequently, mobile content needs to be brief and to the point. If a mobile user only needs a few key data points from a web site, they do not want to sift through mounds of irrelevant content.

User Variables

The characteristics of the user that are specific to the user themselves is also of key importance to defining the content interaction. People can be categorized on a wide range of characteristics, all of which have shown to have an impact on mobile usage. Users with disabilities, such as poor eyesight or motor controls, have a decidedly different experience with the same content. Different age groups prefer different content formats, and certain age groups utilize mobile devices differently. Gender also plays a defining role in how users interact with mobile devices, both physically and contextually (Gold, et al., 2012).

Global and cultural factors contribute to how content is formatted and displayed. Mobile devices are used in different manners by various ethnic and social groups and the manner in which the content is delivered can make a great impact depending upon the audience. For example, Asian users have been shown to use the Mobile Web in a much higher volume, but through specified portals and SMS-based interactions, despite the majority having a non-technical background. European and American users, while less numerous, tend to be more technically literate and rely on the full Mobile Web experience, preferring advanced features and functionality that are normally associated with Mobile Apps (Kaikkonen, 2008).

Content Frameworks

The various factors of the Mobile environment seem to necessitate developing content parallel to traditional methods. In fact, in many cases this can be beneficial to have a unique mobile experience as opposed to a watered-down desktop interface. Web developers may choose to craft new versions of their desktop sites, but this can be time-consuming and labor intensive. The wide range of possible devices requires a way of reaching as many users as possible while still delivering the same content. The process of altering content to enhance the user experience on a particular device is called "content adaptation" (W3C Authoring Challenges for Device Independence). Content adaptation involves the production, selection, and modification of content to satisfy the restraints or take advantage of the characteristics unique to mobile devices. This could include hardware constraints like screen size or software constraints like native browsers. In today's Mobile environment, content adaptation is typically accomplished through a tedious process of manual reauthorization or automatic conversion by third party software. These solutions aren't generally geared towards improving usability, only maintaining consistency and reducing device dependencies.

Mobile content is relayed from a web site provider to the user through a number of steps and transformations that can be identified as a "content adaptation system". A content adaptation system is made up of a delivery framework and an adaptation mechanism (Zhang, 2007). A delivery framework determines where and when the conversion takes place while the mechanism dictates what format or visual changes will be made to the content. Frameworks may choose to apply a number of different adaptation mechanisms, ranging from simple translations to complex algorithms, but the overall purpose is to get content from a server to a client device and display it with minimal loss of clarity.

Server-Side Frameworks

Content delivery frameworks typically fall into server, client, or hybrid types, depending upon the logical setup and location where adaptation takes place. With server-side adaptation, the server detects the user's device and makes changes to the requested content to make sure that it is accessible (Fudzee & Abawajy, 2008). Server-side adaptation may be as simple as a bit of extra code inserted into the original content to handle different phone types, a catalog known as a Wireless Universal Resource FiLe, or WURFL (WURFL). In some cases, server-side adaptation is necessary just to make the content work reachable at all. Server-side adaptation typically depends upon a device detection library to detect the device accessing the web site and return its capabilities. This list of capabilities provides the web developer the ability to fine-tune the resulting page and sync with the device's capabilities. This technique of adapting to the device is sometimes called "browser sniffing" (Bickmore, 1999).

Server side frameworks are common content adaptation technique used by large internet brands and portals that have an interest in maintaining a well-defined mobile presence, including Google, Facebook, Amazon, Youtube, and Yahoo. Most server-side solutions are created specifically for a particular company, but there are several mechanisms designed for the average content developer. For example, dotMobi is a top-level domain framework that provides mobile W3C standards enforcement which assigns correctly designed pages as "mobile ready". Sites that receive this label are allowed to use a special .mobi domain and users can be assured of a certain level of usability and accessibility (Haumont & Siren, 2007).

Server-side adaptation is not without issue. The device detection solution needs to be kept up to date, which requires overhead by the developer to monitor and maintain this list. Secondly, client information is not explicitly collected or utilized in a pure server-side implementation, thus missing out on the contextual and user factors mentioned previously. The need to tailor each page on the server can become inefficient for content providers; and therefore it typically only feasible for large companies with a wide market reach.

Client-Side Frameworks

The increasing power of individual devices and smartphones has led to the ability to shift some of the burden of adaptation to the client side. Smartphones now come with specialized software built in that provides a higher degree of accessibility to web content than in the past. The client-side framework assumes a user downloads a special modified browser that converts and manipulates web sites into a format more accessible by a mobile device. The issue with this framework is that users must install software prior to accessing the content and does nothing to enforce consistency across multiple device types. A pure client-side framework requires that an entire site's content data be downloaded prior to any adaptation, which makes this an inefficient system. In practice, pure client-side frameworks are only used in special circumstances for niche markets. Today's mobile browsers utilize a combination of server and client adaptation frameworks to provide a hybrid solution.

Hybrid Frameworks

Servers and clients have complementary advantages; therefore, a combination of both, or hybrid, framework helps maintain efficiency while maximizing context during content adaptation. In the proxy-client hybrid a proxy server renders web pages into a proprietary, greatly compressed format. This content is then sent to a corresponding browser installed on the client, which may perform further adaptation as needed (Fudzee & Abawajy, 2008). This framework provides faster adaptation than if either framework handled the bulk of the transcoding and manipulation.

Opera Mini is an example of a hybrid framework and stands as one of the most popular non-native mobile Web browsers on the market. Opera Mini works by converting original content into the specified format of the requesting device, creating a copy of the original content. It reduces file size by offloading conversion to a proxy server and then delivering a slimmed down version to the client. In this manner, it provides a framework that can access any website regardless of source (51 Degrees). There are several similar browsers that take the same approach on other mobile devices, notably Apple's Safari browser and Google's Android WebKit browser. All of these browsers suffer from similar issues, such as lack of Javascript support and plug-ins or the focus on device independence over usability.

Adaptation Techniques

Adaptation techniques comprise a varied number of systems and methods in which web content is adapted into a format more usable on a mobile device. The concept of usability stems from a number of metrics, including efficiency, learnability, error rate, user satisfaction, and recall rate (Nielsen, 1994). This process may involve a number of variables and factors, including not only physical established properties, such as screen size, but also underlying intrinsic factors, such as user type, context, semantics, and usage statistics. The following section details a number of methods for web content adaptation that have been or are currently used in adaptation frameworks.

Segmentation

The concept behind segmentation is the idea that mobile content can be subdivided into portions or blocks, which can then be rearranged to better suit the user. These blocks act as segments that when recombined form the whole web site (Chen, Ma, & Zhang, 2003). Blocks might be something like a menu bar at the top of a page, a login form, or an image gallery. Each block is composed of similar, related elements that define it logically within the same functional use. A login input, for example, may be composed of a user name entry, a password entry, a login button, and a missing password link. These elements all reside under a similar functional context and when combined form logical content block. The adaptation mechanism may decide to keep or discard blocks, based on factors such as size or user preference.

Segmentation begins with a process called Document Object Model (DOM) parsing. Hypertext Markup Language (HTML) code is composed of tags that denote sections of content as being related, either by hierarchy or similar function. DOM parsing will build a "DOM tree", a logical relationship of each element on a page, in order to identify content blocks (Chen, Ma, & Zhang, 2003). The DOM tree, combined with other factors, can then be used to partition content blocks, while maintaining the underlying links between them. The content is then reshuffled into a more accessible mobile layout and then displayed to the user (Joshi & Liu, 2009). The technique of block determination can vary between implementations. Several approaches have been tried, including a structured HTML analysis method, natural language processing techniques, machine learning, and ontology based methods (Soderlan, 1999).

An example of DOM tree parsing and blocks is the VIsion based Page Segmentation (VIPS) approach, which was proposed by Deng Cai (Cai, et al., 2003). It relies upon visual analysis and display coordinates to determine the distance between elements and create relationships between content. VIPS extracts blocks from a DOM tree and then finds visual separators between these blocks. Users are presented a complete layout initially and then may focus on individual content blocks as they hover over as seen in figure 1. Unfortunately, the VIPS approach suffers from being somewhat computationally expensive (Pasternack & Roth, 2009).

Segmentation can be used to distribute content across multiple pages rather than just one single page. The cMo browser is a client-proxy approach that splits up large web pages into multiple smaller pages depending upon how content is related (Borodin, Mahmud, & Ramakirshnan, 2007). It uses a context hierarchy and emphasizes content the mobile user might be most interested in. The cMo browser uses machine learning for identifying content blocks (Xiang, Yang, & Shi, 2006). It implements a subcomponent labeled the Support Vector Machine (SVM), which is a statistical machine-learning model used to compute a relevance score for content sections with respect to the overall context. It offloads a portion of the context algorithm to a proxy server to minimize the bandwidth impact and reduce the download requirements on the user.

Figure 1. Vision based page segmentation layout (Pasternack & Roth, 2009)

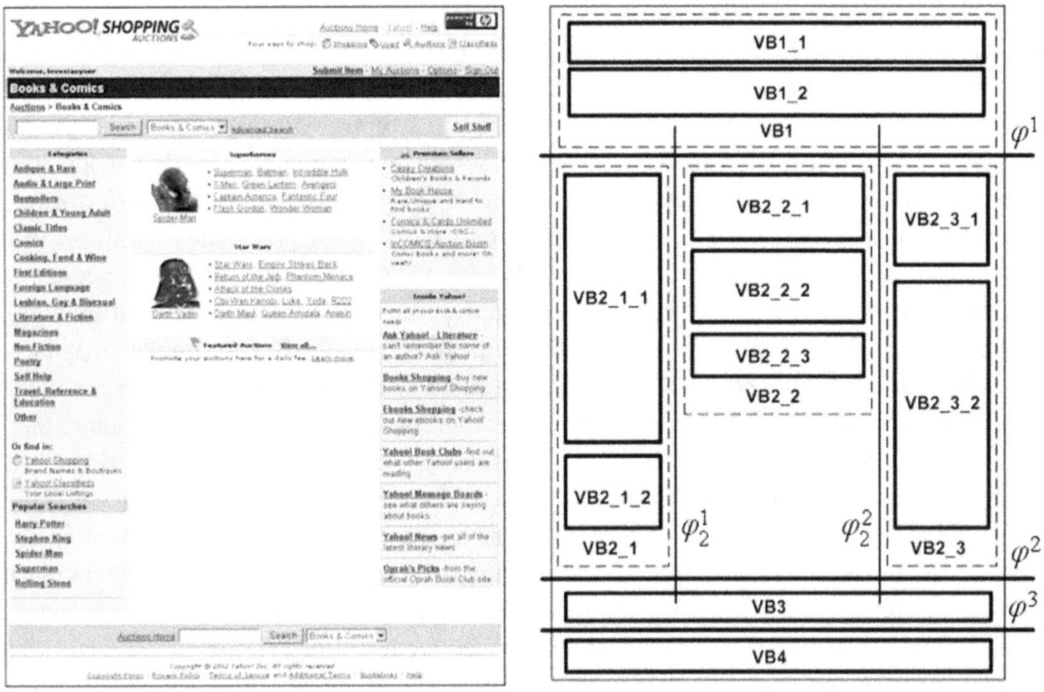

Natural Language Processing (NLP) has been used as a mechanism for content segmentation, identifying semantic relationships between sections of content. NLP has the ability to establish relationships with difficult to recognize elements, such as images. It can determine the content of images based on surrounding textual cues, such as captions or titles. Relevance to other content can then be derived based on this initial analysis. This is a task that would be difficult to accomplish solely based on the document object model (Joshi & Liu, 2009).

Filters

The principle behind mobile Web filtering is to provide users only relevant portions of a page and limit or remove the rest entirely. Filtering involves the extraction and removal of web elements to minimize the amount of information sent over a mobile network. Filter specifications are largely derived from a number of inputs, like keywords, personal web browsing history, contextual relevance, and bandwidth demands. A good example of a mobile content filtering system based on user personal data is the Myngle system. Myngle queries user web history from multiple devices and allows users to quickly find information that they are frequently looking (Sohn, Chun, Li, Battestini, Setlur, & Mori, 2011). This is designed to allow for quick task resumption as users swap between desktop and mobile environments. For example, imagine a user interested in booking a particular flight with a destination set as "Orlando, FL" using a desktop browser. Later, when that user is viewing the same site on a mobile device, the user's browser history from the desktop is relayed to the Myngle browser, indicating that "Orlando, FL" is a high priority keyword. Myngle then filters the flight information to only focus on the Orlando flights in the flight status screen.

Compression

With the demands on bandwidth in the mobile environment, compression techniques have become a mainstay of most modern browser frameworks. Though some adaptation techniques focus on visual presentation, compression focuses on the demands to the system. Compression based strategies are designed to allow multiple devices to display content in a range of screen resolutions, with less bandwidth usage and lower requirements on mobile resources. This technique is designed to optimize the performance of the CPU, memory, and network capacity. Compression minimizes HTTP requests and responses and to limit the total size of content being sent through Mobile Web channels (Paternò, Santoro, & Scorcia, 2008).

Templates provide one method of compression using predesigned site layouts, configurations, and settings for optimal viewing on a given device (Guirguis & Hassan, 2010). A template may be chosen automatically upon initial entrance to the site, using predetermined parameters. Web sites take advantage of the user agent profile, a string of device specific variables that contain such information as the screen size, operating system, and browser. This agent profile string is often cached for an individual browsing session to avoid repeated requests for the profile in the future (Guirguis & Hassan, 2010).

Visual media, such as images and video, are one the biggest demands on bandwidth. In many cases there isn't a need to send a high resolution image to a small display that will either be forced to shrink the media down to a smaller width. Compression models implement media adaptation modules that provide automated resizing of images depending upon the device request. Image compression algorithms can range in efficiency and effectiveness. While some Smartphone browsers, such as Opera mini, do manage image compression, this technique may not always be particularly optimized for every possible web site. Different image compression techniques include color space reduction, a reduction of the palette of available color swatches, chroma-subsampling, removing or averaging out colors values, and fractal compression, a method of identifying texture patterns in imagery (Mai, Hua, Bo, & Wei, 2010).

Scaling

One of the most prevalent methods of adaptation on the market is the use of scaling or "zooming" to deliver desktop content in a mobile environment. Some browsers, generally those on older feature phones, will format a web page within the confines of the mobile device width and maintain the original resolution of the content. This transforms a Web page into a one-column layout with an incredibly long scroll bar to traverse the entire page. The concept of scaling, or magnification, is simply a mechanism by which content's layout is preserved while offering the user the ability to "zoom in" and magnify portions of it. In Opera Mini, as with Android's Webkit and Apple's Safari, a large thumbnail version is created of the entire site, and then shrunk to fit into a mobile screen. Users can zoom to drill down to a particular section and then zoom out to regain perspective of the site as shown in figure 2. This certainly benefits users that are familiar with the desktop version of a web site, while users without prior experience have difficulty hunting for particular items in the zoomed out preview. For complex and content rich sites that aren't designed with mobiles in mind, the original format is not always optimal for small screen viewing.

Scaling has definite advantages from a resource perspective. One particular scaling browser, TutMobi, is specially designed with impressive performance gains, with almost 90% reduction in bandwidth, while retaining content layout (Lehtonen, et al., 2006). Opera Mini and Mobile boast similar performance metrics (Opera Mini & Opera Mobile browsers). Some browsers have also tried scaling individual content elements of a web page, rather than the entire page. The ThunderHawk browser uses font resizing and

Figure 2. Opera Mini browser

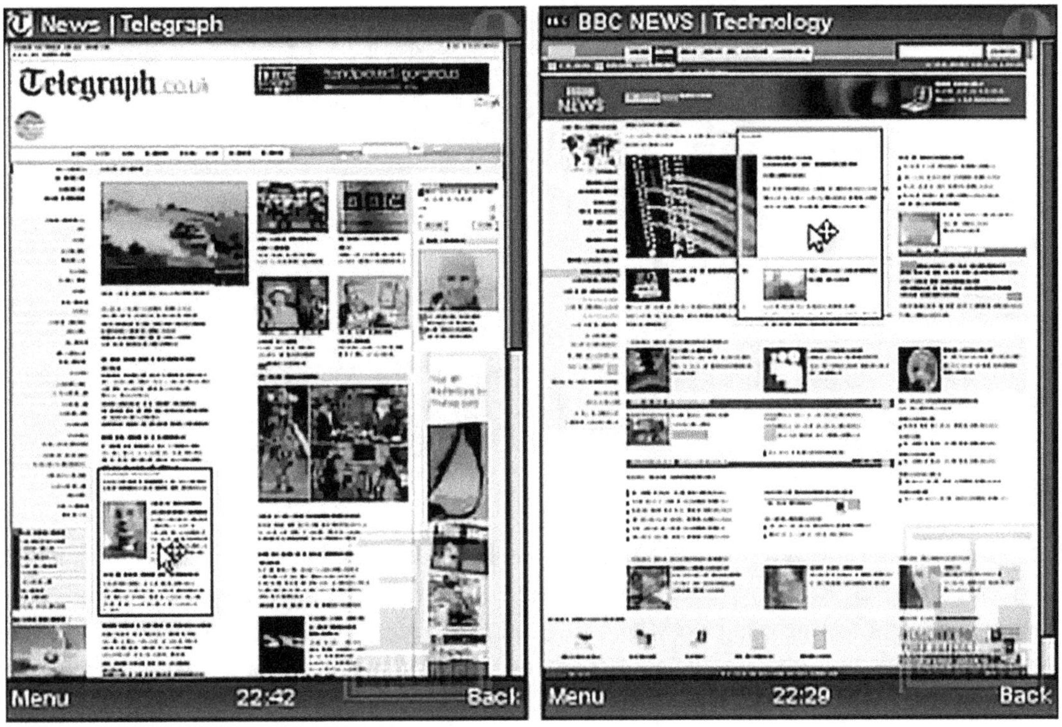

graphic scaling to adapt a web page (Lam & Baudisch, 2005). In order to conform to a small screen, the Minimap browser changes the Cascading Style Sheets (CSS) by modifying the size of the text and limiting the width of the paragraphs (Virpi, Popescu, Koivisto, & Vartianien, 2006). This result is a scaled-down version of the content, called an overview, on top of the browser viewport. These thumbnail previews lets users browse large numbers of web pages, but users still need to zoom and scroll for a more detailed view (Lam & Baudisch, 2005).

Responsive Design

A relatively new construct, Responsive Design is a considered a design principle rather than an automated technique. It requires a certain amount of planning by content developers, but once built allows for a great degree of flexibility and accessibility across many different device types. As seen in figure 3, the basic idea is to create content that has enough leeway and scalability that it can grow or shrink to varying resolutions, including those of mobile devices (Marcotte, 2011). Responsive design is based on three principles. The first is the user of a flexible "grid" where all content pieces are "snapped" to relative rather than absolute positioning. This eliminates the need for specific pixel dimension settings for different mobile devices. Content then scales as needed within the grid as the available screen real estate becomes available on large devices. The second principle is the use of flexible images, meaning that images on the grid must be allowed to relocate, resize, or be removed, if the grid shifts. The third principle is the use of CSS "media queries", special queries to the device that respond with an appropriate style sheet based on the size of the device. This helps ensure a similar style across a broad range of resolutions (Marcotte, 2011).

Figure 3. Responsive web design output (Martin, 2012)

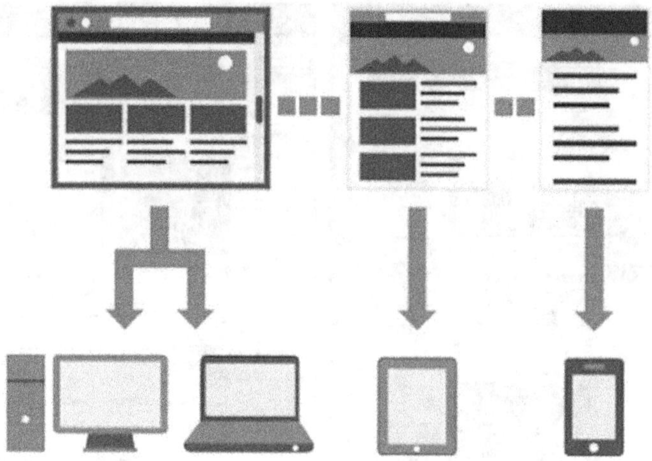

Responsive Design allows a single document to reach a wide range of devices and still retain the same look and feel. However, Responsive Design is not an automated solution; rather it is a set of guidelines to follow when first creating content (Marcotte, 2011). It requires the mobile content creator to plan ahead early on in the development cycle. Responsive Design also ignores other mobile factors beyond screen size and simply creates content that fits into various resolutions. It doesn't take advantage of the features that are unique to mobile devices.

Responsive Design is relatively inefficient from a bandwidth perspective. It still needs to download full desktop-sized images even if they are not used. Second, Responsive Design does not perform well on low-end devices because the underlying HTML remains unchanged, which can leave broken or missing functionality on the page (Marcotte, 2011). Responsive Design doesn't utilize any client-side data to adapt the site for that particular device. It just creates template sites for resolutions that a "family" of devices might use. This can be a very labor-intensive process for web developers, requiring a lot of careful, upfront planning to create a scalable site (Marcotte, 2011). Although this is a reasonable solution towards device independence, it doesn't manipulate the content in such a way as to actually improve usability and performance.

Summarization

The process of reducing the total amount of content shown to a user is a mechanism known as summarization. Summarization tries to manage limited screen size by condensing content information and thus reducing the amount of time a user needs to successfully understand and navigate content on a mobile screen. The process of summarization relies on intrinsic properties of web content, such titles, themes, keywords, metadata, tags, and captions (Ham, et al., 2006). Unlike filtering, which eliminates content, summarization refines content more succinctly without losing the basic intention. Summarization focuses on the written and hidden portions of content, a task that makes it well suited for Natural Language Processing.

NLP techniques involve the creation of "lexical chains", a list of related words and phrases determined by the use of references in the content (Alam, Hartono, Kumar, Rahman, Tarnikova, & Wilcox, 2003). Words that appear to exhibit a relation must retain this connection once the entire summary is complete. Nouns are then extracted and run against synonym sets to determine relations. Depending upon the process technique in use, a scoring device is utilized to determine the optimal arrangement of words. The ultimate determination of the candidate phrase is based on a variety of factors, such as lexical chain depth, distance between terms with content, number of references to a term or term synonym, and the diversity within chains. The final summary is then derived from multiple chains once a strong candidate is selected. The length of content plays a large part in determining the number of chains and thus the length of the final summary (Alam, et al., 2003).

Personalization

Personalization refers to the adaptation of a user's mobile experience to better suit their individual needs. Generally, personalization is "on-demand"; the user actively makes adjustments to a website's configuration on their browser that alters the look and feel. In these sites, a user account is created, which links the user to the site, and then the user selects options to determine which content they wish to view. Many website portals choose this method to allow users the ability to serve up content out of a much larger content source. The problem is that the selection of such options may be difficult or tedious for the user to carry out, especially on a mobile device. Users are often not willing or interested in hunting for personalization options, especially given the limited amount of focus users might give to mobile content.

Some web site portals like Yahoo have begun using learning algorithms that specifically tailor content to individual users without their explicit consent. Autonomous personalization is a method of adapting content without the user's active involvement. While this approach minimizes user involvement, it suffers from scalability, requiring a large number of users to supply data before it can consistently make intelligent recommendations.

Prioritization

Prioritization focuses on determining content's importance based on both the user and site context. This means that not only are user factors considered, such as would be the case in personalization, but site factors like structure, statistics, and data flow are also considered. The result is an alternate version of the original content, redesigned with what the user and the site deem as most important. The MCAP framework was developed as a means of content prioritization that considers all available data points and scales accordingly (Sonnenberg, 2013). The MCAP framework is an autonomous framework that automatically converts web pages into optimized mobile sites. This particular framework parses content into blocks, as with segmentation, then reorganizes and reformats blocks on a stack that delivers high priority content first.

The MCAP framework is a server-client approach; the adaptation is performed on the web host with the aid of site databases and user agent strings. This minimizes user and developer impact, while still providing a mobile optimized solution. It scales with the amount of implicit and explicit site context, analyzing the content, the site hierarchy, the user context, and the user's device at the same time. The prioritization feature of the MCAP framework is handled by the Content Order algorithm, a sorting method by which content blocks are ordered by numerical rank. This numerical rank is derived from a number of individual rank calculations, including distance, function, user, semantic, and novelty weights. As an unordered list of content blocks is provided to the algorithm, a new ordered list of blocks is returned based off the weights and values assigned by each calculation. For example, the novelty weight measures content based on how recently it was updated and how frequently it changes compared to surrounding content blocks. Content providers may choose to emphasize certain weights to better suit their site's overall needs.

Standardization

The World Wide Web Consortium (W3C) stands as the main international standards body for the World Wide Web. This includes all desktop as well as mobile web and application content standards. W3C has documented a list of guidelines and principles into Mobile Web Best Practices 1.0. It details a number of areas including presentation, delivery, security, accessibility, and usability. Many of these guidelines are employed in the W3C mobile readiness checker, a free website that analyzes websites and grades them based on their conformance to the standards. W3C also maintains a list of best practices for mobile application development in a separate document. The following is an example of a best practice guideline taken direct from section 5.2.8 of the Best Practices document:

Do not cause pop-ups or other windows to appear and do not change the current window without informing the user. (W3C, 2008)

Over the past several years, W3C has developed a number of Web technologies that are particular geared towards standardizing mobile devices. These technologies help ensure a seamless mobile experience and eliminate many of the possible issues that traditional content creation on a desktop might create. These technologies include CSS Mobile, a version of the cascading style sheet language that matches the need of mobile authors, SVG Tiny, a scalable vector graphics format suited to the capabilities of mobile devices, and XHTML For Mobile, which defines a subset of XHTML for mobile sites.

SOFTWARE AND RESOURCES

With the explosive growth of mobile content creation, a wide variety of tools and software have been created to help developers. As mentioned, multiple mobile-specific versions of standard technologies like CSS and XHTML exist. Cross-platform mobile development tools are also gaining popularity as the need to develop applications separately for different mobile operating systems becomes laborious and inefficient. The following list includes a number of resources available that can assist content creators in reaching mobile users while maximizing usability and efficiency.

Web Developer Tools

- **goMobi** (http://gomobi.info/) Mobile site builder, providing enforced mobile standards on existing content.
- **jQuery Mobile** (http://jquerymobile.com/) A unified, HTML5-based user interface system for mobile device platforms, built on the jQuery and jQuery UI foundation.
- **jQTouch** (http://www.jqtouch.com/) a Zepto/jQuery plugin for mobile web development on the iPhone, Android and other devices.

- **Mobify** (http://www.mobify.com/) Allows users to create a mobile version of their website through an intuitive and user-friendly graphical user interface.
- **Webflow** (https://webflow.com/) – Responsive Design focused drag-and-drop website builder for designing custom mobile web sites.

Cross-Platform Application Tools

- **Adobe AIR** (http://www.adobe.com/devnet/devices.html) A framework to build applications that run on the Apple iOS, Google Android, and BlackBerry Tablet OS platforms.
- **Appcelerator** (http://www.appcelerator.com/) An SDK for developing native, hybrid and mobile web applications from a single codebase.
- **Corona SDK** (http://www.coronalabs.com) An SDK that aims to deliver, write-once-and-build to iOS, Android, Kindle Fire and NOOK, using Lua on top of C++/OpenGL.
- **Sencha Touch** (http://www.sencha.com/products/touch/) A high-performance HTML5 mobile application framework for iPhone, Android, and BlackBerry.
- **TestPlant** (http://www.testplant.com/): A GUI-based automated test tool for mobile application across all operating systems and devices.

Standards

- W3C Mobile Ready Checker - http://validator.w3.org/mobile/.
- W3C Mobile Application Best Practices - http://www.w3.org/Mobile/mobile-web-app-state/.
- W3C Mobile Web Best Practices - http://www.w3.org/TR/mobile-bp/.

FUTURE TRENDS

A number of dynamic and potentially disruptive technologies have developed alongside mobile devices that could alter the development of adaptation techniques. These include systems that provide additional data to enhance the experience, present data in a new visual fashion, provide alternative input mechanisms, or incorporate data from alternative sources. The following section details some of the areas that will influence mobile adaptation and to what extent this may be a positive or negative contribution.

Location-Based Content

Mobile devices are optimally designed to take advantage of location-awareness, the ability to interact with and interpret a user's immediate surroundings. This is evident with the advent of Augmented Reality technology, but there are other targeted location services available for mobile education. One example, already in use in a number of museums, is the use of contextual audio feedback. These technologies exist as simple applications, scan codes at particular exhibits, or specialized devices that push content to nearby phones. However, new developments are taking the possibilities one step further. One of the most interesting applications of "content push" systems is presented in Apple's iBeacon product. These devices act as distributed nodes that emit localized content in a physical area. Cheap in design and easily configured, the beacon will automatically push content out to any device that is listening in a particular area. The implication of this technology is that as content becomes more narrowly focused and tailored to individual users, the need to localize and provide metadata to contextualize it will become increasingly necessary. This can be a powerful tool for content providers and allow a much more nuanced and granular view of the adaptation and deployment process.

Haptic Feedback

While mobile devices certainly suffer drawbacks, there are some capabilities in development than may give them an advantage in other areas. One such technology is "haptic" feedback, the ability for users to physically sense motion and touch when using a device. Basic tactile feedback is becoming increasingly common in mobile devices. Handset manufacturers like Nokia, LG and Motorola are including haptic technologies in their phones. In most cases, this takes the form of vibration response to touch. In 2013, Apple Inc. was awarded the patent for a much more accurate haptic feedback system designed for multi-touch surfaces. Apple's patent describes a system where at least two actuators are positioned beneath a multi-touch input device to provide vibratory feedback when a user makes contact with the unit (Apple, 2013). The patent provides for one actuator to induce a feedback vibration, while another actuator creates a second vibration to suppress the first from propagating to unwanted regions of the device, thereby "localizing" the haptic experience. This provides much more precise feedback to user's individual actions. Content adaptation will be able to take advantage of this opportunity, providing feedback to user input that would not be possible in a desktop environment. Some applications have even explored the possibility of assisting visually disabled users with haptic-controlled features (María Galdón, et. al., 2013).

Augmented Content

Augmented Reality (AR) presents content in a format that overlays information on top of a user's field of vision. Google Glass is a primary example of this technology in action. While the augmented space is not as mature as mobile space, at least in terms of usability, there has been a handful of interesting developments in the field of content compression and interpretation. One method of compact visualizations for Augmented Reality provides a mechanism to reduce resource demands on the entire system (Tatzgern, Kalkofen, & Schmalstieg, 2013). This research presented a method to transfer visualizations to AR settings and achieve improved frame rates on limited-resource hardware. It utilizes filter and layout optimization to create compact information overlays. The reduction of information generates an improvement in terms of speed of mobile hardware with limited resources.

Another line of research incorporates gesture-based interactions as a means of interpreting and predicting user intention in AR systems (Keil, et. al., 2013). It utilizes a "magic lens paradigm" and introduced gestures that employ natural movements, such as gazing (pointing for long periods of time at the same area), movement patterns, and the distance/proximity to a target to influence the presented information. This provides a certain amount of contextual cues that can be used for prioritizing particular pieces of content in a dense spatial environment. The researchers note that as a preliminary prototype, there is not a substantial quantitative evaluation of the results. The discussion is largely subjective; usability testing would be needed to verify these particular approaches. However, the potential to utilize a user's entire range of physical gestures could provide enhancements to the localization and personalization of content adaptation.

CONCLUSION

The history and progression of mobile content is one of device independence. As the lines begin to blur between what is a desktop machine, a Smartphone, and a tablet, so too does the future of adaptation. Current practices and research point to improved usability when sites are created specifically for a target device (Pernice, et al., 2014). However, the reason for such findings is not only dependent on the technology, but on the entire contextual and sociological experience. The shift from traditional desktop usage to mobile device usage has ushered in a new era of content adaptation for the mobile environment. As the majority of content access has shifted to mobile devices, and as "mobile devices" are further redefined by what the technology brings, the driving influence on all digital content presentation and adaptation will depend on the mobile ecosystem.

Software solutions and adaptation mechanisms have come a long way towards reaching as many people as possible. However, there is only so much a browser can do to perfect the user experience. It can alleviate latency issues and provide accessibility, but when it comes to usability, the developer of the site must be actively involved. They will need to understand the human element: the user interaction that merges the computer elements with the cognitive process. Only by understanding the usability factors and tailoring content specifically to users can adaptation be truly perfected.

REFERENCES

51Degrees. (n. d.). Retrieved from http://51degrees.mobi/

Adams, R. (2007). Applying advanced concepts of cognitive overload and augmentation in practice; the future of overload. In D. Schmorrow, K. M. Stanney, and L. M. Reeves (Eds.), Foundations of Augmented Cognition (2nd Edition) (pp. 223-229).

Alam, H., Hartono, R., Kumar, A., Rahman, F., Tarnikova, Y., & Wilcox, C. (2003). Web page summarization for handheld devices: a natural language approach. *Seventh International Conference on Document Analysis and Recognition*, (pp. 1153-1158). doi:10.1109/ICDAR.2003.1227838

Apple Insider. (2013). *Apple awarded patent for more accurate haptic feedback system*. Retrieved from http://appleinsider.com/articles/13/02/19/apple-awarded-patent-for-more-accurate-haptic-feedback-system

Baluja, S. (2006). Browsing on small screens: recasting web-page segmentation into an efficient machine learning framework. *Proceedings of the 15th international conference on World Wide Web (WWW '06)* (pp. 33-42). New York, NY, USA: ACM. doi:10.1145/1135777.1135788

Bickmore, T. (1999). Web Page Filtering and Re-Authoring for Mobile Users. *The Computer Journal*, *42*(6), 534–546. doi:10.1093/comjnl/42.6.534

Borodin, Y., Mahmud, J., & Ramakirshnan, I. (2007). Context browsing with mobiles - when less is more. *Proceedings of the 5th international conference on Mobile systems, applications and services (MobiSys '07)*. New York, NY, USA. doi:10.1145/1247660.1247665

Budiu, R., & Nielsen, J. (2012). *Mobile Usability*. San Francisco, CA: New Riders Press.

Cai, D., Yu, S., Wen, J., & Ma, W. (2003). VIPS: A vision-based segmentation algorithm. MSR-TR-2003-70.

Chen, Y., Ma, M., & Zhang, H. (2003). Detecting web page structure for adaptive viewing on small form factor devices. *Proceedings of the 12th international conference on World Wide Web,* Budapest, Hungary (pp. 225-233).

Dandekar, K., Raju, B. I., & Srinivasan, M. A. (2003). 3-D finite-element models of human and monkey fingertips to investigate the mechanics of tactile sense. *Journal of Biomechanical Engineering, 125*(5), 682–691. doi:10.1115/1.1613673 PMID:14618927

Deegan, R. (2014). Mobile HCI: Issues Surrounding Cognition, Distraction, Usability and Performance. *International Journal of Mobile Human Computer Interaction, 6*(1), 1–14.

Dharmasiri, C., Jayendranath, R., Ariyarathna, A. L., Perera, P. M., & Weerawarana, S. M. (2013) Web browsers on smart mobile devices: A gap analysis on the state of the art. Proceedings of the Advances in ICT for Emerging Regions (ICTer) 2013 International Conference, 75(79), 11-15. doi:10.1109/ICTer.2013.6761158

Fu, W.-T., & Pirolli, P. (2007). SNIF-ACT: A cognitive model of user navigation on the world wide web. *Human-Computer Interaction, 22*(4), 355–412.

Fudzee, M., & Abawajy, J. (2008). *A classification for content adaptation system.* Paper presented at 0th International Conference on Information Integration and Web-based Applications & Services. New York, NY, USA.

Gallant, L. M., Boone, G., & LaRoche, C. S. (2014). Mobile Usability: State of the Art and Implications. *Interdisciplinary Mobile Media and Communications: Social, Political, and Economic Implications, 344.*

Global mobile statistics. (2012). Retrieved from http://mobithinking.com/mobile-marketing-tools/latest-mobile-stats/d#mobilebehavior

Gold, J. E., Driban, J. B., Thomas, N., Chakravarty, T., Channell, V., & Komaroff, E. (2012). Postures, typing strategies, and gender differences in mobile device usage: An observational study. *Applied Ergonomics, 43*(2), 408–412. doi:10.1016/j.apergo.2011.06.015 PMID:21764031

Guirguis, S. K., & Hassan, M. A. (2010). A smart framework for web content and resources adaptation in mobile devices. *Proceedings of the 12th international conference on Advanced communication technology*, Piscataway, NJ, USA (Vol. 1, pp. 487-492).

Ham, D.-H., Heo, J., Fossick, P., Wong, W., Park, S., Song, C., & Bradley, C. (2006). Conceptual framework and models for identifying and organizing usability impact factors of mobile phones. *Proceedings of the 18th Australia conference on Computer-Human Interaction: Design: Activities, Artefacts and Environments (OZCHI '06)* (pp. 261-268). New York, NY, USA: ACM. doi:10.1145/1228175.1228221

Haumont, S., & Siren, R. (2007). *dotMobi, a Key Enabler for the Mobile Internet.* Paper presented at Mobile Internet User Experience Workshop.

Huang, M.-J., & Lee, T.-L. (2008). *An integrated software processor with autofilling out web forms.* Paper presented at Computer Systems Architecture Conference.

Joshi, P. M., & Liu, S. (2009). Web document text and images extraction using DOM analysis and natural language processing. *Proceedings of the 9th ACM symposium on Document engineering (DocEng '09)*, (pp. 218-221). New York, NY, USA. doi:10.1145/1600193.1600241

Kaikkonen, A. (2008). Full or tailored Mobile Web- where and how do people browse on their mobiles? *Proceedings of the International Conference on Mobile Technology, Applications, and Systems (Mobility '08)*. New York, NY, USA: ACM. doi:10.1145/1506270.1506307

Keil, J., Zoellner, M., Engelke, T., Wientapper, F., & Schmitt, M. (2013). Controlling and filtering information density with spatial interaction techniques via handheld Augmented Reality. In *Virtual augmented and mixed reality. Designing and developing augmented and virtual environments* (pp. 49–57). Springer Berlin Heidelberg.

Lam, H., & Baudisch, P. (2005). Summary thumbnails: readable overviews for small screen web browsers. *Proceedings of the SIGCHI conference on Human factors in computing systems (CHI '05)* (pp. 681-690). New York, NY, USA: ACM. doi:10.1145/1054972.1055066

Lehtonen, T., Benamar, S., Laamenen, V., Luoma, I., Ruotsalainen, O., Salonen, J., & Mikkonen, T. (2006). Towards user-friendly mobile browsing. *Proceedings of the 2nd international workshop on Advanced architectures and algorithms for internet delivery and applications (AAA-IDEA '06)*. New York, NY, USA. doi:10.1145/1190183.1190190

Lira, W., Ferreira, R., de Souza, C., & Carvalho, S. (2014, September). Experimenting on the cognitive walkthrough with users. *Proceedings of the 16th international conference on Human-computer interaction with mobile devices & services* (pp. 613-618). ACM. doi:10.1145/2628363.2628428

Mai, T., Hua, C., Bo, T. J., & Wei, L. (2010). A Study of Image Compression Technique Based on Wavelet Transform. *Proceedings of Genetic and Evolutionary Computing (ICGEC) 2010 Fourth International Conference* (pp. 826-828).

Marcotte, E. (2011). Responsive Web Design. New York, NY: A Book Apart.

María Galdón, P., Ignacio Madrid, R., de la Rubia-Cuestas, E. J., Diaz-Estrella, A., & Gonzalez, L. (2013). Enhancing mobile phones for people with visual impairments through haptic icons: the effect of learning processes. *Assistive Technology*, *25*(2), 80–87. doi:10.1080/10400435.2012.715112 PMID:23923690

Martin, M. (2012). Retrieved from http://searchengineland.com/responsive-design-alone-is-not-mobile-seo-124202

Mobile Accessibility. Targeting the mobile browser. (n. d.). Retrieved from http://www.reflektions.com/miniml/template_permalink.asp?id=508

Mobile Web Best Practices 1.0, W3C Recommendation. (2008). Retrieved from http://www.w3.org/TR/mobile-bp/#d0e253

Mohomed, I. (2007). Correlation-based content adaptation for Mobile Web browsing. *Proceedings of the ACM/IFIP/USENIX 2007 International Conference on Middleware (Middleware '07)*, New York, NY, USA (pp. 101-120). doi:10.1007/978-3-540-76778-7_6

Network, U. W. (2011). *Wireless infrastructure: Can our current system handle bandwidth demands.* Phoenix, USA: College of Information Systems and Technology.

Nichols, J., Hua, Z., & Barton, J. (2008). Highlight: A System for Creating and Deploying Mobile Web Applications. *Proceedings of UIST'2008*. Monterey, CA. doi:10.1145/1449715.1449757

Nielsen, J. (1994). *Usability Engineering*. San Diego, CA: Academic Press.

Norman, D. A. (2010). *Living with complexity*. Cambridge, MA: MIT Press.

Opera Mini & Opera Mobile browsers. (n. d.). Retrieved from http://www.opera.com/mobile/

Pasternack, J., & Roth, D. (2009). Extracting article text from the web with maximum subsequence segmentation. *Proceedings of WWW '09 the 18th international conference on World Wide Web*, New York, NY, USA (pp. 971-980). doi:10.1145/1526709.1526840

Pernice, K., Nielsen, J., Farrell, S., Mizobuchi, S., Ishida, N., Stover, U. A., & Richardson, A. (2014). *Usability Guidelines for Accessible Web Design. Evidence--Based User Experience Research, Training.* Consulting.

Sauro, J., & Kindlund, E. (2005). A method to standardize usability metrics into a single score. *Proceedings of the SIGCHI conference on Human factors in computing systems (CHI '05).* New York, NY, USA. doi:10.1145/1054972.1055028

Soderlan, S. (1999). Learning information extraction rules for semi-structured and free text. *Machine Learning*, *34*(1), 233–272. doi:10.1023/A:1007562322031

Sohn, T., Chun, F., Li, Y., Battestini, A., Setlur, V., & Mori, K. (2011). Myngle: unifying and filtering web content for unplanned access between multiple personal devices. *Proceedings of the 13th international conference on Ubiquitous computing (UbiComp '11)* (pp. 257-266). New York, NY, USA: ACM. doi:10.1145/2030112.2030147

Sonnenberg, C. (2014) Maintaining Thematic Consistency within Mobile Content Adaptation. *Proceedings of the International Conference: Industry, Engineering & Management Systems*, Cocoa Beach, FL.

Tatzgern, M., Kalkofen, D., & Schmalstieg, D. (2013). Dynamic compact visualizations for augmented reality. In Virtual Reality (VR), 2013 IEEE (pp. 3-6). IEEE. doi:10.1109/VR.2013.6549347

Virpi, R., Popescu, A., Koivisto, A., & Vartianien, E. (2006). Minimap: A Web Page Visualization Method for Mobile Phones. *Proceedings of SIGCHI*, *06*, 35–44.

W3C Authoring Challenges for Device Independence. (n. d.). Retrieved from http://www.w3.org/TR/2003/NOTE-acdi-20030901/

W3C mobileOK Basic Tests. (n. d.). Retrieved from http://www.w3.org/TR/mobileOK-basic10-tests/

W3C Web Accessibility Initiative Standards Committee. (n. d.). Retrieved from http://www.w3.org/WAI/mobile/

WURFL - Wireless Universal Resource FiLe. (n. d.). Retrieved from http://wurfl.sourceforge.net/

Xiang, P., Yang, X., & Shi, Y. (2006). Effective Page Segmentation Combining Pattern Analysis and Visual Separators for Browsing on Small Screens. *IEEE/WIC/ACM International Conference on Web Intelligence (WI '06)*, (pp. 831-840). Washington, DC, USA. doi:10.1109/WI.2006.67

Xiao, Y., Tao, Y., Li, Q., & Li, W. (2008). Web Page Adaptation for Small Screen Mobile Device: A New P2P Collaborative Deployment Approach. *Intelligent Networks and Intelligent Systems, 08*, 191–196.

Zhang, D. (2007). Web content adaptation for mobile handheld devices. *Communications of the ACM, 50*(2), 75–79. doi:10.1145/1216016.1216024

KEY TERMS AND DEFINITIONS

CSS: Cascading Style Sheet, A standard markup language construct used for defining elements, location, fonts, colors, and other visual factors of a web page.

DOM: Document Object Model, the base model used by HTML pages to define the logical distribution of content and hierarchy of dependencies.

GUI: Graphical User Interface, a graphic-based interface for human operators to interact with a machine or device.

MCAP: Mobile Content Adaptation and Prioritization, adaptation mechanism framework used to prioritize mobile content.

NLP: Natural Language Processing, technique of interpreting human speech on a device and converting it to machine commands and constructs.

RGB: Red, Green, and Blue, standard color composition for lighting in tablets and smartphones. Colors defined for use with such devices are given a numerical value to correspond to how much red, green, or blue is needed for each color.

SQL: Structure Query Language, a language used to storing and retrieving data from a database.

W3C: World Wide Web Consortium (W3C), a standards body that governs the protocols and best practices for deploying content to the Web.

Chapter 11
Mobility Matters:
Classifying Locative Mobile Apps through an Affordances Approach

Brett Oppegaard
University of Hawaii at Manoa, USA

ABSTRACT

Researchers generally have not been able to keep up with the rapid pace of mobile app innovation in many ways. One of those holes is with typology. Not being able to clearly describe mobile app artifacts, as a way to start a discussion about them, is a key factor holding researchers back. This chapter splinters off a promising but relatively underdeveloped genre of mobile apps, the locative ones that juxtapose interactive digital media with particular physical environments, for closer examination and better labeling. This piece quickly provides theoretical, historical, and contextual overviews to help situate locative apps as a genre of importance. It then provides informal frameworks, as thought exercises, for examining and understanding apps through a locative perspective. Two cases, also led by the author, will then be taken through this process to illustrate what can be learned from it, followed by reflections and suggestions for related future areas of inquiry.

INTRODUCTION

Just before the iPhone's original release, on June 29, 2007, MacWorld magazine published a story about customers waiting in line for the new device. This account featured the proclamation by an early adopter that "Today, I've got three things in my pocket," a cellphone, a PDA, and an iPod; "Tomorrow, I'll have one" (Barylick & Honan, 2007). Combining functions, it turned out, was only a part of the wonders of the smartphone emergence, and convergence was just the start of our understandings of mobile technologies as a new medium. To demonstrate how disorienting this mobile shift has been for society so far, the generally prescient Steve Jobs at first touted the iPhone's music storage possibilities and then its telephone capabilities. He only reluctantly opened the App Store, 18 months later, and stated that he didn't expect "much profit" from it (Kerris & Dowling, 2007; Wingfield, 2008). More than 10 million

DOI: 10.4018/978-1-4666-8838-4.ch011

apps, though, were downloaded on the App Store's first weekend, and, as tallied at the five-year mark, the App Store was paying out $400 million a month to developers, after Apple took its 30-percent cut from each app sold (Friedman, 2013).

On the other side of the spectrum, BlackBerry, formerly known as Research in Motion, seemingly had control of the smartphone market in the early stages, in the mid-2000s, with a focus on business clientele and a carefully controlled system of information management that matched the initial needs and activities of mobile users. As of this writing, though, less than a decade later, BlackBerry's market share has fallen to virtually zero (Silcoff, McNish, & Erman, 2013; Surowiecki, 2012). Nokia – the first global developer of mobile hardware, and at one time, the largest – has slumped as well, selling its mobile division to another teetering technology titan, Microsoft, which, despite enormous resources toward the effort, has been unable to gain much traction with its mobile devices (Ali-Yrkkö, 2013). As demonstrated in these examples, society quickly has been evolving its conceptualizations of mobile technologies, through exposure and use, and of being immersed within the complex, fluid, and varied confluences of communication, people, and place, which mobile technologies enable. Some highly paid and highly educated industry leaders, such as those within Apple and Google, have been deftly responding to this environment, while others – even those similarly paid, educated, and deft, such as those in the once-prominent Palm, Nokia, BlackBerry, and HTC companies – have suffered an inglorious fade from relevance.

Rather than the mobile content domain solidifying around a coherent industrialized vision, and parts falling into orderly place, based on commercial precedents, it has erupted into a heterogeneous and fragmented ecosystem. Classification systems have been proposed, in response, to provide shape to these ideas, and as fuel for development, yet attention to mobilities needs as much of a focus on the future as the present, with novel uses and interactions continually expected to emerge from new products, platforms, and services (Feijóo, Maghiros, Abadie, & Gómez-Barroso, 2009). While much of the early mobile media could be classified simply as shovelware, or repackaged material dumped on mobile screens, mobile-born forms are surfacing through novel device and platform affordances at various points of inspiration ((Feldmann, 2005; Jaokar & Fish, 2006). These innovations will produce both incremental and divergent developments as well as recombinative and convergent ones, that will require regular re-evaluations of content types and characteristics across multipurpose as well as special-purpose devices (Kim, Lee, Geum, & Park, 2012; Rawolle & Hess, 2000).

This dramatic shifting of human relations to mobile technologies, throughout a worldwide population, can be traced theoretically to the idea that, by enabling new sense ratios in people, a new communication medium simply shocks society, and people take a while before they figure out what to do with it (McLuhan, 1962, p. 23). As Google CEO Eric Schmidt (2010) said during his keynote speech at the Mobile World Congress in 2010: "It's like magic. All of a sudden you can do things that it never occurred to you was possible." McLuhan (1962) stated that historically such technology shock eventually is followed by a period of general adjustment, in which the community absorbs the new habit of perception, and a new normal emerges. Yet to make that adjustment, the medium has to settle through the process of people getting their hands on the technologies and thinking about what they would like to do, while researchers observe, document, and analyze what they actually do, when given the chance.

This chapter has been written during that period of tumult and discovery, in which mobile devices are pervasive, but people ambitiously are experimenting with the related technologies, and researchers still are trying to make sense of what's happening. Potential paths of inquiry seem everywhere, and researchers generally have not been able to keep up with this rapid pace of innovation in many ways. One of those holes is with typology. Not being able to clearly describe mobile app artifacts, as a way to start a discussion about them, has been identified as a key factor keeping researchers behind. This chapter therefore will splinter off a promising but relatively underdeveloped genre of mobile apps, the locative ones that juxtapose interactive digital media with particular physical environments, for closer examination and better labeling. This piece will quickly provide theoretical, historical, and contextual overviews to help situate locative apps as a genre of importance. It then will provide informal frameworks, as thought exercises, for examining and understanding apps through a locative perspective. Two cases, also led by the author, will then be taken through this process to illustrate what can be learned from it, followed by reflections and suggestions for related future areas of inquiry.

LOCATION PLUS MOBILITY

Mobility and location awareness are the common, underlying characteristics of contemporary wireless communication systems (Farman, 2012; Gordon, & de Souza e Silva, 2011; Zeimpekis, Giaglis, & Lekakos, 2002). This place-based type of app – often designed to enhance a historic or culturally meaningful place, such as Philadelphia's Independence Hall or Shakespeare's London – has no established home base in the current mobile app development environment. It has no distinct category in the App Store. Instead, these types of apps are spread throughout the market clutter, tossed under such broad labels as education, books, business, lifestyle, travel and local. They not only are hard to find among these general categories, they also often – rarely attempted as profit-making ventures – are lacking resources for promotion. With few business models for efficiently making money, an audience not quite sure what to make of them, and substantial privacy concerns related to location awareness, these locative apps remain generally marginalized and obscured from public view and discourse. As such, they have not been thoroughly explored by users or designers. Yet these apps also arguably could become the quintessential expression of mobile technologies, creating complex new media forms through unprecedented hardware affordances, based on their fundamental *mobileness*. They symbolize the potential power of mobile – and ubiquitous computing – ideals. Representative of net locality, as envisioned by Gordon and de Souza e Silva (2011), they help to create a complicated and blended physical-digital environment, in which identity constructions, social practices, group interactions, security risks, information interpretations, haptic sensations, and many other dynamics swirl wildly from the Internet tether. But if bottom-line and privacy concerns, at least, can be mitigated, locative apps could emerge as a definitive form of mobile-born expression. An initial obstacle in the way, though, is the rhetorical one of how to fundamentally situate them in our minds. If locative apps cannot easily be classified, categorized, and described to potential users, then they will never be fully investigated and developed as a genre. While a few scholars have begun creating mobile taxonomies of different sorts, no categorization system for locative mobile apps has yet surfaced (Handal, El-Khoury, Campbell, & Cavanagh, 2013; Zhu, Cao, Chen, Xiong, & Tian, 2014). This chapter is intended to start such an effort, through a thoughtful exploration of this topic.

This piece also will argue that a critical next step in the development of this field, and as part of the societal sense-making process, is the development of a robust classification system for locative apps, incorporating empirical data of available variations and following a systematic approach toward the creation of a robust taxonomy. This system should allow interested parties to parse and clump as well as filter mobile aspects from the non-mobile in a heavily integrated technoscape. By bringing together similar locative concepts, under unifying labels, deeper thoughts about those isolated ideas can develop. Mobility and ubiquity are deeply intertwined with contextual awareness in this advancement of a framework. Connectivity and location-awareness are at the analytical tool's core, and without both of which mobile technologies are severely limited in offering distinct experiences. Other substantial attempts at mobile classification have been made, incorporating a variety of strategies, including a focus upon mobile spatial interactions (Bilandzic, & Foth, 2009); mobility and creativeness (Feijóo, Maghiros, Abadie, & Gómez-Barroso, 2009); revenue and economic models (Feldmann, 2005); and innovation through convergence and divergence (Kim, Lee, Geum, & Park, 2012). The framework here, instead, will be guided by an affordances approach to mobile technologies, or a classification effort based upon what the devices allow us to do and what we can do with them; this is not from a prescriptive stance of describing what we should do, or need to be doing, but more focused on describing the possibilities and avenues for development. Users and designers of these technologies and apps ultimately will decide and dictate the details; our purpose here, though, is to classify potential courses of action within this complex mobile environment and to give them some rhetorical boundaries and shapes. Theoretically, the inherent location-awareness aspects of such a mixed reality can be thought of as "net locality," or networked locality, in which something different begins happening to "individuals and societies when virtually everything is located or locatable" (Gordon & de Souza e Silva, 2011, p. 2). This piece, from that net locality perspective, will describe that something different in ways that could help to classify and operationalize it for further study.

NET LOCALITY

A debate in mobile spatial theory is the divergence of those who think mobile devices separate people from their surroundings and those who think mobile devices allow people to blend better into our heavily mediated and digitized physical spaces. Burke (1966) puts this rift into historical perspective by arguing that humankind is defined, in part, by being "separated from his natural condition by instruments of his own making." Once a person, no matter how primitive, he said, departs from the needs defined by survival standards of sheer animality, a new "natural" emerges (Burke, 1966, p. 27). In that respect, mobile devices, over time, likely will feel less foreign and more integrated into our spaces, like other ubiquitous technologies that now are accepted as a part of modern life, such as eyeglasses, and books, and television screens. Envisioning space through a lens of technology is not new, either. Barton and Barton (1993) note that a paper map, for example, not only presents a skewed sense of a physical space but also delivers ideological undertones. Yet most people consider the map a "neutral" form, not a colonizing endeavor, or a type of civic boosterism, creating an optimist's view intended to draw interest and visitors (Barton & Barton, 1993, p. 236).

By being relatively novel, at least, mobile devices maintain their hypermediacy, in which users seem well aware that something significant is happening to them and their surroundings when they use mobile devices, even if they cannot articulate what that is. This moment seems to be the ideal time to attempt a classification model; as Hayles observed, the user and medium are becoming cyborg, with each informing, expanding, and limiting the other's capacities (Hayles, 1999). In the process, space, digital information, locative applications, and users are all bound up together (de Souza e Silva & Sutko, 2011). Mobile spatial experience therefore can be thought of as a merging, or as if people are entering a hybrid realm that incorporates physical, digital, and represented spaces. In this process, the realm generates new spatial perceptions and new possibilities for social networks across all spaces. These spaces, though, do not completely overlap with each other. They instead are superimposed and connected through social actions. This, in effect, creates roughly a doubled perception of space, in which there becomes a clear articulation between the physical and the informational, or what traditionally has been classified as the real and the virtual. While in this realm, users simultaneously see and experience their physical surroundings but also access the representation of that space on the mobile device (de Souza e Silva & Sutko, 2008).

This transformation of physical location into digital information and vice versa illustrates the concept of "becoming," based on Gilles Deleuze's philosophy of the virtual (Gordon & de Souza e Silva, 2011), or the "Singularity," as Ray Kurzweil (2005) described it. From a Deleuzean perspective, the real is pure immanence, absorbing both virtual and physical realities, while the virtual is always ready to emerge, to have physical existence. People do not simply operate in space. Space also operates on people (Gordon & de Souza e Silva, 2011, pp. 33-36). Kurzweil predicts that at the time of the "Singularity," about three decades from now, there will no longer be a distinction between humans and technology. He said, "This is not because humans will have become what we think of as machines today, but rather machines will have progressed to be like humans and beyond" (Kurzweil, 2005, p. 40).

An apt metaphor of this evolutionary shift might be the railway, as McLuhan described it in communication terms (1997, p. 152). Railroads did not introduce movement or transportation or wheel or road into human society, but they accelerated and enlarged the scale of previous human functions, creating totally new kinds of cities and new kinds of work and leisure. This societal acceleration took place whether the railway happened to be functioning in a tropical or a northern environment, and it was functioning independent of the freight or content of the railway medium. With a similar extension of scale, the Internet has been amplifying the cognitive abilities of humans, thrashing apart some media institutions while erecting others. When mobile channels – connected and delivered by such devices as the iPhone – melded with the networking capacity of the Internet, a massive and convergent communication layer was created. No wonder that chaos to the status quo ensued. Accessibility to the communication system, as in getting mobile technologies in the hands of most people, meanwhile has been provided at an unprecedented pace (Castells, Fernandez-Ardevol, Qiu, & Sey, 2009). Most people now have a mobile device of some kind, with more than 90 percent of American adults owning such a device, and a rapidly growing number of those being "smartphones," such as an iPhone or 'Droid. From June 2000 to June 2009, wireless penetration rates in the United States rapidly rose from 34 to 89 percent (CTIA, 2011). Those high penetration rates have been mirrored in countries throughout the world, with about 6.8 billion mobile subscriptions in service today for the globe's 7 billion people (International Telecommunication Union, 2013). As all of these mobile users get the devices in their hands, though, they see they have more than just a digital Swiss Army knife. But what, exactly?

With new hardware features, more computing power, and communication capabilities added to each iteration cycle, mobile devices present learners with broad access to informational, social, and cognitive supports. Museums, historic sites, and other place-based attractions meanwhile are created, at their core, to provoke new thought and interaction, which makes them ideal innovation testing grounds for orchestrating and configuring different media and technologies related to learning (Hsi, 2008, p. 125-126). Handheld guides for such inspiration have been traced to 1777, when a housekeeper at Kedleston Hall in Derbyshire in the United Kingdom reportedly offered, as part of her duties as "distinct articulator," a printed catalog of the fine artworks on display (Parry, 2008, p. 180) Even though the audio guides phase of such curatorial outreach began in the 1950s, experiments with interactivity in handheld guides did not emerge until 1973, at the Milwaukee Public Museum, in which the audio narration there would ask a question, and pause, until the learner pressed the key that corresponded with the correct answer. The solicitation and promotion of the learner's voice, on a parallel plane with the curators, started about a decade later, in 1985, at the Metropolitan Museum of Art, when anecdotal interviews with learners in front of specific artworks were spliced together into an audio tour that challenged the museum's authority as a monolithic voice (Tallon, 2008, p. xxiv). The multimodal and interactive mobile app extends such research but also, with the emergence of its net locality abilities, has the potential to break new ground in lifelong learning opportunities, if we can figure out ways to effectively talk about it together.

Such ideas present a picture of a complex intertwining of the personalities and beliefs of the interface, or interpreter, with the individual and personalized interests of the user, and the affordances and constraints of the technology and the space. Foucault characterized the modern epoch, above all, as an epoch of space (Foucault, 1986). Then, he added (p. 22), "We are in the epoch of simultaneity: we are in the epoch of juxtaposition, the epoch of the near and far, of the side-by-side, of the dispersed. We are at a moment. I believe, when our experience of the world is less that of a long life developing through time than that of a network that connects points and intersects with its own skein." People need physical interfaces to reach digital environments, and virtually every space they encounter is mediated in some way. The physical space, de Souza e Silva & Sutko (2011) contend, can be considered the interface for information while information, in turn, becomes an interface for the physical space. In the symbolism of the alternate reality gaming crowds, the physical space serves as a game board for an experience happening underneath ordinary everyday activities. Location-aware interfaces add to reality, rather than negate, replace, or diminish it.

BACKGROUND

Jeremy Hight, one of the pioneers of mobile place-based experimentation, created such a model hub of information related to a particular GPS coordinate, in a 2003 piece called "34 North 118 West." Through that project, he demonstrated that "context and subtext can be formulated as much in what is present and in juxtaposition as in what one learns was there and remains in faint traces." He began to envision that location in the city as "a collection of data and subtext to be read in the context of ethnography, history, semiotics, architectural patterns and forms, physical form and rhythm, juxtaposition, city planning, land usage shifts, and other ways of interpretation and analysis" (Hight, 2003). Michael Epstein, another early experimenter in net locality, looked at his 2005 project, "History Unwired," a walking tour of Venice, with more of an activist mentality, trying to encourage visitors to explore and get to know better the less-traveled neighborhoods of the otherwise centralized tourist area. He noted that learners could be brought closer to issues and locales through a narrative overlay on maps that also could be consulted for geographic orientation and other data (Epstein, 2009, p. 1). Field expanding research has been at the core of Blast Theory's various "shared network narratives," such as the spatial emphasis on "Can You See Me Now?" and the social emphasis on "Uncle Roy All Around You" and the temporal emphasis on "Day of the Figurines," which, focused upon kairos, relates to timing of information delivery (Benford & Giannachi, 2009, p. 443). These studies, and even ones that focused on audio guides and early mobile systems, typically centered on the designers' perspectives, through the puzzles to be solved, without much insight from users. The studies of this type can help to establish the edges of the potential for net locality today. But because user response also can be highly valuable, and because the technical advancements in mobile devices, such as the iPhone and iPad, have been so rapid in recent years, dramatically changing system dynamics, new explorations need to be launched with new questions in mind about how and why people use net locality in place-based attractions.

During the dawn of institutionalized communication studies research, between the 1930s and 1950s, such efforts likely would have been examined from the theoretical perspectives of a "magic bullet" or "hypodermic effects," in which the messages produced, regardless of medium, were intended for targets, like bullets or needles, with the effects of the insertion immediate and evident. Through the later 1950s and the 1960s, Paul Lazarsfeld and associates at Columbia University rejected such simplistic ideas in favor of the "minimal effects" model, in which audience motivations and prior beliefs also influenced the interpretation of messages, as did the ensuing discussion of the messages among peers, friends, family, etc. The polemic "not so minimal effects," or "big effects," paradigm formed as a response, focusing on ways in which media frame discussions, not necessarily affecting what people think but clearly affecting what they think about (Neuman & Guggenheim, 2011, pp. 171-172). Concentrating on cultural and social impacts of net locality at this point, though, might be premature (Perse & Dunn, 1998). The shift from effects-based inquiry to a more functionalist perspective responds to the opaqueness of understanding of computer-mediated communication, particularly in relation to net locality. Therefore it seems prudent to begin with as blank of a slate as possible and, through mixed methods within contextual inquiries, to establish a corpus of descriptive information about the process of identifying a locative mobile app and develop systems from there.

Studies of mobile media and mobilities in communication around such ideas can be traced back to the 1990s, and sociology, psychology, and many other fields share interests in these topics. Some early subjects of academic concern were the types of communication modes embedded into mobile phones (voice, SMS, internet, television, interpersonal communication, etc.), the social questions to which mobile technologies are linked (communication in developing countries, health communication, youth and media), media use (transistor radios, playing cards and dice, etc.) and traditions widely independent of media, such as urbanism or more logistical perspectives of mobility (Jones, Karnowski, Ling & von Pape, 2013). Current areas of scholarly emphasis include clusters of studies around biomonitoring (using mobile devices to computationally track personal behavior), cyber relationships (from bullying to sexting), and investigations of various strains of the privacy-surveillance debate, demonstrating just how intimately integrated mobile technologies have become in modern society. Recent theoretical overviews expand such thoughts to include conceptualizations of the mobile medium as the global network that now connects us all (Gordon & de Souza e Silva, 2011), the ubiquitous interface through which we embody space (Farman, 2012), and so embedded throughout society today that mobile devices broadly create and maintain social cohesion while also fading into the fabric of society, as other technological infrastructures have done, like roads, clocks, and power lines (Ling, 2012).

As this field of inquiry has grown and developed, Campbell (2013) noted that the lines separating mobile communication from other forms of networked connectivity have increasingly blurred. He therefore suggested defining the field through the devices and services that support mediated social connectivity while the user is in physical motion. Such interaction may take place through voice, text, picture, video, or otherwise, and mobile devices do not necessarily need to be used on the go, but, they afford – make possible – mobility during mediated communication.

Affordances only are one paradigm through which to examine mobile technologies, and by choosing such a focus, this chapter does not further address many other significant issues related to mobile use and societal integration, including concerns about privacy, affordability, accessibility, empowerment, commercialization, and the development of mobile etiquettes. Yet by focusing on affordances, and helping elevate them for broader perception, at this point in the field's history, this chapter can highlight some high-potential opportunities for invention in description, particularly related to the typology task.

CLASSIFICATION SYSTEM VALUE

With more than 1 million apps available in the Apple App Store alone at this point, and the top 1.6% of app developers making multiples of the revenue of the other 98.6% combined, and no universal Dewey Decimal-like system in place for adequately classifying and categorizing all apps, the current app market is considered unsustainable. (Wilcox & Voskoglou, 2014). App business models, though, are beyond the scope of this chapter; instead the focus here is upon the value of typology and classification, using locative mobile apps as an illustration of that potential value. The study of the use of mobile apps is critical to generating an understanding of user preferences, and a key initial step in such a study is thought to be the act of classifying apps (Nickerson, Muntermann, Varshney, & Isaac, 2009; Zhu, Cao, Chen, Xiong, & Tian, 2014). Even though context-dependent delivery can be achieved in other forms, such as through brochures, mobile apps offer many benefits that cannot be matched by analog materials or desktop computers (Emmanouilidis, Koutsiamanis, & Tasidou, 2013). Those need to be described and evaluated in groups of similar apps for wider understandings to be reached. So the focus here has been narrowed to locative apps, as a way to make such an investigation manageable.

The descriptive data of projects that the current app market leaders, Apple and Android, provide to the public – with obscured search algorithms, nebulous and inaccurate meta tags, and routinely incomplete search results – help to create a chaotic development environment rife with ambiguity (Zhu, Cao, Chen, Xiong, & Tian, 2014). This problem has emerged as part of the advent and explosive growth of the app market. Users and developers therefore need an analytical framework, through which locative apps can be classified and thereby better understood within their contexts.

When encountering an app for the first time, particularly a locative app, users and developers need to know where the app fits into the larger ecosystem. Mobile research into what people want to do, and what they actually do, has taken many forms so far, including attempts to establish mobile as a medium, and, from various field studies, determining what mobile does best and what limitations it has. Like with the railway example, mobile technologies have had too many complex effects to neatly summarize the field in a few pages. Few studies to date, for example, have looked at mobile apps from this classification perspective (Handal, El-Khoury, Campbell, & Cavanagh, 2013; Zhu, Cao, Chen, Xiong, & Tian, 2014). So this chapter will focus upon first creating a starting point for determining what the "essence" of the mobile medium is, or what the mobile medium does better than any other medium (Smith, 2010).

Beyond just channels of storage and delivery of information, a medium, as Bolter and Grusin (1998) describe it, is that which remediates, indicating a connection to previous media forms but also demonstrating that something new emerges through the different form. Mobile technologies, such as smartphones and tablet computers, store and deliver but they also appropriate the techniques, forms, and social significance of other media and attempt to refashion them in the name of the real, per the extension of the Bolter-and-Grusin definition. Therefore, mobile technologies should be considered an independent medium, with distinct characteristics, including unique affordances (and constraints; although constraints generally are outside of the scope of this piece, too).

AN AFFORDANCES APPROACH

Affordances is a term created by Gibson (1979, p. 127), to describe what an environment provided or furnished its inhabitants, for good or ill, relative to each individual. From a mobile perspective, not all devices and interfaces are created equally, so an activity within an environment with a new smartphone might be altogether different for a person with an older feature phone, which might have had the top-of-the-line capabilities two years earlier. In that respect, even a change in Internet connection speed, or a break altogether, could drastically affect the affordances of a mobile device user in a particular environment, without changing the individual involved, or the hardware, or the software, or the interface. In that respect, affordances are conceptualized along the lines of thought of Norman (1988), who expanded Gibson's term by separating the inherent properties of a thing, such as a mobile device, from the perceived properties, which help each individual person determine what exactly that thing could be used for, and how (p. 9). The actual properties matter less, according to Norman, because they are merely potential uses. Unless a person identifies the use potential, typically derived from the design itself, the affordance might as well not exist. Locative potential, for example, is irrelevant as an affordance if not incorporated into the design.

Beyond just the device-human connection, though, Laouris and Eteokleous (2005) suggest expanding ideas about mobile engagement further to take into account the mixed physical-digital environment in which the mobile user interacts (p. 6). From that perspective, mobile use is not envisioned as just a person and a device interacting in an otherwise antiseptic space. Mobile use instead typically happens in media-rich places, with a lot of distractions, and users often are engaged in more than one activity at the time of use, causing a heavy cognitive load, meaning that the acknowledgement of the physical-digital dynamic is integral to the understanding, while the layers of complexity, caused by the significant affordances and constraints of mobile use, dictate the perceived potential in any particular place. Before describing other key perceived affordances of the mobile medium, a few important foundational points should be made about the distinguishing features of mobile technologies, compared to other communication technologies, as a way to establish the realm in which the affordance exploration takes place.

Mobile technologies fundamentally are personal, personalizable, and part of a social identity. Turkle (2008, p. 121) equated the ubiquity of mobile devices to an appendage, in which mobile devices are "always on, always on you." Time Magazine's (2012) mobility poll, conducted by Qualcomm, showed that half of Americans check their mobile device every hour, while nearly everyone checks every day. Almost 70 percent of respondents slept with their mobile devices next to their beds, and only 1 percent of users kept their mobile device in the garage, with another prominent social identity technology, the car. These mobile devices not only are different brands, with varying levels of power, but they are modified by users to have different apps, in different places, and they typically are skinned with a case that somehow represents the user's social identity, meaning mobile technologies essentially are as close to people today as a technology can be. Mobile technologies meanwhile should be considered highly usable and accessible, at least compared to other recent computing systems. One does not need to go back as far as punchcards to recall how difficult early computers were to use, especially to laymen. The command-line prompt still was common in the early-1990s, requiring the use of obscure computing terms and syntaxes to make the computers perform basic functions, but, even after windows-like interfaces were adopted, many people still did not see a practical use for the machines. With mobile technologies, though, users pretty much just have to be able to turn the device on, and press the fingerprint-sized button, to make the device perform complicated procedures. Kids can do it. Senior citizens can do it. As shown on YouTube, even cats can do it. Mobile technologies also connect people of all types on an individual level. Unlike the first corded telephones or radios or television sets, which usually were shared among an entire family, the mobile device is inherently individualized, with great personal agency, because people now call other people, not places. Societal expectations, in turn, have become high that each person maintains such an open channel.

ESSENTIAL TOPICS

The distinguishing characteristics of mobile apps are largely defined by their functional and non-functional attributes. While functional requirements vary widely, and are application-specific, non-functional requirements are common to mobile devices in general and imposed by technology usage and its associated restrictions. Common non-functional requirements are related to the mobile device characteristics (such as its small form factor), wireless technology usage and connectivity, security, privacy and interoperability considerations (Emmanouilidis, Koutsiamanis, & Tasidou, 2013). The focus in this chapter is on such non-functional distinctions, or the ways in which mobile apps can distinguish themselves from desktops or mediated forms, beginning with these concepts:

- **User Awareness:** As soon as the user powers on the device, and provides the linking login information, the mobile technology knows a lot about that user, including how many times the user has interacted with the device before, what those interactions have been, the precise location of the user, what is around this place, the time of day, the weather, whether the user has been to this place before with this mobile device, and so on. The more the user interacts with the mobile device, particularly in this same place, the more the device knows about what the user does, and wants to do. Sophisticated programming of the device, in turn, can make each of those data points into a precision tailoring system, to give the user what is needed, when it is needed.
- **Device Awareness:** With the hundreds of different handsets and numerous types of tablet computers available, each user likely will have a different experience based solely on the types of functions, and buttons, and physical characteristics available through the device. That said, if the programming is aware of the device, which it can be, and if the programming tailors the content to the device being used, the effect is that of a custom-made and caring experience, even if complex algorithms make all of the adjustments.
- **Location Awareness:** When Selective Availability of GPS satellites was turned off in 2000, because military leaders no longer deemed the technological shrouding necessary, the accuracy of location awareness on mobile devices instantaneously increased at least five-fold, from precision within about 100 meters to within 20, or even 10 meters. Today's devices can pinpoint location down to the meter, and every mobile device is required by law in the United States to have GPS tracking hardware, which means that the devices know where the user is, and from that knowledge, combined with mapping, what is around that person.
- **Spatial Awareness:** If the world can be mapped, including the identification of points of interest, and the user location can be determined, a combination of those affordances allows a broader view of the potential experience in store for the person wielding the device. As with any digital interface, a virtually unlimited number of facets can be presented to this user, based on a theme,

or physical proximity, or personal interests. Instead of working with knowledge of a small space, like a 3-meter halo around the user, spatial awareness allows a broad and varied perspective on the potential use-case scenarios.

- **Contextual Awareness:** Knowing where a person is, and what is around that person, is helpful, but knowing what the user likes to do, or what other similar users have liked doing, and how that person likes to do it, or the people before liked to do it, creates the ultimate opportunity for precisely tailoring an experience to a particular individual and generating optimal results from that activity. The more a person uses a mobile device, the more data the user generates, and the more recognizable the use patterns can become, especially when combined with other user data at the same place.
- **Social Capabilities:** The publishing potential of mobile technologies has few limits and comparables, with the unprecedented abilities to create, edit, and distribute most media forms, dispatching them all over the world at the touch of a button, within the same small, handheld forms. The media messages can be short and intimate or large and complex, as much as mobile cares, without any additional hardware, software, or costs.
- **Search Capabilities:** Mobile apps cannot control or contain the curiosities of users in the ways that other, more closed systems were able. In short, the user, at any point, can connect in the same device with an Internet search engine and reach out to other sources with other viewpoints about the same subject matter the designer is displaying.
- **Hyperlinking Capabilities:** If the Web provides an endless rabbit hole of possibilities, mobile technologies put such potential into the hands of users when they might want it most, when they might need it most, or simply when it is most convenient to explore.
- **Interactive Capabilities:** While mobile devices have many powerful outputs, such as vibrant HD screens, external speakers, and even the ability to noisily vibrate, they also have similarly powerful input devices, such as cameras, microphones, and keyboards. What this means is that instead of one-way communication, like traditional media forms typically offer, mobile technologies can provoke a two-way interchange of media messages, including prompts that initiate new activities and behaviors, generate more engaged interactions, and develop more robust connections.
- **Analytic Exhaust:** A computer can track every keystroke, but a mobile device can record exponentially more data, creating a highly detailed trail of screen touches, location check-ins, file exchanges, sensory inputs, and environmental contexts. As a pervasive technological appendage, the mobile device therefore can provide a thorough accounting of a person's daily activities, and, with long-term use, most of a life's highlights but also its more forgotten moments, through the details captured and stored with virtually no effort.

FUTURE RESEARCH DIRECTIONS

With those general concepts in mind, at least 10 key questions need to be addressed about an app, to determine its place in the locative field, as part of this working and evolving typology. Those are:

- Does this app register the user's location? A locative app inherently must identify and distinguish the user's place in the world, whether that is through a machine-curated response, such as some sort of algorithmic procedure, or a human-manipulated alignment activity, such as pressing a button to indicate presence in a place. If the mobile app does not in any way connect location and user, it does not belong in this locative category, and it cannot be understand through a locative perspective.

- Does this app dynamically align and filter its content to the user's place, for the user's benefit? Mobile apps that record location, and mobile behaviors, as a way to sneakily serve advertisers or manipulate users into consumption also do not belong in the locative category, because the location responsiveness and the integration of mobile media and place do not advance public good. So, at least from this chapter's locative perspective, the question is not just what can the technologies do, but what can they do for the user? This function, for the user's benefit, should be readily identifiable. From that identification, the user then should be able to make an informed decision about the value of the service versus the cost of the locative information provided.
- Does this app dynamically respond to what is around the user? Knowing what is within the sensory range of the user, such as sightlines, makes an app valuable, but providing extra-sensory information, like what is around the bend, could be an invaluable service, in many ways. To accomplish provide such service, the app not only needs to know where the user is, but it also needs to understand the surroundings, and, at least to some degree, the use-case scenario.
- Does this app respond to the changing context of the setting? With location and spatial awareness, a locative app has power, but to extend that potential even further, it needs to tap into contextual awareness. Each use of a mobile device builds a record of user preferences. When a place is developed spatially, through mobile media, an additional layer of context also can be added, including variations in time of day, weather, site activities, and user preferences. These contextual responses should be easily identifiable, and, again, for the user's benefits.
- Does this app dynamically respond to the user's behaviors? When a user interacts with the app, or the place, a locative app ideally should be reconfiguring itself continually to this present and emerging situation while also determining the best potential courses of action for the next moment in the user's experience. If the app is providing only a static experience that does not change, regardless of any factor, it is not taking full advantage of the locative possibilities of mobile. So while the app would be a locative one, if it responded to the user in place, it might not be as effective or engaging as it could be, which this question would help to isolate and identify.
- Does this app allow users to create in-situ multimedia objects? When the user is in place, and engaged, a locative app could prompt the user into various multimedia production modes, such as photo taking, video making, and journaling. If the app does prompt such an activity, what is the response to that activity? The response rate would be interesting, of course, but better yet, the qualitative output of such activities could provide otherwise inaccessible insights about mobile app use, and locative app use in particular.
- Does this app use network connectivity? While many activities could be initiated without network connectivity, an Internet connection combined with previously mentioned locative app abilities could create many benefits, including various kinds of unforeseen innovations. In short, connecting the user and place to the rest of the world, via the Internet, all at the same time, is a major expansion of our sense ratios, worthy of further exploration, particularly in the locative app genre.
- Does this app allow users to share multimedia objects? With connectivity, and the abilities of locative apps to upload and download information on the move, other interested parties can be connected to the place and the media artifacts of the place. Sharing with people not on site gives the user the role of place guide, and those off site could become more connected to the place through interactions with the user on the ground.

- Does this app curate user content? Raw sharing has great value, but any successful system quickly could become overrun with content, without refined sorting and classification options. The Apple and Android markets are prime examples of such organic growth into poorly organized systems.
- Does this app allow users to connect with each other directly? Sharing to a general repository is valuable, but sharing to friends and family is personal. When a user is personally invested, in the place, and in the locative app, a different relationship with the experience develops.

Ultimately, such a questioning process could help to distinguish locative apps as their own category of mobile media, but it also could have a more profound impact on mobile app classification. In short, does answering yes to each of these questions mean a locative app is more "mobile" in its essence? Could such distinctions, labeled through systematic categorization, indicate factors of increased or decreased *mobileness*, and the more of those there are, the more "mobile" the locative app? To explore these ideas in a pragmatic manner, let's compare two of my recent projects through the process previously described.

A COMPARISON AND DISCUSSION

Case No. 1 is the Fort Vancouver Mobile app (Oppegaard & Shine, 2014; www.fortvancouvermobile.net). That project is a national award-winning one, celebrated widely for creating new kinds of history discourse. It has been categorized in both the Apple and Android markets under "education." That nebulous categorization of the app belies its intent to challenge traditional historical interpretation genres. A user could reasonably try this app and at first consider it a new multimedia form of a brochure, or even find similarities with a wayside sign. Or an audio tour, only this one has videos. But the more the user explores the app, the more it appears as a sort of media equivalent of a platypus, with some recognizable parts that generate more questions than answers. For example, this app contains quizzes. Does that make it a lesson? It contains a treasure-hunt activity. Does that make it a field trip or a game? It asks users to create media objects, such as photos and videos and audio recordings. Are people in that process simply frosting the cake of history, or are they actively participating in the history-making process? The app offers several different modules, about different events, in various storytelling styles. Does that mean it's a digital book?

To help to determine if this Fort Vancouver Mobile app should be classified as locative, I could take it through the questioning outlined above, to discover:

- Does this app register the user's location? Yes, it persistently tracks location.
- Does this app dynamically align and filter its content to the user's place, for the user's benefit? Yes, based on location, and the selected story module in use, the app delivers packets of media (text, photos, animations, video) to the user based on where on site that person is. If the user is not within 1 mile of the center of the site, the app automatically provides travel information to encourage the user to go to the site. If the user is within the radius of the site, the app suggests different story modules that might interest the user. Then, the story unfolds as the user walks around the site and passes through predefined hot spots.
- Does this app dynamically respond to what is around the user? No, even though it might appear to the user that the app is responsively reacting to the experience, the design is a premade path that does not change, regardless of the responses of the user.

- Does this app respond to the changing context of the setting? No, night or day, rainy or sunny, in a group or alone, the design does not change based on the environmental changes. It could accommodate such activities, as expansion of the current design, but at this point, it does not.
- Does this app dynamically respond to the user's behaviors? No, again, the design is static. Besides stalling, or stopping, the app does not care what the user does except for proceeding forward through the design path.
- Does this app allow users to create in-situ multimedia objects? Yes, this app has several prompts for creating multimedia objects of all sorts, including photos, videos, and text entries. It typically gives the user a thought-provoking piece of media first, such as a photo or video, and then asks for something similar in return.
- Does this app use network connectivity? Yes, but only for sharing files with the server, such as user photos, and for off-loading local storage, such as using streaming video. This app does not connect to the Internet for any other reason.
- Does this app allow users to share multimedia objects? Yes, but only in limited and undefined ways. The app receives user content, but it does not describe well where that content will go, what that content will be used for, and who has the rights to it.
- Does this app curate user content? No, at this point, the user content is stored, but nothing is done with it.
- Does this app allow users to connect with each other directly? No, early usability testing showed that users were reluctant to connect with strangers on site, for safety and privacy reasons. No further developments were made to create alternatives, such as a secure system of trusted contacts, or anonymized use.

With five yes answers (and some of those qualified) and five no answers, I would classify this app as using middle-of-the-road mobile potential, including reaching some of the low-hanging fruit of the field, such as tracking location. This questioning process helped me to see where the weaknesses of the design remain, from a locative perspective, but also where the app is tapping into the power of mobility in interesting ways through novel combinations of the technology affordances. In short, this app appears that it could be much more "mobile," but it also is taking advantage of more than one affordance, which shows mobile promise.

Case No. 2 is a recent service app created in collaboration with the National Park Service at Yellowstone National Park (www.yellowstonemobile.wordpress.com). This first-generation Yellowstone Mobile app offers helpful logistical information, such as when the six predictable geysers, including Old Faithful, are expected to erupt. There is a map, safety tips, and FAQ. The app is categorized under Travel in the App Store and Travel & Local in the Google Play Store, which seems reasonable for those previously mentioned functions. But the app also offers access to the Upper Geyser Basin webcam, which primarily is used by people who are not at the park, and therefore not in explicit travel mode. They might be thinking about traveling, or recalling their travels, but they likely are not in the local – meaning inside the park – environment, traveling. The app also offers social media feeds from Twitter, YouTube, and Flickr. These feeds provide news and updates about the park, which could be classified at times, because of the content, as a traditional "news" app, such as when there is a missing person in the park. Or, it could be classified as entertainment, such as during a photo contest. The park, through these feeds, posts beautiful professional photographs and videos ranging from practical (traffic during road closures, or snowmobile pass sales) to artistic (geyser steam at sunset, or flowers in bloom). The app therefore could be considered a fine art gallery just as easily as it could be considered a tool for tourists.

More, though, can be understood through the locative app questioning outlined above:

- Does this app register the user's location? Yes, it tracks the user's location when the app is opened. This is done for practical reasons, though, instead of interpretive ones. When crowds are heavy at Yellowstone, the Internet bandwidth gets clogged, so the webcam feature of the app will not work well, and it will degrade the experience of others. Therefore, this app turns the webcam feature off when a user is inside the park, to avoid straining the Internet connection. Besides that background feature, the app does not use the location information for anything else.
- Does this app dynamically align and filter its content to the user's place, for the user's benefit? No, it provides information that the user can align with location, such as the eruption times of the predictable geysers in the Upper Geyser Basin, but the user has to make the connections, of being in the right spot and looking at the right screen.
- Does this app dynamically respond to what is around the user? No, again, this app offers information to the user about the surroundings, such as boardwalk safety tips, answers to frequently asked questions, and geyser eruption times, but it does not know, or care, where the user is when receiving that information.
- Does this app respond to the changing context of the setting? No, whatever the situation, the app responds in the same way.
- Does this app dynamically respond to the user's behaviors? No, the app does not track user behavior or change based on user behavior.
- Does this app allow users to create in-situ multimedia objects? No, at one point in its design, it did offer users the ability to take a postcard image at Old Faithful, but that feature was tabled for later development. Otherwise, users cannot create anything within this app.
- Does this app use network connectivity? Yes, but only in the very limited sense of transmitting and receiving updated eruption times. Because of the limited bandwidth on site, that design decision was made to keep the system from clogging.
- Does this app allow users to share multimedia objects? No, the photo-taking feature was removed during development.
- Does this app curate user content? No, there is no user content to curate.
- Does this app allow users to connect with each other directly? No, except through indirect social media feeds that are displayed on the Yellowstone app but not accessible through it.

So the Yellowstone app, in this process, only generated two yes answers, and eight no answers. From the locative perspective posited, then, the Fort Vancouver Mobile app is more mobile-oriented than the Yellowstone Mobile app, or more suited to earn the locative app designation. To be clear, this evaluation does not mean the Fort Vancouver Mobile app is better, or more useful, or more robustly programmed, or any other such potential qualitative assessment. This process is simply intended to determine how well an app takes advantage of mobile affordances, from a locative perspective, and to somehow use that process to classify the efforts, from the pure porting of a desktop program to a smaller screen, to designs that fully explore the potential of mobile technologies.

CONCLUSION

One aspect of a comparison between these two apps that is not addressed at all in this process, for example, is the compositional style of the discourse. The Fort Vancouver app was designed as a package of interactive stories, and the Yellowstone app was designed as an expositional service tool. That distinction is important, and would seem to greatly affect the quality of the experience, but it does not necessarily relate to the *mobileness* of the app, or its standing as locative or not.

From that perspective, using these examples, this process did seem to tease out and distinguish the mobile characteristics most critical to a locative label. These questions, and thoughts about them, are intended to help initiate a much-needed drive toward mobile app classification. But several significant questions about the process also were generated during this exploration, including the overarching concern of what might be gained from pursuing more detailed app classifications, and from creating more genres? Would such activities end up in generating several ad-hoc systems that never could be aligned or agreed upon or used? Are locative mobile apps a significant genre, worthy of focus? Or are they simply a subset of travel apps, or entertainment apps, the contemporary version of the audio tour? If locative apps should be classified and categorized, and are worthwhile to study as a genre, then are other important genres also being obscured or suppressed because of the current corporate-dictated categorizations?

As part of this process, several other questions were generated related to app classifications and classification systems that were beyond the scope of this chapter. Current app delivery systems, such as the Apple and Android markets, set agendas biased toward business use. That is how the companies that control them make money, in a capitalist system, but that approach is not necessarily the extent of what can – or should – be done with mobile technologies and with the development of mobile apps to better society. The locative genre is a good example of how a possibly rich use of mobile technologies, rife with potential, has been left out of the current conversation. This chapter proposes a path to bringing locative apps more to the forefront. Much more work on this classification topic clearly is needed, and the approach outlined here is not intended to cover all concerns. Yet it is intended to raise the classification topic, and its related issues, as a fertile area of further inquiry worthy of increased attention.

REFERENCES

Ali-Yrkkö, J., Kalm, M., Pajarinen, M., Rouvinen, P., Seppälä, T., & Tahvanainen, A.-J. (2013). Microsoft Acquires Nokia: Implications for the Two Companies and Finland. Retrieved from http://pub.etla.fi/ETLA-Muistio-Brief-16.pdf

Barton, B., & Barton, M. (1993). Ideology and the map: Toward a postmodern visual design practice. In N. Blyler & C. Thralls (Eds.), *Professional communication: The social perspective* (pp. 49–78). Newbury Park, CA: Sage.

Barylick, C., & Honan, M. (2007). iPhone release brings out the crowds. *MacWorld,* retrieved from http://www.macworld.com/article/1058682/iphone_crowds.html

Benford, S., & Giannachi, G. (2009). Temporal Convergence in Shared Networked Narratives: The Case of Blast Theory's Day of the Figurines. *Leonardo*, *42*(5), 443–448. doi:10.1162/leon.2009.42.5.443

Bolter, J., & Grusin, R. (1998). *Remediation: Understanding new media*. MA: MIT Press Cambridge.

Burke, K. (1966). *Language as symbolic action: Essays on life, literature, and method.* Los Angeles: University of California Press.

Campbell, S. W. (2013). Mobile media and communication: A new field, or just a new journal? *Mobile Media & Communication, 1*(1), 8–13. doi:10.1177/2050157912459495

Castells, M., Fernandez-Ardevol, M., Qiu, J. L., & Sey, A. (2009). *Mobile communication and society: A global perspective.* Cambridage, MA: MIT Press.

CTIA. (2011). Wireless Quick Facts. Retrieved from http://www.ctia.org/media/industry_info/index.cfm/AID/10323

de Souza e Silva, A., & Sutko, D. M.de Souza e Silva. (2008). Playing Life and Living Play: How Hybrid Reality Games Reframe Space, Play, and the Ordinary. *Critical Studies in Media Communication, 25*(5), 447–465. doi:10.1080/15295030802468081

de Souza e Silva, A., & Sutko, D. M.de Souza e Silva. (2011). Theorizing Locative Technologies Through Philosophies of the Virtual. *Communication Theory, 21*(1), 23–42. doi:10.1111/j.1468-2885.2010.01374.x

Emmanouilidis, C., Koutsiamanis, R.-A., & Tasidou, A. (2013). Mobile guides: Taxonomy of architectures, context awareness, technologies and applications. *Journal of Network and Computer Applications, 36*(1), 103–125. doi:10.1016/j.jnca.2012.04.007

Epstein, M. (2009). *Moving Story.* Paper presented at the Media in Transition 6: Stone and Papyrus, Cambridge, Mass.

Farman, J. (2012). *Mobile Interface Theory: Embodied Space and Locative Media.* New York, NY: Routledge.

Feijóo, C., Maghiros, I., Abadie, F., & Gómez-Barroso, J.-L. (2009). Exploring a heterogeneous and fragmented digital ecosystem: Mobile content. *Telematics and Informatics, 26*(3), 282–292. doi:10.1016/j.tele.2008.11.009

Feldmann, V. (2005). *Information Age Economy: Leveraging mobile media: Cross-media strategy and innovation policy for mobile media communication.* Heidelberg, Germany: Physica-Verlag.

Foucault, M., & Miskowiec, J. (1986). Of Other Spaces. *Diacritics, 16*(1), 22–27. doi:10.2307/464648

Friedman, L. (2013). The App Store turns five: A look back and forward. *MacWorld*, retrieved from http://www.macworld.com/article/2043841/the-app-store-turns-five-a-look-back-and-forward.html

Gibson, J. J. (1979). *The ecological approach to visual perception.* Hillsdale, NJ: Lawrence Erlbaum Associates, Inc.

Gordon, E., & de Souza e Silva, A. (2011). Net Locality: Why Location Matters in a Networked World. West Sussex, United Kingdom: Wiley-Blackwell.

Handal, B., El-Khoury, J., Campbell, C., & Cavanagh, M. (2013). *A framework for categorising mobile applications in mathematics education.* Paper presented at the Australian Conference on Science and Mathematics Education: Students in transition – The learners' journey, Sydney, Australia.

Hayles, N. K. (1999). *How we became posthuman: Virtual bodies in cybernetics, literature, and informatics*. Chicago, IL: University of Chicago Press. doi:10.7208/chicago/9780226321394.001.0001

Hight, J. (2003). Narrative Archeology. Retrieved from http://www.xcp.bfn.org/hight.html

Hsi, S. (2008). Designing for Mobile Visitor Engagement. In L. Tallon (Ed.), *Digital technologies and the museum experience: handheld guides and other media*. Landham, MD: AltaMira Press.

Jaokar, A., & Fish, T. (2006). *Mobile Web 2.0: The innovator's guide to developing and marketing next generation wireless/mobile applications*. London: Futuretext.

Jones, S., Karnowski, V., Ling, R., & von Pape, T. (2013). Welcome to Mobile Media & Communication. *Mobile Media & Communication, 1*(1), 3–7. doi:10.1177/2050157912471456

Kerris, N., & Dowling, Steve. (2007). Apple Reinvents the Phone with iPhone (press release). *Apple.com*. Retrieved from http://www.apple.com/pr/library/2007/01/09iphone.html

Kim, J., Lee, S., Geum, Y., & Park, Y. (2012). Patterns of innovation in digital content services: The case of App Store applications. *Innovation: Management. Policy & Practice, 14*(4), 540–556. doi:10.5172/impp.2012.14.4.540

Kurzweil, R. (2005). *The singularity is near: When humans transcend biology*. New York, NY: Viking Adult.

Laouris, Y., & Eteokleous, N. (2005). *We need an educationally relevant definition of mobile learning*. Paper presented at the 4th World Conference on Mobile Learning, Cape Town, South Africa.

Ling, R. (2012). *Taken for grantedness*. Cambridge, MA: MIT Press.

TIME Mobility Poll. (2013). *TIME Magazine*. Retrieved from http://www.qualcomm.com/media/documents/time-mobility-poll-cooperation-qualcomm

McLuhan, M. (1962). *The Gutenberg galaxy: The making of typographic man*. Toronto, ON: University of Toronto Press.

McLuhan, M. (1997). *Essential McLuhan*. New York, NY: Basic Books.

Neuman, W. R., & Guggenheim, L. (2011, May). The Evolution of Media Effects Theory: A Six Stage Model of Cumulative Research. *Communication Theory, 21*(2), 169–196. doi:10.1111/j.1468-2885.2011.01381.x

Nickerson, R., Muntermann, J., Varshney, U., & Isaac, H. (2009). *Taxonomy development in information systems: Developing a taxonomy of mobile applications*. Paper presented at the European Conference in Information Systems, Verona, Italy.

Norman, D. A. (1988). *The psychology of everyday things*. New York, NY: Basic books.

Oppegaard, B., & Shine, G. (2014). Going Mobile: Changing the Face of Interpretation in the National Park Service. *Federal History Journal*, *6*(1), 123–142.

Parry, R. (2008). Afterword: The future in our hands? Putting potential into practice. In L. Tallon (Ed.), *Digital technologies and the museum experience: handheld guides and other media*. Landham, MD: AltaMira Press.

Perse, E. M., & Dunn, D. G. (1998). Utility of Home Computers and Media Use: Implications of Multimedia and Connectivity. *The Journal of Broadcasting & Electronic Media*, *42*(4), 435–456. doi:10.1080/08838159809364461

Rawolle, J., & Hess, T. (2000). New digital media and devices: An analysis for the media industry. *International Journal on Media Management*, *2*(2), 89–99. doi:10.1080/14241270009389926

Schmidt, E. (2010). Mobile World Congress keynote address [YouTube video]. Retrieved from http://www.youtube.com/watch?v=YuqiE2lukDM

Silcoff, S., McNish, J., & Erman, B. (2013, November 4). BlackBerry financing aims for a new lease on life. *The Globe and Mail*. Retrieved from http://www.theglobeandmail.com/report-on-business/blackberry-fairfax-deal-dies-thorsten-heins-out/article15240310/

Smith, G. (2010). *What Media Classes Really Want to Discuss: A Student Guide*. New York, NY: Routledge.

Statistics. (2013). *ITU.int*. Retrieved from http://www.itu.int/en/ITU-D/Statistics/Pages/stat/default.aspx

Surowiecki, J. (2012). Blackberry Season. *The New Yorker*. Retrieved from http://www.newyorker.com/talk/financial/2012/02/13/120213ta_talk_surowiecki

Tallon, L. (2008). *Digital technologies and the museum experience: handheld guides and other media*. Landham, MD: Altamira Press.

Turkle, S. (2008). Always-on/always-on-you: The tethered self. In J. Katz (Ed.), *Handbook of Mobile Communication Studies* (pp. 121–138). Cambridge, MA: MIT Press. doi:10.7551/mitpress/9780262113120.003.0010

Wilcox, M., & Voskoglou, C. (2014). *State of the Developer Nation, Q3*, 2014.

Wingfield, N. (2008, July 10). IPhone Calls On Software Developers. *The Wall Street Journal*. Retrieved from http://online.wsj.com/article/SB121565491776341571.html?

Zeimpekis, V., Giaglis, G. M., & Lekakos, G. (2002). A taxonomy of indoor and outdoor positioning techniques for mobile location services. *ACM SIGecom Exchanges*, *3*(4), 19–27. doi:10.1145/844351.844355

Zhu, H., Cao, H., Chen, E., Xiong, H., & Tian, J. (2014). Mobile App Classification with Enriched Contextual Information. *IEEE Transactions on Mobile Computing*, *13*(7), 1550–1563. doi:10.1109/TMC.2013.113

KEY TERMS AND DEFINITIONS

Affordance: Whatever element that exists in an environment that is actionable, in some way, at least to the person who perceives the use of it in that way.

Constraint: Whatever lack of element, or restriction of an element, in an environment that prevents a potential action as perceived by a person considering the action.

Contextual Awareness: The abilities of our devices, especially the more we use them, to know about our individual contexts and to respond to those contexts.

Convergence: The amalgamation of forms, i.e. the ways in which a smartphone can subsume both television and radio and act as a substitute for traditional hardware while also creating novel kinds of media content that transcends the original forms.

Hypermediacy: When a person has a heightened state of awareness about a technology involved in their activity, putting a focus on the technology, rather than it blending into the background and seeming "natural."

Mobile Technologies: Communication software and hardware that is easily portable and often personalized, especially in an effort to enhance sociable behavior.

Technoscape: A technology-enhanced environment, usually featuring many layers of human interventions, intended to alter the landscape in ways small and large.

Chapter 12
Who Brings the News? Exploring the Aggregators Apps for Mobile Devices

João Canavilhas
University of Beira Interior, Portugal

Ivan Satuf
University of Beira Interior, Portugal

ABSTRACT

The spread of mobile devices and digital networks generate new perspectives for the circulation of news. The development of mobile communication technologies and the migration of habits consolidated in personal computers promote the expansion of news aggregators for smartphones and tablets. This work seeks to inquire the operational logic of applications designed to aggregate journalistic content. An immersive analysis shows that the general operations of these apps are based on three dimensions: integration between news and online social media, always-on recommendation systems, and parameterization of outputs. It is argued that the approach between aggregators and social media makes the news extrapolate its information domain and start to circulate as symbolic reference for online interaction. Finally, this work posits that the operational logic of aggregators neglect fundamental aspects of new technologies related to locative information.

INTRODUCTION

The circulation of news content on mobile devices is rapidly growing but at the same time it is facing huge experimental dispersion. The traditional media conglomerates have different approaches when it comes to the search for mechanisms to adjust their outputs to smartphones and tablets (Wolf & Hohlfeld, 2012; Westlund, 2013). Concomitantly, new agents, without any links to news companies, emerge as important intermediaries in the complex mobile ecosystem (Castellet & Feijóo, 2013).

In this paper we attempt to understand the functioning of a specific group of such intermediaries, i.e. news aggregator mobile apps, some of which are very popular among smartphone and tablet users, such as Flipboard, Pulse and Zite. We have worked on the assumption that news aggregators are not mere

DOI: 10.4018/978-1-4666-8838-4.ch012

adjuvants, but rather important stakeholders that draw direct benefits from an ubiquitous and pervasive media system (Deuze, 2012; Ling, 2012) made up of high-speed digital transmission networks and "always-on" mobile technologies (Baron, 2008; Turkle, 2008, 2011).

Despite frequently cited in studies on mobile journalism, aggregating apps still require further consistent research. We believe that this gap in research may conceal a phenomenon which is decisive for our understanding of the present and, especially, of the future of news distribution and circulation, bearing in mind that the search for news is one of the most common habits of mobile digital device owners. (Taylor, Samuels & Ramey, 2011, Pew Research Center, 2012; comScore, 2013; Newman & Levy, 2014).

Some relevant data stress the need to have more in-depth analyses about aggregating apps. In only three years, Flipboard has reached an impressive 85 million registered users (Yeung, 2013). The app interface simulates on mobile devices the habit of reading paper magazines by combining visual and touch features. It is better adapted to tablet size but also has good usability on smartphones. It further allows users to integrate in a single platform their online social media and their favourite news channels. Although exact data about the activity of registered users is not available, suffice it to know that even if only 1% of them access the app on a daily basis that alone represents a contingent of 850,000 people.

The success of aggregating apps is reflected in the existence of several businesses that actively seek a means of adapting to the new distribution platforms. The rapid rise of Flipboard has caught the attention of traditional media such as The New York Times, which signed a contract in order to allow their subscribers to have full access to the newspaper contents through that app. It was the first time that this newspaper implemented a paywall strategy in a platform that does not belong to the group. CNN, The Telegraph, Forbes and other traditional channels also have agreements with Flipboard, but without immediate charging for access to the contents. Instead they explore the advertisements (featuring others and themselves) that pop up in between the news items.

Other businesses in the sector go beyond the actual contents and directly involve the aggregators and media conglomerates. In August of 2011 CNN purchased Zite under the justification that it aimed to expand its presence in the universe of mobile devices. The business value has not been officially announced but values ranging from 20 to 25 million dollars have been reported. Less than two years later, in May of 2014, Zite changed hands again. The buyer was now Flipboard, paying CNN approximately 60 million dollars (Bercovici, 2014). In addition to being in control of a direct competitor, Flipboard announced that it would incorporate part of Zite's recommendation algorithms to enhance its performance and subsequently discontinue their rival[1].

Another important negotiation for the sector took place in November of 2013 when the social media LinkedIn acquired Pulse for 90 million dollars (Geron, 2013). In addition to promoting the integration of two gigantic networks (at the time of negotiation Pulse had 30 million users), LinkedIn was also interested in increasing the number of visits and the users' time online by encouraging greater interaction with news content.

It should be further noted how technology giants shifted towards news-aggregating services for mobile devices. Facebook chose to develop its own aggregator, i.e. Paper, which incorporates news channels into the social media.

These examples show that the circulation of news combines traditional media (CNN, The New York Times) with major companies of the digital sector (LinkedIn, Facebook) and emerging start-ups (Flipboard, Zite, Pulse).

In this paper we analyse what is behind the apps in order to answer a direct but complex question: "Who brings us the news?" The goal is to understand the operations and features of aggregators that are directly related to contents.

In order to answer the question, we shall divide this study into three stages. We depart from a historical perspective that will allow us to explain the reasons that caused news aggregators to migrate to mobile devices. We will then demonstrate that mobile aggregator apps are structured around three features:

1. Integration of news content with the social media;
2. "Always-on" recommender systems;
3. Output parameter definition.[2]

In the third and final part of the paper, we claim that these features essentially rely on one of the many properties of mobile devices, i.e. networked portability, side-lining important functionalities of mobile technologies such as geolocation (with the help of GPS, QR code reader, and motion sensors).

THE EMERGENCE OF AGGREGATOR APPS

Why and how have aggregators reached mobile devices? The answer requires a twofold approach. On the one hand, it has to do with technological evolution, which allowed for the gradual adjustment of news content to cell phones. On the other hand, it is related to the migration of a habit: news aggregators were widely used on desktop computers, and users looked for ways of transporting that habit to the platform to which they grew more accustomed. This twofold analysis is intended to discard any attempt to apprehend the aggregating apps phenomenon in a one-dimensional perspective, be it technological or social.

The Technological Evolution

The expansion of 3G networks in the first decade of the 21st century was a disruptive innovation, capable of causing deep changes to the mobile ecosystem. First and foremost, the significant increase in the speed of data transmission allowed a wider variety of formats to be put into circulation.

The gradually falling prices of data packages offered by telephone companies, the improvement of Wi-Fi networks and the subsequent adjustment of HTML language for mobile browsers have made it possible for initial barriers to be overcome.

The technological evolution in these devices led to the decline of the telephone function in favour of new, more computer-like features (Klemens, 2010), as well as bigger and higher definition screens to display new contents. Standard QWERTY keyboards have been adopted in several models to make it easier to type longer texts. The reorganisation of technology added a new layer to news contents. The "push news" system (Fidalgo, 2009) remained important and it is still present in most apps (alerts), but the expansion of 3G connections was a definitive step towards the consolidation of "on-demand" content in mobile technologies, with consumers actively seeking contents.

Evolution met a turning point upon the iPhone launch in June of 2007, a "game changer" that influenced the entire composition of the contemporary mobile ecosystem. The physical keyboard was replaced by a single touch screen and interaction was now done through applications, i.e. software specifically designed to operate in this device. The gradual replacement of cell phones with smartphones is a paradigmatic change. The "telecom model", focused on the commercial exploration of voice transmission, made way for the "software model", based on the new computational features of devices and on the data-centred approach (Castellet & Feijoó, 2013). It was at this time that mobile devices turned into an exceptional platform for news distribution.

The approach would be incomplete if we failed to mention the remaining elements of the ecosystem: distribution channel for contents and services, operating system and software development kit (SDK). The distribution channels are the "stores" (App Store, Google Play, etc.) that group the apps into sections and provide access to registered users so that they may download them (freely or for a fee). Stores and apps are directly connected with the operating system (iOS, Android and Windows Phone, etc.) that controls all the functions of the device. This connection has technological and economic implications as it imposes constraints on the users, who will have access to a specific range of contents depending on their choice of operating system. Lastly, the software development kit is made available by companies that control the operating systems in order to allow external developers to create new apps.

For Goggin (2011), the release of the SDK was crucial for apps to become contemporary cultural elements, because it allowed thousands of developers to gain creative access to and decisively influence the contents available on mobile devices. Together, the two main operating systems – iOS and Android – offer on their distribution channels over two billion apps that feed the App Economy, a booming market worth thousands of billions of dollars worldwide (Organisation for Economic Co-operation and Development, 2013).

The opening up to external developers helps to explain the heterogeneity and fragmentation of mobile contents (Feijóo, Maghiros, Abadie & Gómez-Barroso, 2009; Scolari, Aguado & Feijóo 2012). News aggregators can be regarded as a relevant sub-group of news apps. However, as we will conclude during this analysis, it is also a heterogeneous sub-group.

From the PC to the Mobile App

Aggregating platforms are quite popular on the internet. Their development was, to some extent, a response to the explosion of contents on the web and to the need to select and organise information without the user spending much time and effort (Calin, Dellarocas, Palmi & Sutanto, 2012). Aggregators work as intermediaries between content producers and consumers in an environment of data superabundance (Cádima, 2013).

Two of Google's main executives, Cohen and Schmidt (2013), ascribe the rise of new platforms to the widespread crisis of the traditional media, which proved incapable of competing with more agile and effective distribution channels. The type of intermediation accomplished on these platforms inflicts collateral damage on historically established business models and has therefore spawned a strong reaction from news companies, which witnessed the invasion of new competitors (Pauwels & Picone, 2012). The misappropriation of material and copyright infringements are but a few of the accusations that reach the courts (Isbell, 2010; Jasiewicz, 2012).

Users have been taking advantage of aggregators since the late 1990s, when Rich Site Summary appeared. This format is present in the web to this day. Commonly known as RSS Feed, the system allows users to subscribe contents from a variety of sources that are automatically gathered on a single graphical interface.

According to the environments in which they run, Fortes *et al* (2007) identify four types of aggregators: desktop-aggregators (run on the user's machine), web-based aggregators (run on a remote web server), widget aggregators (integrated in a web page or in a desktop side bar) and mobile aggregators (run on a mobile device).

Web-based aggregators operate on the basis of two main features: process automation and content customisation. Automation is the software's ability to perform previously defined tasks without human intervention. It is convenient when it comes to selecting different sources of information and receiving contents, avoiding the need to access all the websites where the news was originally published. In addition to being convenient, RSS is based on content customisation, which is one of the main characteristics of digital news contents. On a single platform, it is possible to gather news about one's favourite soccer team, posts from a blog on national politics, videos released on a newspaper website or any other kind of content.

Automation and customisation are the operational pillars of one of the most common aggregators - Google Reader - launched in 2005 and discontinued in 2013. Web aggregators integrate pull and push systems. The demands are set by the users (pull) and the news arrives automatically (push). Other aggregators remain active on the web, such as Feedly and Feedspot, but they are becoming more and more similar to social media.

It seems obvious that news-aggregating apps for mobile devices should have a solid baseline as well as a source of inspiration. Aggregators developed for personal computers have broken new ground and set new habits of online news consumption. The evolution of smartphones was accompanied by the adaptation of that kind of service to the context of mobility. In addition, the arrival of tablets on the market, since the iPad launch in 2010, encouraged the development of apps which aim to fully exploit the potential of screens and of ergonomic advantages in content consumption (Palacios, 2013). Indeed, some news aggregators seek to captivate users by means of attractive graphical interfaces that explore the tactile dimension (Palacios & Da Cunha, 2012).

It must be made clear that the shift from computers towards mobile devices is not a mere transposition. Rather, it represents a process of adjustment to new features. Smartphones and tablets have characteristics that set them apart from personal computers. Portability, touch screen, integrated GPS, motion sensors and other features make it necessary to rethink the operations performed by aggregators on mobile devices.

NEWS AGGREGATING APPS IN ACTION

Assuming that apps are software developed for mobile devices, we seek to answer an apparently simple question: "Who brings us the news?". In fact, the question is not simple and hence neither is the answer. The operations triggered by each piece of software have crucial implications on contents, determining the news that gets released and the manner in which consumption takes place. Through a careful analysis it is possible to uncover three basic features of these apps: integration of news with the social media, "always-on" recommender systems and output parameter definition.

Given the wide variety of aggregators available on app stores, we have selected a small but representative group for the purpose of this analysis. First of all, we selected only those apps that define themselves as news aggregators and are inserted in the "News" section of the App Store or in the "News and Magazines" section of Google Play, which are the two main distribution channels of mobile apps. Subsequently we searched for the most cited news aggregators in scientific papers and technology expert sites. Finally, we excluded those apps whose aggregation is exclusively done by news sites such as Circa, Inside and Newser, since contents are aggregated by professionals instead of software operations that interact with users[3]. This brings us to a group of twelve apps: Feedly, Flipboard, News360, Newscron, NewsMix, Niiiws, Paper, Prismatic, Pulse, SkyGrid, Trove, Zite.

Integration with the Social Media

The union between the social media and mobile devices has generated a force that is still difficult to measure but its consequences are immediate in several areas, such as journalism. The last few years have been marked by the intense approximation between the news and the online social media. The culture of sharing and the re-appropriation of contents produced by others on platforms such as Facebook and Twitter has led to the coinage of the term "spreadable media" (Jenkins, Ford & Green, 2013) to designate the new relationships between content producers, social media and the audiences.

It would be naïve to think that news companies could remain oblivious to such significant changes. Stories are reposted by users on the social media at the same time as newspapers, radios, television and news sites create profiles on those networks to interact with the public in a territory where they do not control the flow of data. Amateur videos posted on YouTube are used to illustrate stories on TV, mass media stories are released on YouTube and on other video-sharing sites only moments after their exhibition.

These phenomena translate into a major shift within the field of news production. We moved from a model focused on contents and their producers, typical of cultural industries, towards a model focused on users and their relationships to other users as well to the contents themselves (Giardina, 2012; Aguado & Güere, 2013). This change is not peaceful and many journalists disagree that the apps launched by the companies for which they work should be open to participation (Westlund, 2012).

At a time when Facebook gets one billion monthly visits from mobile devices (Lunden, 2014) and approximately 80% of Twitter traffic originates from smartphones and tablets (Durrani, 2014), aggregating apps are in a privileged position to explore the connections between the news, the social media and the users. Although optional, one of the first steps of all the apps under analysis is to encourage users to log into their personal accounts on online social media. But why would we want to associate a Facebook or Twitter profile with an aggregator when we are interested in consuming news? From the user's perspective, that decision may be influenced by the wish to make the most of all the app features, which depends on being logged into social media. Additionally, with the growing culture of sharing, it is fair to assume that a significant amount of users choose aggregators precisely because they are so closely interwoven with social media.

On the other hand, from the perspective of apps and their developers, the answer is more diversified. Four out of the twelve news aggregators analysed – Newscron, News Mix, Niiiws e SkyGrid – use integration with social media only for sharing news. This is a basic feature nowadays.

Two apps belong in a second sub-group: Paper and Pulse. The purpose of these aggregators is to create a news consumption environment that stimulates sharing in specific networks. Paper, developed by Facebook, only allows users of the network founded by Mark Zuckerberg to log in. This app allows you to see all the updates associated with your Facebook account. It further offers a news menu, which is automatically grouped into 16 sections. What is the purpose of that approach? Simple: every user has to be connected to Facebook in order to have access to Paper. To simply use the app means that you have to be logged into the social media, which brings up Facebook stats in terms of the number of visits and time spent online. At the same time, users are exposed to a series of news and pick the channels that appeal most to them, according to personal preferences. Since news consumption also takes place in the integrated environment of the social media, the probability of sharing the news on the "Timeline" is also expected to increase. More news circulating on Facebook means more "Likes", more shares and more message exchange.

It is enough to see how the news is grouped to realise that the operation mode of Paper tries to adapt news consumption to Facebook's interaction patterns. Among the 16 sections there are some traditional ones such as "Headlines", "Score" (sports) and "Tech" (technology), which could well be sections of any news site. However, other sections are not quite as conventional if we consider the formal model of journalism. That is the case of the "Cute" section, which aggregates news about animals, mostly pets, e.g. cats and dogs. Illustrated with pictures and videos, most pieces of news get thousands of "Likes" as well as dozens or even hundreds of comments. Every time that type of news is shared, it tends to trigger actions that feed the social media. The same goes for the "LOL" section, named after an acronym that is commonly used as slang in the internet and that stands for "laughing out loud". That section features links to websites such as Buzzfeed and Gawker, which are known for posting viral contents.

Another aggregator of this subcategory is Pulse, which offers the possibility of connecting exclusively with the LinkedIn user profile. Similarly to Paper, the goal is to incorporate a news content platform in order to strengthen the interactions of the social media. As we mentioned in the introduction, LinkedIn, currently the largest professionally-oriented social network online, bought out Pulse in a multimillion dollar transaction. In order to understand the business it is necessary to take a closer look at the social media. LinkedIn is built on a vast repertoire of professional CVs that are all interconnected. However, if the CV is already up-to-date and if connections with other professionals have been established, there remains little to do on the platform. Unlike Facebook, users are not encouraged to post dozens of photos, videos and personal messages. The average LinkedIn user is not an "always-on" type of user.

News circulation may help to stimulate interaction and hence the number of visits and the time spent on LinkedIn. The news itself is not of a personal nature and contents are often linked to topics of professional interest, whether they be specific or general. The app will show you the other users who also liked a given piece of news and allow you immediate access to their profiles. The aggregator experience expands onto other platforms. When going on LinkedIn on a computer, the following message will pop up: "Pulse recommends this news for you". Based on user activity, the system automatically selects some contents that can be read, commented on and shared on the web browser.

Therefore, the Paper-Facebook and Pulse-LinkedIn integrated systems demonstrate that the news-aggregating app may serve as an additional channel of support to online social media. These aggregators are exclusive to the social media with which they are directly associated, and they rely on the news feed as an appropriate means to catalyse interaction. The target is to keep online profiles permanently logged in and active. After all, "always-on" users are the ultimate goal of any social media platform.

Always-On Recommender Systems

The remaining six aggregators – Feedly, Flipboard, News360, Prismatic, Trove, Zite – incorporate the social media and their basic features as a content recommendation system, although they have different approaches.

Before analysing the apps, it is essential to stress that journalism has historically asserted itself as a recommendation system. That perception is present in the seminal study by White (1950) on the operations carried out by a newspaper editor (Mr. Gates) in order to select the group of events that would make the news. Later research has contributed to redefine the focus, which moved away from individual decisions towards the complex power relations of organisational structures (Gans, 2004). Despite that, the gatekeeping theory continues to be one of the most relevant theories in the field of journalism (Shoemaker, Vos & Reese, 2008). It lies on the idea that to select is to recommend, in other words, it is the journalist telling to his audience “this is what truly matters”.

This centralised model, which prevailed well into the 20th century, was made possible thanks to the limited number of news channels and the intrinsic limitations of each one of them (space in the printed newspaper and time on the radio and/or TV). The proliferation of personal computers and the emergence of the World Wide Web have significantly altered the landscape. According to Bruns (2011), the influence of journalists on the circulation of contents stands in inverse proportion to the number of available channels. This is quite visible nowadays, as there are more mechanisms to manipulate the contents directly as well as an ever-increasing exposure to algorithmic and social recommendation networks, which are not necessarily based on the operation modes of journalism.

We are faced with a process of consolidation of “always-on” recommender systems associated with portable communication technologies, which are permanently connected to high-speed digital networks. Aggregator apps have become relevant platforms in this recent configuration. Obviously one should put into perspective the expression “permanent connection” as there are significant global inequalities with regard to technological infrastructures (Castells Fernández-Ardèvol, Qiu & Sey, 2006). However, research shows a steady fall in the prices of smartphones and network access fees. These factors are gradually allowing millions of people to join in the mobile ecosystem.

“Always-on” recommender systems result not only from technology but also from technology-related sociocultural changes. Turkle (2011) argues that today’s citizens are living in a kind of “media bubble” that simultaneously isolates and connects because each one of us “is tethered to a mobile device and to the people and places to which that device serves as a portal” (p.155). This means that we no longer interact with the people around us. Instead, we connect with others who are physically more remote. These “others” demand from us the same as we demand from them: permanent contact.

Social media, such as Facebook and Twitter, are perfect environments for the proliferation of contents attuned to current events. As our emotional relationships are mediated by media technologies, Grusin (2010) contends that we also establish relationships of an emotional nature, maybe even of love, with our technological gadgets. From a similar perspective, Deuze (2012) claims that there is no life “outside the media” because we are immersed in the media. With our smartphones and tablets packed with apps conceived for communication and content-sharing, it becomes even more difficult to discern the boundaries of “inside” and “outside”. In a study about young users of mobile devices, Gardner & Davis (2013) admit to the possibility that we become “app-dependent” if we allow mobile apps to restrict and shape our behaviour, choices and goals.

Life surrounded by the media has a number of consequences, one of which could be the transformation of everyday activities into digital data. With each message, photograph or video that we post on social media, we are giving computing systems an opportunity to take advantage from access to such data. Three of the news aggregators we analysed in this chapter do exactly that: Feedly, News360 and Zite. By logging into Facebook and/or Twitter, the app begins a search that within a few seconds will retrieve a set of keywords linked to content recommendations. In general, recommendations point to wide topics, e.g. “Technology”, “Business”, “Journalism”, and “Music”, or they may also contain direct links to the user profile such as the country of origin or current country of residence.

The analysis shows that the recommendation system is still rather basic if one bears in mind its potential, but it also demonstrates that it could be the initial stage of a trend that may definitively influence the course of news. It is not journalists who recommend the contents, nor is it the user who seeks possible areas of interest, but rather the software programmed to access a database (the user profile on the social media) that cross-checks data and selects "relevant" subjects. Much is said nowadays about the consumption of customised information, but customisation did not use to mean loneliness or isolation. What we post on social media does not concern only ourselves but others as well. A post on an online platform built for the purpose of interaction is always a reflexive action. We publish posts and await other people's reactions, and their feedback may take various forms, such as comments, "likes" or "retweets". The possible absence of feedback may have an influence on future posts, in terms of both content and form.

When aggregators go through our walls to recommend contents, the output could be the result of our own interactions but also of the expectations we transfer to the online universe. Ultimately this output is the reflection of our online experiences, be it real or fictional. The close relationship that we maintain with mobile devices and the constant flow of information that we transfer to the social media may lead to the development of apps that recommend news contents based on feelings and emotions. While this is a highly prospective analysis, it could be argued that the news tends to turn into an extension of the social media, even though consumption does not take place on platforms such as Facebook or Twitter. The future of "always-on" recommender systems seems to be pointing to the strengthening of what Pursell, Rainie, Mitchell, Rosenstiel and Olmstead (2010) designated as "Daily Me" or "Daily Us". The user's media life provides the input for the recommender system, based on the "digital representation of a real-world personality" (Uhlmann & Lugmayr, 2012, p.87). The consumption of news information is more and more attuned to the interests of users, according to their interactions with others through platforms that engage specific communication patterns.

Whereas Feedly, News360 and Zite use the information posted on Facebook and Twitter as a database for content recommendation, Flipboard, Prismatic and Trove take further advantage of the close relationship with the social media.

After all, what defines online social media? It is basically the ability to create a profile and use it as a hub, i.e. as a point of connection with other hubs in a jumble of connections where it is possible to produce, share and receive information conditioned by the user's own choice.

These three aggregators are authentic social media focused on news-sharing, encouraging users to create a profile on their apps. For instance, by allowing Prismatic access to your Facebook and Twitter profiles, the app will automatically inform which ones of your "friends" also have a Prismatic profile and it will suggest contents "from your activity" on the social media. A user profile will also be created automatically, with the same image as that of the Facebook (or Twitter) profile. On this space it is possible to see the users' interests (keywords that point to news topics they are most interested in), whom they follow (following) and who follows them (followers). Some expressions and symbols used on social media are incorporated into news aggregators. To "Follow" is a basic action on Twitter, and below each piece of news there is the famous "Like" button (thumbs up) on Facebook. Prismatic is a platform for news consumption camouflaged as social media. Profile posts do not include opinions, personal photos or videos of your most recent trip. The profile is a collection of news that the user decided to share and through which he/she expects to attract other people with matching interests.

Flipboard allows users to create "social magazines" based on personal interests. All contents are re-appropriations of news and information selected from various sources that the app allows users to aggregate in a single interface, similar to a paper magazine. Anyone can create a magazine about "International Relations" by aggregating news from sources such as CNN, Fox News, Washington Post, etc. The originality of the magazine lies not in the contents themselves, but in how they are gathered and presented. As pointed out by Thorson (2008), "one can no longer assume that an article originally published on the front page of the New York Times will be read in that context" (p.485).

The app recommends a list of "Flips by Friends" with the magazines made by persons that are on the list of personal contacts on Twitter and Facebook. The magazines created by the user, the articles included in these magazines and the number of users that subscribed to the contents, are also displayed on the profile information. In addition, Flibpoard gathers all of the channels selected by the user in the "Cover Stories" section, where the contents posted on the Facebook and Twitter timelines are presented next to news aggregated into magazines created by other users and to material published on news companies' websites. Presented side by side, the interactions on the social media and the news blend into an uninterrupted flow of information.

Trove follows in the same footsteps, generating an automatic user profile as soon as users go on Facebook and Twitter through its app. It aims to encourage users to create "Troves", i.e. news-related sections linked to the user profile that may be followed by other users. This app also searches the user's social media in order to recommend "The best stories picked by people who share your interests". The mechanism behind Trove is basically the same as Prismatic and Flipboard. The goal is to create news-related sections linked to a user profile and connect with other people who take an interest in similar subjects. These people can be old acquaintances from the social media or new contacts established via the aggregator. As mentioned by Aguado & Castellet (2013), "the true strength of aggregators lies in how they incorporate the flow of social media into content consumption" (p. 203).

All three apps are in line with the emerging concept of "digital content curation" (Rosenbaum, 2011; Correa, 2012; Bruns & Highfield, 2012). At the epicentre of the process are the users responsible for selecting the contents. However those users do not act on their own, as they are constantly monitored by algorithms that recommend contents viewed by others. On these apps, the news feed is subject to the operation mode of social media. The "contents of others" turn into "my contents". But the process does not end there because "my contents" are always addressed to "others". Content consumption in the 21st century is definitely taking a very different path than in the broadcasting era. The news extrapolates the realm of information to become a symbolic reference for online social interaction. In essence, the system ceases to be "mediacentric" and becomes "Icentric", with the media serving as mere suppliers of contents that are then selected, commented on and contextualised by networks of users.

Output Parameter Definition

Mobile apps are the most recent phenomenon of a process that has been taking place for approximately 50 years, since computational systems began to permeate the social sphere. During the first years, computer programming was restricted to the execution of specific tasks, generally linked to complex calculations and process automation, and only experts were able to perform such procedures. Since the late 20th century, computer logic came to be used in connection with many daily activities through the continuous utilisation of software (Kitchin & Dodge, 2011; Manovich, 2013).

Software is a fundamental mediator of routine tasks. Even without noticing, the driver of an automobile equipped with an electronic injection system will interact with software every time he/she presses the accelerator pedal. The same is true of a citizen who withdraws cash from an ATM or makes a credit card payment. Even when it is explicit, the software may be concealed from the user. All computer operations are mediated by software, specifically its visible part, i.e. the graphical interfaces that allow for man-machine sensory interaction instead of forcing the user to deal with programming languages.

In the case of news-aggregating apps, all the programming was focused on achieving an intuitive way for the user to access the news, which is selected and organised beforehand. The tasks are performed by algorithms that are previously programmed by developers. But what is an algorithm? Broadly speaking it is a set of steps necessary to accomplish a task. In computing logic, the algorithm is a sequence of command structures aimed at the execution of a given operation (Goffey, 2008; Cormen, 2013).

Up until now we have analysed apps in which algorithms allow for the integration of news aggregators into the social media. It is now time to look at the algorithms used for the definition of output parameters directly by the user. Apps entail a number of parameters that are directly manipulated, e.g. the selection of contents to be aggregated. Choosing a specific blog from a list of options is to act directly on a parameter. The user tells the software "I would like to get updates from this blog only". This is, of course, the most basic level of parameter definition.

On News360, each news item recommended can be rated by the user. A positive rating will cause other news with similar contents to go up in the recommendation ranking. It is also possible to give a negative rating and thus lower the status of a certain type of content. The more ratings, the more the software will understand consumption patterns and produce more relevant outputs. As mobile devices become key items in today's apparel, parameter definition systems gain relevance due to their role in helping us make sense of the contents with which we interact.

Newscron is an aggregator that deals fundamentally with output parameter definition. The app presents a "default" set of news that can be thoroughly customised. Upon selecting the "Settings" menu, a set of 17 horizontal bars is displayed. These bars are equivalent to news categories, such as "Latest", "Knowledge & Culture", "Digital & Technology" and "Politics". By sliding a finger over each bar it is possible to change the pattern. The content layout possibilities are virtually endless. Imagine what it would be like to be able to buy every newspaper, put them in your pocket and, with only a few touches on the screen, to have the freedom of changing the parameters to consume only the news of the sections that interest you the most. What's better, the news is updated in real time. As there are no limits to the changes in parameters, it is possible to increase exposure to politics news during an election period or, if soccer is not your cup of tea, simply lower the level of sports news during the World Cup.

As with most apps under analysis, which firmly believe in the fusion between news and online social media, the definition of parameter outputs seems to be an emerging trend, which is evolving towards an enhanced customisation of information.

FAR BEYOND PORTABILITY

The characteristics of the apps presented in the previous item are deeply related to networked portability. The freedom of movements associated with the ability to send and receive data in high-speed networks promotes the use of social media on an unprecedented scale. The evolution of recommender systems associated with social media and the ability to change the parameters that generate the outputs are also direct consequences of technological mobility. With devices that are permanently connected within hand reach, news contents are now subject to new consumption habits focused on the re-appropriation and the continuous circulation of information. If personal computers were the departure point for the current state of affairs, then smartphones and tablets represent the consolidation of the user as the central element of a mobile ecosystem, as seen above.

Networked portability is only one of the aspects of mobile devices, the most elementary and visible one, but one that lies at the origin of several other features. The integration with the social media, the "always-on" recommender systems and the definition of output parameters are still located on the outermost layer of technology, whereas new horizons remain untapped by aggregating apps. It is in fact surprising that apps should have significant flaws when compared with the potential of the new media. Developed by start-ups and by giants from the digital sector, which are already born free from the shackles of the broadcasting system, they would be expected to, at least, offer experimentation models so that they could be distinguished from the traditional media that are facing evident problems of adjustment.

But what are, after all, the horizons that have yet to be explored by news aggregators? They are features that result directly from the mutual adjustment of software (apps) and hardware (smartphones and tablets), such as geolocation. The development of that feature could represent a new level with respect to the circulation of news information.

Geolocated (context-aware) information is defined as the association between the contents and the microenvironment in which the user is located (Gay, 2009; Gordon & Souza e Silva, 2011). Smartphones and tablets contain an array of resources that allow the data to be associated with their immediate spatial and temporal dimensions, such as the Global Positioning System (GPS), the motion sensors (accelerometer and gyroscope) and the camera (for QR Code reading or shape recognition). Scoble & Israel (2014) claim that mobile devices, social media, data, sensors and locations are the five forces that command the era of context. The present generation of aggregator apps has yet to realise that the exploitation of the last two forces – sensors and locations – is vital to adapt the consumption of news to modern times.

The integrated GPS in mobile devices, for instance, allows satellites to determine the user's exact coordinates, placing news aggregators in a privileged position to take advantage of that feature. If a person is driving to work and comes close to the scene of a road accident, they may start receiving news in time to choose an alternative route. In this case, the advantages of aggregators as opposed to the traditional means are substantial. An app developed by a specific news channel would not issue a news alert unless it had information about that accident. Aggregators are able to program the algorithms to search for last-minute news about road accidents and other obstacles located within a certain distance radius from the user. As the goal of the app is to bring together different sources of information, all it takes to generate the alert is for a channel to spread the information.

The news aggregators analysed in this paper do not have significant context-related features. They focus exclusively on networked portability and thus reinforce the user-device duality, in other words, the subject that moves and interacts constantly with the smartphone. The potential of the user-device-environment triad is neglected. This is consistent with the study conducted by Goggin, Martin & Dwyer (2014) on locative news, in which the authors emphasize the low presence of news contents adapted to the geographic context of the users.

CONCLUSION

The possibility of accessing and distributing contents at any time and place requires new ways of adapting contents to smartphones and tablets. News-aggregating apps for mobile devices play an important role in today's media ecosystem, because they are platforms that draw on the abundance of information resources in order to carry out the selection and organisation of the data. As the broadcasting system and mobile technologies take on new communicational shapes, the consequences for journalism are massive.

The aggregator apps strengthen the bond between the news and the social media. At a time when networks such as Facebook and Twitter control a significant slice of data traffic on mobile devices, it is expected that content consumption apps will take advantage of the situation.

The aggregators analysed in this paper have different approaches to the social media.

Paper and Pulse serve as catalysers of interaction for Facebook and LinkedIn, respectively.

In turn, Feedly, News360 and Zite take the users' actions on the social media as "always-on" recommender systems, suggesting topics that are likely to appeal to them. The user that is permanently connected with other people via the social media becomes the system's input. If ordinary life feeds the system and exerts influence on the recommendation of contents, it can be expected that the news (outputs) will bear an affinity to the users' feelings and emotional states.

The union between aggregators and the social media finds its highest expression in apps such as Flipboard, Prismatic and Trove. These aggregators are authentic social media, encouraging the creation of personal profiles on their own platforms. It becomes increasingly difficult to discern the boundaries between news and other spheres of social interaction. Future studies should extend the analysis about the activity of digital curation in aggregating apps.

The social media are complemented by the possibility of the users setting the parameters on the app, defining the contents according to personal choices. This is one of the consequences of the strong sense of identification between the device and its owner (Katz & Aarhus, 2002), who seeks, in this way, to adjust the information to his/her interests. The sophistication of algorithmic systems of output parameter definition, as is the case of Newscron, represents the ability to change in a few seconds the set of news to be presented. All it takes is to change a few parameters in order to transform the aggregator into a "sports diary" or into a collection of posts from political blogs.

But this strong connection between users and devices cannot be explored only in terms of the customisation performed by the user. This paper has demonstrated that the greatest shortcoming of mobile aggregators has to do with the absence of a relationship between the contents and the consumption contexts. The mode of operation lies at the surface of networked portability, which is the most basic feature of mobile devices. This causes the user's direct actions (the interactions on the social media and the definition of parameters) to be favoured over the potential arising from the immediate spatial and temporal context, such as the georeferencing of the news. This kind of contextualisation, which Canavilhas (2013) titles horizontality, builds "a chain of meaning in which the understanding of the informational element is achieved through the supply of data about the element within its spatial context" (p.7). In other words, other than what is suggested by the social media and the parameters defined by users, it is further possible to add a third layer of customisation that results from the exploration of the technology already available on the devices.

The vertiginous evolution of apps may alter the present-day scenario rather quickly. The existing apps provide periodical updates in order to fix bugs and add new features. Additionally, as the mobile system is open to developers, news apps, with significant bits of innovation, may be launched at any given time.

Another aspect to be monitored is the consolidation of the sector. Transactions like the purchase of Zite by Flipboard seem to be the beginning of a path that may well set the standards for apps and single out the leaders of the sector. The overwhelming number of apps available hinders visibility, causing a relatively small number of players to attract the attention of the public and thus gain greater influence. Recent data indicates that 80% of the apps available on app stores have extremely low download rates. These are known as "Zombie Apps" (Kelly, 2014).

Bringing content aggregation apps to the forefront of research is a valid initiative in order to understand the circulation of news contents in mobile devices, thus contributing to theoretical studies about the news in a context of mobility.

ACKNOWLEDGMENT

This article was financed by FCT- Foundation for Science and Technology - Ministry of Education and Science (Portugal) and CAPES Foundation, Ministry of Education of Brazil, Brasília - DF 70040-020, Brazil.

REFERENCES

Aguado, J. M., & Castellet, A. (2013). Periodismo móvil e información ubicua [Mobile journalism and ubiquitous information]. In J.M. Aguado, C. Feijóo, & I. J. Martínez (Eds.), La comunicación móvil: Hacia un nuevo ecosistema digital. [The mobile communication: towards a new digital ecosystem] (pp. 187-217). Barcelona, Espanha: Editorial Gedisa.

Aguado, J. M., & Güere, H. N. (2013). Comunicación móvil, ecosistema digital e industrias culturales [Mobile communication, digital ecosystem and cultural industries]. In J.M. Aguado, C. Feijóo, & I. J. Martínez (Coords.), La comunicación móvil: Hacia un nuevo ecosistema digital [The mobile communication: Towards a new digital ecosystem] (pp. 57-77). Barcelona, Espanha: Editorial Gedisa.

Baron, N. S. (2008). *Always on: Language in an online and mobile world.* New York, NY: Oxford University Press. doi:10.1093/acprof:oso/9780195313055.001.0001

Bercovici, J. (2014, May 3). Flipboard buys Zite from CNN; founder Mark Johnson leaving. *Forbes.* Retrieved from http://www.forbes.com

Bruns, A. (2011). Gatekeeping, Gatewatching, Real-Time feedback: New challenges for Journalism. *Brazilian Journalism Research*, *7*(11), 117–136.

Bruns, A., & Highfield, T. (2012). Blogs, Twitter, and Breaking News: The produsage of citizen journalism. In R. A. Lind (Ed.), *Produsing theory in a digital world: The intersection of audiences and production* (pp. 15–32). New York, NY: Peter Lang.

Cádima, F. R. (2013). A Google, o sistema de media e a agregação de informação [Google, media system and news aggregation]. *Intercom: Revista Brasileira de Ciências da Comunicação, 36*(1), 19–37.

Calin, M., Dellarocas, C., Palme, E., & Sutanto, J. (2013, February 16). Attention allocation in information-rich environments: The case of news. *Boston University School of Management Research Paper Series.* Retrieved from http://ssrn.com/abstract=2225359 doi: 10.2139/ssrn.2225359

Canavilhas, J. (2013). Jornalismo Móvel e Realidade Aumentada: o contexto na palma da mão [Mobile Journalism and Augmented Reality: the context in your hand). *Verso e Reverso, 64*, 2–8. doi: 10.4013/ver.2013.27.64.01

Castellet, A., & Feijóo, C. (2013). Los actores en el ecosistema móvil [The actors in mobile ecosystem]. In J.M. Aguado, C. Feijóo, & I. J. Martínez (Coords.), La comunicación móvil: Hacia un nuevo ecosistema digital [The mobile communication: towards a new digital ecosystem] (pp. 27-56). Barcelona, Espanha: Editorial Gedisa.

Castells, M., Fernández-Ardèvol, M., Qiu, J. L., & Sey, A. (2006). *Mobile communication and society.* Cambridge, MA: The MIT Press.

ComScore. (2013, March). *Europe digital future in focus 2013: Key insights from 2012 and what they mean for the coming year.* Retrieved from http://www.comscore.com

Cormen, T. H. (2013). *Algorithms unlocked.* Cambridge, MA: The MIT Press.

Correa, E. S. (Ed.). (2012). *Curadoria digital e o campo da comunicação* [Digital curation and the communication field]. São Paulo, Brazil: ECA – USP.

Deuze, M. (2012). *Media life.* Cambridge, United Kingdom: Polity Press.

Durrani, A. (2014, February, 24). Mobile now driving 80% of Twitter traffic. *Media Week.* Retrieved from http://www.mediaweek.co.uk/

Feijóo, C., Maghiros, J., Abadie, F., & Gómez-Barroso, J. L. (2009). Exploring a heterogeneous and fragmented digital ecosystem: Mobile contente. *Telematics and Informatics, 26*(3), 282–292. doi:10.1016/j.tele.2008.11.009

Fidalgo, A. (2009). Pushed news: When the news comes to the cellphone. *Brazilian Journalism Research, 5*(2), 113–124.

Fortes, F., Lv, R., Dehghan, D., & Moraveji, N. (2007). Taxonomies for News Aggregator Interfaces and User Patterns. Microsoft Research Technical Report. Retrieved from http://research.microsoft.com/pubs/70396/tr-2007-02.pdf

Gans, H. J. (2004). *Deciding What's News: A Study of CBS Evening News, NBC Nightly News, Newsweek, and Time* (25 th ed.). Evanston, IL: Northwestern University Press.

Gardner, H., & Davis, K. (2013). *The app generation: how today's youth navigate identity, intimacy, and imagination in a digital world.* New Haven, CT: Yale University Press.

Gay, G. (2009). Context-aware mobile computing: Affordances of space, social awareness, and social influence. *Synthesis Lectures on Human-Centered Informatics*, *2*(1), 1–62. doi:10.2200/S00135ED1V01Y200905HCI004

Geron, T. (2013, November 11). LinkedIn buys pulse newsreader for $90 million. *Forbes*. Retrieved from http://www.forbes.com

Giardina, M. (2012). Emerging media channels: Mobile tablet powered news dissemination. *f the Technology & Emerging Media Track–Annual Conference of the Canadian Communication Association*. Waterloo, Canada: Canadian Communication Association.

Goffey, A. (2008). Algorithm. In M. Fuller (Ed.), *Software studies: a lexicon* (pp. 15–20). Cambridge, MA: The MIT Press. doi:10.7551/mitpress/9780262062749.003.0002

Goggin, G. (2011). Ubiquitous apps: Politics of openness in global mobile cultures. *Digital Creativity*, *22*(3), 148–159. doi:10.1080/14626268.2011.603733

Goggin, G., Martin, F., & Dwyer, T. (2014). Locative news: Mobile media, place informatics, and digital news. *Journalism Studies*, 1–19. doi: 10.1080/1461670X.2014.890329

Gordon, E., & Souza e Silva, A. (2011). *Net locality: Why location matters in a networked world*. West Sussex, United Kingdom: Wiley-Blackwell. doi:10.1002/9781444340679

Grusin, R. (2010). *Premediation: Affect and mediality after 9/11*. New York, NY: Palgrave Macmillan. doi:10.1057/9780230275270

Isbell, K. (2010, August 30). The rise of the news aggregator: legal implications and best practices. *Berkman Center Research Publication*. Retrieved from http://ssrn.com/abstract=1670339. doi: 10.2139/ssrn.1670339

Jasiewicz, M. I. (2012, December 20). Copyright protection in an opt-out world: Implied license doctrine and news aggregators. *The Yale Law Journal*. Retrieved from http://ssrn.com/abstract=2207489

Jenkins, H., Ford, S., & Green, J. (2013). *Spreadable media: Creating value and meaning in a networked culture*. New York, NY: New York University Press.

Katz, E., & Aarhus, M. (2002). *Perpetual Contact: Mobile Communication, Private Talk, Public Performance*. Cambridge: Cambridge University Press. doi:10.1017/CBO9780511489471

Kelly, S. M. (2014, July 10). Report: Nearly 80% of apps in Apple's app store are basically lifeless. *Mashable*. Retrieved from http://mashable.com

Kitchin, R., & Dodge, M. (2011). *Code/space: software and everyday life*. Cambridge, MA: The MIT Press. doi:10.7551/mitpress/9780262042482.001.0001

Klemens, G. (2010). *The cellphone: The history and technology of the gadget that changed the world*. Jefferson, NC: McFarland & Company.

Ling, R. (2012). *Taken for grantedness: The embedding of mobile communication into society*. Cambridge, MA: MIT Press.

Lunden, I. (2014, April 23). Facebook passes 1B mobile users, 200M messenger users in Q1. *TechCrunch*. Retrieved from http://techcrunch.com

Manovich, L. (2013). *Software takes command: extending the language of new media*. New York, NY: Bloomsbury Academic.

Newman, N., & Levy, D. A. L. (2014). *Reuters institute digital news report 2014: Tracking the future of news*. Retrieved from https://reutersinstitute.politics.ox.ac.uk/sites/default/files/Reuters%20Institute%20 Digital%20News%20Report%202014.pdf

Organisation for Economic Co-operation and Development (2013, December 16). *The App Economy*. Publishing. doi: 10.1787/5k3ttftlv95k-en

Palacios, M. (2013). Prefácio - O mundo no bolso e o contexto na palma da mão. [The world in the pocket and the context in the hand]. In S. Barbosa & L. Mielniczuk (Eds.), *Jornalismo e tecnologias móveis* [Journalism and mobile technologies] (pp. 1–5). Covilhã, Portugal: Livros Labcom.

Palacios, M. S., & Da Cunha, R. E. S. (2012). A tactilidade em dispositivos móveis: Primeiras reflexões e ensaio de tipologias [Tactility and mobile devices: First approximations and a typology]. *Contemporanea (Bologna, Italy)*, *10*(3), 668–685.

Pauwels, C., & Picone, I. (2012). The tussle with trust: Trust in the News media ecology. *Computer Law & Security Report*, *28*(5), 542–550. doi:10.1016/j.clsr.2012.07.003

Pew Research Center. (2012, October 1). *The future of mobile news: The explosion in mobile audiences and a close look at what it means for news*. Retrieved from http://www.journalism.org/files/legacy/ Futureofmobilenews%20_final1.pdf

Purcell, K., Rainie, L., Mitchell, A., Rosenstiel, T., & Olmstead, K. (2010, March 1). *Understanding the participatory news consumer: How internet and cell phone users have turned news into a social experience*. Retrieved from http://www.pewinternet.org/files/old-media//Files/Reports/2010/PIP_Understanding_the_Participatory_News_Consumer.pdf

Rosenbaum, S. (2011). *Curation nation: How to win in a world where consumers are creators*. New York, NY: McGraw-Hill.

Schmidt, E., & Cohen, J. (2013). *The new digital age: transforming nations, businesses, and our lives*. New York, NY: Alfred A. Knopf.

Scoble, R., & Israel, S. (2013). *Age of Context: mobile sensors, data and the future of privacy*. United States: Self Published.

Scolari, C. A., Aguado, J. M., & Feijóo, C. (2012). Mobile Media: Towards a definition and taxonomy of contents and applications. *International Journal of Interactive Mobile Technologies*, *6*(2), 29–38. doi: 10.3991/ijim.v6i2.1880

Shoemaker, P. J., Vos, T. P., & Reese, S. D. (2008). Journalists as Gatekeepers. In K. Wahl-Jorgensen & T. Hanitzsch (Eds.), *The handbook of journalism studies* (pp. 73–87). New York, NY: Routledge.

Taylor, C. A., Samuels, N., & Ramey, J. A. (2011). How people really use the mobile web: A framework for understanding motivations, behaviors, and contexts. In J. Lumsden (Ed.), *Human-Computer interaction and innovation in handheld, mobile and wearable technologies* (pp. 195–214). Hershey, PA: Information Science Reference; doi:10.4018/978-1-60960-499-8.ch012

Thorson, E. (2008). Changing patterns of news consumption and participation. *Information Communication and Society*, *11*(4), 473–489. doi:10.1080/13691180801999027

Turkle, S. (2008). Always-On/Always-On-You: the tethered self. In J. E. Katz (Ed.), *Handbook of mobile communication studies* (pp. 121–137). Cambridge, MA: MIT Press. doi:10.7551/mitpress/9780262113120.003.0010

Turkle, S. (2011). *Alone together: why we expect more from technology and less from each other*. New York, NY: Basic Books.

Uhlmann, S., & Lugmayr, A. (2012). Portable personality and its personalization algorithms: An overview and directions. In A. Lugmayr, H. Franssila, P. Näränen, O. Sotamaa, J. Vanhala, & Z. Yu (Eds.), *Media in the ubiquitous era: Ambient, social and gaming media* (pp. 66–93). Hershey, PA: Information Science Reference. doi:10.4018/978-1-60960-774-6.ch004

Westlund, O. (2012). Producer-centric versus participation-centric: On the shaping of mobile media. *Northern Lights*, *10*(1), 107–121. doi:10.1386/nl.10.1.107_1

Westlund, O. (2013). Mobile news: A review and model of journalism in an age of mobile media. *Digital Journalism*, *1*(1), 6–26. doi:10.1080/21670811.2012.740273

White, D. M. (1950). The "Gate Keeper": A case study in the selection of news. *The Journalism Quarterly*, *27*(4), 383–390.

Wolf, C., & Hohlfeld, R. (2012). Revolution in Journalism? Mobile Devices as a New Means of Publishing. In C. Martin & T. von Pape (Eds.), *Images in Mobile Communication: New content, new uses, new perspectives* (pp. 81–99). Germany: Springer VS Verlag für Sozialwissenschaften. doi:10.1007/978-3-531-93190-6_5

Yeung, K. (2013, September 11). Social reader app Flipboard now has 85 million registered users. *The Next Web*. Retrieved from http://thenextweb.com

KEY TERMS AND DEFINITIONS

Algorithm: Set of instructions that enable a computer system to perform specific tasks.

Always-On Systems: Devices permanently connected to Internet.

App: Acronym of application, software developed for mobile devices, typically downloaded from an app store.

Content Curation: Online practice of gathering and (re)organizing text, video, audio, pictures, etc. in order to create new contents.

Contextualisation: A chain of meaning in which the main information is integrated in a set of related facts, figures, places or data.

Mobile Devices: Small computing devices with a touch screen and wireless internet connection, e.g. smartphones and tablets.

News Aggregator: Online platform tailored to select and organize digital news, automatically received from different web sources.

Push News: System to send/receive news alerts on mobile devices.

ENDNOTES

1 "Zite FAQ - Questions About Flipboard and Zite": http://blog.zite.com/228-2/

2 These features emerged from a larger observation, carried out as part of a doctoral research project currently ongoing at the Laboratory of Online Communication (LabCom), a research unit of the University of Beira Interior, Portugal

3 For a discussion on the role of journalists in news-aggregating platforms see Anderson, C.W. (2013). What aggregators do: Towards a networked concept of journalistic expertise in the digital age. *Journalism*, 14(8), 1008–1023. doi: 10.1177/1464884913492460.

Chapter 13
Mobile Journalism and Innovation:
A Study on Content Formats of Autochthonous News Apps for Tablets

Marcos Palacios
Federal University of Bahia, Brazil

Fernando Firmino da Silva
State University of Paraiba, Brazil

Suzana Barbosa
Federal University of Bahia, Brazil

Rodrigo da Cunha
Federal University of Pernambuco, Brazil

ABSTRACT

Using a qualitative methodological approach, this chapter brings to bear a theoretical and conceptual framework with particular consideration to the notions of innovation, journalistic convergence, design and the theory of affordances, as well as mobile communication and mobile journalism. This analysis covers the period 2010-2014 and seeks to examine the specific affordances, the structure, organization and types of narratives produced; the design and use of multimedia features as well as the level of integration with other products in the same news organization. The empirical corpus consists of products selected for their representativeness within a broader sample: O Globo a Mais, Estadão Noite, and Diário do Nordeste Plus (Brazil), La Repubblica Sera (Italy), La Presse+ (Canada) and El Mundo de la Tarde (Spain). The study concludes that - although timid - experimentation continues to prevail regarding formats, content, language and narratives based on the features of mobile devices, the continuous evolution of operating systems and changes in hardware platforms.

INTRODUCTION

Four years after the creation of the first journalistic products for tablets, and the iPad in particular, how have autochthonous – and specially evening news apps – contributed to the evolution and innovation of content formats? This is the central question of this chapter, which is part of the research developed at the Convergent Journalism Laboratory Project[1] hosted at the Faculty of Communication, Federal University of Bahia, Brazil.

DOI: 10.4018/978-1-4666-8838-4.ch013

The aim of this chapter is to take a step forward in the systematization (with previously cataloged experiences as background and context) of a specific type of innovative journalistic product, namely autochthonous news app, with special attention to evening news apps, and a focus on identifying their contribution to the complexification and differentiation of the mobile information ecosystem in an environment of convergent multi-platform journalism.

As elements of the new mobile ecosystem, evening newspaper apps for tablets are specially designed products that form part of the multiplatform information landscape. They are newcomers to the contemporary media ecosystem, characterized by fluid, horizontalized production, resulting in a multimedia continuum (Barbosa, 2013) and they constitute a rich field for the identification and observation of innovative features in journalistic production.

The premise to be explored in this text is that 'autochthonous' apps – *i.e.*, created by specialized teams, producing exclusive materials for mobile platforms, with distinct treatment of contents (Barbosa, Firmino da Silva, & Nogueira, 2014; Barbosa, Firmino da Silva, Nogueira, & Almeida, 2013) – are products which exist and compete in a new innovation cycle with regards to the editorial line, treatment of information, design, formats, and use of multimedia language. As such, they may be considered as prime spaces for experimentation with new affordances such as tactility (Palacios & Da Cunha, 2012) personalization and geolocation, as well as a breeding ground for new business models, consumer dynamics, circulation, recirculation and response to journalistic content.

Brazil has been one of the leading countries in the production of evening news apps. Their emergence, marked by a convergent and multi-platform logic, has been designed to take advantage of the way content is consumed on tablets during the evening period, *i.e.*, in what is known as 'lean-back' mode, when the user is usually at home, more relaxed and with a (pre)disposition to explore longer, more complex/sophisticated textual constructions and/or HD/multimedia content.

The emergence of these journalistic apps for tablets can also be associated with a general boost in the consumption of content on tablets and smartphones, thereby opening new horizons to business models directed towards a mobile ecosystem. Prospects of economic sustainability seem to be rising, based on the innovation and creation of narrative forms more aligned to audiences who have already identified with the platform's potentialities, particularities and distinctions, *vis-a-vis* other media formats.

The basic parameters of analysis were defined in our research project and will be applied to systematize the data collected through structured observation of a *corpus* comprising Brazilian and international evening news apps, namely: *O Globo a Mais, Estadão Noite*, *Diário do Nordeste Plus* (Brazil), *La Repubblica Sera* (Italy), and *El Mundo de la Tarde* (Spain). These apps were selected from a larger universe which have been mapped in our earlier studies and constitute a representative sample for the purposes of the proposed investigation. We have also included *La Presse+* (Canada) in our *corpus;* although not characterized as an evening newspaper, in contrast to the other apps analysed in this chapter, it is an autochthonous tablet product, with distinctive innovative and advanced features that justify its inclusion as a relevant object for our case study.

The methodological approach is qualitative and employs theoretical constructs created by researchers at the Convergent Journalism Laboratory Project, especially the notion of innovation gradients as a derivation of Gartner's Hype Cycle (Palacios, Barbosa, Firmino da Silva, & Da Cunha, 2014). The theoretical and conceptual framework of journalistic convergence, developed by a large contingent of the authors in several countries, will also be central to situate the phenomenon of evening news apps.

INNOVATION: A PREAMBLE

Since the 90s of the previous century, the discourse on innovation as an essential dimension to the survival of news organizations in contemporary society has become more firmly embedded. This is due mainly to the emergence of the digital as a major factor to mobilize the efforts of media companies around strategies, operations, processes and products able to ensure them a place in the complex ecosystem that took shape around communication and specifically journalism. During the 70s and 80s, some innovation was contemplated in the strategic plans of such corporations; however, the scene was more stable and changes were processed at a gentler pace (Machado, 2000).

To define and quantify 'innovation' and 'innovativeness' in media products poses problems that, in some respects, are similar do those one faces in the field of 'quality assessment' in media product studies. One has to confront the crucial questions of "what is to be measured" and "how to measure" whatever one decides to measure. In both cases the researcher faces a maze of definitions, typologies and indicators and an awesome volume of literature, in different areas of knowledge and investigation, which tended to grow considerably in recent decades (Fagerberg, 2005; Cohen, 2010; Crossan & Apaydin, 2010).

Garcia and Calantone (2002) in a critical review of academic literature dealing with innovation and innovativeness indicated that:

No less than fifteen constructs and at least 51 distinct scale items have been used in just 21 empirical studies in the new product development (NPD) literature that model product innovativeness (...) This lack of consistency is operationalizing 'innovativeness' has resulted in the interchangeable use of the constructs 'innovation' and 'innovativeness' to define innovation types. Although this may seem to be an insignificant point, this has led to incongruent categorizations of innovation typology and widespread confusion as to what empirical studies are actually reporting.

Even without a consistent and uniform definition of innovation, academic works dealing with innovation in journalism have tended to cluster around the theme of technological innovation. Grubenmann (2013), in a critical review of 60 English language academic articles dealing with innovation in journalism, concluded that "[E]ven though research in journalism is not limited to technological innovations, around 70% of the identified literature deals with this topic. About 9% of the research goes into design innovation in journalism and the remaining 21% investigates different forms of journalistic innovation" (p.16).

It is also fair to state that although the area of innovation research has a long academic tradition, specific studies on media innovation is a relatively new field, which accelerated with the emergence, diffusion and generalization of digital technologies and received a new boost with de universalization of mobile devices. Available material on media innovation is mainly based on case studies, tends to be descriptive and is poor in comparative analyses.

On the other hand, 'measuring innovation' constitutes yet another battlefield in academic research. The discussion goes at least as far back as Schumpeter´s divergences with his econometrician colleagues as to the possibility of creation of comprehensive models capable of fully accounting for the "little devil" of innovation, in a dispute marked by irreconcilable contrapositions of organic and mechanical metaphors (Louçã, 2014). The subject is open to dispute and qualitative analyses and measurements of 'degrees of innovation' are to this date confined to very specific areas of innovation research (Van der Pannen, 2007).

There seems to be some consensus among academics that measurement of the process of innovation "is critical for both practitioners and academics [but also that] the literature is characterized by a diversity of approaches, prescriptions and practices that can be confusing and contradictory" (Adams, Bessant, & Phelps, 2006). If this is so for the innovation studies field in general, it is even more so for the area of media innovation studies. In this chapter we adopted a qualitative approach without any attempt to suggest quantification models.

As Machado (2010) pointed out, innovation programs that are now being implemented from the 90s of last century began with the assumption that innovation was essential for progress and social change.

(...) Innovation in journalism is a phenomenon that faces journalism as an industry and focuses on finding conceptual or technological solutions capable of, at the same time, maximizing production and meeting social demands for quality, instantaneous information at the lowest possible cost, in keeping with the rigor of the highest professional conduct and accessible by all available means. (Machado, 2010, p. 67)

Therefore, innovation may be understood as a process that provides successive upgrades that can ensure continual improvement of product and production methods, as well as the emergence of new procedures to shape new products, modes of distribution, consumption and business models.

For Machado (2010), the notion of innovation in journalism is related to any change in techniques, technologies, processes, languages, formats, teams, devices and applications, values or business models designed to streamline and enhance the production and consumption of journalistic information. The author asserts that innovation endeavors to maximize production to meet social demands for quality and instantaneous information at the lowest possible cost, in keeping with the rigors of the highest professional conduct and accessible by all available means.

Franciscato (2010) stresses the need to also consider the interwoven aspects of technology and its social use in situations of innovation. Journalistic institutions, in feeding society each day with a stream of news about the world, build, reinforce and modify social and cultural processes. "Beyond its presence in the market, each news product presented to society brings, even implicitly, these sociocultural elements, which alter or add substance to habits, knowledge and social values in the short or long term" (p. 16).

Rossetti (2013) is one more voice to make chorus to the assertion that there is no unified definition for innovation. One is faced with a multiplicity of terms related to the notion of innovation, e.g. new, novelty and change; creation, invention, evolution and rupture; growth and experimentation, renovation, redesign and metamorphosis; unique, differentiated, different, original and unrevised; innovative subject and innovated object. In the field of communication, the innovative act refers to actions present in communicative processes that generate new and innovative communicative products, or that introduce changes in communication theory or language. And additionally: formats, narratives, and the process of their reception and formation of new consumers' habits are also aspects to consider when one searches for innovation indicators.

To build a framework of categories of innovation, in which the various meanings of innovation are classified based on the notion of change (considered central because change brings the new), Rossetti uses the Aristotelian categories of substance, quality, quantity, relation, place, time, action and affection. Based on these, she organizes the terms relating to change in categories of innovation in a matrix which illuminates the amplitude of possible areas of investigation. The problem, once again, is the inevitable multiplication of similar and possibly juxtaposed terms and classifications in the matrix. According to

Rossetti´s categories, innovations in design, for instance, could constitute elements of 'substantial innovation', 'qualitative innovation', 'relative innovation' or even 'active and affective innovation', as major changes in design may also mobilize actions and affections of the user of a product.

In a more restrained way, David Nordfors (2003) distinguishes the concept of innovative journalism as the process of creating and introducing new values, having such components as a brand, a definition, a business model, a narrative, along with other key elements: a technology, a platform, a device, a format, a product, a concept.

A summary of distinct approaches to innovation in journalism in academic literature, with an indication that "the emerging consensus is to reject deterministic explanations and instead propose that technological innovations are mediated and shaped by initial conditions and contextual characteristics" can be found in Michelstein and Boczkowski (2009).

Based on this outline, it is evident that innovation processes in journalism are continuous and to be found in many different areas of product development: design, production processes, content, formats, genres, languages, narratives, business models, circulation and consumption targets etc. From time to time major ruptures do happen, as illustrated by the invention of smartphones, iPad and similar devices. Transposition and remediation of already existing models and formulas constitutes a first movement of occupation of new spaces/platforms; autochthonous solutions come next.

AUTOCHTHONY AND INNOVATION

Amid the tactics employed by news organizations for the distribution of contents for tablets and smartphones, the implementation of autochthonous applications is one of the current trends. In breaking with the initial pattern of simple replication of print editions (transposition), these applications defined themselves by inroads to the platform with more systematic usage of a wide range of features.

At the same time, practices were established that required more from the standpoint of design, programming and thinking about the narrative while taking into account the specificities of the device, especially as it relates to the operation of the tablet, with its surface tactility and 'lean back' mode of usage (García, 2012). The return of the 'evening edition' is in itself an innovation, as the format had been abandoned decades ago. This confirms the assertion that the newness "may be a recombination of old ideas challenging the present order in such a way that it is new to the people involved" (Van de Ven, 1986; Åkesson, 2009).

Thus, 'autochthony' was selected as a promising indicator of innovativeness in journalistic practices and was adopted as the main criterion for inclusion in the constitution of a *corpus* for this chapter, formed by a selected set of 'autochthonous' news apps produced between 2010 and early 2014.

Once established the notion of autochthony as filter and once selected the cases for study, one still confronts the inevitable need to choose which elements to highlight and analyze in the search for innovative features of the products. To aim at 'holistic approaches' may sound like an instigating challenge, but in practice such attempts may prove wild goose chases. Our aim, therefore, was less ambitious, examining some selected traits which we consider as possible indicators of innovation:

1. Novel use or enhancement of specific affordances (such as tactility, personalization, interactivity, geolocation and augmented reality);
2. On-board functions;

3. Structure, organization and types of narratives produced;
4. Design and use of multimedia resources (video, audio, photography and infographics);
5. Level of integration with other journalistic products from the same media company and their respective social networking sites as such integration may result in new, economically sustainable business opportunities.

CONVERGENCE, MOBILITY, AND AUTOCHTHONY

Journalism is going through two main processes of restructuring of its operations, due to digitization and the complexification of data communication networks. The first one refers to the process of journalistic convergence (Quinn, 2005; Salaverría & Negredo, 2008; Salaverría, García Avilés, & Masip, 2010;) and to editorial transformations in aspects of integration or multi-platform distribution and, consequently, in the production process characterized by the horizontal and dynamic flow of the multimedia continuum (Barbosa, 2013). Convergence has been an important element in the process of organizational innovation in digital media production and has "challenged two pervasive notions of traditional and digital media scholarship: the one newsroom–one medium assumption and the trend toward product homogenization in media convergence" (Boczkowski & Ferris, 2005).

A second process refers to mobile communication (Castells et al, 2006; Silva, 2008; Pellanda, 2009; Canavilhas & Santana, 2011; Peters, 2012; Westlund, 2013) – smartphones, tablets, e-readers, gaming consoles, ultrabooks etc. – through the development and ownership of mobile devices within the notion of multi-platform distribution and publication. In order to participate in this space, newspaper companies are repositioning themselves in terms of products, journalistic routines, content production, circulation, consumption and business models.

These twin processes of convergence and mobility are seen in this chapter as triggers for the creation of new products composing the informational supply of newspaper companies, as well as the production of new narratives that employ specific affordances for the shaping of content that follows different formats. Content production and delivery take place in a mobile domain in a heterogeneous and fragmented digital ecosystem:

(...) and the resulting mobile content space can be better categorized, then, into four broad classes not mutually exclusive: "adapted" for existing information (coming from different media) that has been adjusted for display and use in a mobile environment, "repurposed", for creative content that has been re-used and adapted to mobility, "original" or "specific", for creative content that has been designed with mobility in mind, and augmented", for content (of any type) that uses the additional and specific properties of mobile systems (such as location-awareness, for instance) to increase its value and interest for users ". (Feijóo, C. et al., 2009)

With the introduction of iPad and other tablets with different screen-sizes, the media ecosystem became even more complex, as those devices demanded that new specificities and new affordances had to be taken into account in the process of production and circulation of contents. Mobile devices become potential boosters of a new cycle of innovation, because they tend to reconfigure production, publication, circulation and recirculation, modes of consumption and reception of news content, as well as business models in multiple platforms. They may also very likely entail changes in the editorial cycle (routines) and foster new skills for professionals operating in this ecosystem. Consequently, new journalistic applications (apps) emerge, especially autochthonous products designed and produced specifically for mobile platforms with a large percentage of exclusive materials and differentiated treatment (Barbosa et al, 2013). These apps can be targeted equally well for smartphones and tablets.

Scolari, Aguado and Feijóo (2012) suggest that all classes of content ("adapted", "repurposed", "augmented" and "original or specific") are observable in news apps, as autochthonous mobile products increasingly become part of the contemporary media ecosystem.

In this chapter we focus on autochthonous apps for tablets, exclusively for the Apple iPad, because we identified these products as more innovative regarding the use of specific properties for developing narratives and content formats.

In general, the composition of the journalistic material posted on these apps gives precedence to reports produced exclusively by their teams, with longer narratives enriched by the resources cited above, analyses by columnists/writers, slideshows as well as summarized news highlights of the day, in a mix of curated content from the print edition, supplemented by editorial and copy material produced exclusively by their teams.

Moreover, autochthonous apps incorporate the strategies for diversification of media companies' business models, which, based on both internal and external research and as suggested by trends, began to value digital products and the 'digital first' operational strategy. In doing so, they purport to strengthen and expand the power of their respective brands in a much more complex environment and in direct competition with new content producers, as underscored by *The Full New York Times Innovation Report* (2014).

According to numbers released by the Pew Research Center (2014, p.6):

The vast majority of Americans now get news in some digital format. In 2013, 82% of Americans said they got news on a desktop or laptop and 54% said they got news on a mobile device. Beyond that, 35% reported that they get news in this way "frequently" on their desktop or laptop, and 21% on a mobile device (cell phone or tablet).

The growth in the consumption of news through multiple platforms has been indicated and monitored, for example, by surveys conducted by the *Reuters Institute* at Oxford University (Newman & Levy, 2014), which began to include urban Brazil in their studies as of 2013. The results show that mobile devices occupy a prominent position. Whereas in 2013, Brazil ranked third in the frequency of news consumption on different platforms (radio, TV, newspaper or online, with 88%)[2] in 2014, the *Digital News Report* by the *Reuters Institute* revealed that Brazilian digital news readers (6 out of 10) are the most likely to pay for content[3].

As a whole, 41% of users consume news content through digital platforms: 18% said that mobile phone is their primary means of reading online news, while 8% of users consulted replied they read it on a tablet. Distinct types of digital payment were identified: "single payment", 36%; "digital subscription", 31%; and "digital and print subscription package", 15%.

Meanwhile, data collected by the *Brazilian National Newspaper Association* (Associação Nacional de Jornais) indicated (2012) that digital subscriptions to newspaper websites grew 128% over the previous year and showed that digital platforms are used by 120 million people to seek information. Between January 2013 and June 2014, the use of smartphones to read content from the four major Brazilian newspapers grew by 430%[4]. In the first half of 2014, the increase was 88.3% compared to the corresponding average from 2013.

Broader surveys that measure the usage rates of mobile devices also show that smartphones and tablets are ascending rapidly in Brazil (Banks, 2014). The rate of increase was registered as 132%, even higher than that found in the United States.

These surveys confirm the increasing consumption of news on digital platforms, and also indicate that the diversification of products and launching of apps for smartphones and especially tablets, encounters users who are indeed interested and willing to pay for access to differentiated content. Moreover, they detected a clear preference among users for the evening hours to consume journalistic information. Directed towards this potentially new audience group, one would also expect new 'communication contracts' to emerge concerning the framing of news and structure of narratives (Satuf Rezende, 2013) in innovative journalistic mobile products.

Evening news applications were launched in Brazil and in Europe as new products of an informational galaxy (Macri, 2011) redefined by recent developments. They sought to meet convergent strategies: to diversify operations, to fulfill specific demands of audience and to attract new consumers of news on tablets and smartphones. Barsotti (2014), for example, points out that production for evening applications has changed the journalistic routine in newsrooms. Thus, a robotic editing environment turns into a manual one with the formation of dedicated (exclusive) teams to handle the production of content and the new journalistic language to tablets, such as the Brazilian app *O Globo a Mais.*

Tablets have been a driving force for change in production routines, business models, and journalistic language. Products launched for these devices have pushed integration boundaries in newsrooms. They require their own teams and impose limits on the logic of unceasing, 24-hour content production. Tablets have also restored the faith of journalistic companies in the viability of their business, giving new life to the model underpinned by subscription and advertising income. This device has been the engine for a new journalistic language, which draws on the intensification of sensations. The strategy relies on the reception of news by appealing to three senses simultaneously: sight, hearing, and touch (Barsotti, 2014, p.128).

As Barsotti points out, changes to journalistic language indicate aspects of innovation related to what can be considered a grammar specific to journalism for tablets due to the properties that characterize and distinguish such devices. Referring specifically to smartphohes, Feldman (2005, p.30-32) stressed the importance of taking advantage of new possibilities in interactivity an personalization which could "tie media audiences closer to their media brands" and suggested that selection of contents could be improved by the use of 'collaborative filters'. Batiston and Bello Neto (2013) emphasize the importance in news production for tablets of using elements in various formats (photography, video, sound, etc.) applied to the possibilities of touch interaction."The ways in which content is presented are important to keep the reader's attention in any medium and in tablets there are many features to be employed." (Batiston & Bello Neto, 2013, p.106).

NEWS DESIGN AND AFFORDANCES

Newsrooms were slow to understand the importance of the visual aspect of journalistic information. Only since the Modernist Movement of the early twentieth century, when Ben Sherbow developed his first experiments to recast the appearance of American newspapers and John E. Allen created the first manual on news design, did the appearance of the information gain some relevance to journalism. Over three hundred years after the first periodical was created, information was typed out with no thought to graphic design, and all work focused on fitting the news somewhere on the page, sometimes sacrificing a picture or reducing the size of typography (Barnhurst & Nerone, 2001; Díez & Cuadrado, 2001).

Coates and Ellison (2014, p. 10th) define information by the equation between semantic and pragmatic aspects involving information systems, with the objective to incorporate more essential data, interpret them and communicate them in the most fluid manner possible through the graphical interface[5]. This is done, for example, through the production of schematic diagrams, maps, instruction manuals, outdoor signage and in the case of journalism, infographics. A few landmarks were important for the development of information design such as isotype pictograms developed by Otto Neurath, between 1920 and 1940; the schematic plan of the London Underground, by Henry C. Beck, 1933; and the graphical user interface of the Macintosh, from 1984.

The design of news, as per Larequi's definition (1994, p. 26), may be summarized as a technique that enables the organization of a set of printed (text and graphics) and non-printed (white space) elements on the surface of a graphical space (page) in order to structure, prioritize and facilitate the readability of journalistic information. Still focused on traditional means of communication, the concept of design presented by Larequi underscored the importance of diagramming, which ceased to be haphazard and became a defining element of the personality of the publication.

The internet, and more recently, mobile devices have stressed the role of interactivity, considered as one of the defining characteristics of online journalism. Cameron (1998, p. 7) states that the voice of design in this regard is displaced, so that the professional merely furnishes the language and its syntax and the users create their own voices based on these signs. It is the end of the sequential narrative, with the emergence and growing hegemony of a model in which the user navigates freely for information, thereby modifying the relationship between author and reader (Julier, 2010, p. 218).

It is fair to say that mobile devices have forced designers to seek new ways of organizing elements and graphical resources, not only as a result of the interactive factor, but also due to the limited space on a smartphone or tablet screen. Whereas before the reader scanned the news distributed across large pages and was drawn to other stories by peripheral vision, (s)he now accesses the information in a single take which appears momentarily on a small screen. The presentation format, by means of applications, has become another distinct factor in the visual interface, with the addition of new features and programming languages such as HTML5 and Objective-C.

To enhance the understanding of each of these visual transformations of journalistic language and their respective relationships with the reader, it is worth going back to ideas and concepts presented in *The Theory of Affordances*, an article by James J. Gibson (1979).

The idea of affordance may help to explain the phenomenon of interfaces in which the user deals directly with basic commands for navigating and reading the pages of a digital journal from movements executed directly on the screen's surface, which is conventionally called tact-reading (*tactoleitura*) (Salaverria, 2012) or tactility (Palacios & Da Cunha, 2013). Also significant among affordances to be found in news apps are the increased use of multimedia (videos, photos, picture galleries in 3D or 360 degrees), animated infographics, higher degrees of interactivity, non-linear navigation through internal dynamic links and navigation via hidden text. They may also include features to view internal images or infographics with the touch of a finger, as well as details for data, numbers, and geolocation features, among others.

The theory, based on the nominalization of the verb to afford, was presented in the late 1970s by psychologist James J. Gibson (1979) to describe the interrelation between the properties of the environment, with its particular characteristics, and the animal during the process of perception. Gibson formulated these ideas in the previous decade, when he had contact with concepts from Gestalt psychology and functionalism (Santaella, 2012, p. 49th).

Almost ten years later, Donald A. Norman brought the term into the field of design with his book *The Design of Everyday Things*. The relationship between environment and animal is replaced by object, user and action. In this book, Norman explains that "affordance refers to the perceived and actual properties of the thing, primarily those fundamental properties that determine just how the thing could possibly be used" (1988, p. 9). For the author, the objects provide clues that determine their use, even if they are not uses originally intended by the developer.

Murray (2012) states that digital devices provide four affordances or properties that are useful for their representation: procedural, participatory, encyclopedic, and spatial. One digital work may employ any of these properties better than another. In the context of interaction design as a cultural practice, which is the focus of the author, a digital artifact is able to refine or create conventions that best exploit each of these four affordances.

The following is a quick description of each affordance referenced by Murray (2012) to characterize new media:

1. The *procedural property* of new media refers to the ability to represent and perform conditional behaviors. Programmers can write lines of code able to create abstractions, as well as anticipate behaviors from the negotiation of interaction with the user. Design is an important part of this process as a way to create more flexible systems within the range of possible anticipated responses;
2. The *participatory property* is related to the reciprocal relationship between any digital device and the party interacting with it. The design should provide for significant actions by both sides, as well as selecting appropriate conventions to communicate to the user what actions are possible for the interaction to be understood and carried out;
3. With higher processing capacity and transmission of information than any other media previously created, the *encyclopedic property* refers to unmatched computer storage potential, an infinite library that inherits the same time-old traditions of human culture, with the possibility to collect, preserve and transmit knowledge;
4. The *spatial property* refers to the ability of new media to represent, in the graphical interface, physical spaces – a classic example is the WIMP (Windows, Icons, Menus and Pointing Devices). According to Murray, this space is not restricted to a mere literal reproduction of commonplace elements, but may incorporate abstractions that reinforce their roles as fragments of information or programming of codes. Such abstractions need to be easily learned and should be transparent to the user.

In practice, these affordances are mainly represented by icons that refer to each action within the application, whether to guide reading or clarify items of audiovisual content (video, audio, photo galleries, interactive infographics). These icons, also known by the term 'coach marks,' educate the reader in a new, previously unknown device and incorporate themselves in a grammar already defined by the familiar editorial design of printed newspapers and magazines.

Westlund (2013) discussed current trends in production of mobile news for smartphones and proposed a model of analysis focusing on the roles of humans and technology in activities characterised by customising or repurposing. He believed that the tendency was towards treating journalistic content as 'plataform agnostic' with mobile news intertwined with news publishers' cross-media news work and publishing becoming "increasingly synonymous with excelling in technological customisation, harnessing technological assets that enhance the perceived affordances of mobile devices" (p.21). His conclusion was that "despite exceptions (...) this marks a potentially important shift away from established and normative conceptions, suggesting that for any emerging medium journalism must be shaped in a distinct way. However, this obviously depends on the importance given to humans *vis-à-vis* technology when we speak of journalism." (ibid).

By the end of 2011, the emergence of evening newspaper apps reopened the discussion on 'repurposed x original content' as the new apps became a novel space for potential experimentation relating to a set of interactions, in which the design tries to encourage readers' participation through tactility on the screen surface, the activation of the multimedia elements and other device-specific properties/gimmicks. As products created exclusively for smartphones and tablets, their characteristics should express affordances relating to inherent features of mobile devices; or so one could expect.

JOURNALISTIC APPLICATIONS: CASE STUDIES

The analysis of the six journalistic applications selected as the empirical *corpus* – *O Globo a Mais*, *Estadão Noite*, and *Diário do Nordeste Plus* (Brazil), *La Repubblica Sera* (Italy), *La Presse+* (Canada) and *El Mundo de la Tarde* (Spain) – includes observations regarding affordances employed in the narratives in terms of features that characterize the production of content for tablets and smartphones, as well as aspects of innovation and product evolution identified in the period between 2010 and 2014 (Table 2).

The choice of this *corpus* comprises five evening news applications which were identified as innovative products designed specifically for a new platform. As explained before, La Presse+ which is *not* an evening newspaper was also included in the corpus as it is an autochthonous tablet product, with distinctive and advanced features that justify its inclusion as a relevant object for our case study.

Table 1. Development of features in applications for tablets, 2010-2014

2010	2011	2012	2013	2014
• Use of audio and video features; • Sharing of content via e-mail and social networks; • Screen adaptability for reading on various devices; • Use of panoramic images and HTML5 animation.	• Creation of newsstand feature, with the ability to purchase single editions or subscribe within an application; • Use of augmented reality features; • Printing pages over wifi in capable printers.	• Screen adaptability to also read on smartphones; • Use of video galleries; • Automatic archiving of back issues; • Previews of some internal pages of each edition before having to pay (*paywall*).	• Use of banner ads within applications; • Simplified layout, with the adoption of *flat design* style, to be implemented with the interface for iOS7; • Integration with other device hardware, such as camera and accelerometer.	• Ability to stream audio and video; • Insertion of reader commentary within the pages of a publication; • Sharing of content with external services such as *LinkedIn* and *Evernote*.

Table 2. The corpus: a general view

	O Globo a Mais	Estadão Noite	Diário Plus	La Repubblica Sera	El Mundo de la Tarde	La Presse+
Affordances	360 degrees Interactive infographics Image galleries Video Newsgames Hidden text Image carousel Tactile narratives Hyperlinks Pop-up menus	Image galleries Video Audio Hyperlinks	360 degrees Interactive infographics Image galleries Videos Newsgames Hidden text Image carousel Tactile narratives Hyperlinks Pop-up menus	Audio Video Animation Image galleries	Audio Video Hyperlinks	360 degrees Interactive infographics Image galleries Videos Newsgames Zoom Image carousel Tactile narratives Hyperlinks Pop-up menus
Narratives	Narratives based on photos, videos, infographics and interactive elements with magazine-style coverage	Analytical articles. Doesn't offer reporting.	Narratives based on photos and videos with magazine-style coverage	Narratives based on photos and videos with magazine-style coverage	Narratives based on photos and videos with newspaper-style coverage	Narratives based on photos, videos, infographics and interactive elements with magazine-style coverage
Design	Emulates a print magazine (layout and positioning), but with interactive and dynamic features	Similar design to the columnar print edition.	Emulates a print magazine (layout and positioning), but with interactive features, audio, video, infographics, image galleries	Emulates a print newspaper (layout and positioning) with static elements	Emulates a print newspaper (layout and positioning) with static elements	The design is like an app for tablet with horizontal orientation
Media formats	Audio, video, infographics, image galleries	Audio, video, image galleries	Audio, video, infographics, image galleries	Audio, video, image galleries	Audio, video, image galleries	Audio, video, image galleries, infographics
Sections	Giro, Fotos do Dia, Gente do Globo (Colunistas), Artigo, Dicas a Mais, Imagens a Mais	Artigos, Colunistas, Cenas do Dia, NaTV Estadão & na Rádio Estadão, Amanhã no jornal O Estado de S. Paulo	Panorama, Colunistas, Diário em Fotos, Teenagers, Roteiro Cinema, Roteiro Cultural, Reportagens	Primo Piano, Politica, In Redazione, La Giornata in immagini, Esteri, Lo Scatto, Pollole, Intanto net Mondo, Sport, Cultura, Spettacoli, Commenti, La Memoria, Le Scelte di Rsera, Il Libro, Il Film, Stori ai Fornelli, Web Circus	Opinión, España, Mundo, Economia, Deportes, EM2, Perspectivas, Última, Pasatiempos, Videos e Imágens	Actualités, Débats, Arts, Sports, Affairres, Pause Beauté, Pause Repas, Reportagens, Le Monde en Images
Business models	Distribution for tablets and smartphones (iOS and Android). Sold in the Tablet and web catalogs. Single purchase, subscription or digital package.	Distribution for iOS tablets. For single purchase or paired with subscription to other digital or print products.	Distribution for iOS tablets. Sale paired with subscription to the print edition.	Distribution for iOS tablets. Sale paired with the digital print edition.	Distribution for tablets (iOS and Android). Monthly subscription.	Distribution for tablets (iOS and Android). Free, ad-supported distribution

We sought to characterize apps according to the process of innovation and how they currently position themselves after the maturity period. The year 2010 was taken as a point of reference, as it marks the emergence of the Apple iPad tablet and its first dedicated products, although it was only in 2011 that the first evening news apps were created: *La Repubblica Sera* (Italy), released on November 23rd of that year with the introduction of audio, video, photo galleries and animation in HTML5.

In Brazil, the trailblazer was *O Globo a Mais.* Its first edition was published on January 21, 2012, initially offering such features as audio and video, page zoom, HTML5 animation, thumbnail navigation, photo galleries, horizontal and vertical layout, and 360° panoramas. Subsequently, there appeared *Estadão Noite* (April 2012), *Diário do Nordeste Plus* (January 2013), *La Presse+* (April 2013) and *El Mundo de La Tarde* (October 2013).

The first half of 2014 comprises the final period for observation and case analysis, so as to better understand the stages of evolution and development of apps selected for the empirical study. The key question that guided our observation and analysis was: how have these applications evolved since their launch and what innovations have they made?

Following this perspective, we observed the collection of applications identifying changes and aspects that stood out. Initially, we systematized a generic timeline of the features introduced in these applications and kept track of the emergence of evening newspaper apps as of 2011 (See Table 1).

Systematic observation of evening news applications permitted the characterization of the current state of each one of the products in terms of affordances, design, content formats, construction of narratives, sections, business models and integration between platforms in the same media group such as web and mobile (See Table 2). Data collected revealed that there was limited progress in the implementation of new features since the release of the respective autochthonous evening journalistic products. These native apps designed specifically for tablets were timid in exploiting the potential of the platform for the construction of longer and more complex narratives and in experimenting with new content formats.

In most cases there was little concern with innovating and some applications were launched as a finished version, as is the case of *La Repubblica Sera* and *Estadão Noite.* They kept virtually the same structure of design and features suitable to the language of *tablets,* but without adding innovative features afterwards.

La Repubblica Sera

The pioneering evening news application, *La Repubblica Sera,* released few updates since its creation and some of these updates were limited to bug fixes, schematic changes or changes in subscription or price. It offers the same basic features as the first news apps (as detailed in Table 1): audio and video, HTML5 animation, and a map of international correspondents with audio and weather forecasts (See Figure 1).

Aspects such as horizontal viewing are limited to image galleries, but other features adopted by other applications were not introduced, such as navigation by thumbnails of pages (there is only a vertical menu on the left side), printing, page zoom and the ability to change the typographic font. The sharing system is still based on static pages and the layout is similar to the design in print. The print version of *La Repubblica* received a recent make over to its graphic design, but this did not extend to the application.

We could obtain no data concerning the size and/or composition of *La Repubblica Sera´*s staff.

Figure 1. La Repubblica Sera offers the same basic features as the first news apps

Estadão Noite

As in the case of *La Repubblica Sera,* the Brazilian *Estadão Noite* has maintained almost unchanged its initial design structure (a black, static background) and features (audio, video and image galleries), vertical and horizontal navigation, and content based on columnists and political and economic analysis. The audio format predominates since it has appeared on the cover of each issue, as well as videos. The application, in the end, delivers the print newspaper headlines from the following day and the "Latest News" section, which links to a page on the website. In this way, there is an attempt to connect the app with the platforms of print, web and radio. The app staff consists of five professionals: an executive editor, an editor, an art director, a designer and a photo editor, besides the contributions of *Estadão Group* columnists (See Figure 2).

The parent application, *Estadão,* encompasses *Estadão Noite, Estadão Premium* (reproduction of the print edition with some multimedia elements added), *Estadão Light* (version based on iPad features) and *Olhar Estadão* (which circulates on the weekend with pictures from *The New York Times,* photo essays and Instagram pictures sent in by readers). This strategy coincides with the recommendation of García (2012) on three ways to distribute content for tablet[6]: "curated edition", "continuous news aggregator" and "PDF of the print version." *Estadão* offers a combination of these three forms with applications available on its platform. For Garcia, the "curated edition" would be the best suited for customization and the use of audio, video and images. However, in the case of *Estadão Noite,* the object of our analysis, there were no new features added based on innovative journalism.

Figure 2. Estadão Noite emphasizes analytical content. Audio is the predominant feature.

Diário do Nordeste Plus

The *Diário do Nordeste Plus* (https://flic.kr/p/wdGipj) was the third application to be launched in Brazil and followed the model of *O Globo a Mais*. In this context, the app has implemented practically the same features as *O Globo a Mais* and the design itself is similar with sections and features approximating those used in *the O Globo a Mais* app. The application is navigated vertically and changes to horizontal for the image gallery. The team consists of 10 people (plus columnists): a tablet editor and subeditor, a reporter, a photo editor, an art editor, a designer, an infographic designer and three interns.

One attempt to create a unique identity is seen on the app cover, which besides the main photo and caption, delivers three more headlines of internal content. The app shows infographics, reports with photo galleries, plus videos, audio and images in 360°. During the Confederations Cup in Brazil, *DN Plus* offered a virtual tour of the Arena Castelão (a World Cup stadium in Fortaleza, Northeast Brazil). From the point of view of innovation, it offers little novelty compared to other evening news apps.

El Mundo de La Tarde

This app has a different logic to the others regarding access. It is limited to offer the download of editions to access on mobile devices. The content is primarily online, therefore the interface connects to a website. The application (especially the top bar with the logo and access to the menu) is a skin-type interface to access the content. It is reminiscent of the first journalistic applications for tablet, like the initial release from the Brazilian magazine *Epoca*, in 2010. The ability to "download content" allows the edition to be read offline (See Figure 3).

Figure 3. With website-like characteristics, El Mundo app is reminiscent of the first products for tablet

In spite of having a website-like appearance, the content of *El Mundo de la Tarde* has some of the resources that are implemented with greater complexity in other autochthonous applications: sharing to social networks (a link is also accessible from the desktop, as with a website), shrink/enlarge the font size (accessibility), electronic library (archive of content and search by date of publication), fluid adaptation to any type of screen (responsive design).As with the other applications, it features audio and video, hyperlinks, and photo galleries. However, it is limited to a vertical format, even for images.

There are no animations. It has been on the App Store since October 21, 2013. The few updates have included the ability to sign up for the publication on a monthly basis (version 1.1.0 of 05/11/2013), since access was previously limited to subscribers of *Orbyt,* their system for managing access and subscriptions; as well as push notifications, customized ordering of the sections by the user, and the ability to see marks on the calendar of the electronic library in downloaded copies (version 1.1.2 of 10/06/2014).

The app staff consists of a chief editor and seven writers/sub-editors. This same team is also responsible for additional copy material produced for *El Mundo* in its digital version in the group´s Orbyt platform. Concerning the app, the main task for the team is the edition of journalistic material produced by the staff in the newsroom.

O Globo a Mais

The application (https://flic.kr/p/vWJoCJ) from the media company *O Globo* made considerable progress with initiatives in several directions. They invested from the start in a dedicated team composed of 13 people: an editor-in-chief, an editor, an assistant editor, six reporters, three visual designers and a coordinator of multimedia images. The formation of a larger team may have contributed to the approach of continual change, as observed since its launch in 2012.

The application strove for its own business model, breaking away from the parent application (web, *O Globo, O Globo a Mais)* and began to form a product sold separately in the Apple catalog. They have also expanded into other operating systems[7] and devices such as a version for Android OS and customized versions for smartphones of both operating systems. Another relevant aspect of *O Globo a Mais* is having been made available also on the web, although without the features that characterize tactility (Palacios & Da Cunha, 2012). This reinforces the notion of a multimedia continuum and the web as the apex of the convergent processes (Barbosa, 2013.

The *O Globo a Mais* application makes use of hypertext capability starting at the cover, with a table of contents superimposed on photos, leading the reader directly to internal content through the link. There are several pictograms indicating features (tactility and affordances) including the touch screen to interact with content such as images, video or hidden text (which are frequently used at the beginning, as in the Giro section). These pictograms demarcate from the outset the distinctive territory inherent to the tablet and its tactile interface. Accordingly, the interaction is established through touch, based on the design of gestural interfaces for applications. The attributes of narratives in the *O Globo a Mais* application are listed according to the characteristics of interaction based on the touch screen or tactility. Navigation may descend without limit in the vertical orientation. There are signals indicating the presence of further content lower down, such as a scroll bar on a web browser.

La Presse+

The *La Presse+* app (https://flic.kr/p/wemaPB), although not an evening publication, given that it distributes content in the morning (5:30 am in the Eastern Time Zone), was included in our sample as a relevant complement to an analysis of apps that were generated with elements of innovation. Similar to *O Globo a Mais, La Presse+* is another example of an application developed specifically for tablets which offers the greatest amount of affordances. The edition is the most robust among applications, with high density of information, covering a wider range of topics in its sections. Since July 17, 2014, the application has also been available for the Android operating system in tablet format, following the example of *O Globo a Mais,* which was the first among the applications analyzed to also distribute for Android.

As regards features which exploit the nature of mobile devices, it uses tactility to zoom in on videos as a convenient way to view full screen in the application. It allows sections to be customized by significance in order of appearance (*Actualités, Sports, Techno, International, Météo, Blogs, Affaires, Le plus populaire sur la Presse*) and alerts (notifications). The content includes infographics and charts. Image galleries and videos gain particular attention on the app and are used frequently. Regarding the videos, *La Presse+* uses a transparency feature, overlaying a video player on images, so that the user can click to watch in a complementary way. The application's navigation also offers a simplified (horizontal) logic to the user, with sufficient indicators. At the end of the application there is also an "*en direct*" (live feed) section with continuous updates.

The application appears as a confluence between the features of print, web and video, incorporating tactility and interaction with the information presented, which makes it quite appealing.

La Presse+ is produced by a staff of over 300 journalists in the newsroom – jointly producing the print edition, web version and app version – and around 100 professionals in areas of design, informatics and product development.

CONCLUSION

This chapter on autochthonous journalistic applications and, in particular, evening news apps has contextualized the emergence of such products as innovative journalism and as a new business initiative. Even considering their newness (from 2011), it is possible to characterize some aspects of the development and evolution of these applications within strategies of convergence and distribution of multi-platform media companies. In the present case, the prospects for mobility and expansion of the number of tablets and smartphones in the market also bear consideration, in addition to the increasing propensity to pay for information consumption on tablets.

During the interval between 2010 and 2014, journalism geared towards tablets experimented with new formats and interactive features applied to narratives, creating unique products, identified here as autochthonous, to employ the novel affordances that tablets make possible, like tactility, movement via the system's accelerometer, among other factors related to the portability and practicality of the device to consume (multimedia) content in a more interactive and intimate way. The narrative format was also object of some – if limited - experimentation as the new products were launched aiming at a 'lean-back' type of audience more likely to accept 'in-depth', longer narratives, side by side with 'summaries' of the main news of the day.

The results of the study undertaken upon the defined *corpus* – *O Globo a Mais, Estadão Noite, Diário do Nordeste Plus* (Brazil), *La Repubblica Sera* (Italy), *La Presse+* (Canada) and *El Mundo de la Tarde* (Spain) – enabled the innovations introduced in these applications to be mapped and systematically organized in order to make comparisons with respect to the degree of innovation made by each application. With respect to this grading, we found that for some applications the innovations were meager or non-existent after the release of their respective evening news apps (for example *La Repubblica Sera, Estadão Noite* and *El Mundo de la Tarde),* while others were continually adding features integrated with the narratives and business models (adding availability for Android OS or sale from the App store catalog), as in the cases of *O Globo a Mais* and *La Presse+. O Diário do Nordeste Plus* occupies an intermediate point, with modest innovations.

O Globo a Mais[8] comes forth as the application that uses strategies with more focus on innovation and the development of a sustainable business model with regard to smartphones. It is the only one so far in our *corpus* that adapted the application to Android and iOS smartphones, thus meeting demand from a greater pool of users. We can infer from this strategic decision that: 1. Availability for smartphones expands the market for the product in view of the larger quantity of smartphones available for Android OS; 2.The strategy is within the scope of multi-screen, multi-platform and responsive design (or liquid design) as there is a challenge to adapt content to be viewed on mobile devices with an adequate experience of "fit", in different screen sizes. We should also consider that smartphones are undergoing a process of metamorphosis in terms of screen size, with the market opening up for devices with screens larger than 4 inches, or even 5 inches as with the so-called *phablets* (a cross between smartphones and tablets). Therefore, a smartphone with a 5-inch screen is an intermediary which can handle the content of a digital magazine almost as well as a tablet, both in terms of facilitating reading as per its readability and in adapting features that are more apt to the tablet.

Based on these observations and analyses, the result is that one set of autochthonous apps are in transition and may be thought of as occupying a basic stage of development[9]. These products were launched for iPad and seek to add some value to the brand of the news organization and its multi-platform strategy. However most of the effort was focused on launching the product as a way of marking territory from competitors in the competition for "innovation," thus generating marketing and attempting to gain the loyalty of readers/customers.

After the release, it is evident that few made improvements, keeping the app in its "beta" version for a long time, with occasional updates pertaining to system errors or to "add items" related to purchasing subscriptions or one-off sales. Meanwhile, these applications have remained languid without progressing to the addition of new features, aggregators of narratives, or the exploitation of features distinctive to tablets. Both *El Mundo de la Tarde* and *Estadão Noite* are examples of this relative stagnation, with low levels of innovation, as indicated above. It is worth remembering that in the case of *El Mundo de La Tarde,* the application was launched in October 2013, when many of the features and affordances of tablets were already employed in several products available at the time.

The research undertaken with the given indicators of innovation in journalism for tablets demonstrates the need to face challenges to develop a suitable language that increasingly incorporates into the narratives those affordances which tablet interfaces make available in terms of innovation and tactile interaction. The present discussion resembles, excepting particular circumstances, the first moments of digital journalism when it sought to implement its own characteristics applied to the web platform, but still making considerable use of 'transposed' models from former platforms. Amid the process that we witnessed, it is not yet clear what will be the future of autochthonous news applications in journalistic organizations, keeping in mind differences in the degree of innovation and even the quantity of apps – especially for evening news – in the market, equally for Brazil, Europe and in North America.

We believe that affordances as inducers of innovation in journalistic products may be further explored in future research. In this chapter (Table 2) they were only mobilized at a basic descriptive level. Feldman (2005, p. 42) differentiates innovations in general according to two dimensions: means-induced (new means for existing purposes) and purpose-induced (new purposes that need to be served). We suggest that for the specific case of innovations in mobile journalism a third dimension could be added: affordances-induced (new interface properties that need to be put to use). The refinement of this approach and the strengthening of the bridge between affordances and innovation may be a fruitful development in future texts to be produced by our research group.

Finally, from what we observed, there is still much to be done concerning experimentation in formats, content, language and narratives based on the features of mobile devices, the constant evolution of operating systems and changes in hardware platforms. Of course innovation is also dependent and limited by financial capacity to invest in new - and sometimes costly - resources, both human and material. Economic crises have direct effects on R&D expenditure, risk taking investments and experimentation and the "2008 economic crisis has severely reduced the short-term willingness of firms to invest in innovation", although the negative effects were not uniform across the economic spectrum (Archibugi, Filipetti, & Frenz, 2013). The expected European economic recovery, in the near future, may be a decisive factor in defining the shape of things to come in the field of mobility and new journalistic formats.

REFERENCES

Adams, R., Bessant, J., & Phelps, R. (2006). Innovation management measurement: A review. *International Journal of Management Reviews*, *8*(1), 21–47. doi:10.1111/j.1468-2370.2006.00119.x

Åkesson, M. (2009). Digital innovation in the value networks of newspapers. *Gothenburg Studies in Informatics*, 42. Retrieved from http://hdl.handle.net/2077/20887

Archibugi, D., Filipetti, A., & Frenz, M. (2013). Economic crisis and innovation: Is destruction prevailing over accumulation? *Research Policy*, *42*(2), 303–314. doi:10.1016/j.respol.2012.07.002

Banks, A. (2014). *2014 Brazil Digital Future in Focus*. São Paulo: comScore. Retrieved from http://www.comscore.com

Barbosa, S. (2013). Jornalismo convergente e continuum multimídia na quinta geração do jornalismo em redes digitais. In J. *Canavilhas (Ed.). Notícias e mobilidade: o jornalismo na era dos dispositivos digitais* (pp. 33–54). Covilhã: Livros Labcom.

Barbosa, S., Firmino da Silva, F., & Nogueira, L. (2014). Análise da convergência de conteúdos em produtos jornalísticos com presença multiplataforma. In E.C. Pellanda, & S. Barbosa (Eds.), Jornalismo e mídias móveis no contexto da convergência (pp.95-126). Porto Alegre: EdiPUCRS.

Barbosa, S., Firmino da Silva, F., Nogueira, L., & Almeida, Y. (2013). A atuação jornalística em plataformas móveis: Estudo sobre produtos autóctones e a mudança no estatuto do jornalista. *Brazilian Journalism Research*, *9*(2), 10–29.

Barnhurst, K. G., & Nerone, J. (2001). *The form of news: a history*. New York: Guilford Press.

Barsotti, A. (2014). Two screen, two paths: News production for smartphones and tablets on the brazilian newspaper scene. *#. ISOJ Journal*, *4*(1), 112–132.

Batiston, B., & Bello Neto, G. B. (2013). Jornalismo para tablets: a dialética entre pesquisa e prática, experiências desenvolvidas na universidade. In R. Paulino & V. Rodrigues (Eds.), *Jornalismo para tablets: pesquisa e prática*.

Boczkowski, P., & Ferris, J. (2005). Multiple Media, Convergent Processes, and Divergent Products: Organizational Innovation in Digital Media Production at a European Firm. *The Annals of the American Academy of Political and Social Science*, *597*(1), 32–47. doi:10.1177/0002716204270067

Cameron, A. (1998). The medium is messy. *Eye (London, England)*, *30*, 6–7.

Canavilhas, J., & Santana, D. (2011). Jornalismo para plataformas móveis de 2008 a 2011: Da autonomia à emancipação. *Revista Líbero*, *14*(28), 53–66.

Castells, M., Fernández-Ardèvol, M., Qiu, J. L., & Sey, A. (2006). *Mobile communication and society*. Cambridge: MIT Press.

Coates, K., & Ellison, A. (2014). *Introducción al diseño de la información*. Badalona: Parramón.

Cohen, W. (2010). Fifty Years of Empirical Studies of Innovative Activity and Performance. Handbook of the Economics of Innovation (Vol. 1, pp. 129–213).

Crossan, M., & Apaydin, M. (2010). A Multi-Dimensional Framework of Organizational Innovation: A Systematic Review of the Literature. *Journal of Management Studies*, *47*(6), 1154–1190. doi:10.1111/j.1467-6486.2009.00880.x

Díez, L. G., & Cuadrado, P. P. (2001). *Principios básicos sobre diseño periodístico*. Madrid: Universitas.

Fagerberg, J. (2005). Innovation: A guide to the Literature. In J. Fagerberg & D. C. Mowery (Eds.), *The Oxford Handbook of Innovation*. Oxford: Oxford University Press.

Feijóo, C., Maghiros, I., Abadie, F., & Gomez-Barroso, J. (2009). Exploring a heterogeneous and fragmented digital ecosystem: Mobile content. *Telematics and Informatics*, *26*(3), 282–292. doi:10.1016/j.tele.2008.11.009

Feldman, V. (2005). *Leveraging Mobile Media*: Cross-Media Strategy and Innovation Policy for Mobile Media Communication. Heildeberg: Physica-Verlag.

Franciscato, C. E. (2010). Uma proposta de incorporação dos estudos sobre inovação nas pesquisas em jornalismo. *Estudos em Jornalismo e Mídia*, *7*(1), 8–18.

García, M. (2012). iPad design lab: storytelling in the age of the tablet. (Kindle Edition).

Garcia, M., & Calantone, R. (2002). A critical look at technological innovation typology and innovativeness terminology: A literature review. *Journal of Product Innovation Management*, *19*(2), 110–132. doi:10.1016/S0737-6782(01)00132-1

Gibson, J. J. (1979a). The Theory of Affordances. In J. Giesekink, & W. Mangold (Eds) (2014), The People, Place and Space Reader. New York: Routledge.

Gibson, J. J. (1979b). *The ecological approach to visual perception*. Boston: Houghton Mifflin.

Grubenmann, S. (2013). Innovation in Journalism – Towards a Holistic Approach. Retrieved from http://www.researchgate.net/publication/235799323_Innovation_in_Journalism__Towards_a_Holistic_Approach

Julier, G. (2010). *La cultura del diseño*. Barcelona: Gustavo Gili.

Larequi, J. C. (1994). *El diseño periodístico en prensa diária*. Barcelona: Bosch.

Louçã, F. (2014). The elusive concept of innovation for Schumpeter, Marschak and the early econometricins. *Research Policy*, *43*(8), 1442–1449. doi:10.1016/j.respol.2014.02.002

Machado, E. (2000). *La estructura de la noticia en las redes digitales: un estudio de las consecuencias de las metamorfosis tecnológicas en el periodismo* [Doctoral thesis]. Universidad Autónoma de Barcelona, Spain.

Machado, E. (2010). Creatividad e innovación en el periodismo digital. *Proceedings of II Congreso Internacional de Ciberperiodismo y Web 2.0*, Bilbao, Universidad del País Vasco (pp. 64-72).

Macri, P. (2011). Per una teoria copernicana delle informazioni. *European Journalism Observatory*. Retrieved from http://it.ejo.ch

Mitchelstein, E., & Boczkowski, P. (2009). Between tradition and change: A review of recent research on online news production. *Journalism, 10*(5), 562–586. doi:10.1177/1464884909106533

Murray, J. H. (2012). Inventing the medium: principles of interaction design as a cultural practice (Kindle Edition). Cambridge: MIT Press.

Newman, N., & Levy, D. A. L. (2014). Reuters Institute Digital News Report 2014: tracking the future of news. Retrieved from http://www.digitalnewsreport.org

Nordfors, D. (2003). *The concept of innovation journalism and a programme for developing it.* Stockholm: Vinnova. Retrieved from http://www.innovationjournalism.org

Norman, D. A. (1988). *The design of everyday things.* New York: Doubleday.

Palacios, M., Barbosa, S., Firmino da Silva, F., & Da Cunha, R. (2014). Aplicativos jornalísticos vespertinos para tablets. Cartografia do fenômeno ante o desafio de uma produção original e inovadora. *Sur le Journalisme, 3*(2), 40–55.

Palacios, M. S., & Da Cunha, R. (2012). A tactilidade em dispositivos móveis: Primeiras reflexões e ensaio de tipologias. *Contemporanea (Bologna, Italy), 10*(3), 668–685.

Pellanda, E. (2009). Comunicação móvel no contexto brasileiro. In A. Lemos (Ed.), *Josgrilberg. F. Comunicação e mobilidade aspectos socioculturais das tecnologias móveis de comunicação no Brasil.* Salvador: EDUFBA.

Peters, C. (2012). Journalism to go: The changing spaces of news consumption. *Journalism Studies, 13*(5-6), 5–6. doi:10.1080/1461670X.2012.662405

Pew Research Center. (2014). Key Indicators in Media and News. *State of the News Media 2014.* Retrieved from http://www.journalism.org/files/2014/03/Key-Indicators-in-Media-and-News-2014.pdf

Quinn, S. (2005). *Convergent journalism: the fundamentals of multimedia reporting.* New York: Peter Lang Publishing.

Rossetti, R. (2013). Categorias de inovação para os estudos em comunicação. *Comunicação & Inovação, 14*(27), 63–72.

Salaverría, R. (2012, May 8). Se ve... y se toca: hacia un nuevo lenguaje periodística multimedia. Retrieved February 15, 2014, from http://www.salaverria.es

Salaverría, R., García Avilés, J. A., & Masip, P. M. (2010). Concepto de convergencia periodística. In: López García, X. & Pereira Fariña, X. (Orgs.), Convergencia digital: reconfiguración de los medios de comunicación en España. Santiago de Compostela: USC, 41-64.

Salaverría, R., & Negredo, S. (2008). *Periodismo integrado: convergencia de medios y reorganización de redacciones.* Barcelona: Editorial Sol90.

Santaella, L. (2011). *Percepção: fenomenologia, ecologia, semiótica.* São Paulo: Cengage.

Satuf Rezende, I. (2013). Jornalismo para tablets e a emergência de contratos comunicacionais, paper present at *XIII Congreso Internacional Ibercom.* Retrieved from https://www.academia.edu/7848328/Jornalismo_para_tablet_e_a_emerg%C3%AAncia_de_contratos_comunicacionais

Scolari, C., Aguado, J., & Feijóo, C. (2012). Mobile Media: towards a definition and taxonomy of contents and applications. *iJIM*, 6 (2), 29-38.

Silva, F. F. (2008). Jornalismo reconfigurado: tecnologias móveis e conexões sem fio na reportagem de campo, paper presented at *Intercom – Sociedade Brasileira de Estudos Interdisciplinares da Comunicação, XXXI Congresso Brasileiro de Ciências da Comunicação*. Retrieved from http://rtietz.com/posimagemfeevale/page0/files/Fernando_firmino_intercom.pdf

Innovation report. (2014, March 24). *The New York Times*. Retrieved from http://visualdays.no/files/2014/05/224608514-The-Full-New-York-Times-Innovation-Report.pdf

Van de Ven, A. H. (1986). Central Problems in the Management of Innovation. *Management Science*, *32*(5), 590–607. doi:10.1287/mnsc.32.5.590

Westlund, O. (2013). Mobile News: A review and model of journalism in an age of mobile media. *Digital Journalism*, *1*(1), 6–26. doi:10.1080/21670811.2012.740273

KEY TERMS AND DEFINITIONS

Affordance: The concept was defined by psychologist James Gibson to represent the properties and potential of objects. With respect to graphical interfaces, it refers to design features that induce intuitive navigation and usage of elements of the device content.

Autochthonous News App: Application created by a specialized team, producing exclusive material for mobile platforms, with distinct handling of content. Autochthonous news apps incorporate the strategies for diversification of media companies business models, which, based on both internal and external research and as suggested by trends, began to value digital products and the 'digital first' operational strategy.

Convergent Journalism: A non-consensual concept that for the purpose of this article is defined as "a multidimensional process that, fueled by the widespread deployment of digital telecommunication technologies, affects the technological, business, professional and editorial spheres of the media, providing an integration of previously separated tools, spaces, working methods and languages, allowing (...) content to be distributed across multiple platforms using their specific languages/codes" (Salaverría R.; García Avilés. J.A.; Masip P.M., 2010).

Evening News App: Application for a mobile device characterized by autochthonous products (originals) and the concept of 'lean back'. It is developed exclusively for tablets, with its own business model and differentiated narratives.

Innovation in Journalism: A concept based on Machado (2010, p. 67), "Innovation in journalism is a phenomenon that faces journalism as an industry and focuses on finding social demands for quality, instantaneous information at the lowest possible cost, in keeping with the rigor of the highest professional conduct and accessible by all available means."

Mobile Journalism: The production, editing, distribution and consumption of journalistic content to and from connected mobile devices such as tablets, and smartphones.

Mobility: State of displacement, of motion. In mobile communication, mobility is expressed as physical and informational displacement through the portability of devices and the ubiquity of wireless connections that enable information mobility.

Multimedia Continuum: Processes and products with streams for production, editing and distribution of content in horizontal and dynamic forms and with multi-platform characteristics.

Tactility: A significant feature of a mobile device interface employing touch and the sensorial to facilitate tactile interaction with the content.

ENDNOTES

1 The first phase of the Convergent Journalism Laboratory Project (http://www.labjorconvergente.info/) was financed by the Research Foundation of Bahia (FAPESB, 2011-2014). Now in its second stage, is financed by the National Council for Scientific and Technological Development (CNPq, 2014-2016) and hosted at the Faculty of Communication, Federal University of Bahia, Brazil <http://www.ufba.br>.

2 Second place went to Denmark, with 89%, and first was Japan with 92%.

3 In 2014, the survey was conducted in ten countries: Germany, urban Brazil, Denmark, Spain, USA, Finland, France, Italy, Japan and the UK, using populations *samples.* Finland was included for the first time in 2013,

4 Data from the IVC in June 2014 showed the *Folha de S. Paulo* (São Paulo) as leading in circulation and audience on different platforms, putting *O Globo* (Rio de Janeiro) in second and, in third place, *O Estadão* (São Paulo). In fourth, is *Zero Hora* (Porto Alegre).

5 The popularization of audiovisual communication, computerization of newsrooms and the development of the GUI (graphical user interface) in the 1980s, promoted a new impulse in the visual formatting of journalistic products. The competition with television encouraged color printing of newspapers, as well as better distribution of texts in notebooks and sections, with the support of features from design to attract readers and help them visualize the news. At least two subareas received evidence of design in this period, whether in the professional or academic field: information design, shown mainly in the use of infographics and news design, which is reflected in the visual organization of pages as a whole.

6 To Garcia (2012) there are different classifications to delimit tablet apps as native applications (which we call autochthonous), HTML5 application, in addition to sites for tablets ("native app", "HTML5 app", "web app", " Tablet-optimized website").

7 The adoption of the Adobe Publishing Suite Design – DPS, the main tool for production applications for tablets and smartphones, contributes to the production platform adapted to the specificity of each screen. The DPS remains the primary tool for implementing innovations in production applications.

8 *O Globo a Mais* app has been discontinued. The final version was published on May 15, 2015. In a "Letter to readers", editor Pedro Doria announced that "very shortly a new product for tablets and smartphones will go live".

9 Applications like *Folha10*, the *Folha de S. Paulo*, Brazil, also belong in this stage. This application, although not part of the empirical corpus for analysis, took part in the research sample of the Project Convergent Journalism Lab in his first stage.

Chapter 14
The Role of the Profile and the Digital Identity on the Mobile Content

Ana Serrano Tellería
Beira Interior University, Portugal

ABSTRACT

The (re)construction of the user profile and the digital identity resulting from both conscious and unconscious activity on the Internet is directly linked to the process of creation and diffusion of content. Many times, neither the users-prosumers are aware of this process nor are the authors cognizant of the original content, as new authors – humans as well as machines - cover the relations established between the users, their content, their activity online and the combination of these. Big data and information economy are not just consequences of the possibility of collecting as much information as possible about users. Instead, they provide a quasi unlimited means of mapping and shaping every movement of our behavior and lives through devices and technologies. Therefore, the aim of this chapter is to understand how users handle this liquid ecosystem and its relationship with the evolution of mobile content, considering the different rhythms and dimensions.

INTRODUCTION

The brain directs us but it also fools us when perceiving the world, explained neuroscientist Susana Martínez Conde (2013). Between sensory and attentional limitations, what we perceive represents 0.01% of the information that surrounds us. Thus, the brain makes up a simulation of reality. This is similar to what happens in the movie *Matrix*, where people live in a virtual reality and have no means of distinguishing it from the real world. But this is of no consequence, provided we are still capable of finding food and interacting with the environment. As a matter of fact, it goes back a long way, not only to Pre-Darwinian days but as far back as to Plato's cave. These discrepancies are used in everyday life as more or less evil motives in advertising, marketing, politics, magic, etc. (Martínez Conde, 2013).

DOI: 10.4018/978-1-4666-8838-4.ch014

Similarly to the *Matrix* comparison, Mark Deuze (2012) mentioned *The Truman Show* when explaining how we are increasingly living 'in' the media, like a fish in water, rather than 'with' the media, as we did before. We consider that similarity is a suitable parameter to describe the process of technological appropriation that McLuhan labels as 'ecological'. In this respect, we refer to Vilém Flusser's *Homo Ludens* and his thoughts on how "for the first time, technologies are simulating the nervous system" (1988). Bauman's liquidity (2005) flows into technology as well, which is specifically studied in connection with mobile devices and media (Aguado, Feijóo, and Martínez, 2013; Serrano Tellería and Oliveira, 2015). Thus, a challenging ecosystem surrounds users, who already exhibit some alerting behaviors like the 'fear of missing out' (Turkle, 2011; Rosen, 2013), continuous partial attention (Stone, 2002-2014), lack of rationality in some attitudes and performances, limitations in the extension between knowledge and action, and strong circumstantial pattern behavior. The ambiguity and volatility of most 'terms and conditions' of apps, developers, platforms and software environments should be further considered (Serrano Tellería, Oliveira, 2015).

Therefore, the aim of this chapter is to delve into how users deal and understand their online profiles and digital identities management through mobile devices. In relation to content evolution, the objective is to describe both their attitudes and consciousness about the level of interaction between big data, the media ecology, the social media and their role as users-prosumers. In the frame of the European FEDER project 'Public and Private in Mobile Communications' (LabCom, Beira Interior University), different methodologies have been implemented: online surveys, content analysis, focus groups and interviews.

Background

Previous approaches can be divided into four main groups. Firstly, national and international surveys focused on users' actions and preferences concerning mobile adoption and content access. Secondly, content analysis about users' performances on apps and platforms, establishing relationships, in many cases, with the users' awareness about the logic of the devices and the media involved. The third approach relies on focus groups and interviews to better understand the perception of users' attitudes and motivations. The fourth and last is dedicated to analyzing the 'terms and conditions' of apps, developers, platforms and software environments both from a legal and technological standpoint.

MANAGING MEDIA, IDENTITY, NETWORKS, AND TECHNOLOGY

User Paradoxes from Survey Data Analysis

The first surveys about general uses and habits showed the rapid implementation of mobile devices as well as the adoption of a multitasking performance of media consumption that places smartphones in the center of users' lives, reflecting the previous boom of the mobile phone and the incorporation of the tablet into our daily routines (Meeker, 2014; Nielsen, Pew Research, WAN-IFRA, GEN, etc.). Economic resources and technical availability may differ from country to country, depending on their level of development and market implementation. For that reason, fewer studies have been found concerning the concrete issues of this study. Our project is dedicated primarily to the Portuguese population, drawing on the experience of entities that have conducted this type of survey in Europe and America.

1. Europe

The EU27 survey of electronic identity, privacy and data protection (Lusoli, Bacigalupo, Lupiañez, *et al*; 2012) focused on four key Digital Agenda areas: Authentication and Identification, e-Commerce, Social Networking Sites, and Medical Information as Personal Data. The report shows that more than one third of Europeans (34%) uses Social Networking Sites (SNS) to keep in touch with friends and business partners and almost four out of ten to shop online. In both of these contexts, people disclose vast amounts of personal information and also manage a large and growing number of electronic identities. Among Member States, there are equally significant differences and considerable digital exclusion, mainly due to socio-demographic differences in affluence, education and age.

An attitude to highlight is the overall 'knowing, understanding, or even consent' that if they want to benefit from using the Internet to its full potential, they have to disclose their data (biographical, social, financial or medical) and manage online identities. Therefore, almost three in four accept that revealing personal data, so as to benefit from online services, is part of 'everyday life' - access a service (61%), connecting with others (54%) -. Users are more likely to be younger and well educated, are heavier Internet users and are still studying or are unemployed. More than half also utilized websites to share pictures, videos, movies, etc. (68%); instant messaging, chat websites (57%) and have purchased goods or services online (57%). The more the Internet is widespread, the more SNS are used. Age plays a key role at national level, which means that younger people in most countries use the Internet very little outside SNS, while older people who use SNS are practically the same as the percentage of Internet users.

With respect to risk behaviors, SNS users are less cautious about sharing their social information (friends, activities, etc.) since they think that disclosure is unavoidable in today's life, even though they consider it personal. In addition, they are less concerned with being 'observed' online – more comfortable with online profiling – but more cautious in sharing their sensitive (medical, financial, etc.) information. The report highlights the complicated issue of informed consent in SNS. Four groups are identified: 19% claim to have not been informed of either conditions or consequences; 29% say they have been informed about the conditions of data collection but feel unhappy with the degree of information about possible consequences; 40% report having been informed about both collection conditions and consequences; and 11% are happy about having been informed of the consequences but have hardly been given information on how the data collected will be used. In policy terms, the report states that significant work is required to enforce informed consent and enhanced information about what may happen with people's personal data once it is disclosed in SNS.

In terms of e-Commerce, the most prominent aspects detected are the high percentage of personal information disclosed - except for social activities which are not normally asked - and the lack of a relation between perceived self-responsibility and Internet protection. Overall, there is no apparent relation between considering one's data personal and disclosing it. So even if people consider information personal, they will still disclose it. Even more surprising is the fact that, for many items (name, address, nationality, financial information), the more people consider this information personal, the more they disclose it. Thus, there is no relation between perceptions of self-responsibility and most other regulatory perceptions: desire for the possibility to delete one's data, to move one's data and awareness of identity theft and data loss. As found in previous surveys, even people who feel responsible do little to protect their personal data once it has been disclosed. This may be due to the lack of tools that enable people to take protective measures. However, when such tools are available - privacy notices -, people read them if they feel responsible. Therefore, better tools may be required if people wish to take care of themselves.

Users shopping online also provide address and financial information, while users of SNS share more social information but not financial. Both in e-Commerce and SNS, the responsibility for personal data safety is mainly attributed to the user, followed by the site owners and public authorities.

The report stresses the need of raising awareness regarding the identity-related personal data that users routinely provide to online service providers via their accounts, without being aware. Personal experience of identity theft and data loss is very low, affecting only 2% of EU27 population - identity theft alone affected 3.5% of US residents in 2010. Largely, identity theft and data loss affect managers and other office workers as well as their families. Sensitivity to these phenomena is relatively high, as more than half of Internet users are aware of the issue via different or multiple sources, which increases to two in three in most northern countries, where Internet access is higher. Thus, general Internet skills alone do not provide an answer to these problems, and other more specific skills may be required. Also, concerning remedies, media awareness appears intertwined with calls for enhanced regulation, including greater harmonization of data protection rights, request for information if/when data lost or stolen and the possibility to delete personal data. The media may thus be playing a role in generating support for a more vigorous and more articulated response to the challenge.

The majority of Internet users take one or more actions to protect their personal identity data (average is 2.3) while a minority does not minimize disclosure: they do not withhold bank details, provide information to controllers they do not trust and disclose usernames and passwords. All in all, this is in line with the widespread perception that disclosure is unavoidable. However, lack of protection is not caused by resignation. Rather, it is strongly linked to propensity to disclose personal data, which one in three users happily does. Specifically, those who are happy disclosing personal data, those who trust companies as well as those comfortable with online profiling are much less likely to minimize data, as may be obvious, but are also less likely to withhold sensitive information and to use software measures to protect.

Protection behavior rests on passive use of existing tools (e.g. to limit unwanted emails – 40%) rather than on active strategies of information control (e.g. changing the security settings of your browser – 22%). There is a strong correlation between the overall number of internet activities carried out (a proxy for internet skills), and online protection behavior. Nevertheless, people in some countries tend to stand more protected regardless of the number of activities. These deviations from the trend hint at the importance of variables others than internet use; these may have to do with national technical culture and with maturity of the market. This all implies that where simple tools are not available, or are cumbersome to use for the average user, people are unlikely to take proper care of their personal identity data online.

Specifically, data minimization is strongly correlated with regulatory preferences and data protection principles. In short, people who minimize the information they disclose also tend to have particularly strong feelings regarding the needs for stronger protection of their rights and on enhanced control of their personal data, such as deletion on demand and data breach notification. Also, existing rules and principles engender greater self-protection. Namely, those who think they had to disclose more than they wished compensated by using reactive, proactive and deception strategies. Information about data collection conditions is associated positively with reactive and proactive behavior and with minimization. The concern about re-use of one's data is associated with significant minimization of the data disclosed.

2. Portugal

The scope of the main surveys varied, focusing more on general uses and habits as well as access to contents and devices. A broad survey conducted by the Portuguese Institute of Statistics (I.N.E Portugal, 2014) reports that the Portuguese use (70%) SNS more than the European average (57%). More than half access the Internet from home (63% via broadband) and work. Twenty-six percent use services for storing and sharing files. Photographs are the most common type of file (85% of users of these services), followed by text, spreadsheets, electronic presentations (79%) and music files (53%). Sixty-five percent of people aged between 16 and 74 access the Internet, and 17% use it to place orders. The age group with the highest proportion of Internet users is between 16 and 24 years, decreasing with increasing age.

In e-Commerce, people with 25-34 years report a higher frequency of use (33%). Using the Internet to place orders is below average in the EU-28, by more than 20 percentage points. Among Internet users, more than half (57%) access the Internet on the move, i.e., outside their homes and workplaces, in portable devices (mobile phone, smartphone, laptop or other portable device). This type of access has increased by 19 percentage points last year, reaching in 2014 similar values to the EU-28 average ratio in 2013. They also use the Internet to make more calls or video calls: 36% of people say they do so, compared with 33% in the rest of the EU. The mobile phone is the device most used by users to access the Internet on the move (48%), followed by the laptop computer (38%).

3. North America

A survey of Pew Research (Madden, 2014) suggests that Americans consider certain kinds of data to be more sensitive than others. Media tastes and purchasing habits are among the least sensitive categories of information. Their perceptions of Privacy are varied. Thus, the term "Privacy" evokes a constellation of concepts. A large segment of respondents associated it with security, safety and protection.

They feel more secure using landline phones and less surfing SNS. There is a broad concern about government and third-party access to information on SNS. An overwhelming majority senses a loss of control over how their personal information is collected and used by companies. However, many are willing to share some information about themselves with companies so that they may use online services for free. Approximately one in three adults says they value the greater efficiency of online services because of the increased access they have to personal data. In this sense, mobile internet users are more likely to value the efficiency of services that rely on personal data. Most would like to do more to protect their personal information online.

Content creators employ multiple strategies for managing their identities when posting online, whereas few adults think it is "easy" for them to be anonymous. Nearly nine in ten adults agree that if inaccurate information was posted, it would be very difficult to have it removed. Few call for the withdrawal of information, which includes mostly photos and videos. Many still do not assume that people they meet will search for information about them.

Seventy-four percent of adults are active on SNS (Pew Research, 2014). Fully 40% of cell phone owners use a SNS on their phone, and 28% do so on a typical day. Young people, blacks, Hispanics, the highly educated and those with a higher annual household income are more likely to use SNS on their phones than other groups. SNS are increasingly used to keep up with close social ties and the average user has more close ties and is half as likely to be socially isolated as the average North American. Forty-six percent of adult users post original photos or videos online that they themselves have created – creators – while 41% take photos or videos that they have found online and repost them on sites designed for sharing – curators. Overall, 56% do at least one of the creating or curating activities and 32% do both.

The average Facebook user gets more from their friends than they give to their friends. This is because of a segment of "power users" – interesting concept -, who specialize in different activities and contribute much more than the typical user does. The Pew Research team conducted a new study that for the first time combines server logs of Facebook activity with survey data to explore the structure of friendship networks and measures of social well-being. The findings (over a one-month period) show that: 40% of users made a friend request, but 63% received at least one request; users pressed the like button next to friends' content an average of 14 times, but had their content "liked" an average of 20 times; users sent 9 personal messages, but received 12; and 12% of users tagged a friend in a photo, but 35% were tagged.

4. Worldwide

The EMC Privacy Index (2014) research on consumer attitudes about online behavior and information privacy in 15 countries establishes a ranking of nations based on consumer willingness to trade digital privacy for greater convenience. The regions surveyed were 'Asia Pacific and Japan': Australia, New Zealand, China, Japan, India; 'Europe, Middle East and Africa': France, Germany, Italy, Middle East (UAE, Saudi Arabia and Qatar), Netherlands, Russia and UK; 'Latin America': Brazil and Mexico; 'North America': Canada and USA. Prior to questionnaire development, 'six personas' were created to encompass people's online exposures. These were: '1 Consumer': Defined by interaction with online stores; '2 Employee': Defined by interaction with employment-related systems & websites; '3 Social': Defined by interaction with social media sites, email programs, text/SMS & other communications services; '4 Medical': Defined by interaction with doctors, medical institutions, & health insurers; '5 Financial': Defined by interaction with banks & other financial institutions; and '6 Citizen': Defined by interaction with government institutions.

Four metrics were then asked of each persona: '1 willingness to' trade privacy for greater convenience; '2 confidence in' institutions' 'ethics' & transparency when protecting my privacy; '3 confidence in' institutions' 'skills' & abilities to protect my privacy; and '4 confidence in my level of future privacy'. Key findings point out a 'world of paradoxes' in users' behaviors: 1 "we want it all", consumers say they want all the conveniences and benefits of digital technology, yet claim they are unwilling to sacrifice any of their personal privacy to get them; 2 "take no action", although privacy risks directly impact many consumers, most take virtually no action to protect their privacy – instead placing the onus on government and businesses; 3 "social sharing", users of social media sites claim they value privacy, yet freely share large quantities of personal information – despite lacking confidence in institutions to protect that information. As a final remark, it is pointed out that global confidence in privacy is extremely low and confidence in future levels of privacy is even lower.

Yet many consumers do not take the necessary action required to protect the Privacy of their Data and are mostly passive in their behavior. There is a general consensus (87%, from 71 to 94%) regarding the need for Legal Protection for Data & Privacy but there is concern about those who are perceived to be responsible for protecting it. In this respect, the top risks identified for the future of privacy security were: 64% financial fraud and identity theft; 51% "businesses using, trading or selling my personal data for financial gain without my knowledge or benefit"; 35% lone/crazy hackers, hacker groups or anarchist types; 35% incompetence on the part of those who should protect privacy and secure personal data; 31%

lack of government attention (laws, regulation, funding, enforcement) when it comes to the privacy of personal data; 26% "strangers learning things about me, including those I'd rather be kept private"; 21% "my government spying on me and learning things about me, including those I'd rather be kept private"; 15% "a person or people I know learning things about me, including those I'd rather be kept private"; and 11% "lack of personal oversight and attention from regular people like me".

According to another survey by CIGI-Ipsos (2014) in 2014 across 25 countries, two thirds (64%) are concerned about their online privacy compared to 2013; 78% are concerned about a criminal hacking into their personal bank accounts; 77% about someone hacking into their online accounts and stealing their personal information like photos and private messages; and 74% about a private company monitoring their online activities (such as my Internet surfing habits) and then selling that information for commercial purposes without their explicit consent. Sixty-two percent are concerned about Government agencies from other countries secretly monitoring their online activities and 61% about the police or other government agencies from their own country secretly monitoring their online activities. One third (36%) believe that private information on the Internet is very secure and 64% do not believe it to be so. Thirty-seven percent share personal information with private companies online all the time and think "It's no big deal" to do so, while 63% disagree with that opinion.

Many have changed things about how they behave online compared to one year ago: Avoiding certain Internet sites and web applications 43%, changing your password regularly 39%, self-censoring what you say online 28%, changing who you communicate with 18%, closing Facebook and other social media accounts, etc. 11%, and using the Internet less often 10%. Seventy-three percent want their online data and personal information to be physically stored on a secure server, while 72% do it in their own country. Only half (48%) believe their Government does a very good job of making sure the Internet in their country is safe and secure. Only 41% believe the chance they would ever have their personal information compromised on the Internet is so small that it is really not worth worrying about. Concerning the importance of the Internet for their future: 'Accessing important information and scientific knowledge (very/somewhat)' 91%; 'personal enjoyment of recreation' 87%, 'social communication' 85%, 'free-speech of political expression' 83%, and 'my own economic future and livelihood' 81%.

As far as apps are concerned, more than half of mobile media users are unhappy with the fact that they shared personal information with them, according to an international survey conducted by the MEF in partnership with AVG (2014). The survey reveals that 65% of smartphone owners are not comfortable with app privacy settings. The research was carried out in 13 countries and 10,000 smartphone users were interviewed. The dissatisfaction regarding privacy in mobile apps feature the highest rates in developed countries like the UK (79%) and the USA (76%). Despite high levels of unhappiness, only 24% of respondents protect themselves somehow – the majority ignores the lengthy terms of service and privacy available while they download apps. Surprisingly, transparency recorded a decline: 42% of respondents consider it essential to know whether an app collects and shares their data (49% in last year's survey).

Mobile has become the leading digital platform (Mashable, 2014), with users spending 60% of their digital media time on mobile. Of this time, 52% is spent on app usage with more than 1.3 million Android apps in the market and 1.2 million apps in the Apple app store. The average smartphone user spends more than 80% of their phone time on "non-voice activities" and 50 billion MIMs (Mobile Instant Messages) are sent every day. The most used apps are Google maps (54%), Pandora (44%), YouTube (35%) and Facebook in the USA with 86.1 million monthly visits. Thirty-eight percent of Internet users go online using mostly their phones instead of their desktop computers or laptops; 15% of all online video viewing across the globe occurs via smartphones and tablets and 40% of YouTube's global watch time is mobile.

The SNS that are more commonly accessed via mobile are Facebook (68%), Twitter (90%), Instagram (98%) and Vine (98%). Twenty percent of SNS users check their phone every 10 minutes; 46% of seniors check it for messages even though they did not hear a notification; and 22% of seniors sleep next to their phones. Ninety percent of 18-24-year-olds sleep with their smartphones and 95% of people use their phones for something just before going to bed. Eighty-four percent cannot go a single day without their cell phones and 29% of cell phone owners and 36% of smartphone owners "couldn't imagine living without" their mobile devices ('Nomophobia'). Sixty-seven percent of cell owners find themselves checking their phones for messages, alerts or calls even when they do not notice their phone ringing or vibrating ('Phantom vibrations'). Fifty-eight percent cannot go one hour without checking their phones.

Reflections on the Online Mobile Ecosystem

The main reflections about the scope of our study highlight the era of 'big data', also described as 'dataism' (Van Dijck, 2014), where media and technologies shape the users' behaviors and even identities (Wellman, 2001; Fortunati, 2002; Stald, 2008; Turkle, 2011; Marichal, 2012; Marwick, 2013; Van Dijck, 2013; Boyd, 2014; Brake, 2014; Fuchs, 2014; Grosser, 2014; Hermida, 2014; Horning, 2014; Walker Rettberg, 2014). To date, outstanding research projects developed in the field of communication have focused essentially on 'mobile identity' and the social media. Following the concept of 'technology of intimacy' introduced in a previous state-of-the-art review (Boyce and Hancock, 2012 in Fidalgo; Serrano Tellería; Carvalheiro *et al.* 2013), four dimensions have been established to explain the relationship between humans and technology: internal perspective, external extension, interpersonal interactions and societal reflection. However, as noted in the above-mentioned review (Serrano Tellería, Oliveira, 2015), traces of liquidity have been found to blur the boundaries between these dimensions, both from a technological perspective and from the standpoint of users' habits and performances.

Therefore, we will group the four dimensions in two with the aim, on the one hand, of respecting the natural intersections between them, and, on the other hand, of maintaining a categorization scheme that is suited not only to our particular issues and controversies, but also to the methodological procedure, reflecting clearly the liquidity observed between them.

1. The Technology of Intimacy and the Liquid Media

The line between man and machine is increasingly blurred. The concepts of human *qua* human and machine *qua* machine are no longer located in the extreme: on the one hand, human spectrum; on the other, automation. Instead, man and machine are a converging dyad that has been developed towards a hybrid "commonality". The relationship between the user and the technology can have different degrees. This relationship gives rise to the construction of the 'technology of intimacy', which is the identification of the emotional and physical connection between a human and a technological system (Carnegie, 2002; Bennett, 2011; Boyce and Hancock, 2012; in Fidalgo, Serrano Tellería, Carvalheiro *et al.* 2013).

Specific features have been observed regarding the mobile ecosystem. General and outstanding characteristics of the online content and media are hipertextuality, interactivity, multimediality and usability. The mobile ecosystem adds ubiquity to the fluidity, transience, reticularity and dissolution of boundaries defined by Bauman (2005) in the modern liquid life. It is further characterized by 'spreadability' and the tension between mass and collaborative culture within itself (Jenkins, Ford and Green, 2013). Aguado *et al.* (2013) drew a parallel between these features and the modern liquid life. Bauman's metaphor of liquid life, consistent with the logic of consumer society, provides a useful counterpart to address some of the characteristics of mobility. Beyond the correlation between the impact of the digital technology and the digital features of the liquid society (which refers to the reflections on acceleration, dislocation, consumption and the role of identity), the mobile media particularly fits the fluidization parameters of the technological, institutional and cultural media dimensions previously described by McQuail (2006; Aguado *et al*, 2013).

The dissolution of the link between content and support, which had been the basis for the definition of genres and formats, finds its highest expression in the distribution models based on storage services and cloud sync. The mobile environment is, in essence, a multi-device environment, whose core lies on the access and consumption mode of contents and services. The ubiquity, the diversification and the intertwining of the consumption scenarios, with a marked tendency towards transversal use of the media and access modes (multi-use), as well as their insertion into social dynamics where real identity games become objects of consumption, redefine the users' value perception of the contents and convert them into the new valuable resource of the digital economy: personal information (Aguado *et al.* 2013).

In connection with this, Rheingold (2012) discriminates five dimensions of digital literacy that users should be aware of and develop consciously: attention; ability to detect trash; participation; collaboration and clever use of networks. The reasons may be found, mainly, when the same author highlights the four properties of any social network: persistence; replicability; scalability and ability to be searched, warning of the importance of being aware of what we share and with whom. By this the author means that behind the networks there are invisible and variegated audiences, moving unsuspected between public and private spheres (also Boyd, 2014). From a similar perspective, Brake (2014) emphasizes the relevance of time and memory in social media, where accessibility, durability, comprehensiveness, inequality and evolution are the main features to bear in mind.

2. Identity, Subjectivity, Content, and Performance

A profile is structured information about the users of digital services. Its concrete structure depends on the way the information has been gathered, the technology involved, the types of data or what the law allows. In the mobile arena this profile is usually richer than in the generic case of the Internet and is also linked more clearly to a particular individual, through a mobile identifier device connected to the IP address. Moreover, in case of ambient intelligence (context-awareness), the user profile may even include biophysical parameters, charge level of the battery and a rich set of variables on the environment, which may include, in addition to the location and the orientation, information such as temperature, humidity, noise, etc. - challenges regarding the use and abuse of personal information - (Aguado *et al*, 2013).

The concept of Digital Identity (ID) arises from the combination of three main factors (Cerra, James, 2012): the concrete manifestation of the self-image of the person on the digital service in question (e.g. a professional social network); the elements arising from the protection, limitation, modification and/or concealment of certain information that users define according to their preferences (e.g. to highlight a certain employment); and the implementation of policies regarding provider data (creating a unique user ID and log all activities that occur when accessing the social network in question). Therefore, a certain individual may have multiple digital identities.

Since Wellman (2001) employed the concept of 'portal', academics have been reflecting on different aspects and dimensions of our performance as users-prosumers on the online mobile ecosystem. Fortunati's 'Cellular and Nomadic Intimacy' (2002), Stald's 'Fluidity of Identity' (2008), Turkle's 'Tethered Self' (2008, 2011) and Walker Rettberg's 'Quantified Selves' (2014) are major examples of the tendencies observed and that reflect the character and liquidity of the online mobile ecosystem.

Remembering the process of McLuhan's ecological approach method, Ling (2008) concluded that the mobile phone, multidimensional as it is, generates a mediated interaction that occupies the same place, if not even a top position, in the minds of individuals (Ling, 2008, p.168). The concept of managing privacy in networks was referred to by Campbell (2014) as 'networked privatism'. Marwick and Boyd (2014) have in turn employed the similar term 'networked privacy' to describe the negotiations emerging within specific environments (media-devices) and constellations (Serrano Tellería, 2015), liquid spheres (Serrano Tellería, Oliveira, 2015).

Who we are - our personal identities - and who we think we are - our self-conceptions – flourish if they support each other in a mutually healthy relationship. Things get more complicated because our self-conceptions, in turn, are sufficiently flexible to be shaped by who we are told to be, and how we wish to be perceived, i.e. a third sense in which we speak of 'the self.' It is the social self, described by Marcel Proust, that Floridi (2014) evoked in order to describe the growing influence of Information and Communication Technologies (ICTs) in shaping our personal identities. In his opinion, these are the most powerful technologies of the self to which we have ever been exposed. Never before in the history of humanity, continued Floridi, have so many people monitored, recorded, and reported so many details about themselves to such a large audience. The impact of short and eclectic micro-narratives is also changing our social selves and hence the way we see ourselves, representing an immense, externalized stream of consciousness.

Academics' reflections on these 'mediated memories' - "normative discursive strategies" - (Van Dijck, 2007, p.22) and 'terministic screens' - the terms in our language through which our understanding of the world is filtered - (Markham, 2013) bring attention to how filters allow us to express ourselves in certain ways but not in others. "In today's algorithmic culture the filter has become a pervasive metaphor for the ways in which technology can remove certain content and how it can alter or distort texts, images and data. Filters can be technological, cultural or cognitive, or they can be a combination of these" ('Filtered Reality', 'Filtered World'; Walker Rettberg, 2014, p.22). Marwick's (2013) argument in her ethnography of developers of social media in Silicon Valley explains that, in general, people who are not effective neoliberal subjects are filtered out - fashions, self-improvement, purchasing goods and services to achieve self-realization. They fail to integrate market logics into many aspects of life, including education, parenting, and relationships. The ideal neoliberal subject is an entrepreneur (Marwick, 2013, p.13). Fuchs (2014) also criticizes its neoliberalism but offers a description of its potential to generate a new type of communism.

Technology has been described as 'architecture of intimacy' (Turkle, 2011) and social media as 'architecture of disclosure' (Facebook: Marichal, 2012), where metric quantification prescribes social interaction ("Desire for more" - Facebook: Grosser, 2014). Both appear to be powerful players in the shaping of normative behavior (Linkedin and Facebook: Van Dijck, 2013). Social media profiles are not a reflection of one's identity, or neutral stages of self-performance, nor are they a means of self-expression (Horning, 2014) - they are the very tools for shaping identities (Van Dijck, 2013). "The conflation of self-expression, self-communication and self-promotion into one tool, which is subsequently used for personality assessment and manipulating behavior, should raise the awareness of users in their different roles as citizens, friends, employees, employers and so on" (Van Dijck, 2013, p.213).

'Dataism' is Van Dijck's term (2014) for the common assumption that people and behaviors can be adequately represented by quantitative means and 'big data.' In this regard, Walker Rettberg (2014) alerts to the danger of falling into 'a fantasy of knowing' and 'subjective data visualisation' if one completely trusts 'algorithms to find meaning' in these 'cumulative self-presentations'. Digital self-presentation and self-reflection are cumulative rather than presented as a definitive whole (Walker Rettberg, 2014, p.5).

The freedom to construct our personal identities online is no longer the freedom of anonymity (Floridi, 2014; Fidalgo, Serrano, Carvalheiro *et al.* 2013; Serrano Tellería, Oliveira, 2015). Rather, it is the freedom associated with self-determination and autonomy as far as users can manage it, as well as the uncontrolled search for big data. We may certainly try our best to show off who we may reasonably be, or wish to become. In the long run, that will also affect who we are, both online and offline. Therefore, as compared by Floridi, the onlife experience is a bit like Proust's account-book, but with us as co-authors.

From another perspective, Preston (2014) reflected about the need for privacy to defend the very possibility of creative and meaningful life. There is a barrier/check on our behavior when we know we are being watched: deviancy needs privacy – according to Jeremy Bentham's Panopticon. This is where we find ourselves now, explained Preston, under a form of surveillance that may seem benign enough but which nonetheless exerts a dark, controlling power over us, the watched. The message seems to be that if we really want to keep something private, treat it as a secret, in the age of algorithmic analysis and big data, it is perhaps best to follow Winston Smith's bitter lesson from Nineteen Eighty-Four: "If you want to keep a secret, you must also hide it from yourself." Thus, some academics defend a different perspective from the panopticon Foucault described, like Vaidhyanathan (2011), who contends that "we don't know all the ways in which we are being watched or profiled – we simply know that we are. And we don't regulate our behaviour under the gaze of surveillance. Instead, we don't seem to care" (p.112).

Preston (2014), however, picks up on the words of Don DeLillo: "You need to know things the others don't know. It's what no one knows about you that allows you to know yourself." By denying ourselves access to our own inner worlds, we are clogging up the well of our imagination, that which raises us above the grind of mere survival, that which makes us human. Preston then asked Josh Cohen why we need private lives and his answer was a rallying cry and a warning. "Privacy," he said, "precisely because it ensures we're never fully known to others or to ourselves, provides a shelter for imaginative freedom, curiosity and self-reflection. So to defend the private self is to defend the very possibility of creative and meaningful life."

Finally, to add a subtle nuance, recent studies have confirmed the long-time psychology–oriented hypothesis of "episodic memory formation", which states that memories are inextricably tied to a specific location in the brain (Meyer, 2014). Also worthy of consideration is the conclusion raised by Hermida (2014) about the natural and inner disposition of humans to share, in her analysis of user behavior through social media.

PRELIMINARY MAIN RESULTS OF THE 'PUBLIC & PRIVATE' PROJECT

In the frame of a broader project, which aims to analyze users' perceptions and performances concerning the delimitation of the public and private spheres with mobile devices, we will present the main preliminary results related to the objective of this chapter. The methodologies applied – and which we continue to apply – were online and offline surveys, content analysis, focus groups and interviews.

Overall Users' Habits and Privacy

Several times a day (based on an exploratory survey: ES), users mainly 'make calls' and 'send and receive text messages,' then surf the web -'visit sites, browse the Internet, searching information' and still 'visit social networking sites'. Half might live without mobile phones, 32% would acknowledge the lack of the devices and 18% considered that they did not have such a powerful role in their lives. Mobile phone use was deemed strongly circumstantial, since there were no general rules stipulated by the actors depending on different situations - different contexts elicited different behaviors.

Sociability, coordination and its maintenance (from exploratory focus group: EFG) were outstanding in mobile communication while issues related to personal and professional life were also considered very important, in some cases essential. They recognized (from Focus Group) "anxiety" when "having to be always available". Retired people specifically criticize the "abusive" use of mobile devices and online communication by young people. Participants tended to consider that data was always private, but in some cases (images) did not contradict the possibility to share/publish it (EFG).There was great similarity between motivations and uses (content analysis: CA), with only one respondent mentioning the definitions of privacy. Thirty-four (out of 44 BA students) considered that private life was at risk, only one stated "no" and the rest explained that it would depend on their attitude: the value of one's responsibility and willingness to decide was thought to be determinant. Worries about what could be published by others without their consent were also highlighted. Hence, the relationship between losing privacy and social media as well as the awareness that everything published ceases to be private.

The issue of privacy appeared in the level of content/data and in the level of interaction contexts (FG). Moreover, the risks were not so evident because they continued to act as if they had privacy. The reluctance to speak in public near unknown people, to constrain sociability or to interrupt mobile phone conversations and the strategy to minimize risk of nuisance (silence mode/vibrator) are pointed out as well. The main theme (thematic analysis) within adolescents FG was *Privacy in the Digital Space*. Owing to its complexity, this included several subtopics: *Control Strategies of Privacy* -weak-; *Capacity Choice of the Actor; Negotiation of Actors* as to what is Public and Private; *Invasion of Privacy; Violation of Privacy; Alienation of Control* and *Surveillance*. From the views expressed by participants, there seems to be a false sense of control of their privacy. They emphasized the danger of social networks, including some serious situations that result in violation of privacy, which depend largely on the choices of the subjects. This allowed us to classify them as 'digital naïve' (Boyd, 2014). In this respect, they showed a lack of awareness of invisible or imagined audiences as well as of the characteristics of social media (Rheingold, 2012; Boyd, 2014, Brake, 2014).

Managing Accounts and Privacy

There was a wide range of definitions for 'profile' (CA). In connection with the different accounts in digital services and/or social media, Hotmail was considered both to be and not to be a 'profile': "Account, personal page, platform, image, portrait, space, local, a description of oneself, a characterization of us, an identification, a method, a way to quickly systematize basic information, where I expose moments and events of my life". Most configurations of 'profiles' were presented as voluntary and without risks and only one answer showed awareness of their commercial purposes. They included the results of their actions with the 'profile' as a part of it and did not specifically mention the digital footprint. Participants have from one to fifteen 'profiles', with an average of four profiles. Furthermore, there was considerable variety of timings as well as perceptions about the time spent on construction and maintenance: eleven emphasized the definition of "short" and one emphasized "the required time". The ratings ranged from 15 minutes until 3/4 hours a day. It seemed that daily publications were not taken into account. All students used a real photograph of themselves in their 'profiles', stressing the importance of remaining visible and recognizable to their friends in order to achieve proper interaction with them.

Fifty-five percent (ES) did not allow apps to access the lists of contacts or information and 57.7% (S) did not synchronize data with apps, 41.3% did it, and a mere 1% "do not know / no answer". Twenty-six synchronized their accounts (CA), nine did not answer, four specified the link with the Hotmail account, which is required, but only one was aware of it and actually gave an explanation. Six said it was between Facebook and Instagram, and one specifically referred to the "instant upload". Among these answers, a couple responded a general "yes" and another couple "no", four reported doing the synchronization of all their accounts, two of "most of them", three of "some", one did it "rarely" and one responded "not necessarily".

More than half (60%) use different passwords for different accounts (from data survey: DS), while 23.1% reported not having a diverse set of passwords, using the same one to access the various accounts. Fifty-one point nine percent changed them, and 48.1% did not. As for altering the privacy settings, 51.9% did it, 31.3% had different settings for different parts, 13.5% had never changed the default settings and 1.9% did not know about the existence of privacy settings. When inquired (CA, 44) if different levels of privacy settings for the several 'accounts-profiles' were established, why and in which of them these configurations varied, 28 answered affirmatively, 5 did not reply and 4 stated "the same". It ought to be highlighted that twelve mentioned Facebook was only visible to "friends". They define the different levels in accordance with the 'accounts-profiles' and their functionalities. They equate "having privacy" with content that only your "friends" can see and setting one's own privacy levels. Forty-two altered privacy settings, and the 26 who changed the content did so in the following order: photographs, other content, localization and comments – from the status of "public" to the status of only visible to "friends".

Control Strategies of Privacy (AFG): most also changed the rules of privacy, revealing minimal knowledge of it, in order to control who had access to their data. Some explicitly mentioned changing "friends of friends" to "friends". The use of access codes to mobile phones was cited by some, motivated partly by curiosity and partly to evade surveillance of mothers. *Alienation of Control*: also regarding the use of access codes, although aware of its importance (the participants did it in the past), many confessed they ignored them because of the work involved. Nearly 40% read 'privacy policies (DS), terms and conditions', 58.7% did not and 1.4% chose "do not know / no answer". They "were aware" (CA, 44 students) that once something is published, it is difficult to erase it forever. Eight directly related to the Internet and 12 specifically to the social media. Eleven "were aware" that, despite the privacy policies, data can be held by others. These "others" could be "friends of friends" and not necessarily our direct friends, hackers, "malicious persons", "third parties" (i.e. business) and Facebook. Only one stated that "it is not as protected as it should" (alienation, accustomed to this situation as normal).

Only a couple read terms and conditions, four did not answer, one did not own a smartphone, twelve argued that these statements were too long and required much time, a couple checked them in case of additional costs, another couple justified themselves by explaining that reading them will imply spending more time in the installation, five recognized "lack of patience" and "laziness", the rest answered "no".

Some situations could be plausibly described as *Alienation of Control* (AFG), particularly with regard to reading privacy policies before installing a new app. All respondents admitted to not reading them, merely clicking to accept. Simultaneously there was awareness that this behavior may carry risks and result in manipulation. *Control Strategies of Privacy* (AFG): the concept of friendship was very broad, including all acquaintances. It also referred to other conclusions that pointed to adolescents perceiving social media as an extension of their social relationships (Boyd, 2014). In terms of *Negotiation of the Actors* as to what is 'Public and Private', the idea prevailed that the availability of content was based on an agreement between the people involved, and care was taken not to offend sensibilities.

Managing Media and Privacy

Seventy-one percent of respondents used the phone for taking photos or videos between one and three times per week (ES). Forty (CA) used the mobile phone to take pictures, one did not, one did not have a camera phone and, finally, five only made use of it when lacking a camera. The moments selected for pictures were in the company of family, friends, "moments to remember", "important dates", culture, travel. Thirty-one did not take pictures or films of strangers; two did so rarely, upon finding a ridiculous or interesting situation; one did so on a relevance basis; and, finally, six took into account the rights of/respect for others. Seventeen saved images taken both for personal purposes and/or to share on social media, and two preferred to take pictures for personal reasons. The concern about the dissemination of pictures of themselves (EFG) had to do with embarrassing situations and aesthetics of the images. Their publication in open or closed circles was based on common sense. *Control Strategies of Privacy* (AFG): the strategy of not posting photos and pictures was suggested and reminded by parents.

Sixty-eight percent (ES) checked if the app offered "app permission" but 61% did not read those "permissions" before installing, only 32% did that. Sixty-five point four percent (DS) have abandoned apps due to a detected misuse of data, 21.2% choose "I will consider this possibility in the future", 12.5% did not, and 1% "do not know, no answer". *Control Strategies of Privacy* (AGF): participants developed a set of weak strategies. Most discarded the use of apps in the social media that enable the identification of where people are at any given time. Also, the choice to download an app depended on the knowledge of others who had already done so. The importance of peer behavior at this stage of the life cycle is also reflected in the behavior assumed when entering into the digital world, Boyd (2014).

Ninety-two point three percent (DS) were registered in a SNS, with the following data sharing distributions: 73.6% basic information; 13.9% contact information; 43.8% personal interests; 58.7% education and training; 36.1% work/job and 34.6% travels. Sixty-six point three percent set data visibility and sharing in SNS to be visible only for friends, 15.4% had different data visible for different people, 8.7% were not sure who could see their data, and 8.2% allowed everyone to see it. Thirty (CA) felt that the information they shared was not secure, three felt it was and the rest stated it depended. A gradation of awareness and knowledge was noted – users are aware of some risks but do not delve into the reasons or even know them.

Control Strategies of Privacy (AFG): the interviewees were aware that access to SNS should be done on trusted devices, taking care to leave when they did. They also showed concern for what they published, giving preference to trivial things or things that do not compromise them. Some of them only did it for a very restricted circle of friends. The sending/accepting of invitations was addressed to people they knew, at least by sight. The existence of a consciousness was a prevailing notion, but it was not linked to the risks inherent in the use of social media. Thus, this awareness did not lead to the mobilization of all appropriate behaviors. The verbs and nouns as well as the syntax employed are indicative of normalization and of an acquired routine in the process of exposing personal information (CA).

There was a notion of risk in exposing personal aspects that may stay online indefinitely, but the practices seemed to be unconcerned (EFG). Questioned if they regretted disclosing personal data (DS), 65.9% answered no, 15.4% yes on SNS and 17.8% yes in online shopping. Regarding the use of tools for data protection, 85.1% answered no, 13.9% yes and 1% "do not know, no answer". *Invasion of Privacy* (AFG): the majority referred to photos posted by parents and in which they were tagged.

Concerning the level of preoccupation with the lack of control about data (DS), people were rather concerned (31.3%), worried (30.3%), a little worried (28.2%) and not worried (9.19%). The reasons for this were: 54.8% believe the inner SNS structure hinders control over personal data, 16.3% had difficulties setting up the privacy of their personal data, 16.3% did not have time to set up privacy of their personal data and 12% produced other reasons. Asked if companies should and for how long they should storage data, 66.3% said they should not save data, 24% answered within one year, and 18% indefinitely. Twenty-four percent made bank transactions online, 75.5% did not, and 0.5% "do not know, no answer". On access to data by security agencies, 54.3% agreed only with a legal foundation, 38% did not, 7.2% agreed on a case-by-case basis, 0.5% "do not know, no answer". Twenty-two (CA) felt they did not control the information available about them, whereas eight did. They showed concern about who controls the information - themselves or others - and especially about what others can publish about them and about the appropriation of information without their permission.

Facebook (CA) was mentioned several times as the 'profile' that displays the most data, because it requests users to provide it or due to its interaction-oriented nature. One respondent thought Facebook was more secure since it had more definitions of privacy settings. Only one person established a relationship between publications and personal data. They showed (EFG) high tolerance to invasion/harassment and unauthorized commercial use of personal data. At this point, they showed an initial awareness of the different sources to collect their personal data. However, the possibilities are wider as it was described in the theoretical frame concerning the profile and the digital identity. Therefore, is it important to circulate the required knowledge in this area.

FUTURE RESEARCH DIRECTIONS

'Education for Privacy' and 'Digital Literacy' specifically focused on managing identity on the Internet are research fields that will require an in-depth analysis. Apart from developing clearer and user-centered policies as well as terms and conditions – which ought to involve app and platforms developers, governments, citizenships, etc.-, a concrete syllabus incorporating digital abilities, attitudes, capabilities and knowledge should be implemented in all stages of Education State plans. Combined with interesting and outstanding areas in 'Child Protection and Safety' and 'Digital Literacy', the proposed new lines would cover identity construction and management both from technological and communicational perspectives. This would require an interdisciplinary approach mainly within Education, Communication, Psychology and Sociology, as well as an intersectional collaboration with technology. Aspects to investigate further would include raising awareness about the liquid characteristics of the online mobile ecosystem, about social media and technologies as tools that shape behaviors and identities, as well as the potential dimensions of identity to manage within this ecosystem (personal, professional, etc.).

CONCLUSION

While users worldwide are fully aware of the existence of 'big data' and consequently concerned with personal information and Internet safety, most of them, however, lack the proper attitudes and knowledge to manage their accounts in this ever-changing, and hence challenging, online mobile ecosystem. Despite differences between countries due to technological development and market implementation, which in most cases seem to be in accordance with the users' habits and mobile incorporation into their daily lives, some overall tendencies have been identified. Users recognized some of the existing and potential risks of surfing the web, but there is nonetheless a gap between their notions and their final actions.

In this respect, it is striking to discover that half of these users allow their personal information to be used in order to access different kinds of services for free, even if they acknowledge the information they shared as personal. Also surprising is the fact that half of Internet users have concerns about the terms and conditions as well as data policies of apps, which are difficult to understand properly, but, at the same time, they admitted to not reading them. In the specific case of our project, the preliminary results show that half the users accept an app installation, firstly because others have done it before, and secondly because they trust the companies or simply do not read them. Clearer and user-centered policies as well and terms and conditions must be developed.

According to the differences between countries regarding mobile appropriation-uses and performances, from the preliminary results of our project, basic functions like making calls and sending/receiving messages are still the main functions while surfing the web for secondary purposes within the Portuguese population. Other world tendencies show that most mobile uses are "non-voice activities". However, the Portuguese exceed the European average in terms of accessing and using SNS. Overall and worldwide, there is also a manifest tendency toward the increased use of SNS, where the youth plays a leading role. Internet users share the concern about who has access to their content and the related lack of control. In the specific case of our project, they have proved unconscious about potential invisible audiences to be dealt with and the possibility of misunderstanding the complexity and subjective interpretations of 'big data' and 'dataism'. They take preliminary steps to avoid data synchronization between accounts and contacts but mix the concepts of profile and digital identity, as described by a technological perspective. In addition, most users are not completely aware of the digital paths they produce.

Thus, and as hinted at with the need to develop further research lines, academic work should thoroughly analyze users' perceptions and performances as well as raise awareness about 'dataism' and the consequences of apps, media, platforms and technologies as tools that shape our behaviors and online identities (even offline as a consequence thereof) to best suit their interests, mainly commercial ones.

As for the users' perceptions and understanding of 'privacy', the preliminary results from our project match those from Pew Research. A constellation of concepts has emerged among users, which translates into a correlation between Bauman's liquidity and our hypothesis about the liquidity of the mobile ecosystem - partly previously stated by Aguado *et al* -, and following the relationship with the fluidization parameters of the technological, institutional and cultural media dimensions described by McQuail.

REFERENCES

Aguado, J.M; Feijóo, C; Martínez, I.J. (cords.). (2013). La comunicación móvil. Hacia un Nuevo ecosistema digital. Barcelona: Gedisa.

Bauman, Z. (2005). *Liquid Life*. Cambridge, UK: Polity Press.

Boyd, D. (2014). *It's Complicated. The Social Lifes of Networked Teens. Yale New Haven and London.* University Press.

Brake, D. R. (2014). *Sharing our Lives Online. Risks and Exposure in Social Media.* Basingstoke, UK: Palgrave Macmillan.

Campbell, S. W. (2015). Mobile Communication and Network Privatism: A Literature Review of The Implications For Diverse, Weak, and New Ties. *Communication Research*, *3*(1), 1–21. Retrieved July 7, 2015, from http://www.rcommunicationr.org/index.php/component/jdownloads/finish/16/55?Itemid=0.

Cerra, A., & James, C. (2012). *Identity Shift. Indianapolis*. IN: John Wiley & Sons.

CIGI-Ipsos. (2014). *83% of Global Internet Users Believe Affordable Access to the Internet Should be a Basic Human Right.* Retrieved November 18, 2014 from http://www.ipsos-na.com/knowledge-ideas/public-affairs/ipsos-global-advisory/?q=internet-basic-right

Deuze, M. (2012). *Media Life*. Cambridge, UK: Polity Press.

EMC Privacy Index. (2014). Retrieved from http://portugal.emc.com/campaign/privacy-index/index.htm?pid=home-emcprivacyindex-120614?fromGlobalSelector

Fidalgo, A., Serrano Tellería, A., Carvalheiro, J. R., Canavilhas, J., & Correia, J. C. (2013). Human Being as a Communication Portal: The construction of the Profile on Mobile Phones. *Revista Latina de Comunicación Social,* 68, 545-565. Retrieved November 8, 2014 from http://www.revistalatinacs.org/068/paper/989_Covilha /23_Telleriaen.html

Flusser, V. (1988). *On writing, complexity and the technical revolutions. Interview in Onasbrück, European Media Art Festival.* September 1988. Retrieved November 8, 2014 from https://www.youtube.com/watch?v=lyfOcAAcoH8&app=desktop

Foridi, L. (2014). The Facebook-ification of everything! Sex, authenticity and reality for the status update era. *Salon*. Sunday, Aug 31. Retrieved from http://www.salon.com/2014/08/31/the_facebook_ification_of_everything_sex_authenticity_and_reality_for_the_status_update_era/?utm_source=twitter&utm_medium=socialflow

Fortunati, L. (2002). The Mobile Phone: Towards New Categories and Social Relations. *Information Communication and Society*, *5*(4), 514–528. doi:10.1080/13691180208538803

Fuchs, C. (2014). *Social Media: A Critical Introduction*. London, UK: Sage. doi:10.4135/9781446270066.n2

Grosser, B. (2014). What Do Metrics Want? How Quantification Prescribes Social Interaction on Facebook. *Computational Culture: A Journal of Software Studies*. Retrieved November 17, 2014 from http://computationalculture.net/article/what-do-metrics-want

Hermida, A. (2014). *Tell Everyone: Why We Share and Why it Matters*. Toronto: Doubleday Canada.

Horning, R. (2014). Social Media Is Not Self-Expression. The New Inquiry, November 14. Retrieved November, 17, 2014, from http://thenewinquiry.com/blogs/marginal-utility/social-media-is-not-self-expression/

Jenkins, H., Ford, S., & Green, J. (2013). *Spreadable Media. Creating Value and Meaning in a Networked Culture*. New York, NY: New York University Press.

Ling, R. (2008). The Mediation of Ritual Interaction via the Mobile Telephon. In J. Katz (Ed.), *Handbook of Mobile Communication*. Cambridge, MA: MIT Press. doi:10.7551/mitpress/9780262113120.003.0013

Lusoli, W., Bacigalupo, M., Lupiañez, F., Andrade, N., Monteleone, S., & Maghiros, I. I. (2012 June 4). EU27 survey of electronic identity, privacy and data protection [Final Report]. *IPTS of the EU Commission*. Retrieved from http://is.jrc.ec.europa.eu/pages/TFS/eidsurvey.html

Madden, M. (2014). Public Perceptions of Privacy and Security in the Post-Snowden Era. Pew Research. November 12, 2014. Retrieved November, 18, 2014, from http://www.pewinternet.org/2014/11/12/public-privacy-perceptions/

Marichal, J. (2012). *Facebook Democracy. The Architecture of Disclosure and the Threat to Public Life*. Farnham, Surrey; Burlington, VT: Ashgate.

Markham, A. (2013). Undermining "Data": A Critical Examination of a Core Term in Scientific Inquiry. First Monday, 7 October, 18 (10). Retrieved November, 8, 2014, from http://uncommonculture.org/ojs/index.php/fm/article/view/4868/3749

Martínez-Conde, S., & Macknick, S. L. (2013). *Los engaños de la mente*. Barcelona: Destino.

Marwick, A. (2013). *Status Update: Celebrity, Publicity, and Branding in the Social Media Age*. New Haven, CT: Yale University Press.

Mashable. (2014). *Mobile-Minded*. Retrieved November 18, 2014 from http://mashable.com/2014/11/17/human-smartphone-relationship/#share-action:eyJzIjoidCIsImkiOiJfejQ5OXB4eDY0bnExZ3kwMyJ9

McQuail, D. (2006). *Mass Communication Theory*. London, UK: Sage.

Meeker, M. (2014). Internet Trends 2014 – Code Conference. KPCB. 28 May 2014. Retrieved November, 8, 2014, from http://www.kpcb.com/internet-trends

MEF & AVG. (2014). *Global Consumer Trust Report 2014*. Retrieved November 18, 2014 from http://www.mefmobile.org/activities-and-analytics/analytics/global-consumer-trust-report

Pew Research. (2014). *Social Networking Fact Sheet*. Retrieved November 18, 2014 from http://www.pewinternet.org/fact-sheets/social-networking-fact-sheet/

INE Portugal. (2014, November 6). *Sociedade da Informação e do Conhecimento. Inquérito à Utilização de Tecnologias da Informação e da Comunicação pelas Famílias*. Retrieved from http://www.ine.pt/xportal/xmain?xpid=INE&xpgid=ine_destaques&DESTAQUESdest_boui=211422735&DESTAQUESmodo=2

Portugal, I. N. E. (2014). *Sociedade da Informação e do Conhecimento. Inquérito à Utilização de Tecnologias da Informação e da Comunicação pelas Famílias.* 6 de Novembro de 2014. Retrieved November, 18, 2014, from http://www.ine.pt/xportal/xmain?xpid=INE&xpgid=ine_destaques&DESTAQUESdest_boui=211422735&DESTAQUESmodo=2

Preston, A. (2014). The Death of Privacy. The Guardian. Sunday 3 August 2014. Retrieved September, 11, 2014, from http://www.theguardian.com/world/2014/aug/03/internet-death-privacy-google-facebook-alex-preston?CMP=twt_gu

Privacy Index, E. M. C. (2014). Retrieved November, 18, 2014, from http://portugal.emc.com/campaign/privacy-index/index.htm?pid=home-emcprivacyindex-120614?fromGlobalSelector

Rheingold, H. (2012). *Net smart. How to thrive online.* Cambridge: The MIT Press.

Rosen, L. (2013). Always On, All the Time: Are We Suffering From FoMO? *Psychologytoday*. Retrieved November 17, 2014 from http://www.psychologytoday.com/blog/rewired-the-psychology-technology/201305/always-all-the-time-are-we-suffering-fomo

Serrano Tellería, A., & Oliveira, M. (2015). Liquid Spheres on Smartphones: The Personal Information Policies. *International Journal of Interactive Mobile Technologies*, *9*(1). Retrieved July 17, 2015, from http://online-journals.org/index.php/i-jim/article/view/4065

Serrano Tellería, A. (2014). Interface Design on Mobile Phones: The Delimitation of the Public and Private Spheres. In Paiva, F; Moura, C. (Orgs.) DESIGNA: INTERFACE International Conference on Design Research (pp.87-108) Covilhã, Portugal: Labcom Books, University of Beira Interior. ISBN 978-989-654-141-5. Retrieved December 12, 2014, from http://www.livroslabcom.ubi.pt/book/111.

Serrano Tellería, A. (2015). Liquid Spheres or Constellations: Reflections Towards Mobile Devices. In Carvalheiro J.R., Serrano Tellería A. (eds.) Mobile and Digital Communication: Approaches to Public and Private. 173-198. Covilhã, Portugal: Labcom Books, University of Beira Interior. ISBN: 978-989-654-235-1 (paper) 978-989-654-237-5 (pdf) 978-989-654-236-8 (epub). Retrieved July 17, 2015, from http://www.livroslabcom.ubi.pt/book/141.

Stald, G. (2008). Mobile Identity: Youth, Identity, and Mobile Communication Media. In Buckingham D. Youth, Identity, and Digital Media (pp. 143–164). The John D. and Catherine T. MacArthur Foundation Series on Digital Media and Learning, Cambridge, MA: The MIT Press.

Stone, L. (2014). *Continuous Partial Attention.* Retrieved November 8, 2014 from http://lindastone.net/qa/continuous-partial-attention/

Turkle, S. (2008). Always-On / Always-on-You: The Tethered Self. In J. E. Katz (Ed.), *Handbook of Mobile Communications Studies* (pp. 121–138). Cambridge, MA: MIT Press. doi:10.7551/mitpress/9780262113120.003.0010

Turkle, S. (2011). *Alone together: why we expect more from technology and less from each other?* New York, NY: Basic Books.

Vaidhyanathan, S. (2011). *The Googlization of Everything (and Why We Should Worry).* Berkeley, CA: University of California Press.

Van Dijck, J. (2007). *Mediated Memories in the Digital Age*. Stanford, CA: Stanford UP.

Van Dijck, J. (2013). You have one identity: performing the self on Facebook and LinkedIn. *Media, Culture & Society*, 35(2), 199-215. Retrieved November 14, 2014 from http://mcs.sagepub.com/content/35/2/199

Van Dijck, J. (2014). Datafication, Dataism and Dataveillance: Big Data between Scientific Paradigm and Ideology. *Surveillance & Society*, *2*(2), 197–208.

Walker Rettberg, J. (2014). *Seeing Ourselves Through Technology*. Basingstoke, UK: Palgrave Macmillan. Retrieved November 8, 2014 from http://www.academia.edu/8482366/Seeing_Ourselves_Through_Technology_How_We_Use_Selfies_Blogs_and_Wearable_Devices_to_See_and_Shape_Ourselves

Wellman, B. (2001). Physical Place and Cyberplace: The Rise of Personalized Networking. *International Journal of Urban and Regional Research*, *25*(2), 227–252. doi:10.1111/1468-2427.00309

KEY TERMS AND DEFINITIONS

Architecture of Exposure: Reflects both the interface design and the inner information architecture of content and users performance, including interactivity and usability dimensions, observed mainly in Social Media and in mobile devices (see more at Serrano Tellería, 2014).

Constellations of Spheres: Describes how users manage their public, private, personal and intimate spheres within the online and mobile media ecologies as well as technological ambient / environments (see more at Serrano Tellería, 2015, pp.190-191).

Digital Identity: The data employed to represent entities (persons, organizations, machines or software processes) in information systems and it may have three dimensions: social, legal, and computational.

Liquid Media Ecology: The hypothesis that relates the online and mobile media ecology as well as its technological ambient/environments with Bauman's liquidity and the fluidization parameters of the technological, institutional and cultural media dimensions described by McQuail (2006. See more at Aguado, Feijóo, Martínez, 2013; Serrano Tellería, Oliveira, 2015; Serrano Tellería, 2015).

Liquid Spheres: Describes the blurring boundaries between the public, private, personal and intimate spheres within the online and mobile media ecologies as well as technological ambient/environments (see more at Serrano Tellería, Oliveira, 2015; Serrano Tellería, 2015, pp.190-191).

Liquid Technological Ambient/Environment: The hypothesis that relates the online and mobile media ecology as well as its technological ambient/ environments with Bauman's liquidity and the fluidization parameters of the technological, institutional and cultural media dimensions described by McQuail (2006. See more at Aguado, Feijóo, Martínez, 2013; Serrano Tellería, Oliveira, 2015; Serrano Tellería, 2015).

Online Profile: Structured information about the users of digital services. Its structure depends on the way the information has been gathered, the technology involved, the types of data or what the law allows.

Section 3

Experiences and Performances on Mobile Content Evolution

Chapter 15
Research, Development and Creativity in Ubiquitous Technologies in University: CIDUS Contribution

Enrique Castelló-Mayo
University of Santiago de Compostela, Spain

Antía M. López Gómez
University of Santiago de Compostela, Spain

Roi Méndez Fernández
University of Santiago de Compostela, Spain

Julian Flores González
University of Santiago de Compostela, Spain

Eduardo Sánchez-Vila
University of Santiago de Compostela, Spain

ABSTRACT

In 2009 the University of Santiago de Compostela (USC) proposes the creation of a truly transdisciplinary research platform, suitable for the development of integral projects of excellence in R&D, technology transfer and professional training in all phases of the digital contents evolution cycle: CIDUS Platform for Innovation and Experimentation and Innovation in Digital Contents (www.usc.es/es/Servizos/cidus). Although this chapter covers some of the finest CIDUS exponents in the field of ubiquitous technologies, we need to emphasize that we believe that the main innovation of CIDUS is not only its innovative projects but also the very experience of interdisciplinarycollaborationbetweeneighty-onereferenceresearchers in the field of digital content (assigned to seven groups with such a pure technological base as a social communication profile), respectively located in Faculty of Communication Sciences, Singular Research Centre in Information Technologies, Technical School of Engineering and Institute of Technological Research at USC.

DOI: 10.4018/978-1-4666-8838-4.ch015

INTRODUCTION

Although it seems obvious that large-scale penetration of ubiquitous technologies in contemporary society should tally with a special attention and tracking from the academic environment, the examples of best practices in that fieldwork have not yet reached full efficiency, at the teaching and research levels (Wilson & McCarthy, 2010).

In 2009 and as a response to this social commitment, the University of Santiago de Compostela (USC) proposes the creation of a truly transdisciplinary research platform, suitable for the development of integral projects of excellence in R&D, technology transfer and professional training in all phases of the digital contents evolution cycle: CIDUS Platform for Innovation and Experimentation and Innovation in Digital Contents (CIDUS, 2015).

Although this chapter covers some of the finest CIDUS exponents in the field of ubiquitous technologies, we need to emphasize that we believe that the main innovation of CIDUS is not only its innovative projects but also the very experience of interdisciplinary collaboration between eighty-one reference researchers in the field of digital content (assigned to seven groups with such a pure technological base as a social communication profile), respectively located in the Faculty of Communication Sciences (FCCC, 2015), the Singular Research Centre in Information Technologies (CITIUS, 2015), the Technical School of Engineering (ETSE, 2015) and the Institute of Technological Research (IIT, 2015) at USC.

But to better illustrate our efforts, some of the most iconic projects in CIDUS ubiquitous technologies are included below, some of them carried out so far and others still in progress as we write these lines, covering such diverse areas as telemedicine, tourism promotion, distribution of digital cinema in OV, the impact of the ecological footprint, cultural content generation, dissemination and enhancement of historical and artistic heritage, etc.

BACKGROUND

Before focusing on the main objective of this report (which is to contribute to a better understanding of innovative experiences on the CIDUS Mobile Content Evolution), this will start with a look at a quotation from Vladímir Ilich Uliánov—more commonly known as "Lenin"—, who pointed out that it is sometimes necessary to take one step back in order to take two steps forward. However, it might be necessary to take two steps back —as long as three decades ago— in order to achieve the current scope of ubiquity technologies.

First step: A controversial paradox was raised by Furio Colombo (1974) about media influence more than three decades ago. According to the Italian author, politically mature citizens want to decide for themselves between an actively interact with the social reality, or to maintain a permanent link —Plug-In & Stay-Tune— with the various levels of media information. If it is true that the main goal in a media democracy seeks to have informed citizens making informed choices, the hidden face of this edifying goal is the isolation of people: social struggle and media consumption spaces are incompatible, due to the very characteristics of electronic media.

However, as Rheingold (2004) predicted, changes in ubiquity technologies were not accompanied by structural political changes, but were exploited by social protest movements. In fact, right now, the obsolescence of Colombo's paradox was reflected in technological deployment in the Puerta del Sol Square during its peaceful occupancy by 15-M "indignados": while thousands of people stayed in Puerta del

Sol to avoid the square carried out by the police, one of the key aspects in "militant mobile technology" is keeping global audiences informed —in real time, wherever they are in the world—while the Spanish movement continue their political struggle on the front lines: thus, social struggle and media consumption spaces concurred now in (real) time and (virtual) space, thanks to the advanced ubiquity technologies.

More recently, it was possible to find another similar example in "Podemos" last electoral experience. In fact, this political group —which has become the country's third political force— brings together voters' widespread dissatisfaction with austerity measures. After unexpected results of the European elections, the Spanish party led by Pablo Iglesias prepares a specific programme for the 2015 municipal elections in agreement with all of his voters: Appgree, a mobile application, allows a choice between 16,000 proposals for Community action.

At the time of writing, more than eleven thousand people have participated in this initiative, with a total of 260,179 votes. Appgre is complemented with other virtual tool, "Plaza Podemos" (http://www.reddit.com/r/podemos): nothing could be more laudable for "Podemos" than the defence of freedom of expression, reflecting on the issue of citizens' participation and empowerment."Plaza Podemos" overcomes 33,000 chat followers, at the time of writing, on themes as diverse as unemployment, social exclusion, precarious jobs, development financing, political corruption and access to health or education, among other social problems that take place in the present.

Second step: Joshua Meyrowitz published in 1985 his celebrated and schismatic work in the electronic media field: "No sense of place. The impact of electronic media on social behavior". In his study —awarded by the prestigious National Association of Broadcasters and the Broadcast Education Association (http://www.nab.org and http://www.beaweb.org) as "Best Book on Electronic Media"—, Meyrowitz asserts the existence of a change on the "situational geography" of social life, specifically focusing on the television influence.

Indeed, until the outbreak of electronic media, the situation in the social scale of each of every individual would limit his own access to collective knowledge that is found and created in the social group. Through new patterns of social action, feeling and belief, and with a view to stimulating the debate on the influence of media changes in social environments, Meyrowitz penned his famous remark: «Television takes our kids across the globe before parents give them permission to cross the street».

In this way, the incessant information bombardment targeting different social class dissociates access to collective knowledge of the situational geography; in other words, the specific characteristics of television media impose virtuality as a particular kind of social relation; or, to put it differently, the television potential amounts to calling into question the time-space laws. This is completely at odds with the sociological theory of Pierre Bourdieu's concept of «Habitus» (2005).

As pointed out above, social relationships and media consumption are mutually incompatible prior to the ubiquity technologies proliferation: perhaps, on communication and new media fieldwork, we are confronted with the challenge of that cultural change raised by Hirst, Harrison and Mazepa (2014), from broadcast to narrowcast… That is probably a turning-point period when it comes to the way we consume films or not in theatres.

Without falling into excesses of technological panegyrics, ubiquity technologies usage has rapidly proliferated with millions of customers, while they're on the go, consuming information and entertainment content (Oliver & Raney: 2014).

In the field of tourism, mobile Recommender Systems (RS) have emerged relativity recently providing richer tourist experiences, personalizing multimedia content and including context as a source of key information to improve the overall satisfaction of users (Gavalas et al., 2014). These new mobile RSs take advantage of the myriad of sensors currently deployed in smartphones in order to manage all types of information about user context. Therefore, new context variables like position, heading, and even emotional cues, could be estimated and later used to improve the accuracy of generated recommendations. As a result, different recommendation strategies and algorithms could be deployed by combining user preferences, purchase history and user context.

It is a development model that emerged as a historical consequence of a number of significant choices: in the late twentieth century, successive digitizing waves hit the technology industry and communication markets, under the motto of "integration" among otherwise incompatible technology architectures (Moreno, 2002), would reach its zenith years later with the emergence of the "ubiquitous technologies" competently analysed by Featherstone (2009).

Even though, as followers of Gartner (1995) and his famous Cycle of Over-Expectation of technology emerging through the time axis ("Release" - "Oversized expectations peak" - "Disillusion abyss" - "Consolidation ramp" - " Productivity plateau"), we are aware of the need to temper any technological-loving panegyric (both as any technological-hater diatribe), it is clear that, despite its historical, socioeconomic and cultural significance, this "integration" remained largely alien to the contemporary university education and research.

Yet, as in many other moments of historical disruption, it corresponded to the Alma Mater—while "nurturing mother"—, to provide the social body of those broader, more inclusive approaches that descry the magnitude of these emerging technologies (whose ubiquity reaches nowadays, with respect to the dawn that Mark Weiser peeked in 1991, biblical proportions), transcending the prudish industrial and economic notion of "integration" in favor of "consilience", as it was lit by William Whewell in 1837 and updated by Edward Wilson in 1998.Further, more recently in the field of inclusive education, it also needs to be mentioned Featherstone (2009), El-Hussein and Cronje (2010) or De Souza and Sheller (2014), among many other authors who write about the need to make education more disruptive, consilience and creative.

On a final note, there are, fortunately, some references to similar experiences in other countries would enrich this background. The following list does not claim to be exhaustive, but rather to transmit in a sensitive manner an artistic and technological environment, values we identify with.

It is therefore now becoming relevant for us to start describing how the HIT Lab NZ focuses its works in Interaction Design offering I+D+I and educational programs as it is presented in their web page (http://www.hitlabnz.org/). The key to their courses and projects is the multidisciplinary focus, collaboration with external companies and team work. It mixes people with expertise in philosophy, economics, sociology or engineers to create better human-computer interaction.

The Human Interface Technology Laboratory New Zealand (HIT Lab NZ) developed and commercialized technology that improves the human computer interaction. It conducts research with new emerging technologies such as Augmented Reality, Next Generation Video Conferencing, Immersive Visualization and Human-Robot Interaction. Interaction Design techniques are used to adapt these technologies to the needs of end users and solve real world problems. Its end goal is to improve the user experience with technology.

In a similar vein, the Digital Worlds Institute (http://www.digitalworlds.ufl.edu/) includes programs in Academic, Research and Service areas. Their activities develop and inform the convergent spaces between the Arts, Communications, and Engineering. Through the use of interactive tools and technologies, it promotes transdisciplinary creativity across classrooms, continents and cultures. The diversity of projects attests to the wide range of domains that can be substantively supported and enhanced with interactive digital media. These projects vary from training for different purposes, arts or virtual tours to virtual museums, virtual heritage, etc.

We must also need to cite the institute for film in the Netherlands, *EYE Film Institute Netherlands* (Eye Film Institute, 2015). Since 2010, EYE has become an institution of international reference in the field of film restoration, research, and education, managing more than 40.000 films from all genres. EYE unites the Filmmuseum(cinematography museum founded in 1952, with regard to cinematographic and audiovisual preservation and protection of Dutch films and foreign films), Holland Film, the Filmbank (with materials of the film industry in the Netherlands from 1895), and the Netherlands Institute for Film Education: Media Literacy is one of EYE's main aims, by teaching the various activities of film education and developing materials for primary, secondary and higher education), the Filmmuseum, and EYE International, the international face of Dutch films, to enhance the perception and visibility of Dutch filmmaking worldwide.

To conclude this section, as could not be otherwise, we will focus on a summary description of Canadian Hexagram-CIAM (Canadian Hexagram-CIAM, 2015), an international network that open up a wide range of possibilities for research and work in a broad range of interdisciplinary domains relating to new media art, design and technology, with rules and operating conditions very similar to those applied by Cidus Platform(CIDUS, 2015).At present, more than one hundred members of Hexagram-CIAM (Centre interuniversitaire des arts médiatiques)works in the fieldworks of media arts, design, technology and digital culture, collaborating with industry, arts and cultural organisations in Québec, across Canada and around the world.

Hexagram-CIAM has two main pillars on which its nationally and internationally fame as Canada's largest pole for research-creation in the many intersecting points between art and technology was built: Hexagram-Concordia, Centre for Research-Creation in Media Arts and Technologies (Centre for Research-Creation in Media Arts and Technologies, 2015), and its partner centre HexagramUQAM at l'Université du Québec à Montréal (HexagramUQAM, 2015), with additional researchers at l'Université de Montréal, l'École de technologie supérieure, l'Université du Québec à Chicoutimi and McGill University. Since 2002, Hexagram-Concordiahas evolved into a dynamic environment for research-creation brings together artists and researchers from Concordia University's faculties of Fine Arts, Arts and Science, and Engineering and Computer Science.

MAIN FOCUS OF THE CHAPTER

Research, Development, and Creativity in Ubiquitous Technologies in CIDUS

As indicated, this section is dedicated to the most iconic projects in CIDUS ubiquitous technologies, covering such diverse areas as telemedicine, tourism promotion, distribution of digital cinema in OV, the impact of the ecological footprint, cultural content generation, dissemination and enhancement of historical and artistic heritage, etc. Issues, controversies and problems, and likewise solutions and recommendations, will be treated together at each of the following experiences on ubiquitous technologies field.

Project Nr. 1: Meet the Pórtico, a Mobile Photorealistic Museum in the User's Hand

It is futile to exalt the Portico de la Gloria of the Cathedral of Santiago de Compostela as one of the masterpieces of Spanish Romanesque. Thanks to the financial support of the Barrie Foundation and in order to disseminate among the pilgrims the Porticos heritage significance and some details of its restoration it were developed a set of applications that allow users to explore the history and meaning of each particular element in the monument. Meet the Pórtico is one of these applications.

The original Barrie Foundation project, inside the "ProgramaCatedral" (Fundación Pedro Barrié de la Maza, 2015) main project, involved a virtual exhibition that went one step further in the concept of interactive photorealistic museums proposed by Chun-Rong H., Chu-Song C. & Pau-Choo C. (2005). It was formed by a virtual visit to the monument —Virtual Pórtico (Otero, Méndez & Flores, 2012)—, a specific application for the observation and interaction with the instruments around its middle arch —The virtual instruments of the Pórtico) (Méndez, Otero, Jarque& Flores, 2012)— and some information kiosks that, through a touch screen, allowed the visitor to explore documents and sounds related to the porch. These three installations, which compose the exhibition The Virtual Pórtico de la Gloria, are placed since 2011 beneath the monument itself, which is being studied in a long term restoration process that started in 2010 and shall finish in 2017.

Once the development of these applications was finished and after one year of successful exhibition inside the Cathedral the idea of a itinerant version arised in order to make the importance of the Pórtico visible abroad and attract tourists and pilgrims to visit Santiago de Compostela. The three main components were kept with minor adaptations and the exhibition started travelling around the Europe: Berlin, Lisbon, London, etc. Following this idea of making the monument visible and taking into account the pedagogical power of mobile devices presented in Roschelle, J. (2003), it was proposed to create a new version of The Virtual Pórtico for mobile devices in order to take the already successful museum experience to users worldwide, regardless of where they were, taking advantage of new technologies to preserve and spread this important part of Santiago's cultural heritage and outreaching the external world from a museum, a need that has been presented in Roccetti, M., et al.(2013). The first proposal was to create an application for iPod, iPad and iPhone called Meet the Pórtico.

The main objective of this application was not only to be an advertising medium that encouraged the user to visit the exhibition, but also try to reproduce in the most accurate way the user experience that a visitor feels in the real museum.

This concept is far away from the experience offered by other museistic applications that exist at the moment which are more like a guide and help during the visit —like the one presented in Hou, H.-T. et al. (2014)— than an independent application to download and enjoy at home. For example the MAN MuseoArqueológico application is a virtual guide for the visitors of the National Archaeological Museum in Madrid, and only makes sense to use it as a support tool during the visit. Similar to this application the Vusiem British Museum Lite gives information about the exhibition, tours, events, etc. Second Canvas Museo del Prado tries to empower the visitor to learn about the story of some painting masterpieces present in the museum, but none of them tries to simulate part of the experience of actually being in the museum with a device that is in the user's hands.

The resulting application is divided in four parts: Virtual Pórtico, Restoration, Image gallery and Exhibition (See Figure 1).

Virtual Pórtico is the part that tries to simulate the original Virtual Pórtico installation giving the same information and reproducing the user's interaction paradigm.

Figure 1. Global and detailed view of the monument in the application

Both The virtual Pórtico and The virtual instruments of the Pórtico computer applications are based on real time rendering of huge 3D models, needing an amount of graphics power that an actual mobile device is not able to provide. Thus, when the mobile version was proposed, the viewing and interaction paradigm had to be reviewed. There were two main options: The first one involved the simplification of the complex 3D model, losing all the visual impact of the high resolution original one but keeping the interaction and viewing mode intact. The second option was to use a render of the original model and simulate in 2D the interaction paradigm of the 3D computer application. The decision was made to take the second option as losing the detail of the sculptures was not an option in an informative and visual application.

The original non-mobile application interaction was based on the free movement of the user around some points of interest of the monument using an independent tactile screen as a controller. As this kind of interaction was unreachable on an iPod, iPhone or iPad, the original 3D mesh was substituted by a very high resolution render, resulting in a 2D exploration movement that simulated the 3D original experience.

The user is able to move through the image performing the usual gestures on this kind of devices —pinch to zoom and drag to pan— giving him the opportunity to move freely around the monument. Besides, as in the original Virtual Pórtico application there are 11 points of interest that the user can visit with detail by clicking on the identification number (figure 1). When the user focuses on one particular area the application starts playing an explanation of the meaning of the sculptures and their history, including the display of real photographs, taken from the monument itself, and allowing zoom and pan focusing just in the particular point of interest that is being explored in order to facilitate the interaction.

There is also available an on rail visit that is a recorded version of the museum one. It consists on a video that goes through all the points of interest of the 3D model, explaining them and moving along the whole monument, taking advantage of the communicative power that this kind of representation has. This makes it possible for the user to have a similar 3D experience as he would have visiting The Virtual Pórtico in a museum.

The restoration part of the application intends to explain the complexity of the restoration process being developed by the Barrié Foundation on the Pórtico through the years making it visible for the people the need of this long term solution in order to prevent future deterioration and the correspondent loss of cultural treasures.

One of the most important things to take into account about cultural heritage is its preservation, keeping it as immutable as possible through the time. To achieve this goal it is necessary to recognize deeply the origin and reasons of the problems it has —moisture, wind erosion, etc.— in order to avoid them and preserve the piece in the best possible manner.

This module explains to the user the problems that were found in the Pórtico when the curators of the Barrié Foundation started studying the origin of the preservation problems of the monument. It also explains the process that has been developed since then and the future process that will be followed to restore the figures of the masterpiece.

For this purpose it has been designed a studies map (figure 2) where some images and texts are depicted, presenting the work that is being carried on the Pórtico in an attractive arrangement. This design has being made specifically to make it interesting and easy for the user to learn about the restoration process problems and objectives, showing real photographs and their localization in the monument as examples of the different processes, actions, and problems presented. It is also presented, in a more classical approach, a video where some experts explain the whole process in a purely informative way.

Finally, the application offers also a image gallery to show details of the real monument and an advertising module that presents the exhibition that is being moved all along Europe, composed by the Virtual Pórtico and the The virtual Instruments of the Portico among other images and information, encouraging people to visit the exhibition if it is near to their places, to travel to Santiago and see it or even for organizations to bring it to their respective locations

Meet the Pórtico is not yet available in the appStore but this new concept of museum applications where the main goal is to spread the knowledge and simulate the user experience of actually visiting a exhibition, being more than a guide inside a museum or just an advertising application, can be applied to many other museums worldwide, specially to those that already use virtual or audiovisual resources. Even more, the possibilities of this kind of applications are increasing daily with the unstoppable technical advance of mobile devices, allowing the museums to spread their cultural heritage worldwide in a new and more attractive way.

Figure 2. Studies map of the monument and detailed image of the process

Project Nr. 2: Imotion Comic, mixing Media Formats for an Enriched Story Transmission Experience

It is an innovative narrative, interactive and transmedia format developed by CIDUS in collaboration with Continental Games and the spin-off of the USC Paralaxe Multimedia. It takes advantage of the capabilities of mobile devices —iPone and iPod touch in this case— to create a new way to transmit stories, adding complementary media to the classical comic format.

The idea came across when thinking about taking comic stories to everyone's pocket through the actual multimedia devices. Making a pure copy or adaptation of the paper story to the screen format and resolution was a solution that didn't make use of the communicative power of the new platform.

The market is full of reader applications specifically designed for comics such as the Perfect Viewer, Marvel Unlimited, Comic Flow, YACReader, etc. There are also many applications that make it easier for any user to create its own digital comic such as Comic Life (Comic Life, 2015), Toondoo (Toondoo, 2015), Strip Generator (Strip Generator, 2015) etc. but professional authors have to use non-specific programs such as Corel Draw or Photoshop. So, the main idea of this project was not to create just an application to create comics and another to read them but also a whole new format that allowed the creators to add different multimedia components to their stories in order to enhance their creative possibilities.

Taking into account that this project doesn't intend to substitute the original comic but create a whole new narrative scheme that adds interactivity to classic story telling, it doesn't threaten the current comic world environment. Moreover, it enriches this environment allowing the artists to incorporate its entire potential, just a part of it or nothing at all, preserving as it is the original comic.

The main concept is to have a usual digital comic that integrates animation graphics, audio and video, creating a more enriching experience for the user, adding new capabilities such as selecting language, listening to the voice of the characters, having ambient sounds, animations etc.

The new format is designed to be created in layers. There is a background layer that is the biggest one and goes even out of the screen for animation purposes. For example, if there is a cloudy day, the background will move to generate the feeling of wind and rain (figure 3). Over this background the artist can add several layers to include the characters. These characters are associated to their animations (if they are animated), their text (including the text bubble) and their sound (voice or noises that they make).

Figure 3. Vignette of "Quotidianía Delirious"

Having this scheme makes it easy for the editors to change the text and sound language making it trivial to translate this format and change on the fly from one language to another as every character in every vignette will have as many texts and sounds associated as languages. Even more, it empowers the artist to change parts of his creation as every character is depicted in his own layer and can be removed or replaced easily.

From the users point of view the format is similar to any digital comic that presents the story using one vignette per screen, being this the most adequate presentation style for small screen devices, avoiding the reading user experience problems that are presented in Arai, K. and Herman, T. (2010) when reading comics in their common multiple vignette presentation. As usual, the transitions from one page to the next are fired by a swipe gesture and simulate, using an animation, the behavior of a real paper comic. The main difference is that it is interactive so that the viewer is able to choose what element of the story —plot, characters...— he wants to deepen in, combining traditional comic statism with kinetic images.

The first installment of this series was based on the comic "Quotidianía Delirious" from the renowned cartoonist Miguel Anxo Prado. This comic was selected because, as the author said, it is a story that is very adaptable to this new format as it is composed of very "specific and small stories". The artist was a part of the development team that gave to the beautiful vignettes a number of transmedia elements that allowed users to view the story from a closer point of view to the perspective of its creator.

This first story was distributed through the appStore for free in more than 70 countries and 50 million people, entering in just a week in the top 20 of the most downloaded free books in the Apple Store. It can be downloaded at (Quotidiania Delirante, 2015).

The main conclusion of this project is that the new multimedia mobile devices that are spreading worldwide hold a completely new set of possibilities both for creators and editors to transmit knowledge and information in many new ways and that, keeping traditional formats, new ways of communication can be developed and explored to reach a bigger audience and enrich the communicative process.

Project Nr. 3: The European Digital Space and the Role of Small Cinemas in Original Version

From ten years ago, Audiovisual Studies Group (GEA), a consolidated research group of the University of Santiago de Compostela (http://www.estudosaudiovisuais.org/), has shifted the focus of its research strategy towards detection of those barriers that small cinemas in original version have to surpass for the consolidation of cinema as a fundamental factor of social change (Ledo & Castelló, 2012 & 2013).

As evidence of their interest, all of the GEA research projects have been supported for many years by Galician, Spanish or European institutions, and always granted by competitive tendering: "Lusophone space: communication, interculturality, citizenship and interactivity" (Xunta de Galicia, INCITE08PX-IB212124PR), "Cinema, Diversity and Networks" (National R&D Plan of the Spanish Ministry of Science and Innovation, CSO2009-13702), "eDCINEMA: Towards the European Digital Space. The role of small cinemas in original version" (National R&D Plan of the Spanish Ministry of Science and Innovation, CSO2012-35784), and a long etcetera.

The Audiovisual Studies methodology is based on cooperative and multicentric research (without eurocentric prevalence), with researchers from Argentine, Brazil, Denmark, France, Finland, Guinea, Holland, Portugal, Spain, United Kingdom, Uruguay... To date, several international investigations were developed in three strategic geographical areas: Lusophony (over 223.000.000 people), Hispanophony (over 470.000.000 people) and European Union (over 501.000.000 people).

That is how the Audiovisual Studies methodology's largest potential is also its greatest problem. Thus, in the face of the dispersed location of researchers, multifocal target groups and objects of study, there is no option but to have a strong and solid alliance with specific tools based on ubiquitous technologies (Coiro, Knobel, Lankshear & Leu, 2014). It is in this sense that an increase in the number of interactions was noted after the adaptation of the website elements for mobile devices (at the present time: Android, IPhone and IPod Touch), following the significant work of Conti and Giordano (2014).

However, another problem arises with the above adaptation: in fact, investigative and analytical processes are hosted by an original online instrument, designed ad hoc, based on a Web Content Management System (WCMS). From the researcher's point of view, the implementation is always the biggest challenge of research on mobile communication based on multimedia technology (Xiong, Wang & Li, 2014), but how constructing and hierarchalising complex technological systems…? What's the GEA researching practice kernel and its ramifications…? Which of the terminals will be enabled to hold a feedback process with the kernel…?

Therefore, GEA uses a simplified diagram to help explain what is really happening on its research routine (See Figure 4):

As is evident from the figure above, the WCMS is for GEA the technological true operating nucleus: responsible for managing a centralized database, supports the online platform for the survey, different access profiles, discussion forums and the samples of analysis, as well as the extraction of the resulting data —particularly to the typical syntax of Predictive Analytics SoftWareor PASW— and the interactivity between the participating groups and the matrix group. However, it should be clarified that, despite WCMS position at central level, GEA project was implemented on an entirely decentralized basis, encouraging an interactivity interface between all the elements to work synergistically with contributions from all the system fields.

Figure 4. WCMS, technological nucleus for GEA's research routine

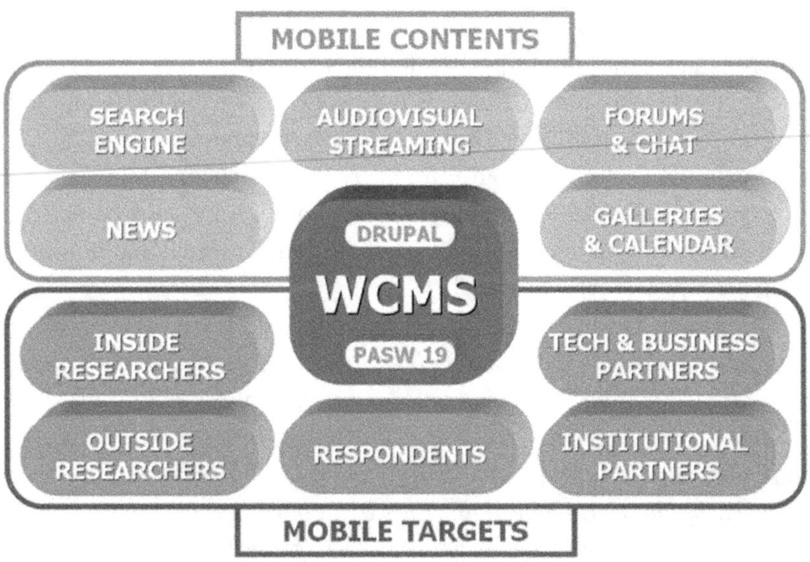

In this regard, Ledo and Castelló (2012: 95) values GEA methodological application of two technological instruments, designed to collect and systematise information in synergetic combination, as key assets: on the one hand, the already mentioned WCMS designed with a versatile superstructure to manage any digital content in its entire chain of value, design, creation, dissemination and interaction. WCMS system design was supported and developed internally as a patented tool for the analytical and taxonomic analysis of the reception of audiovisual media contents. This manager takes advantage of the current potential of multiplatform streaming making use of the growing implementation of the mobile technology throughout the research work.

On the other hand, due to the large amount of data generally involved, a predictive analysis and data management software was implemented: PASW Statistics 19. Why? PASW is proven effectiveness as a processor of reports on non-parametric algorithms or its capacity to correct errors and enable cross-platform collaboration, but its versatility in research design through its innovative automatic potential to adapt other open source programs, such as R and Python.

The troubling question is find aversatile superstructure —at the same time powerful and flexible— to manage any digital content in its entire chain of value, design, creation, dissemination and interaction, could be adapted to the ubiquitous reality. In GEA particular case, the solution calls for current free and open-source content management framework par excellence: Drupal.

Why Drupal…? As is well known, Drupal architecture is based on PAC (Presentation Abstraction Control) which has demonstrated its versatility and polyvalence mobile remote care, with a broad spectrum of applications: geriatric and health care, kindergartens, rescue services… Aligned to the above-mentioned, the GEA priority in the design and development of its ubiquity tools aims to increase an intuitive response capacity of very heterogeneous reception groups: a full-blown example of ubiquitous learning website (Chen, Chang & Wang, 2008).

In this respect, each new tool added was always preceded by an extensive conceptual, ergonomic or procedural testing (beta test), and therefore the modular and layered character of the Drupal was seen from successive stages in the design, development, access and digital interactivity.

Yet the advantages do not end here: as mentioned before, Drupal facilitated the upward integration between very diverse instruments and platforms, as an online WCMS, a predictive analysis and data management software, and a series of ubiquity technology rollouts. But what is most important is that, because of diverse socio-cultural context in each fieldwork, all of the GEA online and mobile tools should be a decisive support for multilingualism, especially with regard to actual research about the role of small cinemas in original version through European Digital Space: while GEA react negatively to prevalence of the monolingualism or the linguistic homogeneity, care must be taken in an EU with 24 official languages —in addition of a myriad of co-official, semi-official and non-official languages— without it becoming a confused tower of Babel:

Let's say it: The linguistic diversity is an indisputable reality in Europe, and the EC must operate with it. It should be noted, in any case, that it is not so much a question of the relative prevalence of the monolingualism or the linguistic homogeneity, but rather the idea of how the multilingualism is a necessary condition for mutual understanding both in the European and global economic context. On the contrary, the very concept raised by the European authorities is a controlled and limited multilingualism, and exactly tailored to the demands of Europe where the diversity is therefore a right, but it is a right that should not be institutionalized (Ledo, López & Castelló, 2014: 14).

For this purpose, specific tasks were designed for best suited to each access profile for single or group users—as inside and outside researchers, respondents, technical, business and institutional partners...—structuring tool processes and the provision of extensive information in different languages. According to GEA researchers Ledo and Castelló (2013: 186), the restricted access to users provided with protected accounts ensures data integrity and security, reducing the potential for error and increasing the reliability of statistics results, after identification in the login page. This research was financed via competitive bidding by the National R&D Plan of the Spanish Ministry of Science and Innovation (CS02009-13702).

In particular with regard to mobile terminals, several profile questions should be simplified by the use of significantly fewer options. It should be noted that, in online case, users had to respond to a group of 16 questions that investigated their user profile: 13 closed, 1 open and 2 mixed questions, and a second block about consumer of 19 closed and 8 mixed questions. In any case, at each data extraction level, users are fully informed of the GEA responsibility to safeguard the confidentiality of the information obtained, using this information squarely and exclusively for ensuring proper research processing, in accordance with an extraordinarily stimulating reflection of Park and Mo Jang (2014) on the many different consequences, or the mechanisms that promoted popular participation (Foth, 2011), or the institutional structures that facilitate student participation and develop model projects (Rius, Clarisó & Masip, 2014).

In this context, it has to be noted that the majority of the cooperating users —from the initial design stage, or Beta test— were active processors, and not only end-user. That is precisely why design of the mobile environment should ensure an intuitive operation by the reception groups which had extraordinarily high and low levels of digital literacy, while ensuring its continued and safe functioning and end-to-end solutions.

Besides, and not less important, the levels of data management were articulated in a treelike way through a series of sections such as news, search engine, galleries, forums, chat, calendar, video/audio streaming and surveys. The case of video/audio streaming would have been particularly problematic in small displays of mobile devices, it becomes necessary to design content for a smaller screen, as well as the different stages of the research work and the main activities involved.

Thus, according to the perspective of the audiovisual analysis adopted by GEA, the methodological evolution of the research included the following stages (Ledo & Castelló, 2012: 96-99):

1. A careful selection of the "unit of audiovisual analysis", to avoid introducing undesirable deviations on reception universes.
2. An effort to simplify the methodological scales used to measure complex variables with multiple indicators: semantic differential scale or bipolar rating scale or Markham's scale, Guttman scale or scalogram analysis in its quadripartite morphology, Likert scale or summative scale and situational scalogram.
3. A training of encoders and pilot testing at the individual and collective levels, to effectively test an application's functionality and usability, optimize the design of the tool, to detect possible errors and to establish the main conceptual and procedural difficulties to answer the questionnaire.
4. An interpretation of data, to ensure the extrapolation of results through four types of reduplication: literal, of operations, instrumental and constructive.

This type of analysis —with the aim of establishing the influence of certain thematic, formal and linguistic variables in the assimilation and acceptance of certain contents— also works towards among the groups as part of the content analysis tradition(Krippendorf, 2004), for evaluating the impact of audiovisual narratives in intercultural relations (Bautista, 2009) and detecting stereotypes and prejudices (Rodríguez, 2011).

To conclude, it would be worth noting some of the main solutions to the problems raised in proactively implementing the latest ubiquitous technologies and features:

1. Ubiquitous technologies make it easier for diverse targets to answer the questionnaires, at any time and in any place, with a very high level of replies of survey respondents.
2. Responding the requirements of the users in real time enhance interactivity with targets, playing a key role within effective research process.
3. In order to streamline the screening and analysis by our reception universe, we limited the length of the film samples to a maximum of fifteen minutes, and selected those that had the largest number of representative items for our research.
4. Improves the design of survey instrument to do not allow of respondents to continue the survey if they do not respond the current question in order to decrease missing responses, or including notes to respondents stating the responses are compulsory.
5. Improves the surveys adding a time control tool to control the time they spend in a question in order to distinguish between its difficulty.
6. Allow the respondents to start the survey and continue in another time for avoid dropout rates.

Project Nr. 4: A Mobile Recommender System for the Gastronomy Domain

The main goal of a tourist interested in gastronomy is to taste excellent food in quality restaurants. As long as the offer could be overwhelming on certain tourist destinations, a recommender system (RS) would be handy to find suitable restaurants and eating places according to each tourist's tastes and personal context. An RS is defined as a tool to provide relevant personalized alternatives to users in large and complex information spaces (Burke et al., 2011). Three main approaches of RS exist (Adomavicius & Tuzhilin, 2005): (1) collaborative-based, (2) content-based, and (3) hybrid. Collaborative-based RSs are the most popular nowadays. They take advantage of the degree of satisfaction previously reported by users in order to predict the satisfaction for a new user faced with an alternative that was not previously experienced. Content-based RSs generate recommendations by comparing the alternative attributes with the user preferences that were learnt beforehand. Finally, hybrid RSs combine collaborative and content-based approaches weighting their contribution depending of the problem at hand. Recently, new paradigms based on complex networks and processes of classical physics have arisen (Lü et al., 2012). The aim of these new approaches is to model the impact of social networks as well as to take into account temporal dynamics.

In the field of gastronomy, the firm GeoWhiz has developed a collaborative-based RS in which the alternatives are previously defined as points of interest (POIs) (Horozov et al., 2006). By collecting previous ratings provided by tourists on each POI, a similarity measure is computed between the target tourist and those tourists who rated each POI. Finally, the POI with higher expected rate for the target user is recommended. A different approach based on Bayesian networks and probabilistic inference is proposed by Park et al. (2007). A dataset is created including information of user profiles, context, and previous experiences. Network parameters and conditional probabilities are then computed. After that, recommendations can be generated by picking the alternative with highest probability.

The gastronomy domain can also be modeled by describing restaurant attributes in a precise way. This was the line followed by Yang and Wang (2009), who developed an algorithm that first searches on the attribute space and then applies attribute constraints to rank alternatives by distance measures.

On the other hand, Internet companies, like Google, have also implemented their own recommender engines. Hotpot by Google (2011) relies on Android OS and Google Maps Services to first gather and then make public tourist opinions about relevant places. User feedback is obtained through appropriate widgets installed on Android OS. Recommendations are computed by aggregating user ratings as well as considering ratings and comments of friends in the social network of the target user.

In order to integrate collaborative-based with location-based approaches in a unified framework, we have developed FieldRec, a recommender engine that supports mobile users by means of algorithms inspired on the electromagnetic field theory. The engine collects the degree of satisfaction of tourists that have previously visited the restaurants of any given city. Then, it gathers the geographic location of the target tourist and computes the distance to restaurants located in the neighborhood. Finally, the algorithms generate recommendations by estimating the attraction field at the location of the target tourist. In what follows, the algorithm as well as the implemented distributed system (engine plus mobile client) is described.

The algorithm is inspired on the concept of electromagnetic field, and concretely on the notion of electric field. Coulomb's Law states that the electrostatic force between two charges varies proportionally to the product of their charges and inversely proportional to the square of the distance between them (Wangness, 1990). The mathematical formulation is as follows:

$$F = k\frac{q_1 q_2}{d^2} \tag{1}$$

where d is the distance between the charges, q_1 and q_2 represent the magnitude of the charges, and k is Coulomb's constant. The electrostatic force is the result of the influence of any given charge q on the space around it. That influence could be quantitatively described by means of the concept of electric field:

$$E = \frac{1}{4\pi\varepsilon_0}\frac{q}{r^2}\hat{r} \tag{2}$$

which is a vector whose magnitude depends on q and r^2, and its direction on the sign of the charge Q.As we are looking for ways to represent the energy of attraction perceived at any given point of space, neither the electrostatic force nor the electric field are suitable concepts. There exists a related concept, the electric potential, which seems to be the best choice for our purposes. It determines the energy that a unitary point charge would have when located at distance r from charge q:

$$V = \frac{1}{4\pi\varepsilon}\frac{q}{r} \tag{3}$$

The electric potential will be the basis of the proposed recommendation algorithm. The algorithm follows the concepts of electric field and electric potential, as it is shown in Figures 5 and 6. Figure 5 shows the electric potential with five negative charges:q_1=-1, q_2=-2, q_3=-3, q_4=-4, q_5= -5, and a positive charge q=1 in the middle. It can be observed that chargeq_5, the highest one, is the one exerting the highest attraction force on charge q. Figure 6 illustrates the application of the electric field concept but using

Figure 5. Electric potential with five negative charges: $q_1 = -1$, $q_2 = -2$, $q_3 = -3$, $q_4 = -4$, $q_5 = -5$; *and one positive charge:* $q = 1$

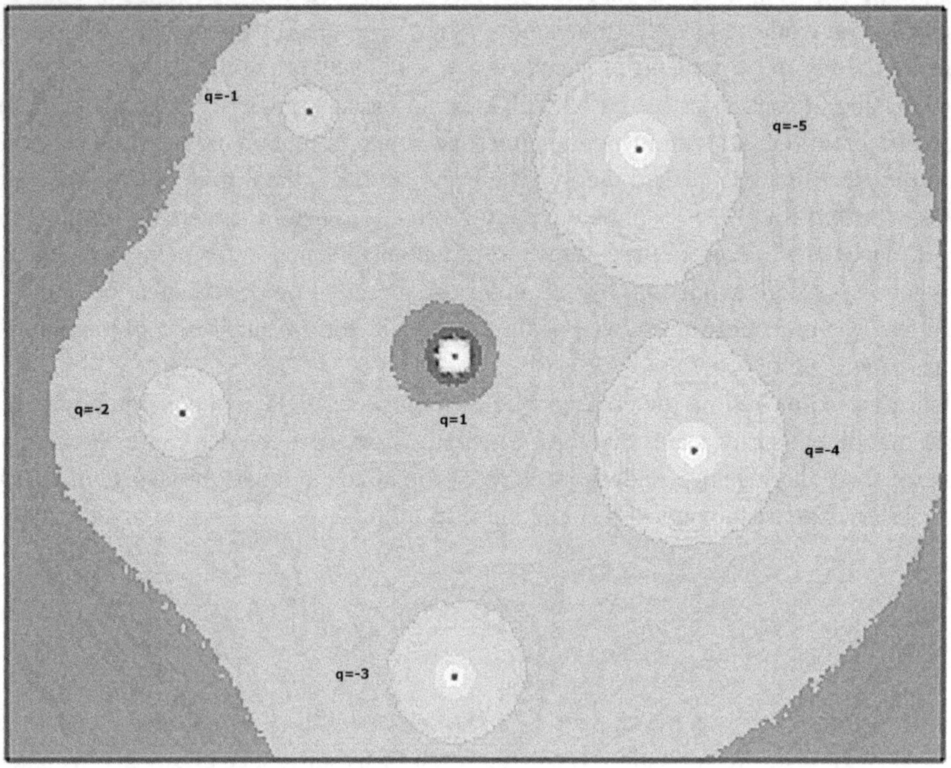

restaurants and users instead of charges. In this example, chargesq_1-$q_5$would represent restaurants, and charge q the target user. Our goal is to choose the recommended restaurant by computing the attraction energy, and its corresponding attraction force, generated by each "charged" restaurant. In our approach, the charge of each restaurant is proportional to the average of the ratings previously provided by other users. In order to compute these charges we have resorted to the data obtained in a real experiment in the context of the Santiago(é)Tapas contest, which was held in year 2011 in the city of Santiago de Compostela in 2011 (Santiago(e)Tapas, 2011).

Simplifying the electric potential equation by letting k=1, we get the following expression:

$$V = \frac{q}{r} = \frac{\text{average of restaurant's ratings}}{\text{distance}} \quad (4)$$

The distance between the target user and each restaurant is required to compute the electric potential generated by each restaurant. To estimate the distance we have used the Gellert's equation (Gellert et al., 1989):

$$d = R \times 2 \arcsin\left(\sqrt{\left(\sin^2\left(\frac{\Delta\Phi}{2}\right) + \cos\cos\phi_5 \cos\cos\phi_f \sin^2\left(\frac{\Delta\lambda}{2}\right)\right)}\right) \quad (5)$$

Figure 6. Map with five restaurant locations and a target user. Each place has been rated before the target user is located on their neighborhood.

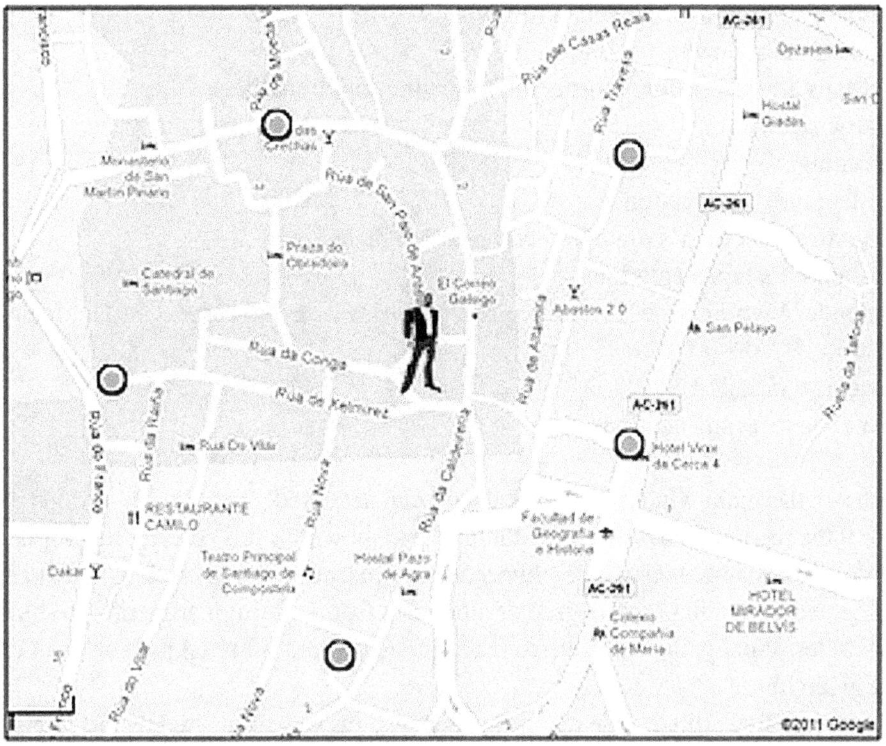

On the basis of these elements, the recommendation algorithm is described as follows:

- First. The data of available restaurants is gathered from a database and its location plotted by using the Google Maps API.
- Second. The location of the target user (latitude and longitude) is obtained from the user's mobile device.
- Third. For each restaurant, the average of ratings previously provided by the users is computed.
- Fourth. The distance between the target user and all available restaurants is estimated.
- Fifth. For each restaurant, the electric potential (attraction potential) at users' location is computed.
- Sixth. A ranking of the restaurants on the basis of their generated attraction potential is elaborated.
- Seventh. The top 10% of the ranking is chosen for recommendation.

The recommendation algorithm was wrapped into a recommender system, which in turn was integrated in the system architecture illustrated in Figure 7. The mobile device, equipped with Android Os, plays the role of the client, while the web services implement the business logic. Two different services were developed: the Recommender Web Service and the Google Maps Web Service.

In what follows, a detailed description of each component is presented:

- Mobile client. It was implemented on Android OS. Its main functionalities are:
    - Restaurant list.
    - Visualization of recommended restaurants.
    - Evaluation of new restaurant.
- Google Maps Service. It implements the following functionalities:
    - Map visualization.
    - Zooming.
    - Restaurant's geolocation.
    - Visualization of car's route to recommended place.
    - Restaurant´s representation.
- Recommender Web Service. It is implemented in Java EE on Tomcat 7. The main tasks of this service are:
    - Save new ratings.
    - Generate recommendations.

Figure 8 shows the main window where the user can access all the relevant options of the application. In Figure 9 the recommendations of restaurants are shown. In this screen, the user is located on its current position, the restaurants in the neighborhood are placed (colored circles) and the recommended ones are highlighted indicating their degree of attraction from red (high attraction) to blue (low attraction) colored circles. Finally, the car route to reach the restaurant with the highest level of attraction is shown with a green line.

The research regarding with mobile recommender systems was sponsored by the Ministry of Science and Innovation of Spain under grant TIN2011-22935.

Figure 7. System architecture including the Recommender System as a Web Service

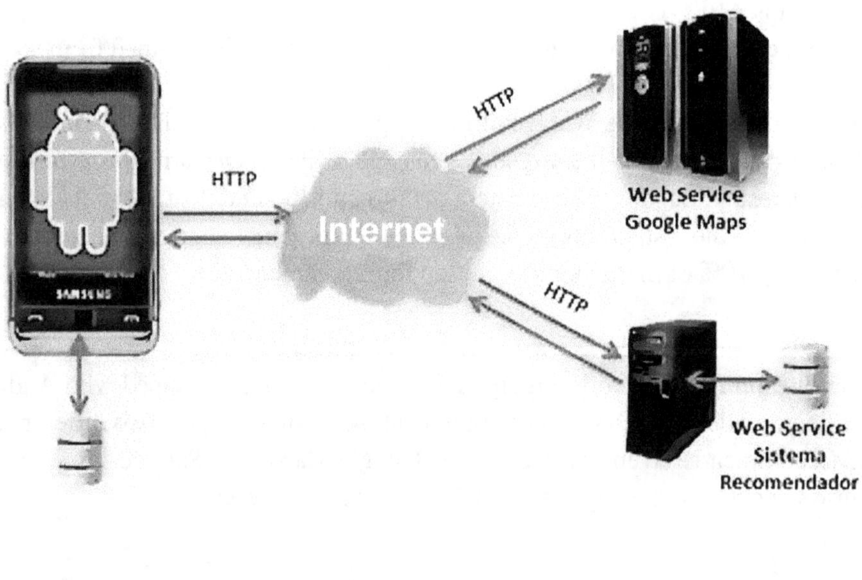

Figure 8. Main window

Figure 9. Recommended places

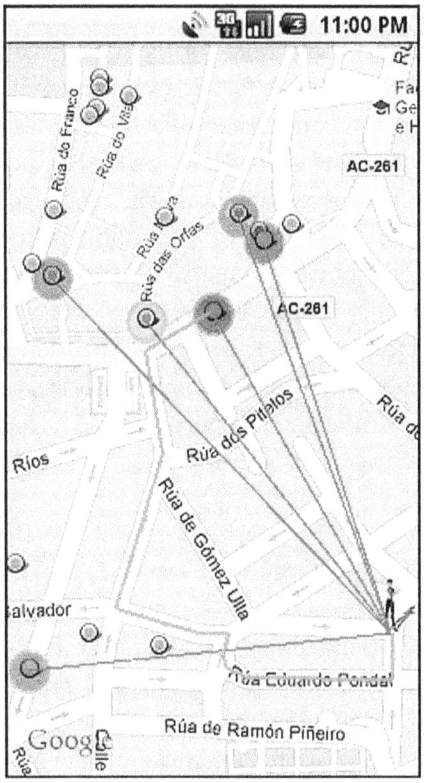

FUTURE RESEARCH DIRECTIONS

It is clear that ubiquitous technologies have come to stay. Every single action that can be solved by a computer, is tending to go mobile. The fact that, nowadays, connected mobile devices have spread reaching a high percentage of people's pockets opens a whole new world of possibilities in the fields of communication and computer science.

The collaboration scheme that CIDUS represents is placed in an unbeatable position in this field as it represents the union of how to present information to people and how to technically make it possible.

The future must make mobile applications more attractive and easier to use for the people. Using them must be fun independently of their objectives and the use of new techniques such as virtual reality (VR) and augmented reality (AR), which are currently booming, will help to achieve this goal and reach a bigger number of users.

Also, new techniques of localization such as Wi-Fi tracking that improve the accuracy of GPS tracking will be very useful in the fields of tourism promotion, gamification, etc. Making it possible to create a whole new type of applications that take advantage of this precision like tourism guides, educational games, telemedicine (controlling the movements of patients), etc.

By way of example, in the field of recommender systems for mobile applications, current trends focus on considering context (social, spatial or temporal) as a key source of relevant variables when generating suggestions for the user. Recent studies demonstrate the impact of context on conditioning user preferences. In fact, relevant research contest are now open looking for efficient algorithms to solve the computational issues related with this topic.

As a second example, following the success of Drupal implementation on GEA research routines, one of the lines of research is devoted to the assessment of Near Field Communications (NFC) potential. In actual fact, the chances of success of this contactless technology depend on so many imponderables, which should not prevent so that Wi(Fi)lite style technology could already be in GEA routines, and could soon be a regular feature of its research work: security procedures to users IDs through contactless authentication to the host server, access to different levels of information according to their responsibilities…

Either way, technological, historical, sociological, cultural and economic changes will ultimately concretise the possible purposes and its materializations to the research community to consider strategic research agendas relevant to social context and to have in promoting our future.

CONCLUSION

Since its foundation, CIDUS actively works with institutional and business networks, developing research, technology and knowledge transfer projects applied to multiple socioeconomic sectors: telemedicine, tourism promotion, distribution of digital cinema in OV, the impact of the ecological footprint, cultural content generation, dissemination and enhancement of historical and artistic heritage, etc.

The Platform also generates entrepreneurial initiatives from the University and develops training activities aimed to satisfy both the demands of the academic curriculum and the continuous education for companies in the digital sector.

As the best verification of its sustainability, it is sufficient to note that, in its still fledgling career, CIDUS has led or participated in more than thirty R + D projects with a collection of resources that far exceeds one million euros.

In the field of mobile virtual museums CIDUS has joined the expertise of communication experts, human-computer interaction experts and historians to achieve the creation of a new type of mobile app that is not only a complement for a museum exhibition but a complete experience in itself. It brings to the users hands the chance to have a similar experience to the one they would have visiting the virtual museum even though they can be far away from the exhibition. Moreover it becomes an advertisement of the monument itself, encouraging people to know more about its history and visit its real emplacement.

This multidisciplinarity has also lead to the creation of a whole new transmedia format for storytelling. It is based on the comic format but it adds new kind of multimedia components to enrich the user experience. It also gives tools for the creators to use the whole potential of the tool adding audio, animations, etc. or just use some components as it better suits his needs. This kind of development would have been impossible if the team that created it hadn't been formed by people from so different fields as arts, computer engineering or communication and this is what CIDUS does, joining together people with very different expertises.

Likewise, another matter for action and investigation aimed at evaluating and reviewing the effectiveness of research methodologies, currently characterized by dispersed location of researchers, multifocal target groups and objects of study, which requires intensive interaction with all relevant tools based on ubiquitous Technologies. In this respect, it should be noted that ubiquitous technologies make it easier the survey process and improving researchers and partners dialogue (any time, any place): and this is without prejudice to the evaluation of each new tool added to find a powerful and flexible superstructure to manage any digital content adapted to the ubiquitous reality in its entire chain of value, design, creation, dissemination and interaction.

The contribution of CIDUS in the field of mobile recommender systems relies on thinking recommendations as a problem made up with entities, charges, and processes, electrical fields. This vision follows the trend initiated some years ago in which other ideas coming from physics have been applied to predict the satisfaction of users with available products.

But the most important conclusion of these years of multidisciplinary research is that unifying efforts and knowledge always leads to richer results and though to a bigger advance and return to society. The difficulties of collaboration that appeared at the beginning between researchers of such different fields have been overpassed, and a change of mentality from the “doing it yourself” to the “someone could enrich this project” has arisen.

The quality and diversity of technological infrastructures available to Cidus platform also facilitated a collaborative work between inside and outside researchers (as well as between technological, business and institutional partners), to develop new methodologies, concepts, tools or work routines of research-creation based on the anticipated case load.

Consequently or as a corollary, the creation of this collaboration network besides the bigger coordination between different fields of knowledge can lead in the future to more important projects that mix two of the most important concepts in society nowadays: Communication and technology.

REFERENCES

Adomavicius, G., & Tuzhilin, A. (2005). Toward the Next Generation of Recommender Systems: A Survey of the State-of-the-Art and Possible Extensions. *IEEE Transactions on Knowledge and Data Engineering*, *17*(6), 734–749. doi:10.1109/TKDE.2005.99

Arai, K., & Herman, T. (2010). Method for Automatic E-Comic Scene Frame Extraction for Reading Comic on Mobile Devices. *Seventh International Conference of Information Technology New Generations (ITNG)* (pp. 370-375). doi:10.1109/ITNG.2010.22

Balez, M. (2011). Personally yours on the go. Retrieved from http://googlemobile.blogspot.com/2010/11/personally-yours-on-go-personalized.html

Bautista, A. (2009). Relaciones interculturales en educación mediadas por narraciones audiovisuales. *Comunicar*, *33*(33), 149–156. doi:10.3916/c33-2009-03-006

Bourdieu, P. (2005). *Capital cultural, escuela y espacio social*. México: Siglo XXI.

Burke, R., Felfernig, A., & Göker, M. H. (2011). Recommender Systems: An overview. *AI Magazine*, *32*(3), 13–18.

Canadian Hexagram-CIAM. (2015). Retrieved from http://www.hexagram.ca/

Centre for Research-Creation in Media Arts and Technologies. (2015). Retrieved from http://hexagram.concordia.ca

Chen, G. D., Chang, C. K., & Wang, C. Y. (2008). Ubiquitous learning website: Scaffold learners by mobile devices with information-aware techniques. *Computers & Education*, *50*(1), 77–90. doi:10.1016/j.compedu.2006.03.004

Chun-Rong, H., Chu-Song, C., & Pau-Choo, C. (2005). Tangible Photorealistic Virtual Museum. *IEE Computer Graphics and Applications*, *25*(1), 15–17. doi:10.1109/MCG.2005.22 PMID:15691165

CIDUS. (2015). Retrieved from: www.usc.es/es/servizos/cidus

CITIUS. (2015). Retrieved from: https://citius.usc.es/Coiro

Colombo, F. (1974). *Televisione: la realtà come spettacolo*. Milano: Bompiani.

Comic Life. (2015). Retrieved from http://plasq.com/products/comiclife3

Conti, M., & Giordano, S. (2014). Mobile ad hoc networking: Milestones, challenges, and new research directions. *Communications Magazine, IEEE*, *52*(1), 85–96. doi:10.1109/MCOM.2014.6710069

De Souza e Silva, A., & Sheller, M. (Eds.). (2014). *Mobility and Locative Media: Mobile Communication in Hybrid Spaces*. New York: Routledge.

El-Hussein, M. O. M., & Cronje, J. C. (2010). Defining Mobile Learning in the Higher Education Landscape. *Journal of Educational Technology & Society*, *13*(3), 12–21.

ETSE. (2015). Retrieved from http://www.usc.es/etse/en

Eye Film Institute. (2015). Retrieved from https://www.eyefilm.nl/en

FCCC. (2015). Retrieved from http://www.usc.es/en/centros/cc_comunicacion/index.html

Featherstone, M. (2009). Ubiquitous media: An introduction. Theory, Culture & Society, 26 (2-3). Nottingham, SAGE & TCS Nottingham Trent University, 1–22.

Fenn, J., & Linden, A. (2005). Gartner's Hype Cycle Special Report for 2005. Retrieved from http://www.gartner.com/DisplayDocument?doc_cd=130115

Foth, M. (Ed.). (2011). *From social butterfly to engaged citizen: Urban informatics, social media, ubiquitous computing, and mobile technology to support citizen engagement*. MIT Press.

Fundación Pedro Barrié de la Maza. (2015). Retrieved from http://www.programacatedral.com

Gavalas, D., Konstantopoulos, C., Mastakas, K., & Pantziou, G. (2014). Mobile Recommender Systems in Tourism. *Journal of Network and Computer Applications*, *39*, 319–333. doi:10.1016/j.jnca.2013.04.006

Gellert, W., Gottwald, S., Hellwich, M., Kästner, H., & Küstner, H. (1989). *La Enciclopedia Concisa de las Matemáticas* (Vol. 2). VNR.

Hexagram UQAM. (2015). Retrieved from http://www.hexagram.uqam.ca/

Hirst, M., Harrison, J., & Mazepa, P. (2014). *Communication and new media: From broadcast to narrowcast*. Don Mills: Oxford University Press.

Horozov, T., Narasimhan, N., & Vasudevan, V. (2006). Using location for personalized POI recommendations in mobile environments. Saint. *International Symposium on Applications and the Internet (SAINT'06)*, 124-129. doi:10.1109/SAINT.2006.55

Hou, H.-T., Wu, S.-Y., Lin, P.-C., Sung, Y.-T., Lin, J.-W., & Chang, K.-E. (2014). A Blended Mobile Learning Environment for Museum Learning. *Journal of Educational Technology & Society*, *17*(2), 207–218.

IIT. (2015). Retrieved from https://www.usc.es/en/institutos/iit/index.html

Knobel, J., Lankshear, M., & Leu, C. D. J. (Eds.). (2014). *Handbook of research on new literacies*. New York: Routledge.

Krippendorf, K. (2004). *Content Analysis: An Introducing to its Methodology*. Beverly Hills: Sage Publications.

LedoAndión, M. &Castelló Mayo, E. (2012). Identification of barriers and motivations in the circulation of audiovisual media contents in the Lusophone space: Communication, interculturality, citizenship, and interactivity. *Communicatio Socialis*, *XXV*(2), 85–116.

LedoAndión, M. &Castelló Mayo, E. (2013). Cultural Diversity across the Networks: The Case of National Cinema. *Comunicar. Scientific Journal of Media Education*, *XX*(40), 183–191.

LedoAndión, M., López Gómez, A., & Castelló Mayo, E. (2014). Creative Destruction as Economic Pattern of Europe's Small-Scale Audiovisual Media Markets and Minority-Language Film Industries. *China-USA Business Review*, XIII(1), 12-19.

Lü, L., Medo, M., Yeung, C. H., Zhang, Y. C., Zhang, Z. K., & Zhou, T. (2012). Recommender Systems. *Physics Reports*, *159*(1), 1–50. doi:10.1016/j.physrep.2012.02.006

Méndez, R., Otero, A., Jarque, S., & Flores, J. (2012). Exploración en tiempo real de la reconstrucción virtual de los instrumentos del Pórtico de la Gloria. *VAR. Virtual Archaeology Review*, *3*, 49–53.

Meyrowitz, J. (1985). *No sense of place: The impact of electronic media on social behavior*. Oxford University Press.

Moreno Sánchez, I. (2002). *Musas y nuevas tecnologías: el relato hipermedia*. Barcelona: Paidós.

Oliver, M. B., & Raney, A. A. (Eds.). (2014). *Media and Social Life*. New York: Routledge.

Otero, A., Méndez, R., & Flores, J. (2012). Pórtico de la Gloria Virtual, aplicación interactiva en tiempo real para la exploración virtual del Pórtico de la Gloria. *VAR. Virtual Archaeology Review*, *3*, 18–25.

Park, M., Hong, J., & Cho, S. (2007). Location-Based Recommendation System Using Bayesian User's Preference Model in Mobile Devices. In Ubiquitous Intelligence and Computing (pp. 1130-1139). Springer Berlin Heidelberg. doi:10.1007/978-3-540-73549-6_110

Park, Y. J., & Mo Jang, S. (2014). Understanding privacy knowledge and skill in mobile communication. *Computers in Human Behavior*, *38*, 296–303. doi:10.1016/j.chb.2014.05.041

Quotidiania Delirante. (2015). Retrieved from http://itunes.apple.com/es/app/quotidiania-delirante-1-imotion/id356356294?mt=8

Rheingold, H. (2004). Mobile media and political collective action. In J. Katz (Ed.), *The handbook of mobile communication studies*. Cambridge, MA: MIT Press.

Rius, À., Clarisó, R., & Masip, D. (2014). Student projects empowering mobile learning in higher education. *RUSC. Universities and Knowledge Society Journal*, *11*(1), 192–207. doi:10.7238/rusc.v11i1.1901

Roccetti, M., Marfia, G., Varni, A., & Zanichelli, M. (2013). How to Outreach the External World from a Museum: The Case of the Marsili's Spirit App. *Arts and Technology Lecture Notes of the Institute for Computer Sciences. Social Informatics and Telecommunications Engineering*, *116*, 25–32.

Rodríguez, M. P. (2011). Imágenes del otro. Identidad e inmigración en la literatura y el cine. *Journal of Spanish Cultural Studies*, *3*(3), 375–377. doi:10.1080/14636204.2011.669256

Roschelle, J. (2003). Keynote paper: Unlocking the learning value of wireless mobile devices. *Journal of Computer Assisted Learning*, *19*(3), 260–272. doi:10.1046/j.0266-4909.2003.00028.x

Santiago(e)Tapas Contest. (2011). Retrieved from: http://www.santiagoetapas.com

Strip Generator. (2015). Retrieved from http://stripgenerator.com

Toondoo. (2015). Retrieved from http://www.toondoo.com

Wangsness, R. (1986). *Electromagnetic Fields*. Wiley.

Wilson, S., & McCarthy, G. (2010). The mobile university: From the library to the campus. *RSR. Reference Services Review*, *38*(2), 214–232. doi:10.1108/00907321011044990

Xiong, Y., Wang, H., & Li, Y. (2014). Research on Mobile Communication Based on Multimedia Technology. *Proceedings of the 9th International Symposium on Linear Drives for Industry Applications* (Vol. 4, pp. 677-684). doi:10.1007/978-3-642-40640-9_85

Yang, F., & Wang, Z. M. (2009). A mobile location-based information recommendation system based on GPS and WEB2.0 services. *Journal WSEAS Transactions on Computers*, *8*(4), 725–734.

ADDITIONAL READING

Adomavicius, G., & Tuzhilin, A. (2005). Toward the Next Generation of Recommender Systems: A Survey of the State-of-the-Art and Possible Extensions. *IEEE Transactions on Knowledge and Data Engineering*, *17*(6), 734–749. doi:10.1109/TKDE.2005.99

Ally, M. (Ed.). (2009). *Mobile learning: Transforming the delivery of education and training*. Athabasca University Press.

Anderson, E. F., McLoughlin, L., Liarokapis, F., Peters, C., Petridis, P., & de Freitas, S. (2010). Developing serious games for cultural heritage: A state-of-the-art review. *Virtual Reality (Waltham Cross)*, *14*(4), 225–275. doi:10.1007/s10055-010-0177-3

Besora, I., Brunet, P., Chical, A., Morales, D., & Moyes, J. (2008). Real-Time Exploration of the Virtual Reconstruction of the Entrance of the Ripoll Monastery. *Proceedings of the CEIG, 08*, 1–6.

Bruno, F., Bruno, S., De Sensi, G., Luchi, M., Mancuso, S., & Muzzupappa, M. (2010). From 3D reconstruction to virtual reality: A complete methodology for digital archaeological exhibition. *Journal of Cultural Heritage, 11*(1), 42–49. doi:10.1016/j.culher.2009.02.006

Burke, R., Felfernig, A., & Göker, M. H. (2011). Recommender Systems: An overview. *AI Magazine, 32*(3), 13–18.

Castelló, E. (2012). Cidus, de la integración digital a la consiliencia. *Estudiossobre el MensajePeriodístico, 18*, 233–242.

Cheon, J., Lee, S., Crooks, S. M., & Song, J. (2012). An investigation of mobile learning readiness in higher education based on the theory of planned behavior. *Computers & Education, 59*(3), 1054–1064. doi:10.1016/j.compedu.2012.04.015

Chinnery, G. M. (2006). Emerging technologies. Going to the mall: Mobile assisted language learning. *Language Learning & Technology, 10*(1), 9–16.

Crook, C., & Barrowcliff, D. (2001). Ubiquitous computing on campus: Patterns of engagement by university students. *International Journal of Human-Computer Interaction, 13*(2), 245–256. doi:10.1207/S15327590IJHC1302_9

David, K., Bieling, G., Bohnstedt, D., Jandt, S., Ohly, S., & Robnagel, A. et al. (2014). Balancing the Online Life: Mobile Usage Scenarios and Strategies for a New Communication Paradigm. *Vehicular Technology Magazine, IEEE, 9*(3), 72–79. doi:10.1109/MVT.2014.2333763

Goodall, S., Lewis, P. H., Martinez, K., Sinclair, P. A. S., Giorgini, F., & Addis, M. J. et al. (2004): SCULPTEUR: Multimedia Retrieval for Museums. *Proc. of the International Conference Image and Video Retrieval CIVR* (pp. 638-646). doi:10.1007/978-3-540-27814-6_74

Herrington, J., Herrington, A., Mantei, J., Olney, I., & Ferry, B. (2009). *New technologies, new pedagogies: Mobile learning in higher education.* Wollongong: University of Wollongong.

Humphreys, L., Von Pape, T., & Karnowski, V. (2013). Evolving mobile media: Uses and conceptualizations of the mobile Internet. *Journal of Computer-Mediated Communication, 18*(4), 491–507. doi:10.1111/jcc4.12019

Hwang, G. J., Tsai, C. C., & Yang, S. J. (2008). Criteria, Strategies and Research Issues of Context-Aware Ubiquitous Learning. *Journal of Educational Technology & Society, 11*(2), 81–91.

Hwang, G. J., & Wu, P. H. (2014). Applications, impacts and trends of mobile technology–enhanced learning: A review of 2008–2012 publications in selected SSCI journals. *International Journal of Mobile Learning and Organisation, 8*(2), 83–95. doi:10.1504/IJMLO.2014.062346

Järvelä, S., Näykki, P., Laru, J., & Luokkanen, T. (2007). Structuring and regulating collaborative learning in higher education with wireless networks and mobile tools. *Journal of Educational Technology & Society, 10*(4), 71–79.

Jones, R. (Ed.). (2014). *Discourse and creativity*. Routledge.

Jones, V., & Jo, J. H. (2004). Ubiquitous learning environment: An adaptive teaching system using ubiquitous technology. *Proceedings of Beyond the comfort zone: the 21st ASCILITE Conference* (Vol. 468, p. 474).

King, L. J., Gardner-McCune, C., Vargas, P., & Jimenez, Y. (2014). Re-discovering and re-creating African American historical accounts through mobile apps: The role of mobile technology in history education. *Journal of Social Studies Research*, *38*(3), 173–188. doi:10.1016/j.jssr.2013.12.005

Liu, C., Zhu, Q., Holroyd, K. A., & Seng, E. K. (2011). Status and trends of mobile-health applications for iOS devices: A developer's perspective. *Journal of Systems and Software*, *84*(11), 2022–2033. doi:10.1016/j.jss.2011.06.049

Looi, C. K., Seow, P., Zhang, B., So, H. J., Chen, W., & Wong, L. H. (2010). Leveraging mobile technology for sustainable seamless learning: A research agenda. *British Journal of Educational Technology*, *41*(2), 154–169. doi:10.1111/j.1467-8535.2008.00912.x

Lü, L., Medo, M., Yeung, C. H., Zhang, Y. C., Zhang, Z. K., & Zhou, T. (2012). Recommender Systems. *Physics Reports*, *159*(1), 1–50. doi:10.1016/j.physrep.2012.02.006

Mata, F., Claramunt, C., & Juarez, A. (2011): An Experimental Virtual Museum Based on Augmented Reality and Navigation. *Proceedings of the 19th ACM SIGSPATIAL International Conference on Advances in Geographic Information Systems* (pp. 497-500). doi:10.1145/2093973.2094058

Mcconatha, D., Praul, M., & Lynch, M. J. (2008). Mobile Learning in Higher Education: An Empirical Assessment of a New Educational Tool. *Online Submission, 7*(3).

Mortara, M., Catalano, C. E., Bellotti, F., Fiucci, G., Houry-Panchetti, M., & Petridis, P. (2014). Learning cultural heritage by serious games. *Journal of Cultural Heritage*, *15*(3), 318–325. doi:10.1016/j.culher.2013.04.004

Mörtberg, C., Bratteteig, T., Wagner, I., Stuedahl, D., & Morrison, A. (2010). Methods that matter in digital design research. In *Exploring digital design* (pp. 105–144). Springer London. doi:10.1007/978-1-84996-223-0_4

Motiwalla, L. F. (2007). Mobile learning: A framework and evaluation. *Computers & Education*, *49*(3), 581–596. doi:10.1016/j.compedu.2005.10.011

Noguera, J. M., Gutiérrez, M. V., Castillo, J. C., & Segura, R. J. (2012). Arqueología Virtual en Dispositivos Móviles. Un Caso Práctico: Patrimonio Defensivo Medieval. *Virtual Archaeology Review*, *3*(7), 109–115.

Peng, H., Su, Y. J., Chou, C., & Tsai, C. C. (2009). Ubiquitous knowledge construction: Mobile learning re-defined and a conceptual framework. *Innovations in Education and Teaching International*, *46*(2), 171–183. doi:10.1080/14703290902843828

Rau, P. L. P., Gao, Q., & Wu, L. M. (2008). Using mobile communication technology in high school education: Motivation, pressure, and learning performance. *Computers & Education*, *50*(1), 1–22. doi:10.1016/j.compedu.2006.03.008

Roschelle, J., &Sharples, M. (2010). Guest editorial: Special issue on mobile and ubiquitous technologies for learning. *IEEE Transactions on Learning Technologies, 3*(1), 0004-5.

Rossing, J. P., Miller, W. M., Cecil, A. K., & Stamper, S. E. (2012). iLearning: The Future of Higher Education? Student Perceptions on Learning with Mobile Tablets. *Journal of the Scholarship of Teaching and Learning*, *12*(2), 1–26.

Rua, H., & Alvito, P. (2011). Living the past: 3D models, virtual reality and game engines as tools for supporting archaeology and the reconstruction of cultural heritage – the case-study of the Roman villa of Casal de Freiria. *Journal of Archaeological Science*, *38*(12), 3296–3308. doi:10.1016/j.jas.2011.07.015

Schreiber, T. (2013). Information Literacy, Mobile Technologies and Study Practices. In *Worldwide Commonalities and Challenges in Information Literacy Research and Practice* (pp. 317–323). Springer International Publishing. doi:10.1007/978-3-319-03919-0_41

Sharples, M. (2002). Disruptive devices: Mobile technology for conversational learning. *International Journal of Continuing Engineering Education and Lifelong Learning*, *12*(5), 504–520. doi:10.1504/IJCEELL.2002.002148

Shih, J. L., Chu, H. C., Hwang, G. J., & Kinshuk, . (2011). An investigation of attitudes of students and teachers about participating in a context-aware ubiquitous learning activity. *British Journal of Educational Technology*, *42*(3), 373–394. doi:10.1111/j.1467-8535.2009.01020.x

Styliania, S., Fotisb, L., Kostasa, K., & Petrosa, P. (2009). Virtual museums, a survey and some issues for consideration. *Journal of Cultural Heritage*, *10*(4), 520–528. doi:10.1016/j.culher.2009.03.003

Sylaioua, S., Maniab, K., Karoulisa, A., & Whitec, M. (2010). Exploring the relationship between presence and enjoyment in a virtual museum. *International Journal of Human-Computer Studies*, *68*(5), 243–253. doi:10.1016/j.ijhcs.2009.11.002

Taipale, S., & Fortunati, L. (2014). Capturing methodological trends in mobile communication studies. *Information Communication and Society*, *17*(5), 627–642. doi:10.1080/1369118X.2013.862562

Vodanovich, S., Sundaram, D., & Myers, M. (2010). Research Commentary-Digital Natives and Ubiquitous Information Systems. *Information Systems Research*, *21*(4), 711–723. doi:10.1287/isre.1100.0324

Wang, D., Xiang, Z., & Fesenmaier, D. R. (2014). Adapting to the mobile world: A model of smartphone use. *Annals of Tourism Research*, *48*, 11–26. doi:10.1016/j.annals.2014.04.008

KEY TERMS AND DEFINITIONS

CIDUS: Platform for Innovation and Experimentation and Innovation in Digital Contents (www.usc.es/es/Servizos/cidus) created in 2009 by the University of Santiago de Compostela, between eighty-one reference researchers in the field of digital content. Truly transdisciplinary research platform, suitable for the development of integral projects of excellence in R&D, technology transfer and professional training in all phases of the digital contents evolution cycle.

Cooperative and Multicentric Research: Engaging with students and researchers worldwide, without overarching perspective or eurocentric prevalence.

Drupal: Free and open-source content management framework. Its architecture, based on PAC (Presentation Abstraction Control), has demonstrated its versatility and polyvalence mobile remote care, with a broad spectrum of applications: geriatric and health care, kindergartens, rescue services, etc.

Multifocal Target Groups: Research methodology focalised on prospective targets tested in different places around the world, with very different sociodemographic characteristics, however with the same object of study.

Predictive Analytics SoftWare(PASW): Predictive analysis and data management software with a high degree of efficiency as processor of reports on non-parametric algorithms or its capacity to correct errors and enable cross-platform collaboration, but its versatility in research design through its innovative automatic potential to adapt other open source programs, such as R and Python.

Recommender System (RS): Tool to provide relevant personalized alternatives to users in large and complex information spaces.

Transmedia Formats: Aggregation of a variety of formats to create a whole new narrative structure.

Ubiquitous Research WCMS: Web Content Management System responsible for managing a centralized database, supports the online platform for surveys, different access profiles, discussion forums and the samples of analysis, as well as the extraction of the resulting data. This kind of WCMS takes advantage of the current potential of multiplatform streaming making use of the growing implementation of the mobile technology throughout the research work.

Virtual Museums: Digital entity that substitutes, complements, enhances, or augments the traditional museum experience through personalization, interactivity and richness of content.

Chapter 16
Mobile Content and Walking Documentary:
Teaching and Learning Science Step-by-Step with Smartphones

Frédéric Adam
supercluster.eu, Spain

ABSTRACT

This chapter presents creative strategies to teach Science using Smartphones and focuses on projects by researchers and teachers from the international "Big History" network. They teach scientific based knowledge from a cosmic point of view starting from the theory of the Big Bang. A large number of contributors believe that connecting learning to the physical, natural world is the key to stimulating the interest of students before studying the theory in the classroom. This is where Mobile Content is invaluable. A general selection of interesting outdoor learning projects are presented, together with experimental, hypermedia creations achieved by the author in collaboration with scientists. The original concept of a "Walking Documentary" is advanced as a way for students to experience large scales of space and time by walking with a smartphone synchronized with every step of the walker. Finally, some emerging technologies are mentioned, announcing new ways to teach and learn Science with technology.

INTRODUCTION

The First Step on the Moon: A New Scientific Perspective for Humankind

On July 20, 1969, the first human step made on the lunar surface was transmitted to a worldwide audience. This incredible technical achievement was to have very deep consequences in the imagination of humankind for decades. For the first time we saw the Earth from the outside and the images broadcast on TV will be forever engraved in the minds of millions of people.

DOI: 10.4018/978-1-4666-8838-4.ch016

The most important fact was not brought back in 25 kilos of lunar rock but was planted, like a seed, in the collective imagination. This profound experience was acting, as Richard Dawkins (2012) describes in his book *The Selfish Gene* as a "Meme," a process of viral information transmitted from mind to mind like a shockwave. This stunning technological success by the space industry was followed by another major revelation explained in 1990 by the scientist Carl Sagan. Deeply inspired by a picture of the Earth taken by the space probe *Voyager 1*, he describes the Earth taken from a distance of 6 billion kilometers as a tiny "pale blue dot" lost in the dark. This insignificant 0.12 pixel sized dot is our home. This radical point of view of Earth, compared to the size of the solar system will also mark a new perspective in the way we think and experience our position in the Cosmos (Sagan, 1994). These two facts from 1969 and 1990 will have deep consequences in a generation of scientists born in the 20th century. One of the major repercussions of this phenomenon is the theory advanced by James Lovelock in 1970, with the help of Lynn Margulis, of a living Earth. In which they describe the Earth as a single living organism composed of millions of interconnected species of plants, fungi and animals in equilibrium and where humans are just another form of life. The consequences of this new perspective are still expanding, the shockwave is becoming a large and long wave of energy, a necessary tsunami of consciousness, if we want to preserve the pale blue dot.

Today several key fields of knowledge are influenced by the exploration of the Cosmos. Probably the most important knowledge is that we live on a finite Earth with a limited number of resources, and it is absolutely necessary to preserve them. This redefines how we understand ourselves and the place of humanity in the story of space and time. Several scientists like Fred Spier, Rich Blundell, Elisabeth Sahtouris and Lynn Margulis, among others, understood that teaching Science today is about explaining things in a expanded space-time context and in an inter-disciplinary way. Lynn Margulis (1998) was able to call into question the most established concepts of evolution, arguing that multicellular organisms are the results of symbiogenesis, that is a form of collaboration between single cells sharing their DNA (Eukaroytes). This point of view gives more weight to collaboration than competition in the process of natural selection. Collaboration could be the key to humanity's future, and the survival of the "pale blue dot".

The new worldview of Earth as a self-regulating complex system has been embedded in our culture since the 19th century. The Russian scientist Vladimir Vernadsky was the first one in 1926 to coin the term "Biosphere" meaning all living things, bacteria, multi-celled organisms, animals, plants, and fish are all part of a single living system. Vernadsky (1926) identified Homo sapiens as an intelligent life form, who with our minds, created the "Noosphere" (minds sphere) distinctly separate from the "Biosphere." This idea of a global human intelligence network will be retaken from another angle a few years later by the Christian theologian Pierre Teilhard de Chardin (1955). His concept of a single nervous system composed of all human beings, creating a common thinking layer has now seen its formation in the World Wide Web. It is the hybridization of the "Noosphere" with the "Technosphere".

The biologist Elisabet Sahtouris explained in a very nice metaphor called "The Butterfly Story" what open minded scientists are trying to communicate to the new generation. In Sahtouris' metaphor humanity is symbolized by a caterpillar eating three hundred times its own weight in a single day, and metamorphosing into a completely new and distinct life form, a butterfly. Sahtouris suggests that it's time for humanity to convert itself in a distinctly new life form, considering that we already have all the information, knowledge and technology to do so (Sahtouris, 2000). This important message is embedded in the content of some interesting audiovisual and multimedia experiments described below.

A New Orbital Point of View for the Creation and Distribution of Media

The landing on the moon in 1969 was the most significant live TV broadcast event in the history of television at that time. Obviously, the live transmission of the first step on the moon would not have existed without the invention of video and the ability to capture, in real time, the images of the crew. Gene Youngblood (1970) is a theorist of media arts and politics and in the early 70's, he took the ideas of Pierre Teilhard de Chardin and applied the vision of the human network as a decentralized system of production and distribution, to media. He was one of the first to consider the video as a form of art because of the democratization of this medium as a portable audiovisual recording technology for everyone. This visionary understanding of the future of the distribution of audiovisual media by the people announcing the Internet peer to peer technology, and the social networks in the digital age coincided with the moon landing and the video retransmission. He announced a deep change in the field of Cinema envisioning the end of what he calls the "perceptual imperialism", the unilateral distribution of formatted content to the public through movie theaters. A new generation of video artist from the 70's tried to apply the idea of a decentralized distribution of independent movies and documentaries. Today, we can see that this strategy didn't work, there was no real powerful parallel decentralized system of distribution of videos, fictions and documentaries, no parallel social environment for autonomous production and distribution of media capable of competing with the distribution chain of the movie industry[1]. But the idea was in the air and the fusion of the Technosphere and the Noosphere was running forward. Today the technical possibilities offered by the Internet and the telecommunication technology are endless and the production of audiovisual media by the public is continually increasing. The rise of the Smartphone as a micro-computer, a million times more powerful than the *Apollo 11* onboard computer[2], now in the pockets of billions of people is a great opportunity to rethink the relation between the audiovisual production, the social networks, and our understanding of Earth as a complex evolving ecosystem.

We will explore a range of audiovisual and multimedia practices from scientists, teachers and researchers, attempting to teach how Earth works by using mobile technology and non-fictional storytelling. We will also describe projects based on the use of Smartphones as real time environmental sensors potentially combined with mobile learning apps. All the mobile projects described have a common theme, the search for ways to perceive our position in time and space, a new kind of "outerspace", designed to be experienced outdoors. We will especially speak about two major projects, the *Walk Through Time* and *Cosmosis1* I have been working with scientists and members of the *Big History*[3] international network. My innovative contribution as Art Director of these two projects is based on my previous experience working on the idea to synchronize smartphone contents with the walking steps using the pedometer features of the smartphone. Finally we will analyze some experimental practices involving a complex set of innovative resources like the *Google Glass* and wearable sensors as upcoming technology for emerging teaching practices. The "Walking Documentary" terminology will be potentially the best way to define an audiovisual, interactive, scientific experience. We will see how teaching and learning by walking is a key component for future projects in the context of *Big History*.

BACKGROUND

The famous professor of communication, Henry Jenkins, in his report for the MacArthur Foundation, *Confronting the Challenges of Participatory Culture* states that according to a 2005 study conducted by the *Internet & American Life* project, 57% of teens who use the Internet could be regarded as media creators. He describes how educators can teach a number of fundamental skills with digital technology to better integrate society. The new skills include play, performance, simulation, appropriation or multitasking and more. This communication reveals without any doubt that the education system is facing major changes, students are not only receptors, but need to learn to be emitters (Jenkins, 2009). The integration of technology in the classroom is a challenge, which has to be taken very seriously. It's more than just an opportunity to produce media, it is a new way to conceive the exploration of the environment, and get access to knowledge in real physical places, expanding the educational process in time and space. This new perspective of outdoor teaching can be achieved thanks to the exploration of the physical environment with Earth science courses like geology, physics or biology and several other disciplines. This opportunity for expanded learning is the subject of this chapter, investigating creative opportunities for the communication of Earth science with mobile technology. Several of the projects mentioned below have been presented during the *International Big History Association* (IBHA)[4] events in the US since 2012.

There is some controversy within the Big History network about the use of the term "Story" or "History" for the description of the knowledge transmitted by teachers and scientists. This fuzzy frontier, between what is a story and what is history, has to be taken into consideration when creating interactive audio-visual content on digital devices. Sharing and contextualizing scientific information in the outdoors is influenced by the point of view of the user in time and space. Part of the work for the creator of digital content on a mobile device relative to a physical place must take into consideration the way we perceive the world through our unique organs of perception. What we see through our technology and the filters of our beliefs can't be declared as an unchanging description or perception. This should be taken into account for any digital content designed to communicate a view of our world and outer space.

The projects at the University of Murcia are listed on the website *Lab de Creación Hipermedia* (um.es/arisgames/). Other projects from the author, and a database of creative locative media and mobile learning projects are listed on the *GPS Museum* portal (gpsmuseum.eu).

COMMUNICATING EARTH SCIENCE WITH DIGITAL TECHNOLOGY IN THE OUTDOORS

The Early Times of Science Cloud Computing and Social Participation

In 1970, NASA launched the *SETI* project, - the Search for Extraterrestrial Intelligence – Using the very large radio telescope at the Arecibo Observatory in Puerto Rico they started to search outer space for signals. Even though the likelihood of detecting signals was slim, Carl Sagan strongly supported the project. In 1999 the *SETI@home*[5] project was created, based on the idea of using the interest of amateur personal computer enthusiast to search through millions of pieces of data collected from the radio telescope. Software was created to enable any person interested in processing the data. Over 5.2 million participants, worldwide aggregated 2 million years of computing time in just a few years. If Carl Sagan saved the *SETI* project in 1982 after being removed from the NASA budget in 1981, it is probably not because he believed in finding extraterrestrial signals, it is because it was a great project to keep in mind that we are not the center of the Universe and to reinforce the idea that we are all in the tiny spaceship Earth, ready for a global collaboration and it worked! Thanks to this project the participation of millions of people gathered to create the biggest computer project ever done in the history. This project pinpointed the potential of public Cloud Computing and how human networks can work for a common goal. With the help of computers, telecommunication technology and satellites we can understanding and learn how to preserve the environment. As of today, the *SETI* project has not found any positive signs of intelligent life on other planets, but it has shown the great power to process and exchange scientific data on a global scale.

The Smartphone, Sits between the Internet and the Satellites

Approximately 3000 satellites are orbiting Earth each day and night, capturing information from the Cosmos but also gathering data from Earth. In the book *Observer la Terre depuis l'Espace, enjeu des données spatiales pour la société* the French authors Dubois, Avignon and Escudier (2012) explain clearly how some international emergency protocols permit the synchronization of data from private and public satellites to produce, in a few hours, a large amount of public environmental data about a specific area of the territory. Satellites are able to read in a broad spectrum of frequencies from X-ray, UltraViolet, Visible Light or Infrared and beyond, to detect the composition of the atmosphere, the surface and the subsurface of Earth. They also share valuable information, including our spatial coordinates, by triangulation between several satellites and our smartphone or GPS. The smartphone sits in between the network of home computers, connected to the Internet and the satellites, and is able to catch information from both sources of data, making this piece of technology one of the most powerful devices we can carry. Because of its mobility, smartphones are the best technological vehicle to experience the world by the exploration of nature, studying directly the physical elements with the relevant information downloaded from the network. But the experience of teaching Earth Science with digital technology doesn't start with using a smartphone, not at all. The common way to teach science is to explore the Internet, because it is the easiest way to get access to the largest database of information in the classroom. We will not describe here the infinite list of great learning content we can find on the Internet. Every day a large number of people publish audiovisual information for teaching Earth Science. The common access to the information is through a computer in a classroom because we are teaching the outside from the inside, favoring the theory over the experience, disconnected from the environment. The American philosopher and educational reformer John Dewey (1938), in the 19th century, built on the teaching philosophy of the day by emphasizing the idea that education and learning are social and interactive processes, engaging students in practical, high quality outdoor, experiences. Despite the fact that a large majority of public and private schools are unconcerned by the lack of outdoor education with mobile technology, a great number of experiments are going on all over the world.

Contextual Learning of Earth Science with Smartphones

The Global Positioning System "GPS" is a navigation system created by the US army in the early 70's, a network of more than 20 GPS satellites designed to control the location and trajectory of planes and missiles. It's because of the high precision of the technology, below 1 meter, that the system was deliberately degraded to 150 meters accuracy for public access. This was to deny our potential enemies from using the system. It was impossible to know our exact coordinates until the year 2000 when the full capabilities of the GPS system was available for civil purposes. Thanks to the liberation of the precision, a group of hikers created, in 2001, one of the most exciting locative games called "Geocaching," which is played by millions of people today. It's about hiding a box somewhere and sharing the coordinates through the

Internet or a smartphone app. Walkers find it and exchange objects inside the box, it's an adaptation of the typical "Treasure Hunt" game with an environmental and social value. There is a large variety of wonderful modalities, like the community of Earth Caching, that "let the Earth be our teacher!"[6]. The Earth Cache application pulls the walkers to a special place to unfold the story from the box. It can be text, pictures, objects, or anything useful to tell the geologic story of the place. The GPS is simply used to direct people to the right place and learn something captivating about the environment.

There is a long tradition of naturalists collecting information from outside the laboratory, in the field research, especially since the 17th century. The naturalist is passionate about the exploration of the environment, collecting information and samples from a large array of sources, an interdisciplinary approach combining Biology, Botany, Zoology, Mineralogy, Astronomy, and other disciplines. The place to represent the field research before the 19th century was called the "Cabinet de Curiosité", a room full of mixed objects of any kind, a miniature museum of curiosities. When arriving at a place the next step for students and teachers is to explore and collect information. That's what the project *NOAH*[7] (Networked Organisms and Habitats) is about. Any teacher can create a account on the website and set up a working group in a specific area, with the aim of creating an outdoor pedagogic mission for kids, what they call a "Bioblitz"[8]. A smartphone app enables the kids to geolocate species, take pictures, write a description and share them with the community. Because of the very large number of species on Earth, it is very hard to know the name of the observed specie. The community of citizen scientists, teachers and students are potentially there to put the right name on the pictures. Project *NOAH* is supported by *National Geographic*, gaining a great number of users each day. It is surely a solid model to follow and in a certain way comparable with the *SETI@Home* project in the sense that a group of users are investigating and collaborating to improve the results of their research (See Figure 1).

The project *NOAH* mobile app is basically the combined use of the GPS function, the camera, text writing and social networking of the smartphone. Any information produced during the process of exploration is controlled and generated by the user. But what if the camera itself could help users analyze the images? Pattern recognition is very well developed today and several software packages analyze and recognize forms or faces. This technology is tested and already working for security and control purposes in airports or at cash machines. In 2010 *Google* created the image recognition mobile application *Google Goggles*[9], it is based on text recognition and on the comparison between a photo taken by the user and a database of images from the *Google Search Engine*.This technology is quite useless for a field research trip on a natural trail. There's no way of recognizing life forms like an insect because there is no exact picture in the online database fitting the exact position, form or color. It can only be an approximation of the subject. Developers at the universities of Columbia and Maryland have produced an algorithm to help recognize species of trees. The project is called *Leaf Snap*[10] .After taking a picture of a leaf, on a white background, the smartphone app compares the contour of the leaf with the 184 species of trees in the database and identifies the name of the specie (See Figure 2).

The *Leaf Snap* app also locates where the leaves can be found on a map, which is potentially a powerful extension for the project *NOAH* application. The combination of the geo-coordinates and the recognition of the environment, by the camera of the smartphone, is very close to what we call Augmented Reality technology. The A.R technology pastes on top of a picture of the physical world a digital layer of information. This can be a text, a sound, a 3D object or any digital media we can imagine. The recognition of the geometry of space allows the object to follow the angle of rotation of the screen. A good example is the astronomy app that uses the terrestrial coordinates and the rotation of Earth to show the exact position of celestial objects. The app *Celeste*[11] detects and follows the trajectory of planets in our solar

Figure 1. The Project Noah by the Columbia University, the University of Maryland, and the Smithsonian Institution

Figure 2. Screens of the app Leafsnap by researchers from Columbia University, the University of Maryland, and the Smithsonian Institution

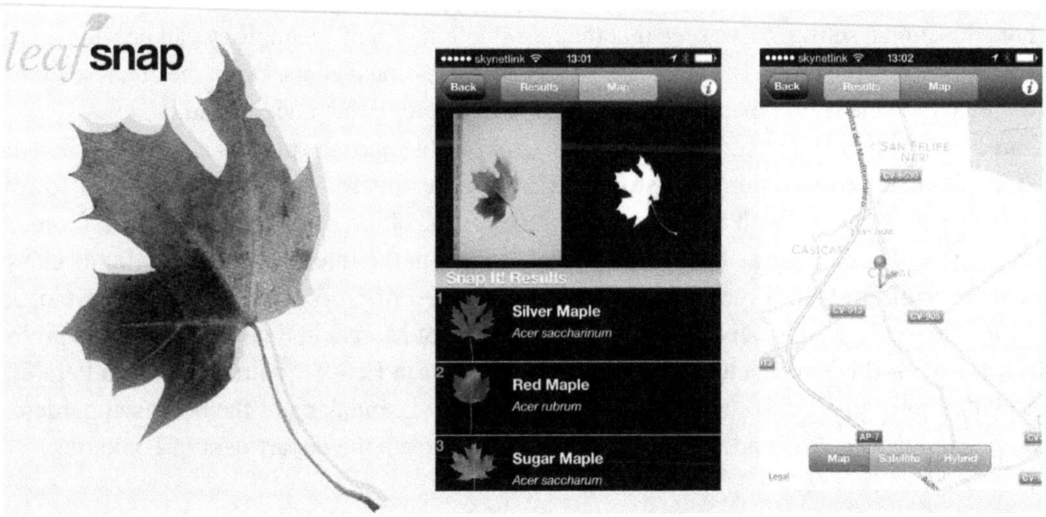

system. Another example is the project *WeBird* based on the framework *ARIS Games*[12] which detects bird sounds, compares with a database, and names the specie of the bird. The ARIS Games framework is described later.

Delivering Compelling Earth Science Stories with Smartphones

In a certain way the Augmented Reality acts as a Head-Up Display, what is called in the videogame culture, an "H.U.D.", showing a set of information on the screen to the player during a game like a mini map, health status and game progression. The displayed information, which is related to a real environment, and is meta-information of the landscape, helps the walker go deeper into the exploration of the physical space, just like a virtual world 3D game does. But in a game there is a story, and until now, the technology we described doesn't integrate a narrative. Storytelling is the next step in improving the experience of Earth Science during an outdoor walk. The advantages of a story are multiple. It helps to fix the information inside a timeline of events, or an historical context. It could be a walk through evolution with Charles Darwin or a biology walk with Lynn Margulis. The story also helps to convert information into an organized and logical sequence, which is easier to memorize showing connections between parts, revealing several levels of interpretation. The strategy for teachers is to capture the attention of students, like a video game, by converting a learning process into what is called a "Serious Game." There are many benefits to embedding a story inside an interactive mobile app. But in the last analysis it is the manifestation of a collaborative and evolutionary process, in shared imaginary that really matters.

The Research group *ARIS Games* led by David Gagnon from the University of Wisconsin have programmed a digital framework to create interactive locative games for smartphones in the field of education. Their idea was to insert a game engine inside a digital interactive audiovisual mobile app that permits students and teachers to design and create interactive locative outdoor activities. To play an outdoor game, students and teachers have to previously elaborate the plot and the interactivity, record and produce the final audiovisual media and geolocate them on the online graphic editor. Most of the work is not in playing the game, it is in building it by the study of the media and the elaboration of a strategy to convert them into an interactive multiplayer game. And most of all it is necessary to study in detail how knowledge is linked with the environment, applying a subject to the world.

According to David Gagnon (2012) "After nearly four years of studying and designing video games in higher educational settings, I've seen that the introduction of new technology can provide a reason to rethink a course from the ground up and reassess its core educational goals. Often the greatest educational benefits seem to come from this process, not just the technology that encouraged it."

A game based on the interaction and feedback between the app, the players and the teachers, all immersed in the same locative interactive story, builds a motivation for outdoor learning. It also connects the physical space of nature with the later study of the results of the game back in the classroom. All the game activity is recorded during the event and retrieved from the Internet for a study in the classroom.

One wonderful example is the Earth Science game called *Microcosmos Explorer*[13] developed with *ARIS Games* in the Fine Arts Department at the University of Murcia in Spain in 2012. This project was a collaborative effort between a teacher and the media artist in Eco-Art named Verónica Perales. They established a connection between the artistic approach of the caretakers of the botanical garden, of the university, and the scientific understanding of the students from the department of Biology.

Figure 3. The project Microcosmos Explorer by the Dpt. of Fine Arts from the University of Murcia

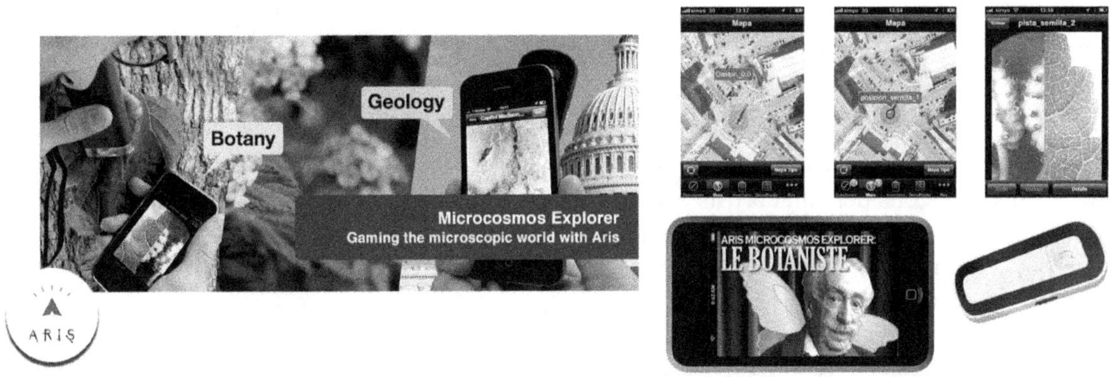

Students and teachers created during a five days Game Jam organized by the University of Wisconsin an interactive game, searching for virtual seeds and real plants in the Campus. The story was told by Gaston Bonnier a famous French botanist from the 19th Century. During a walk on campus seeds fell from Bonnier's pocket and players had to find the corresponding plants or the seeds themselves, within a perimeter of a few meters from where they have been retrieved. Players had to follow the tracks on the map and identify tiny details of plants surrounding them, pointing the camera of the smartphone paired with a portable monocular, a *Keeploop* microscope[14] to identify the relations between virtual seeds and real species. Each player had to report their discovery from the smartphone game app and modified the digital information of the map seen by other users. For example, when a user collected a virtual seed in a specific place the seed disappeared for the other players, playing at the same time. Each team had to collaborate to have more chances to win, encouraging the collaboration during the field research (See Figure 3).

Locative games need to be constructed around a solid story to be really interesting. It's not about going to random places and collecting the greatest amount of virtual objects. The game really needs a historical and/or a physical description of real places visited by players. It is necessary to find a balance between factual information, like plants and animals, and fictional components like the presence of a character from a past century. Characters are used in *ARIS Games* to present to players a list of alternatives in a form of conversation, which modify the continuity of the story depending on the choices made. This mechanism challenges the knowledge of students by asking for concrete information, like "what is the name of a plant" and also more subjective questions. Both will modify the story depending on the answers.

To be really interesting the story has to be linked with real facts and built as a non-fictional story. Going deep into the real historical facts and the lives of the great figures of Science as a basis for the interaction with players is really more powerful than creating unreal characters. It's like building a bridge between the objective study of nature and the history of science to explain the context of human discoveries. Just grouping information is not sufficient to communicate knowledge. The most advanced technological tools, like image recognition or geolocation within the smartphone, needs to be incorporated into the experience to increase interest.

Holding disciplines and fields of knowledge in separate silos does not help explain the history of Earth. Geology needs to be connected with chemistry, and chemistry needs to be connected to biology, and biology connected to cosmology and so on. That's the point where such technology like *ARIS Games* can be a place to combine a larger spectrum of disciplines. The ability to produce and format the audiovisual material into a smartphone experience is where artists, storytellers, designers and programmers can collaborate with people who know but not necessarily have the keys to communicate their stories to the world.

In his talk *Shakespeare in the Cave*[15] the scientist Rich Blundell points to the fact that the collaboration between artists and scientists is a powerful way to go deeper into the meaning of what we learn from Science. He considers that artists and scientists can discover similar results by subjectivity or of by objectivity. Such transformational collaboration between scientists, artists and programmers is happening today, in the *Big History* network.

THE BIG HISTORY NETWORK IS AN IDEAL ENVIRONMENT FOR EXPERIMENTAL TEACHING AND LEARNING WITH TECHNOLOGY

A Short Introduction to the Big History Network

Outdoor education is not new. Aristotle founded in Athena the early Peripatetic school in 335 B.C. It means ambulant or itinerant school, linking the act of learning with motion in space. The etymology suggest that Aristotle probably walked during his lectures. Since then several creative educational reformers like the American John Dewey or the Italian Maria Montessori at the beginning of the 20th century envisioned school in a very new way, a method well ahead of its time. Today, hundreds of schools that follow the *Montessori* method have been created all over the world. The method is based on the interaction with their environment, which enhances the natural development of kids, follows their intellectual abilities and their psychological self-construction.

Times have changed, and science has made great progress, and so has our work; but our principles have only been confirmed, and along with them our conviction that mankind can hope for a solution to its problems, among which the most urgent are those of peace and unity, only by turning its attention and energies to the discovery of the child and to the development of the great potentialities of the human personality in the course of its formation. (Montessori, 1948, p.1)

Several organizations around the world, like the *Association for Science Education*[16], promote outdoor activities. Teaching science, and especially Earth science, is a key pathway to gain a better educational methodology, because it's about our story as specie and all life in the largest context of space and time. Maria Montessori called this methodology "Cosmic Education", and the *Montessori* schools became a pioneer in education. The international *Big History* network, born in the 80's, is clustering several alternative ways to teach the story of the Universe. *Big History* is growing as an international web of teachers, students, researchers, pedagogues, artists, and scientists, giving birth to an academic multidisciplinary approach. The target is to teach what they call the "Big Picture" by creating educational content that tells the story of everything from the theory of the Big Bang to the present. The *Big History* Project, which operates in several countries, is driven by great scientists and academics like the Australian historian David Christian (Christian, 2012) and philanthropist Bill Gates. Teachers and students can sign into the *Big History Project* (bighistoryproject.com) website and have access to large amounts of video tutorials and didactic material.

Some digital resources like the *Chronozoom*[17] interactive timeline have been programmed by the *Microsoft Research Group* to navigate into a growing collaborative database of information from 13.7 billion years until today. *Chronozoom* is a very sophisticated interactive computer program that zooms over enormous scales of time from billions of years to tens of years.

The Big History Network of Creativity

Some very creative people from the *Big History* network are going deeper into the exploration with digital technology, like the researcher and media artist David McConville, chair of the Buckminster Institute and designer of immersive virtual learning environments. His Geo-domes are inflatable planetariums where the public can dive into a visual experience of the 180º projections screens, explaining the latest cosmological theories (See Figure 4).

This way to communicate science with digital projection systems that fill your peripheral vision immerses the viewer into a deep experience of time and space. David McConville points out that in the early 60's the polymath Buckminster Füller's concept of an Earth system visualization, what he called the control room for Spaceship Earth "Geoscope", predated the framing of "Big History" by many decades. The Geoscope is the project of a spherical environment for displaying satellite data of the dynamic flows of Earth's interconnected systems. David McConville (2011) directs his research to present and make real Buckminster Füller's concepts through the prism of the most advanced audiovisual and multimedia technology. Science mobile technology is starting to spread inside the *Big History* network thanks to scientists like Rich Blundell. Mobile learning has the potential to contextualize information in a new way. Complex notions like space and time, the nature of light, or the geometry of the Universe are really hard to understand and it is necessary to develop strategies that enable people to get it. Rich proposes to create physical experiments he calls "Epiphany" transmitted to the users through a set of smartphone apps. Each app is aimed at understanding a "threshold" in the evolution of the Universe. These thresholds include the appearance of light, the formation of a star, the emergence of life, the awareness of reflective thought in humans or many very important moments in evolution.

Figure 4. View of the Geodome by David McConville and The Elumenati, immersive projection design company

Figure 5. Screens of the app Cosmosis1 by Rich Blundell in collaboration with the author

The first of a series of "epiphany" smartphone apps *Cosmosis1*[18] is an audiovisual interactive guide to the electro-magnetic spectrum, which he uses to explain the Cosmic Background Radiation (CBR), the electromagnetic footprint of the first moments of the universe 13.7 billion years ago. Rich starts his introduction by walking in the woods, speaking to the user as if he was a companion. A person to person relationship is established for a 20 minutes experience. New users have to follow a sequence of video introductions before gaining access to the tools and necessary information to complete the experience. At a certain point in the narration, Rich invites people to use the camera of the smartphone to see through these various wavelengths of light by simulating the X-ray, Ultraviolet and Infrared vision on the screen. The user can jump back and forth from the camera mode to the video mode at any time, seeing their surroundings in different wavelengths. This process of capturing the environment and simulating the view through different wavelengths is one of the most important moments of the experience, because the traditional point of view of the spectator is shifted to the point of view of the narrator. Smartphones can enable participation of the user, just like recording with the camera and microphone. This new perspective is reflecting the transition between the 20th century mode of passive consumer to the 21st century mode of active creator of media. In 1996 the teacher and researcher, Isidro Moreno (2002), coined the term "lectoautor" (readwriter), meaning the fusion between readers and writers. The deep significance of the capture of space makes the experience not inside the app but around the user and everywhere in the outdoors. The world is the playground of the wonders of the Universe, whether pointing at a flower or a star (See Figure 5).

At the end of the experience with the app, the discoverer is invited to point at the sky and tune to the Cosmic Background Radiation. Suddenly, the CBR thermo picture is revealed and mapped in the sky, all around the user as a reality never seen before in the real context. The *Cosmosis1* app is linked to the website where people are invited to share their screen captures of the CBR seen from their place on Earth. The first time user is a "Discoverer", the second-time user is called an "Explorer". This change of status of the user means that now it's time for Explorers to teach to other people the meaning of the Cosmic Background Radiation with the help of the app.

The CBR app could be better described as an interactive lecture or a documentary, inviting people into the story hoping they have an "epiphany" which deepens their understanding of the universe and their position in it. The contribution is not really the technology or the large resource of information but creating meaning with the right data at the right time. When the app has been used one time and people understand it then it's time to study more information and go to a theoretical level in the classroom. At the moment any content on a smartphone is represented by an icon on the home screen and is described as an "app." There is no established name to describe this new type of interactive audiovisual media, it is more like a hybridization of genres. *Cosmosis1* is much more interesting than an app, it's an experience, a "Meme" ready to be transmitted over a network of "Explorers" through the Internet or cellphone networks.

An Intimate Relation between Science and Technology

Is technology absolutely necessary to teach *Big History*? — of course not. Some creative scientists like Larry Edwards while teaching at Genesis Farm[19], represents the huge scales of space and time by drawing on the floor a large spiral for members of the group to slowly walk around. *The Cosmic Walk*, the name he gives the walk, traces cosmic evolution from the Big Bang to the present. It's a type of ritual in which each member of the group lights a candle at threshold points while listening to a story teller describe the emerging "gifts" of life on Earth.

The most important thing is to be able to create a deep understanding of Big History, it can be by a geo-dome theater, a smartphone app, a ritual or whatever provokes an effective memory print. Two major parameters, the experience before the theory and the contextualization of the information in the real places are great benefits to provoke a good understanding of the big picture. The history of science moves parallel with the invention of technology. Galileo Galilei would never have discovered the satellites of Jupiter without the invention of the telescope and an orbiting artificial satellite is far more complex than any computer or mobile device we use in our daily life. Great scientific discoveries can be demonstrated or activated with the help of technology. There is an intimate relationship between the evolution of science and the history of scientific instruments. An interesting balance has to be found between what can be a simple and direct human to human relationship, and the use of technology. It is a natural desire to improve the transmission of knowledge according to each situation. But it is clear that it is necessary to communicate scientific knowledge with audiovisual resources as much as possible because our cosmic story is so complex and reveals humans humble place as part of a large horizontal network of life forms. The fact is that teaching science with smartphones is a great opportunity to experiment with the most interesting potential the telecommunication technology can offer.

The smartphone is understood here as a sensor helping us to improve our organs of perception, just like a microscope is for a biologist, and a telescope is for an astronomer. The personal computer entered into our daily life half a century ago now and we have finally the opportunity to look outside with new viewing tools and digital wearable sensors. The people involved in *Big History* represents of a new generation of scientists who could change our worldview, by communicating what they know from outside the laboratories using a renewed and enlarged vision. It is an appropriate time for a constructive combination between the *Noosphere* and the *Technosphere,* like Teilhard de Chardin and Gene Youngblood envisioned, as described in the introduction. It's about finding the good balance between our immersive and sometimes invasive virtual connection to the network with the direct physical and emotional experience of Earth. As the storyteller Jennifer Morgan (2002), active member of the *Big History* network, is accomplishing with kids, the most important thing is to tell the story of the Universe and our own story in a comprehensive and engaging way. That's the most important challenge for mobile science teaching - creating great experiences with great non fictional stories right from the audiovisual screen, with the sensors of the smartphone and in harmony with direct physical and emotional interactions.

EXPERIMENTAL SCIENCE WITH BODY AND MIND THROUGH MOBILE TECHNOLOGY: WALKING NON-FICTIONAL STORIES

A *Walk Through Time*, from an Outdoor Installation to a Smartphone App

In 1996 the advanced research group *Hewlett Packard Laboratories* in Palo Alto, California was one of the major places running innovative, breakthrough projects. They were ahead of their times, Geoff Ainscow, employee and independent business consultant, reported in a conversation that 20 years ago engineers were working on a project very similar to the *Google Glass*[20], trying to project digital information on the surface of a pair of glasses. Today they are working on a new revolutionary concept of photonic computing based on the exploitation of electrons, photons and ions in a single unit with a project they call *the Machine*. In the interest of challenging engineers and scientists to think outside the box, they created an outdoor exhibit called the *Walk Through Time,* which described the evolution of life on Earth along a one mile walk, where each foot represented one million years. A set of 95 illustrative large panels were placed to scale along a track starting from the formation of Earth 4,600 million years ago until today. One of the biggest challenges in teaching *Big History* is to convey the understanding of the tremendous spans of time and space at a scale relevant to us. This outdoor installation converts literally time into distance. The walker, with every cell of their body, not only the mind, gets a visceral sense of time. For the first three quarters (3/4 mile) of the walk the only form of life is various mutations of bacteria. Surprisingly hominids appear just four steps from the end of the walk, and Homo sapiens appear just two inches from the end. A human life span is represented by the thickness of a human hair. This understanding of scale by the physical effort of walking is a shock to the person, and indelibly printed on the heart. This is the power of timelines. The time spans are converted into a body and mind experience making real the distances between events.

In 2012 Geoff Ainscow decided to adapt the *WTT* walk to a smartphone app[21]. It's quite easy to detect the number of steps with a smartphone. The integrated accelerometer in the mobile device senses the movements of the person. A large number of fitness apps count steps during a walk or a run. But what about synchronizing audiovisual media on the screen with the pedometer? In the *Walk Through Time* mobile app each step corresponds to one million years. A skilled narration with graphics synchronized to the distance walked give the person the unique experience of walking along a time scale (See Figure 6).

A team of media artists, programmers and scientists was formed with the goal to create a first prototype of the app. The two sources of information were the book and the outdoor installation. Obviously it was not possible to include all the contents of the book inside the app. First of all, there was too much information, and secondly when walking, we can't read a screen full of tiny characters, and the time to walk one hundred million years is less than one minute. Until today and as far as I know, no such scientific narrative linked to a walk has ever been released at a large scale, it is an absolutely new area of creativity.

Figure 6. Screens of the app Walk Through Time by Geoff Ainscow in collaboration with the author

The first prototype published in the summer of 2012 is a compromise between a classic vision of an audiovisual narration and an experimental walk. At many points along the one mile walk the walker is invited to watch a short motion graphics, animated video explaining in detail what was happening at the corresponding moment in time. During the walk itself, a set of audio voices and still images are synchronized with a specific number of steps. This information can be scientific facts or illustrations of how the landscape and life was evolving on Earth. The results of the prototype are very interesting for several reasons. We discovered that the most important thing is the way the story is told to the walker. The walker needs to feel the narrator is a guide walking by their shoulder. This mode of narration is not the same as a TV documentary. It has to be also a physical effort, breathing in the ear of the walker. As Rich Blundell in his *Cosmosis1* walk did, it has to be a person to person relationship, an augmented emotional experience. The second parameter specific of this new kind of creations is that the story has to be formulated in a linear succession of facts in time and space. It means that the story can't be disrupted, going back and forth in time, it has to be progressive in one direction at the exact speed of the walker. The ground has to be taken into consideration as an alive membrane. In the case of the *Walk Through Time*, walkers have to feel that below their feet the earth is a, boiling, cracking, dry, wet, moist, hard or soft surface, changing at the speed of 1 million years per step. This can be achieved by an augmented soundscape synched with the number of steps and the speed of the walker.

The Concept of a Walking Documentary

The walk in time and space is a marvelous exercise of creativity for storytellers. Non-fiction story is the most logical way to share information because science is about facts and not about imaginary events. The term "documentary" as documenting reality seems to be the most appropriate for the description of this style of story, even though it involves mechanism where the walker has to complete actions to go further in the storyline. The nuclei of a Walking Documentary as described in this text, includes three fundamental components which are (1) a non-fiction story, (2) an outdoor activity and (3) the media driven by the steps. The story can be based on a timeline of events, as in a walk through time, or on the description of a group of objects at different scales or distances, as in a walk in space (See Figure 7).

Figure 7. Diagram of the concept of walking documentary by the author

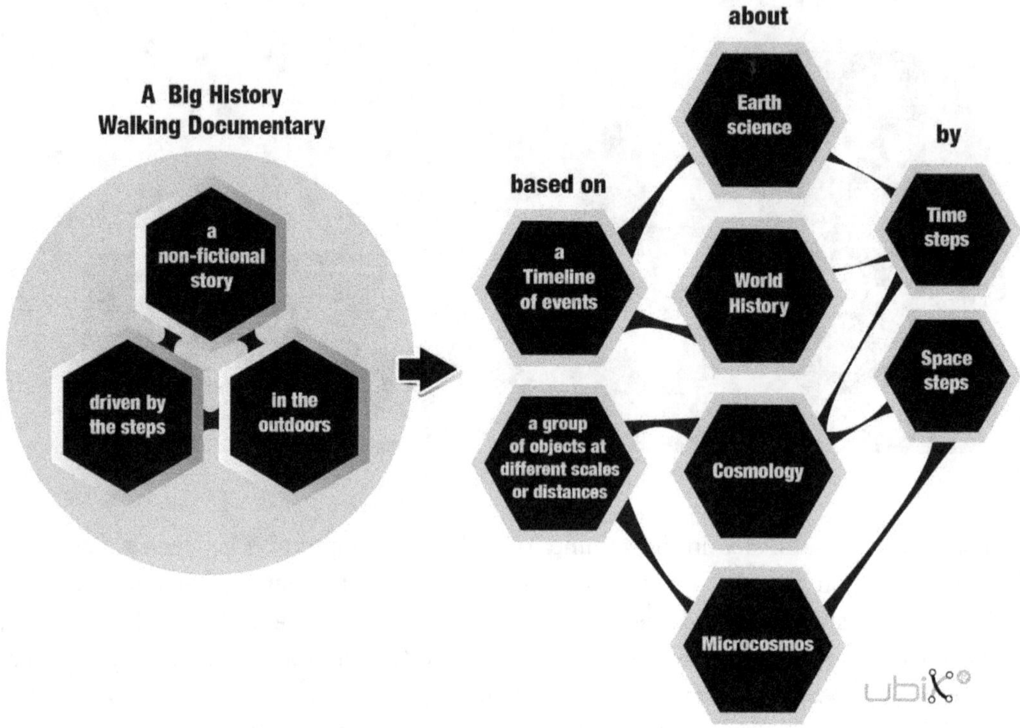

Many timelines can be extracted from Earth Science or World History, which could be converted into a walking documentary, like the *Walk Through Time* project. Different timelines would have different scales. For example, walking the story of hominids could be 1,000 years per step and the history of our occidental civilization 10 years per steps. A walk between objects can be a biological walk into matter at several scales. But one of the most interesting possibilities is a walk in combined space and time. Looking at the night sky is looking into space and time because of the finite speed of light and the huge distances between celestial objects. A journey into space beyond our solar system is an interesting progression from a space-walk to a space-time walk. Let's look at some case studies (the following descriptions are some cosmological proposals by the author not yet released as mobile apps). It's not too hard to figure out our distance from the Sun and the appropriate scale to use. Some simple trick, like representing the Earth as a grain of sand orbiting at 9 meters from an orange, as the Sun, gives us an idea of the scale.

Solar Walker[22] is the prototype of a walk with the smartphone where each step is by default one million kilometers. In this project, the user deploys the Sun with a throwing gesture of the smartphone. The strength of the throw determines the position of the sun in the model, then the solar system unfolds at that scale and the walk is ready. The fact is that if we only have to walk 150 steps to go from the Sun to the Earth, on a scale of one million km per step, it is far more difficult to escape from the solar system. To reach the Oort cloud walking on the same scale would take more than 7500 km of walking. It is then necessary to walk on a scale of one step equals one light year of distance to explore our galaxy. Crossing the Milky Way on this scale would take a very long 120,000 step walk (See Figure 8).

Figure 8. Graphic illustrating the project Solar Walker by the author

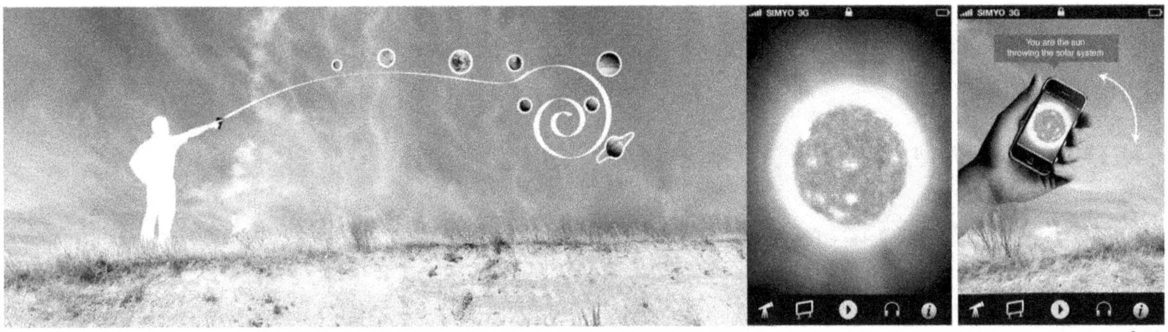

The virtual exploration of a galaxy by walking can be a long process. To cover large distances the mobile app in the pocket should work continually in the background counting the steps of our lives. One such app is called *Moves*[23], which continually records your movements. This app is working as an engine for other developers who want to connect the data retrieved to their own apps, opening the area for a new kind of digital content expanded in time. It is based on the automatic detection of the speed of displacement classified in categories like walk, run, bike or motorized transport. Walking at a realistic speed to the moon would require ten years. Such "Life Walk" apps are challenging the way we consume bits of information, and the way we process learning at large time scales. The "cosmos" is a space-time walk where each parameter can be manipulated separately during a walk. Changing our relation to the space, for example shrinking our size to a very tiny life form could permit us to virtually convert the walk around an apple into an interminable adventure. *Atom Walker*[24] is a prototype that explores relative scales of the inner space of molecules and atoms. It's a long walk from the nuclei to the orbits of the electrons. The pace of the walk can also be changed depending on the amount of information. Describing the first second of expansion after the Big Bang would need the equivalent of super slow motion animation. The upcoming wearable technology like the *Google Glass* is perfectly adaptable for animating these new parameters and scales. Using this technology to explore scientific facts stimulates learning. Imagine walking from the nucleus of an atom to the electrons, then changing your size and walking across the Milky Way galaxy.

Traveling in space and time is a universal dream, such mobile content stimulates our imagination, and expands the way we perceive the universe. We gain the perspective described beautifully by Carl Sagan in his words about the "Pale Blue Dot". Today technology is offering a very large spectrum of new user experiences, but it is very hard to escape from the tiny environment of common practices. It is very hard for each one of us to integrate new ways to use technology. We still process infinite fragments of information like columns of ants transporting food. Each day it gets harder and harder to escape from the feeling of being totally submerged by too much information. There is no time for a walk to ponder what we do and what we effectively learn from the perspective of the big picture. That's why it is a question of education. How can Earth Science and outdoor learning help us consume the new digital information? The list of emerging technology is quite long and the distance between the tiny ecosystem of established platforms and software used by the majority is widening, because a large amount of users are clustering around a narrow, small list of technologies. Paradigms like *Facebook* or *Twitter* will have to collapse before other similar services have a chance to become a standard and change the way we use technology. And more seriously, the traditional educational system will have to change before we can expect the real development of outdoor mobile learning, beyond the few experimental niches described here.

ENHANCING WALKING DOCUMENTARY BY EMERGING TECHNOLOGY

Wearable Hardware for an Earth Science Walk

The smartphone is a common device and a very large majority of people from the industrialized countries carry one in their pocket. The miniaturization of electronic components is enabling the development of wearable smartphones mounted on a pair of glasses like *Google Glass*, the optical head-mounted display (OHMD) developed by *Google*. It is still an experimental project and it will probably attract a lot of users. The problem of privacy is a concern because the camera and the screen is mounted on top of the glasses and is controlled by touch or voice command. From my point of view, the initial audience for the *Google Glass* is in the professional or educational arena. It is clearly an interesting technology for a "Walking Documentary". A number of virtual objects can appear on the screen and change their size according to the number of steps in their direction. But "wearable technology" is not only a smartphone embedded into a pair of glasses, it can be any type of sensors carried on the person. A bracelet can be for example an environmental sensor checking the composition of the atmosphere or a ring on a finger could check the composition of water. The collected data can modify, in real time, the narrative about our environment or create a new story. Haptic wearable technology[25] is also a way to interact with the environment by tactile signals into the body. *Supershoes* is a project by the artist Dhairya Dand[26], his shoes are equipped with vibro-tactile motors sending micro vibrations to each foot. The shoes are paired with a smartphone app and used to guide people to go from one place to another indicating the right direction to walk by sending vibrations to one foot or another. Wearable technology deserves serious consideration for the learning of Earth Science, especially regarding the environmental sensors. Considering a smartphone from this perspective is far more exciting for education and the classroom.

The Body as Cursor during the Walk

Moving in the outdoors with wearable technology is a way to retrieve information from the digital network. But the movements tracked by the sensors of our device, potentially contain another major opportunity for the user interface more interesting than just recording our travels on a map. The combination of parameters, like the inclination of the device, our speed, our coordinates and trajectory can be used to modify the way the information is displayed on the screen of a smartphone. The GPS Drawing[27] is the act of drawing with the GPS function of a smartphone in the pocket. A large community of people use the real time tracking of coordinates with apps like *Map My Tracks*[28] to virtually write words or draw figures on the ground and at very large scale. Beyond the act of simply drawing, it could be possible to communicate with the software of any app by drawing geometric figures. For example a classic play, pause, rewind interface of an audio player can be converted into a circular walk to rewind, a jump to pause or a short run to select the next media. This looks like a lot of fun but when well calibrated it can be a new way to interact with a more complex story, where the plot changes depending on the physical reactions.

The Environment as Part of the Story

The environmental data retrieved by sensors can also change the storyline of a walking documentary. Detecting the time of day or night, the composition of the air, the temperature and humidity offers the walker a deeper involvement into the story. A specific app explaining the composition of the atmosphere could use the direct data of the user to construct an introduction by the description of the air breathed at this time. But it could also capture the sounds of the environment and inject them into the dynamic soundtrack of the story. This is demonstrated very well in the augmented soundscape app Inception[29]. The simple detection of the level and the nature of the noise surrounding the walker identifies the type of space, for example; city, urban, farmland, mountains or the beach.

Several projects like *Smart Citizen*[30] are good examples of the introduction of the concept of "environmental citizen sensors" where a large community of people shares in real time their environmental data from a wearable open source tiny electronic device. The expansion of "Spimes" or space and time traceable objects, will constitute the majority of electronic devices connected to the Internet. The project *Urban Cursor*[31] by the artist Sebastian Campion is based on a very simple idea, he created a rolling giant arrow made of wood representing the typical computer mouse and inserted a GPS inside. The sculpture is situated in an urban setting where people spend a good time just moving the mouse around. The movement is recorded and drawn in real time on the web page of the project. This installation reveals something more than just the body as cursor, the Earth itself is presented as the ultimate big Earth digital screen similar to the computer screens where we move our arrow. The surface of the Earth itself is the place where we write digital information with our displacement as we do on a computer with the keyboard and the mouse.

In the 90's the famous game designer Will Right created the software *Sim Earth*[32], an Earth simulator inspired by the vision of James Lovelock. Players could manage from their PC the equilibrium between the needs of every species and the supply of life giving support systems of the environment to maintain life on the planet for all future generations. Now we are immersed with computers in the middle of real places. The project *Driverless car*[33] by *Google* is based on a technology that constantly scans the space around the car creating a one to one digital replica of the environment. This is not science fiction, it is real. We should now use dynamic 3D modeling for environmental purposes. The serious game *Sim Earth* is not anymore on the computer's screen it is outside accessible by mobile technology and obviously Earth is not a playground, it's a real environment. The *Rain Forest Connection*[34] has created a network of sensors with old mobile phones in the rain forest to detect the noise of chainsaws, which has enabled authorities to act instantly against the illegal traffic of natural resources. Coupling technology with the education of environmental values, the world will benefit from the intelligent use of the incredible expansion of interconnected digital devices.

CONCLUSION

The distance between our body and digital devices is changing, it is getting closer and closer and more transparent. The way we interact with these devices is also changing, our body itself is becoming what the mouse was to the desktop computer. The physical space is now part of the media landscape, the information we receive is the sum of the digital layer of meta-information on top of the physical space. We are the sensors of the environment capturing real information as nutrients for a digital model build in the cloud. Every digital sensor we carry, like our smartphone, is part of a bigger network of interconnected traceable objects. The social network is where we exchange, in real time, our personal perception of how we understand our environment. The fusion between the digital network and the physical space is like an informational skin covering the ground ready to be converted into knowledge. What drones and satellites can achieve today is what we are achieving virtually, a bird's eye-view of the world in space and time. Carl Sagan's "pale blue dot" (Earth) is reflecting on itself by satellites and the digital network.

The creative and technological possibilities for better science education are very promising. Teaching science with mobile technology, in the context of *Big History* is revealing a deep respect for all life forms, their habitats and the wondrous beauty of Nature. The evolution of technology grows in parallel with the evolution of our understanding of our 13.7 billion year story. From the point of view of the creator of digital content, investigating emerging technology to teach science is a strategy for finding a good balance between the digital world and the physical world. Expanding learning in the outdoors and in everyday life with the help of mobile technology can potentially trigger an evolutionary leap to a new educational paradigm by breaking the barrier between learning and experimentation. The act of walking is a great way to expand scientific knowledge through an embodied experience of cosmic time. Synching the steps with time scales help us to better understand our place in time and space and appreciate, with a renewed perspective, our presence on Earth.

The possibilities for the concept of a Walking Documentary are becoming a reality with the emergence of upcoming wearable devices like *Google Glass*. The potential easy access of such augmented reality devices for education could open up deep Earth Science experiments for a larger public. The author is working on a PhD investigation into the evolution of documentaries using mobile technology and telecommunication networks. Any digital media art director or designer who wants to be prepared to collaborate with scientists or historians for the creation of didactic content related to Earth science or cosmology is invited to explore the Big History network as a great place for a dialogue between creative people and scientists. The description by the author of the "Walking Documentary" can also be a good resource for any researcher who wants to communicate science with mobile devices.

REFERENCES

Blundell, R. (2009). *Communicating Science: a novel frame of reverence* [Master's Dissertation]. Retrieved from ProQuest Dissertations and Theses database (UMI No. 1461771).

Christian, D. (2007). *This Fleeting World, a Short History of Humanity*. Great Barrington, MA: Berkshire Publishing Group.

Dawkins, R. (2006). *The Selfish Gene*. Oxford, United Kingdom: Oxford University Press.

Dewey, J. (2007). Experience and Education (Original edition 1938). New York, NY: Simon and Schuster.

Dubois, C., Avignon, M., & Escudier, P. (2012). *Observer la terre depuis l'espace, enjeu des données spatiales pour la société*. Paris, France: Dunod.

Gagnon, D. (2010). *Mobile Learning Environments*. Retrieved from http://www.educause.edu/ero/article/mobile-learning-environments

Jenkins, H. (2009). *Confronting the Challenges of Participatory Culture: Media Education for the 21st Century. The John D. and Catherine T. MacArthur Foundation Reports on Digital Media and Learning*. Cambridge, MA: MIT Press.

Lovelock, J. (1995). *The Ages of Gaia: A Biography of Our Living Earth*. Oxford, United Kingdom: Oxford University Press.

Margtullis, L. (1998). *Symbiotic Planet: A New Look at Evolution*. New York, NY: Basic Books.

McConville, D. (2011). Reflexive Ecologies: Visualizing Priorities. In M. Lima (Ed.), *Visual Complexity: A visual exploration on mapping complex networks*. New York, NY: Princeton Architectural Press.

Montessori, M. (2004). The Discovery of the Child (Original Edition 1948). Delhi, India: Aakar Books.

Moreno, I. (2002). *Musas y nuevas tecnologías: el relato hipermedia*. Barcelona, Spain: Paidós Comunicación.

Morgan, J. (2002). *Born with a Bang: The Universe Tells Our Cosmic Story*. Nevada City, CA: Dawn Publications.

Sagan, C. (1994). Pale Blue Dot: A Vision of the Human Future in Space. Ann Arbor, MI: University of Michigan.

Sahtouris, E. (2000). Earthdance: Living Systems in Evolution. Bloomington, IN: iUniverse.

Spier, F. (2011). *Big History and the Future of Humanity*. New York, NY: John Wiley & Sons.

Teilhard de Chardin, P. (1959). *The Phenomenon of Man*. Paris, France: Éditions du Seuil.

Vernadsky, V. (1998) The Biosphere (Original edition 1926). Berlin, Germany: A Peter N. Nevraumont book Copernicus Series. Springer Science & Business Media.

Youngblood, G. (1970). *Expanded cinema*. New York, NY: Dutton.

KEY TERMS AND DEFINITIONS

Atom Walker: It is project of Walking Documentary by the author where the walker is walking at an atomic scale.

Life Walk: In the context of a Walking Documentary, it is a walk connecting a series of events distributed on an imaginary line that needs the time of a human life to be walked.

Solar Walker: It is a project of a Walking Documentary by the author where the walker throw virtually with the accelerometer of the Smartphone a digital model of the Solar system at a variable scale regarding the strength of the gesture.

Space-Time Walk: In the context of a Walking Documentary, it is a walk along a description of physical objects or events distributed on an imaginary line, separated from us by light years, covered at an astronomical scale per step.

Space Walk: In the context of a Walking Documentary, it is a walk along the description of a set of physical elements distributed by relative distances or relative scales in a progressive order on an imaginary line, covered at a relative scale per step.

Time Walk: In the context of a Walking Documentary, it is a walk along a description of historical events distributed in a chronological order on an imaginary line, covered at a relative time span per step.

Walking Documentary: A Non-fiction audio-visual narrative displayed on a digital mobile device synchronized to the distance, speed or trajectory covered by the walker. The walk is organized along a linear distribution of events or objects distributed in time and/or space.

ENDNOTES

1 Lucas, G., & Youngblood, G. (1971). Watch videocast retrieved from http://youtu.be/nUFQetgs-VNg

2 *Apollo 11* - 2048 words ram memory computer. Retrieved from the Wikipedia: http://en.wikipedia.org/wiki/Apollo_Guidance_Computer

3 More information on the *Big History Project* can be found at https://www.bighistoryproject.com (accessed September 06, 2014).

4 More information on the *Big History Association* at http://ibhanet.org/ (accessed September 06, 2014).

5 More information on the *Seti@Home* project at http://setiathome.ssl.berkeley.edu (accessed September 06, 2014).

6 More information on the *Earth Cache* project at http://www.earthcache.org (accessed September 06,
2014).

7 More information on the *Project Noah* at projectnoah.org (accessed September 06, 2014).

8 Retrieved from the Wikipedia: http://en.wikipedia.org/wiki/BioBlitz

9 More information on the *Google Goggles* project at https://support.google.com/websearch/answer/166331 (accessed September 06, 2014).

10 More information on the *Leaf Snap* project at http://leafsnap.com (accessed September 06, 2014).

11 More information on the app *Celeste* at https://play.google.com/store/apps/details?id=com.t11.celeste.se&hl=es (accessed September 06, 2014).

12 More information on the *ARIS Games* project at http://www.arisgames.org (accessed September 06, 2014).

13 More information on the locative game *Microcosmos Explorer* at http://arisgames.org/featured/microcosmos-explorer/ (accessed September 06, 2014).

14 More information on the *Keeploop* company at http://www.keeploop.com (accessed September 06, 2014).

15 Blundell, R. (2014) Watch videocast retrieved from http://www.youtube.com/watch?v=VoisuLgCDr0

16 More information on the Association for Science Education at http://www.ase.org.uk/resources/outdoor-science/ (accessed September 06, 2014).

17 More information on the *Chronozoom* project at http://www.chronozoom.com (accessed September 06,
2014).

18 More information on Rich Blundell and the *Cosmosis1* app at http://cosmosis.omniscopic.com (accessed September 06, 2014).

19 More information on *Genesis Farm* at http://www.genesisfarm.org (accessed September 06, 2014).

20 More information on the *Google Glass* project at https://www.google.com/glass (accessed September 06, 2014).

21 More information on Geoff Ainscow and the *Walk Through Time* app at http://www.walkthroughtime.org/

22 More information on the S*olar Walker* project at http://www.bighistoryapps.org (accessed September 06, 2014).

23 More information on the *Moves* app at http://www.moves-app.com (accessed September 06, 2014).

24 More information on the A*tom Walker* project at http://www.bighistoryapps.org (accessed September 06,
2014).

25 Haptic Technology. Retrieved from the Wikipedia: http://en.wikipedia.org/wiki/Haptic_technology

26 More information on the artist Dhairya Dand at http://dhairyadand.com (accessed September 06, 2014).

27 More information on GPS Drawing at http://www.gpsdrawing.com (accessed September 06, 2014).

28 More information on the *Map My Tracks* app at http://www.mapmytracks.com (accessed September 06, 2014).

29 More information on the I*nception* app at http://inceptiontheapp.com (accessed September 06, 2014).

30 More information on the S*mart Citizen* project at http://smartcitizen.me (accessed September 06,2014).

31 More information on the *Urban Cursor* artwork at http://www.urbancursor.com (accessed September 06, 2014).

32 *Sim Earth* By Will Wright. Retrieved from the Wikipedia: http://en.wikipedia.org/wiki/SimEarth

33 More information on the D*riverless Car* project by *Google* at https://plus.google.com/+GoogleSelfDrivingCars/videos (accessed September 06, 2014).

34 More information on the *Rain Forest Connection* project at https://rfcx.org (accessed September 06, 2014).

Chapter 17
The Message Is the Medium: Ecology, Mobility and Emergent Storytelling

Verónica Perales
Murcia University, Spain

ABSTRACT

The aim of this chapter is to examine in depth the potential for ecological awareness that emerging artistic and creative narrative practices involving mobility may possess. In order to do so, this chapter will analyze the current situation in reference to several projects that exemplify this potential. Our time sees the coming of the Creative Activist in a practice that uses art and media to scale up the ecological message to a global level, to reach a lot of people and to move them.

INTRODUCTION

From an ecological viewpoint, we live in a situation of obvious emergency. Global warming and the alterations that human beings have imposed on the planet (biodiversity losses derived from agricultural and cattle raising processes, building, pollution, extreme exploitation of resources, etc.) have produced a lethal blend with outcomes that seem increasingly difficult to foresee.[1] A feeling of uncertainty spreads as a fine low-density layer through our consciousness and into our projections for the future.

Citizen involvement is crucial in addressing any possible change of conscience or perception of who we are at a planetary level. A change that might allow us to live in harmony as part of the shared planetary body of which we are a part and which we belong to. As Daniel Quinn explained in his novel *Ishmael*, as long as humans are convinced that the world belongs to them and that their destiny is to conquer and rule it, then they are heading for destruction. Changing this attitude by means of legislation is impossible – what is needed is to change peoples' minds (1990). Hypermedia, locative media and transmedia creation may be able to tackle one of the key issues for a shift of environmental paradigm, defined by Donna Haraway (1991) as the invention and reinvention of nature: "perhaps the most central arena of hope, oppression, and contestation for inhabitants of the planet earth in our times" (p.1). This paper is about the role of transmedia and mobile content in ecological art and ecological thinking.

DOI: 10.4018/978-1-4666-8838-4.ch017

BACKGROUND

In ecological thinking everything is interconnected (Morton, 2010). It is not difficult to join the dots and reach a clear picture of the crisis we face. As expert in ecological critique Timothy Morton (2012) asserts, "ecology isn't just about global warming, recycling, and solar power – and also not just to do with everyday relationships between humans and nonhumans. It has to do with love, loss, despair and compassion. It has to do with depression and psychosis. It has to do with amazement, open-mindedness, and wonder. It has to do with doubt, confusion, and skepticism. It has to do with concepts of space and time. It has to do with consciousness and awareness. It has to do with ideology and critique. It has to do with reading and writing. It has to do with race, class, and gender. It has to do with sexuality. It has to do with ideas of self and the weird paradoxes of subjectivity. It has to do with society. It has to do with coexistence" (p.2). From this position, ecological thinking has to do with art, philosophy, literature, music, and culture. It has as much to do with the humanities as with the sciences, and it also has to do with factories, transportation, architecture and economics. In this way, "[h]uman beings are each other's environment" (p.4).

As far as art is concerned, recent decades have witnessed a progressive shift in strategies that address the environment, migrating from the production of art objects to the development of participative proposals. New approaches give priority to the creation of synergies and the articulation of practical actions in the field, involving collaborative strategies and people participation. These practices are linked to social rather than aesthetic considerations, and in some cases take advantage of, or rely on, emerging technologies and formats. Infiltrating the technological mainstream in order to convey a message to the highest possible number of people is a strategic issue. The new narrative forms imply bidirectionality, inviting not only interaction but also participation and writing.

The concept of Media Ecology was first theorized by McLuhan (although it has also been attributed Neil Postman). The intellectual background of this current includes Lewis Mumford, Harold Innis, Walter Ong, and Jacques Ellul among a vivid set of resources (Fuller, 2005). The concept-metaphor operates according to the following formula: medium/media = environment(s); and like environments, media mutations depend on the shifting dynamics of the global system. Scolari, Aguado and Feijoo (2013) assert that media operate like species that coexist in the same communication ecosystem. But this communication/ecosystem exists in the context of a bigger one – planet Earth. At the Inaugural Media Ecology Association Convention in 2000, Postman declared that the term 'media ecology' was a good choice "since we wanted to make people more conscious of the fact that human beings live in two different kinds of environments. One is the natural environment and consists of things like air, trees, rivers, and caterpillars. The other is the media environment, which consists of language, numbers, images, holograms, and all of the other symbols, techniques, and machinery that make us what we are" (p.11). This essay attempts to address the dialog between these two environments in the belief that – as Postman (2000) concluded– media ecology exists to further our insights into where we stand as human beings and how we are faring morally on the journey we are taking.

The essay will not tackle issues of responsible, ecological and sustainable production of the technologies involved in creation (computers, smartphones, GPS and so on). This question has already been studied by Sy Taffel from the University of Bristol who has put forward an ecological conception of media and materiality stemming from the work of Gregory Bateson and Felix Guattari (2012) "arguing that if we are to take generational responsibilities regarding social and environmental justice seriously, we need to look beyond content in order to critically engage with the materiality of the technologies that digital media is predicated upon" (p.1). This materiality is what Baruch Gottlieb (2009) has called the irrefutable facts of the origin of the technological surface, facts that are a key to understanding –in a radical new way – our technologized society and the things we produce.

MOBILITY TO GO GREENER

Making art that addresses the present environmental situation is challenging. Attempts to bridge art and activism often fall short. Vague abstractions tend to arouse suspicion among activists, while connoisseurs find fault with almost any art that assumes a didactic or informative tone (Transnational Temps, 2007). But, as Lucy Lippard (1981) affirmed in the 80s, there is no choice, and new kinds of art practice will have to emerge –at least partially– beyond the art world; it may be that these new forms are only to be found buried in social energies as yet unrecognized as art (p.17). Spreading the Medium as a Message (the title of this chapter obviously refers to the famous McLuhan assertion that "the Medium is the Message") is a task for artists, media experts and other disciplines. In fact, this is only possible for multidisciplinary working groups and must be approached from different directions. Ecological questions no longer deal solely with perception but also with participation, movement, and synergies (Ohlenschlager, 2007).

Today, smartphones connect us to the Internet from almost anywhere in the world. This interface with the world transforms our everyday experience. As in the title of the Ohlenschlager text (2007), biosphere and infosphere are close together, real space and virtual space constitute a unique experience of multiplicity. The thumb generation, as Sadie Plant has called it, is born of the ubiquity of mobile practices. Sadie Plant (2001) explains that one of the earliest uses of the word 'mobile' was in association with the Latin phrase *mobile vulgus*, the excitable crowd, "the mobile introduces new senses of speed and connectivity to social life, establishing new kinds of relationships between individuals and with the urban crowd" (p. 61). The strength of many of the projects described here lies in the mobilizing, organizing and strategic potential derived from the use of mobile devices linked to creation.

In 2005, Natalie Jeremijenko created *Feral Robotic Dogs*. These dogs are 'released' *en masse* into a community space to sniff out pollution, ozone, and other environmental toxins. The artist designed packs of feral robotic dogs to 'patrol' sites of public interest such as parks, schools, neighborhoods and industrial sites, in order to generate community and media attention to the issue of environmental contaminants. To do this, artists and engineers worked together to hack commercial robots designed to explore territories and to report the findings. The components include sensors, solar panels, and object detectors. These robots react in front of the toxic substances that surround us, acting temporarily as 'our eyes, our smell', and so boost perception by moving through space to 'go greener'. As Jeremijenko says, the dogs provide "an opportunity for displaying some evidence that certainly experts can interpret, but local community members can interpret, as well"[2]. Primarily, Natalie Jeremijenko explores the interface between society, environment, and technology. She may not be the best eco-design-engineer on the planet but her artworks do meet the challenge of generating social impact. She uses a wide range of media – robots, art, online clips, animals– to communicate her ideas creatively and to reach a large audience. In collaboration with German artist Chris Woebke, she developed *Bat Billboard* (2008), an interactive work designed to create "novel social interactions between bats and humans"[3]. The work uses the structure of a billboard and provides luxury housing for bats, addressing the environmental health emergency faced by New York bats known as white nosed syndrome (a mysterious disease that has killed more than a million bats since 2007). Bats, like bees, play a key role in many ecosystems, pollinating plants and providing insect control, but in urban habitats they are often misunderstood. Inside the standard billboard structure, urban bats found a safe space to live and hibernate, helping to counteract the effects of white-nose syndrome and other threats to their population. Monitoring equipment inside the billboard used voice-recognition software to map and translate the calls of the resident bats, matching them to archives of various call patterns and meanings currently being compiled by biologists. The resulting messages are displayed on a screen. In this way, the billboard has reclaimed urban infrastructure as an animal habitat, and also given the bats a public face, translating their habits and activities into a form that humans can understand[4]. In earlier work, Jeremijenko produced *Housing Birds* and other artworks conceived to open urban spaces to animals. All these animal-related artworks form part of *Ooz*, a place to see and observe animals. *Ooz*, like a traditional zoo, "is a series of sites where animals and humans interact. However, the interactions at an *Ooz* site differ from that of a Zoo. *Ooz* is interactive in that it provides humans with a set of actions, the animals provide reactions and these couplets add to a collective pool of observations"[5]. The artist proposes a redefinition of what health means, not just internal or individual, but also external and communal (air and water quality, sound pollution...), venturing a concept of 'shared health', which is the main issue of her *Environmental Health Clinic*[6]. In this way, Natalie Jeremijenko offers an exemplary model of the 'art of the eco-mindshift', interpellating on the relationships and interactions between space and living beings.

The relation between environmental health and individual health has much in common with Donna Haraway's concept of natureculture, and the way she has considered species or companion terms. "I love the fact that human genomes can be found in only about 10 percent of all the cells that occupy the mundane space I call my body; the other 90 percent of the cells are filled with the genomes of bacteria, fungi, protists, and suchsome of which play in a symphony necessary to my being alive at all, and some of which are hitching a ride and doing the rest of me, of us, no harm. I am vastly outnumbered by my tiny companions; better put, I become an adult human being in company with these tiny messmates. To be one is always to become with many" (Haraway, 2008a, p.4). Another similarity between these eco-activists/artists is how they act as 'disruptors', whereby their discourse disrupts dominant dynamics. Donna Haraway calls upon her audience to work, play and think in terms of multispecies cosmopolitics, a new approach to recuperating the 'Terrapolis' in which we live. She suggests that humans turn to *SF*, a sign (coined by her) that stands for *string figures, science fiction, science fantasy, science facts, speculative fabulation, speculative feminism, so far...* as mechanisms for envisioning the future. Her main goal: to 're-do' ways of living and dying[7]. She explores –and sets out to merge– trans-disciplinary biologies, which might be called EcoEvoDevoHistoTechno, whereby string figures tie together human and non-human ecologies, evolution, development, history, technology, arts, and affect. Infection and indigestion, rather than production and reproduction, are the name of her 'game', as creatures render each other capable both of inheriting the terrible trouble of our mortal, corporeal, sensuous Terra and of shaping ongoingness[8].

In her book *When Species Meet* (2008), Haraway analyzes the *Crittercam*, a research tool to be worn by wild animals. It combines video and audio recording and also collects of environmental data: depth, temperature, acceleration. Its goal is to allow scientists to study animal behavior without interference by a human observer. *Crittercam* "brings the animal's point of view to the scientific community and a conservation message to worldwide audiences"[9]. We might think that this type of device would help us to understand the animal world better and broaden our awareness of the value of ecosystems. But Haraway shows us the perverse side of technology usage. Rather than focus on the tool, she focuses on the ways TV shows are advertised, how they address children, how online advertising operates, and so on, asking: "Have you ever wondered what it would be like to be a wild animal? ... You can experience their life the way they do" (2008a, p. 252). Sometimes entertainment and the banalization of animal life converge. That is what happens in the ad "Sharks Now Tweet When Approaching the Beach"[10] for a data transmitter created by the group Surf Life Saving Western Australia. The app warns surfers and beachgoers about the size, breed and location of sharks that come within a kilometer of the shore. We can 'see better' into the water with this tool. In this way, warnings such as 'shark-infested waters' are no longer necessary. But as Daniel Quinn commented, the world does not belong to humans alone – who *infests* the waters? The ecological crisis demands an equally powerful and urgent opening up of perception of who we are and where we are (Morton, 2012).

One way to Go Greener using smartphones and locative strategies is simply to drop a 'green message' at a key site. This is the basis of *As if by Magic*, a project initiated by the group Active Ingredient. It consists of a series of short interactive films for mobile phones produced with a group of filmmakers in Leicester during 2009. The films focus on real locations in the area. Presented on mobile phones using GPS, each film is triggered when the user passes by the location where it was filmed, merging fictions with physical reality. Funded by Rachel Jacobs and Matt Watkins, Active Ingredient is as an artist led collective that develops projects and conducts research of artistic intent. The collective is based in Nottingham and works closely with the University of Nottingham's Mixed Reality Lab and Horizon Digital Economy. Active Ingredient won the Nokia Mindtrek Ubimedia Award in 2007 for the internationally renowned project, *Heartlands*[11] ('Ere Be Dragons) a mobile phone game controlled by heart rate and GPS.

Rachel Jacobs researches the different contributions made by artists to engage the public with climate data, and how artists work with scientific data. This led to *The Prediction Machine*[12] (2012), an interactive installation developed in collaboration with and through conversations with local residents, engineers, computer scientists, climate scientists and researchers. The main purpose of the project was to encourage community dialogue using local climate data and machine prints outs. These 'climate fortunes' are made using live temperature, precipitation, wind-speed and air pressure readings, which the machine processes to predict both the near and distant future (30 years ahead). The predictions then spark exchanges in response to the environmental problematics they intimate, for example, climate change. This is an issue dealt with by numerous artists, lately associated with the use of locative media. As Claire Colebrook argues in *Extinctions*, the 'discovery' of climate change needs to be considered as an epochal event: it does not simply occur within time but also points to a time after humans, an end-time that humans themselves will be able to witness at the same time as contributing to its acceleration. Rachel Jacobs has also worked on *Timestreams Platform*[13], artist-mediated participatory sensors for environmental discourse, encompassing art practice and public engagement for a better understanding of energy and climate change.

Leila Nadir and Cary Peppermint are EcoArtTech collaborators, whose social art practice focuses on the visualization of scientific data. Since 2005 they have been designing a series of solar-powered mobile platforms: sound art transmission *Wilderness Information Network* (2006), or the *Environmental Risk Assessment Rover ERAR-AT* (2008), an all-terrain mobile station that uses GPS networks and solar power to project real-time data based on local coordinates. *Indeterminate Hikes+* (2011-2012) is a mobile media app developed by the EcoArtTech collective. Its goal is to transform everyday landscapes into sites of bio-cultural diversity and wild happenings. *Indeterminate Hikes+* (IH+) reappropriate smartphones, "which are generally used for rapid communication and consumerism, and turns them into tools of environmental imagination and meditative wonder, renewing awareness of biological, cultural, and media ecologies and slowing participants down at the same time"[14]. The Android app works by importing the rhetoric of wilderness into virtually any place registered by Google Maps, encouraging users to treat these locales as spaces worthy of the attention accorded to sublime landscapes, such as canyons and waterfalls. Participants are incited to 'look for flying creatures' or 'consider car alarms as bird calls'. They are invited to take photos of the 'urban wild' with their phones and upload the images to an online public database through the Android app.

CHANGING PERSPECTIVES

Non-Human Point of View

One of the outcomes of art production with new narrative formats has been a shift of perspective from human to non-human. This can be seen in the book by Daniel Quinn cited above, but this is just one example among a considerable body of literature. Giving voice to non-human animals allows us to place our feelings in their bodies, to represent them as significant elements of the biosphere, and so generate empathy. As specialist in Animal Studies Lucile Desblache (2011) states "[t]he animals of today's literature speak to us, sometimes through very structured logical discourse, sometimes through the utmost whimsical writing. Whatever the medium, they bring wisdom, as they allow our own limits to burst and make us aware of the dangers and restrictions of a pyramidal vision of creatures topped by human beings"[15] (p.51). The transmedia work *Bear 71*[16] is a good example; co-created by Leanne Allison and Jeremy Mendes, the film premiered at the Sundance Film Festival in January 2013. This transmedia documentary provides not only insight into contemporary humankind's relationship with nature, but also into human relationships by means of location technologies (mostly related here to security control and methods of tracking nature). The film leads us into the amazing, eye-opening, intricately detailed life of a grizzly bear. It was inspired by over one million black and white photographs taken in Banff National Park (Canada) by automated cameras set up to track wildlife, and raises the complex issue of (in the words of the narrator) 'where the wired world ends and the wild one begins'. An issue that is not easily resolved. The camera places human and non-human animals on the same level, sharing the same territory. Nature-lovers equipped for leisure pursuits unconsciously tramp over wild animal tracks. Meanwhile the animals must adapt to recent changes to their habitat –fences, railways, boundaries– that split their natural territory to create difficulties that must be overcome. The images show, on the one hand, the innocence of animals whose habitat has been incomprehensibly limited, and on the other, humans seeking the means to traverse these vast territories. One of the key values of the documentary is that camera takes and narration remain objective and unbiased. The film neither denounces nor accuses, but simply portrays events (as an organ of global vision). In this way, it provokes a curious 'swapping of minds' by alternating our point of view between feeling what it is to be inside the skin of the bear and feeling what it is to be inside human skin. By doing so, *Bear 71* provides a new platform from which to discuss a complex relationship without undermining the relevance of either human or non-human beings.

Swedish born artist Lisa Jevbratt has explored the expressions and exchanges generated by the protocols and languages of the Internet and the Web, often manifested as visualization software. She applies her understanding of these unintentional collaborations to exchanges with animals of other species and their experiences of the world around them. She invites participants to imagine the world 'through the eyes of another species'. The goal of her project *ZooMorph* is to produce software filters that generate simulations of how non-human animals see (Jevbratt, 2012). The filters are applied to image and video software (Photoshop, Final Cut), in an online Flickr-based interface, and as an augmented reality application in smart phones (iPhone). When a *ZooMorph* filter is applied to an image it changes its hue, sharpness, brightness, contrast and other aspects, providing an estimation of how the selected species sees. The project hopes to encourage people not only to use the filters on existing images, but to take images with the filters in mind, asking them to imagine themselves as non-human animals in different situations and environments (idem). The iPhone and iPad app applies the filters in real-time to provide a more direct experience of animal vision, an interface for imaging the world from the point of view of another species. In Lisa Jevbratt's words, *ZooMorph* is simple, naïve: "That is its potential strength. It is breaking complex things down to utter simplicity, and then by the sheer amount of these simple parts, statements, and experiences, it hopes that maybe something complex will emerge. Some knowledge" (2012, p. 91).

All animals have their own languages, and they leave their marks and voices on the territory, even if humans cannot hear them (Desblache 2011). As John Gray (2002) says, what is distinctively human is not the capacity for language but the crystallization of language in writing (p.56). Writing allows human beings to analyze the past and influence the future. For Gray, it is clear that technology and science increase human power; we can live longer and have higher living standards than in the past. But at the same time, it allows us to wreak destruction on a larger scale than ever before, on each other and on the earth.

Fiction as a 'Key' to a Point of View

Most people today think they belong to a species that can be the master of it destiny; this is not science but faith (Gray, 2002). This faith is projected in many ways in the fictions we create.

Meredith Drum is a digital artist and has produced the Augmented Reality walking tour of Manhattan. Her work *Oyster City*, developed with Rachel Stevens, reveals hidden narratives of the ecological and social history of New York, producing alternative views for the future: "*Oyster City* invites participants to explore ecological and social histories within the context of the long relationship between oysters and humans across this land and waterscape. The piece also invites speculation regarding the benefits of bioremediation, in the form of oyster reefs as smart city planning with serious potential to mitigate current and future problems, especially flooding. These elements of social documentary unfold as locative media, available only when participants move their bodies and engage their senses in exploring the often-neglected urban water's edge" (Drum, 2013, p.18). Reality is 'augmented' not only by images but also sound and narrative. This is the case of *A machine to see with*, by the group Blast Theory. This machine for seeing uses mobile phones: "just listen to the voice on the phone. The voice tells you what to do. The voice says you're playing the lead in a movie. Hide in the toilets, find the getaway car, stake out the bank and take a deep breath. You're going in"[17]. It is easy to imagine ways of influencing what users perceive using mobile devices as significant sound agents.

In April 2010, the Transnational Temps art collective launched the first edition of its *Safari Urbis*[18] in Poitiers (France). Versions for other cities followed –Madrid, Inca, Saint Jean D'Angely. This artwork combining workshop, performance, and mapping, uses communication technologies (mobiles) and game strategies in order to raise citizen awareness of the real situation of animals in danger of extinction. These alternative urban cartographies locate not only wild, endangered fauna (represented in everyday elements and advertising in the city) but also citizens (as consumers of all these connoted images). Transnational Temps demands that companies help animal conservation, for example the brand Puma, 'making fast product designs for the fastest athletes on the planet', could help 'real' pumas. This work also has to do with fiction. After exploring the city by means of the *Safari Urbis* game, we see the real animals behind the representations (daily fiction). Transnational Temps is an international art collective concerned with ecology, sustainability and media. Since its formation in 2001, it has produced a series of critically acclaimed works and exhibitions under the banner *Earth Art for the 21st Century* (Transnational Temps, 2007). Working primarily in Europe and the United States, works to date have emphasized participation, tactical media, spanning the sometimes awkward divide between activist advocacy and aesthetics.

Recent years have seen the emergence of documentaries related to environmental concerns through a preoccupation with trash. As Karl Schoonover (2013) has called it, 'a wave' of ecological documentaries made in the United States and Europe appear to confer with this sense that waste is something we are blind to. Amongst these are *Flow: For Love of Water* (Salina, 2008), *Crude* (Berlinger, 2009), *The Cove* (Psihoyos, 2009), *GasLand* (Fox, 2010), *Waste Land* (Walker, Harley and Jardim, 2010) and *Trashed* (Brady, 2012). Trash is a fundamental global issue and these films reveal that "the trash heap's real threat lies in an imperceptible danger: a festering toxicity as hard to track down as to contain. The problem of waste today is not only that which we are conditioned not to notice in our field of vision, but also what is simply impossible for human perception and photographic images to register" (p. 487). Obviously, a documentary in the traditional sense is not fiction. But the main objective of these audiovisual works is not to create (build), but reveal. It does not matter if the stories inside are fictional, as long as their overall output describes a reality. Some of these proposals, such as *Highrise: The Towers in the World. The World in the Towers*[19] are local (or hyperlocal) visions explaining the global phenomenon of urbanization. Launched in 2009, the film explores vertical living in the global suburbs. It was a multi-year, many-media, experimental, collaborative documentary backed by the National Film Board of Canada. As the director Katerina Cizek explains, "it takes several years using different platforms and media to experiment on documentary, to explore what the documentary world can learn from the digital"[20]. Along the way, the initial proposal has generated a number of other projects, including mixed media, interactive documentaries, mobile productions, live presentations, and installations. The main goal for creators here is to see how the documentary process can drive and participate in social innovation rather than just document it. In this way, the project aims "to help re-invent what it means to be an urban species in the 21st century"[21]. After *Highrise: The Towers in the World. The World in the Towers,* Katerina Cizek presented *A Millionth Tower* (2011), another interactive documentary experiment[22].

Alternative Cartographies

Analytical cartography reaches beyond the existential territories to which it is assigned (Guattari, 1989). Artists and writers working in some way as cartographers (for example working with mobility) can seek, develop, and innovate to create new perspectives. These new visions, or re-envisions, are like a jump-start to thinking. New relationships between people and between people and the planet can emerge, giving way to new links between environment, social relations and human subjectivity.

Furtherfield is an Art Center based in London (UK) that believes that through creative and critical engagement with practices in art and technology, people are inspired and enabled to become active co-creators of their cultures and societies. In 2009, the center set up Furtherfield's Media Arts Ecologies program. Its focus is on stimulating practice and debate about ecological approaches to work situated at the intersection of art, technology, and environment. The Media Arts Ecologies program engages in a critical view of growth economics and patterns of consumption, that is, emerging ecological media art practices. One of the projects in the program was *Sound Ecologies: Listening in the City*. It consisted of an exhibition and a series of events related to urban sound, networked sound, locative media, and acoustic ecology. The main idea was to explore the relationship between living beings and their environment, as mediated by sound. Here we can see a link between these sound-space-practices and what is called Deep Listening: connecting with the totality of sound in the space/time continuum. A part of *Sound Ecologies:* Ximena Alarcón's *Listening in the City, Sounding Underground*[23] is a 'virtual environment that invites users to interact with the soundscapes of three underground public transport systems: London, Paris, and Mexico City, Commuters', memories and imaginations are represented by sounds and images selected by volunteer commuters through an ethnographic process. These sounds are assembled in a sound score that acts as a multimedia user interface. The score contains sonic features unique to each metro, as well as some shared sonic spaces. *Sounding Underground* invites users to write memories in text, produced by listening. These 'public memories' are published randomly on a 'Listening and Remembering' page.[24] Recently, Ximena Alarcón has participated in another sound-performance based on deep listening: *My Voice Means* (2014). This artwork/workshop, developed with Ron Herrema, gives participants an opportunity to listen mindfully to the sound of their own voice. Using sound (instead of vision) to look around us is another way of interacting with the environment.

Teri Rueb has pioneered a form of GPS-based interactive installation. Her research addresses constructions of landscape and subjectivity in mobile network culture. She has founded and now directs the Open Air Institute, a platform for field-based learning and collaboration at the intersection of media, ecology and culture. Teri Rueb's *Drift* (2004) is a location-specific sound walk set along an expanse of tidal flats in northern Germany. In this work, the shore becomes an interface where participants move with mobiles and head-phones connected to a GPS unit. Within a sound circle, users hear recordings of literature and poetry on the experience of being lost, of drifting. The work invokes "psychogeography, non-utilitarian walking as a refutation of hegemonic economic engines" (Drum, 2013, p.18). In 2012, Rueb developed *Places with Names*, another GPS sound walk. In this case, the walk is through the high desert surrounding the Institute for American Indian Art in Santa Fe, New Mexico. *Places with Names* is about the history of the word 'wilderness' in which participants with mobile devices explore the sound narrative, "an inquiry into how language, native and imported, represents, limits and shapes our relationship to cultural and physical landscapes" (Drum, 2013, p.18).

Acoustic Networks of Birds is a three-year research project led by evolutionary biologist Charles Taylor in collaboration with a research group, including artist Victoria Vesna. The goal of this remarkable sound project is to understand the language of birds. Although we long to understand birdsong, to date this has been impossible. However, modern advances in computing, in linguistic analysis, and a new-found appreciation of how sophisticated other creatures can be, make it possible to study the grammar and possible meaning of birdsong. It is expected that the results of this investigation will emerge in the coming years. In 2006, Beatriz da Costa (1974-2012) released *PigeonBlog*, in collaboration with Cina Hazegh and Kevin Ponto. The project was "a collaborative endeavor between homing pigeons, artists, engineers and pigeon fanciers engaged in a grassroots scientific data gathering initiative designed to collect and distribute information about air quality conditions to the general public" (da Costa, 2008, p.377). The collected data –pollution levels– were visualized and plotted in real-time over Google's mapping environment and open to anyone with an Internet connection. *PigeonBlog* was developed and implemented in Southern California, which ranks among the ten most polluted regions in the country. Its aims were to reinvoke urgency around a topic that has serious health implications, to broaden the notion of citizen science, while building bridges between scientific research agendas and activist-oriented citizen concerns, and to develop mutually positive work and play practices between situated human beings and other animals in technoscientific worlds. The project opens a necessary discussion about possible new forms of cohabitation in urban ecologies between humans and non-humans, whereby da Costa has described the pigeons as her communicative objects and "collaborators" in the co-production of knowledge.

Such projects can and do inspire audiences by generating alternative visions through creativity and collaboration. How we represent space (sound, images, other media) has everything to do with how we embody that space. Mapping, as a means of representing and practicing space, serves as a key example in the exploration of what 'space' means in the embodied practices of mobile technologies (Farman, 2012). As Meredith Drum (2013) asserts, emerging place-based projects with a critical focus have the capacity to activate physical sites as places of social engagement in the public sphere; "Somewhat paradoxically, the same mobile devices that normally remove us from our environment and our immediate community can be used to connect us somatically with sites and encourage dialogue with the lives, human and non-human, occupying these places with us" (p.19). Obviously, these artworks are like social interfaces and the art world itself is changing as a result of such cultural influences, ones that raise new questions focused less on what art is and more on what it can do (Lovejoy, 2011).

The Map and the Territory

Landscapes have become information interfaces much like the graphic user interface of a computer screen (Farman, 2012). The landscape around us can serve as a type of interface in which data reside. In this sense, map and territory are close, the object and the representation of that object are in touch. We could say that cities have always used the urban landscape as a site of information, but today these urban spaces are open to a bi-directional exchange so that users participate in the creation of that information. With locative and mobile media, "the readily available virtual information 'up there' allows new possibilities for configuring experiences in the physical, public space 'down here' that can become an opportunity to reshape experiences in which viewers see things differently on the ground" (Levine, 2014).

In 2009, ecoartivists Kayle Brandon and Heath Bunting published their *Food for Free* map. They mapped the edible plant organisms living within the public realms of Bristol since 2004. Plants that are rooted in private domains yet branch into public space are included in this survey. Potential planting sites are also investigated, with the aim of increasing the life spectrum of the city. It is an unusual map in the sense that it does not include road names and yet must be navigated by following the contours of the roads. The project *Garden Registry* (2008) in San Francisco operates in a similar way; it is an online map and social networking tool created to connect urban gardeners and to locate open spaces in the city that are suitable for growing food. *Garden Registry* is a Futurefarmers[25] creation, led by Amy Franceschini. Since 1995, Futurefarmers have been practicing a form of cultural activism that exploits the interactive potential offered by new media and public space. These territorial practices are capable of carrying bigger ideas into the consciousness of a town, and beyond.

Activist and environment worker Pamela Warhurst[26] is known for founding the voluntary gardening initiative, Incredible Edible, in Todmorden, West Yorkshire. She describes these "public space food plantings as "propaganda gardens" – something to use as a tangible expression of a set of bigger ideas – including growing local, eating local and fresh, eating seasonal, and knowing the provenance of food" (Paull, 2011, p.28). Already, there are copy-cat projects in England, Ireland, Spain, and Canada.

Then, there are geocaching narratives, linked with concepts of territory. These GPS-enabled modern treasure hunting games were first initiated in the spring of 2000 by a man from Oregon, who hid a box of toys among the woods, posted its geographic latitude and longitude on a website, and called on anyone interested to find it (Zeng, 2011). Today, geocaching games are played by people of all ages and countries with over a million 'treasure' boxes hidden worldwide. It has become one of the most prominent applications of locative media in every day usage and it has great potential for generating empathic narratives. These narratives are described by Paula Levine (2014) as those that allow an individual to feel a deep connection to an external feeling or event, as if it was happening to him or herself. While sympathy allows a person to feel a sense of mourning or loss for another, empathy puts the person in another's shoes. Empathic narratives develop stories that engender a level of identification with experiences and feelings. In this sense, ecological issues might be considered suitable content for empathic narrative. Geocaching makes an "explicit claim to link virtual with physical space, it distinguishes itself with its particularly high degree of connectedness to the actual world" (Zeng 2011, p.120). *Earthcache*[27] is a term used to define special geocaches that people can visit to learn about a unique geoscience feature of the Earth. *EarthCache* pages include a set of educational notes along with cache coordinates. This engages with issues that include geological processes, resource management, and how scientists gather evidence to learn about the Earth. Another 'green' type of geocache is *Cache In Trash Out*[28]. This ecocache focuses on litter clean-up, removal of invasive species, revegetation efforts, or building trails. Users are invited to bring a trash bag along with them on their walks in the woods and pick up the occasional piece of trash they see on the trail.

Ecocaching[29], is an individual project developed by Finnish media artist Tapio Mäkelä as an artistic locative game. Players search for boxes called 'ecocaches' hidden in city space according to GPS location and sets of clues. Ecocaches range from green spotting (observing and collecting animal and plant life) to waste spotting, and special ecocaches. Each ecocache contains a task related to the local environment.

As Jason Farman (2010) asserts, through spatial discussions and map overlays, users become interactors involved in the representation of the social space of the global village. The territory is defined through this interaction. The project *Sauti Ya Wakulima*[30] ("The voice of the farmers" in Kiswahili) is an example of how locative media can help communities living in poor economic conditions or in need. This networked communication tool for farmers allows the publication of images and voice recordings on the Internet using smartphones. It provides farmers with access to pertinent agriculture-related information, advice and research to help them make better decisions about their crops, increasing productivity and potential income[31]. The farmers, men and women in Chambezi (Tanzania), struggle not only because of insufficient infrastructure and unreliable markets for their products, but also because they face the challenges of a changing local climate. The project, based on an Android app called *ojoVoz*, has been developed by Eugenio Tisselli, and is being applied in other artistic/social projects, such as *Reporteros ambientales* in Colombia, *Los ojos de la Milpa* and *Red Residuos Sólidos del Valle de México* in Mexico, all of them related to territories and community memories.

MOBILE AWARENESS

Mobile Action and Wildlife

Recent years have seen the development of a considerable number of mobile apps aimed at raising awareness of environmental situations. Most of them are social-action apps. In this way, *Wildlife Witness*[32], developed by Taronga Conservation Society Australia in Partnership with TRAFFIC, the wildlife monitoring Network, allows users to report illegal trade. The illegal trade in animal or animal parts is a highly lucrative area of environmental crime. Iconic species such as elephants, rhinoceroses, tigers, and great apes are the most affected. With this app, users can report wildlife they see captured for commercial gain, consult reports on a world map, and learn about how wildlife is affected by trade. In this way, *Wildlife Witness* represents a global community action tool to tackle illegal trade.

WildObs[33] is a partner of the National Wildlife Federation's Wildlife Watch, which works with a number of other scientific studies to extract citizen science from recorded encounters. *iNaturalist*[34] is a geolocated recording app that marks species locations in the field and uploads associated images. It is not an activist-specific tool but users can set up specific projects and establish networks of observers. In this direction, *Noah*[35] app is another nature-mapping tool. It allows users to create online communities where they can learn more about wildlife and contribute to scientific research. *BirdsEye*[36] registers entries that record hundreds of the most frequently-seen North American bird species including images and bird sounds. It helps to guide bird watchers to birds, and is based on sightings submitted online to eBird.org. This kind of app, called a SciPhone app, is not a social-action tool but a community (citizen scientists) research tool.

From an ecoartivist perspective there is *SimioMobile*[37] (2008) created by Transnational Temps. The project relates telephony to the stressed habitats affected by mobile phone manufacturing in Africa. Partnered with various environmental organizations (Great Ape Project amongst others), *SimioMobile* seeks to spread awareness of coltan's malign effects using fun and pop sensibilities aimed at youth culture. *Digital Jungle Simiomobile* seeks a contractual commitment on the part of the telecommunication industry to guarantee ethical commerce with respect to coltan, including greater efforts to procure the mineral through recycling. With similar concerns, *iMine*[38] is an experimental art-app/game for iPhone conceived by Baruch Gottlieb in 2010. In the game, the user takes on the role of a miner 'digging for one of the precious ingredients for today's electronics: tin, tungsten or tantalum'. As the artist asserts, hardware is made from materials which all started life in the earth; the closer we get to the origin of the materials used in digital technology, the more difficult the conditions become.

Trash Track (2009) is an initial investigation into understanding the 'removal-chain' in urban areas and it represents a change that is taking place in cities: a bottom-up approach to managing resources and promoting behavioral changes through pervasive technologies. SENSEable City Lab[39] has created *One Country, Two Lungs* (2014), "[a]midst a matrix of towering skyscrapers and dense streets, citizens inhabiting two vertical cities are connected by nothing but air… the breathing, invisible, and interconnected landscapes of Hong Kong and Shenzhen. Is it possible to visualize the ebbs and flows of the cities' atmosphere? How do humans fit into this binary urban ecosystem? Can we map what people breathe and absorb, showing the dynamics of air pollution? How can these results empower citizens to transform the urban space around them?"[40].

The *Marine Debris Tracker*[41] (2011) app can be downloaded free for use on iPhones and Android phones. This app allows users to report the type of debris and its location using GPS features pre-installed on a cell phone. Given that the majority of debris tracking might take place in remote areas or even at sea, where there may be no Wi-Fi or even a cell signal, citizen scientists can log and track as many items as they want and store this information in their smartphone until they return to a place where they can submit their findings online.

Global Forest Watch[42] is a dynamic online forest monitoring and alert system that empowers people everywhere to manage forests better. Launched in 2014, this ambitious project links satellite technology, open data, and crowdsourcing to guarantee access to timely and reliable information about forests. It is free to use and follows an open data approach in putting decision-relevant information in the hands of governments, companies, ONGs, and the public. Users can explore forest clearings worldwide over the last six years, submit a report of forest change, contribute or share data, suggest improvements, contribute open source codes, and validate data by participating in citizen science. Another interesting forest-based project is *Rainforest Connection*[43] (2014), which uses recycled mobile technology to monitor and protect remote forests in real-time. It is the first real-time monitoring system for tracking deforestation and illegal logging, and uses discarded Android smartphones to send alerts of rainforest destruction to first responders, giving them the ability to actually intervene (by contrast, current monitoring often relies on aerial surveys or satellite surveillance, which only detects logging days or even weeks after the event). The *Rainforest Connection* system was first tested in 2013 against illegal loggers in Western Sumatra, Indonesia. Using highly sensitive microphones, each autonomous, low-cost device can protect one square mile of rainforest, often home to over a thousand species of plants and animals. The devices, built to operate for many years, employ a patent-pending solar-panel, which generates sufficient electrical power despite the shade of the tree canopy[44].

"StoryBeing": Participating, Acting

One of the necessary conditions for 'creating' within the confluence of entertainment and culture, and for forming a critical and participative social mass, is the generation of inclusive synergies. This is not an easy task. In 2013, writer and performer Nathan Penlington created *Choose Your Own Documentary*[45], in collaboration with filmmakers Fernando Gutierrez De Jesús, Sam Smaïl and Nick Watson. It is a live show described as a unique experience performed in a theatre or cinema space. Audiences receive remote controls at the start of the show and are asked to vote in response to a series of questions and decisions throughout the show. The 'system' is quite simple and based on the *Choose Your Own Adventure*[46] book series. It too starts with a book, the story of someone buying a set of the first 106 volumes of *Choose Your Own Adventure* novels for sale on eBay. At the start-point –*The Boy in the Book*– there are two parallel stories: "there was another story being written– in the margins were scribblings by the little boy who had originally owned them, a boy called Terence". The show moves between different time-lines: fictional and real-time; the novels are already written but the decisions made by the audience have a direct impact on the unfolding story. As the posters and flyers for the show announce "there are 1,566 versions of this show", but the key to its success goes beyond the number of versions, and lies in the fact that the audience writes, in real-time, another story. As the press release declares, this is "decision making to keep you glued to your seat"[47]. Although, *Choose Your Own Documentary* is not an eco-art project, it is a good example of active participation and decision making. Another example is the Augmented Reality Game *Ingress*[48], a science fiction action game created by Google's Niantic Labs. The game images a mysterious global energy, one that occupies space and is projected from different points in the city. It is not difficult to see how the game could easily become an environmentally engaged project. To start with, it sets up two teams or factions on the planet (the 'Enlightened' who seek to embrace the power that this energy may bestow upon us, while the 'Resistance' struggle to defend and protect what is left of our humanity). In the field of ecology, it is easy to identify a similar division between those who take on the environmental emergency and who believe we can repair or relieve it, and those who are not prepared to consider any change. But the point is not to imagine turning the game *Ingress* into environmental activism; what are of interest are the features that offer potential for engaging and mobilizing a remarkable number of players. The game uses geo-location technology in mobile devices to transform the world into a virtual game. Real world locations such as sculptures, parks, landmarks, historical places, become 'portals' for participants (Agents) to capture and control. This Augmented Reality experience encourages groups of players to leave home and discover and interact with the world around them. In this way, *Ingress*, like most games, has a proactive format.

In recent years, games like *Spore* (2008) or its predecessor *Sims Earth* (1990) are examples of player involvement in environmental play. *Spore*, developed by Maxis and designed by Will Wright, simulates the development of a species on a galactic scale. The player molds and guides a species across many generations, growing it from a single-celled organism to a more complex creature. *Spore* uses procedural generation for many of the game's components, providing vast scope and open-endedness. When he developed the game, Wright hoped to address the topic of astrobiology (Bogost, 2008, p.135), the study of life throughout the cosmos. Wright argues that the spread of life in the universe is most likely caused by sentient beings transporting other creatures from star to star – the meaning of 'spore'. *SimEarth: The Living Planet*, also designed by Will Wright, is a life simulation game in which the player can vary the atmosphere, temperature, landmasses, etc., then place various life-forms on the planet and watch them evolve. *SimEarth* models the Gaia hypothesis of James Lovelock, who assisted with the design and wrote an introduction to the game's manual. The new version of *SimCity* (2013) presents a sophisticated toy in which players can lose themselves in fantasy worlds, but also offers users "tough choices about energy generation, environmental costs, and the responsibilities shouldered by inhabitants of a planet with finite resources"[49]. With multiple players, the decisions that a user makes in his/her city –for instance regarding the pollution of water streams– can impact on the decisions made in their neighbor's city, and so the game may open onto a deeper understanding of our own world.

Storytelling is a fundamental action which has to do with how we experience the past and the future. It is not just fictional but also about reality, and this is what makes it essential. Let us reconsider Gray's assertion that writing allows human beings to analyze the past and influence the future. Contemporary writing is more than ink on paper, it is 'Media' or a 'Space of Possibilities'[50]. Media experiences become multiplatform, participatory, content providers, forming connective tissue to the real world. Latitude, a strategic insights consultancy working with the world's foremost media, technology, and advertising companies, affirm in their innovation study *The Future of Storytelling* that audiences want to have experiences that not only allow them to delve deeper into stories, but that also bring stories out of the screen –both literally and figuratively– to mingle with their actual lives. Their actual lives on planet Earth of course. So, the key dimensions of good storytelling experiences should be: immersion, interactivity, integration, impact (does the story inspire me to take action in my own life?)[51].

Solutions; Recommendations; Future Research Directions

The concept of progress and the relations that human development establishes for and with the biosphere is a key issue for debate. The aseptic, machinic future reflected in science fiction literature is far from sustainable and desirable in real terms. In *Symbiotic Planet*, the remarkable and distinguished Lynn Margulis describes the first time she ever saw an episode of *Star Trek*, "I was struck by its silliness. The lack of plants, the machinate landscape, and in the starship, the absence of all non-human life-forms seemed bizarre. Humans, if someday they trek in giant spaceships to other planets, will not be alone. In space as on Earth, the elements of life, carbon, oxygen, hydrogen, nitrogen, sulfur, and phosphorus and a few others, must recycle. This recycling is no suburban luxury; it is a principle of life from which no technology can deliver us" (2001, p.132). *Star Trek,* created by Gene Roddenberry and debuted in 1966 was subject to diffusion techniques that can be classified as crossmedia, generating a large number of novels, comic, video games, books, and other associated products. Star *Trek*'s cultural impact goes far beyond its longevity and profitability. Its fans have developed a global subculture and its creators have spun off eleven movies and four television series. The world of *Star Trek* represents the ideals that are at the core of American public culture and the key to success in the context of the show is warp technology (Crothers, 2013). But when we analyze the series form an environmental or ecological viewpoint, it is, as Margulis has described, truly absurd. "Space – the final frontier" says the voiceover in the title sequence. If we stop to consider this phrase we already have reasons for concern. The landscape of the future as portrayed in *Star Trek* is clear: technologically-empowered people in a world with no place for biodiversity (beyond human-animals). There is no definite link between fiction and real space, the fiction operates inside a hermetically sealed bubble. Star trek does not generate a dialogical space but fiction operates like a blanket smothering but not interacting with reality. Janet Murray (2012) in her latest book writes "When we expand the meaning-making conventions that make up human culture, we expand our ability to understand the world and to connect with one another"(p.2).

From a critical, ecological, and social point of view, a 'desirable' narrative is dialogical. It is a fictional or non-fictional narrative with content related to context; a fiction that permits and recognizes the participation of its users, which aids citizen coordination and drives or makes action possible with an impact in real-time. The desirable narrative is not the representation of a better world but one which helps to improve the reality that we share with other living beings on planet earth. Happiness and well-being is more than the sum of superficial pleasures. We must reflect on this in our writing (regardless of the technological complexity of the 'text'). The philosopher Alicia Puleo (2011), one of the main representatives in the struggle for equality and environmental justice in Europe mentions the "paradox of hedonism" (p.412), known since Antiquity as the state of dissatisfaction of persons seeking a state of happiness through the many pleasures which are not marked out in a project transcending selfish motives. There is no doubt that "[t]he contemporary nihilistic consumer is the most comprehensive historical realization of this" (Puleo, 2012, p.84). And it is obvious that meanwhile millions of people have been trapped in this hedonistic paradox, many others lack the necessary resources to survive. If this is the situation of human beings, "what is to be said of the woodland creatures which are literally being wiped out by hunting, herbicides, pesticides, and alterations due to climate change, in other words the disappearance of the world? Is this the Earth we want?" (Puleo, 2012, p.84).

As Scolari and Ibrus (2014) explained in a recent publication, "to reinterpret transmedia as an important outcome and as a source of contemporary cultural and social complexities –not only as new forms of cultural texts and media institutions or practices but also as new forms of scarcity, inequality, and power struggles. That is, it is necessary to become explicitly concerned about all the manifestations of social power that have conditioned the emergence of transmedia practices and about the new forms of dominance that these practices may have enabled" (p. 2193). They conclude that transmedia is unquestionably an important trend within contemporary media culture. And so it is hoped that intellectual work within transmedia, as well as production (strategies, methods, and also content), will be more critically oriented. This is the narrative we desire.

As Fuller (2005) has declared, "a technology is a bearer of forces and drives, is indeed made up of them (...) it is composed by the mutual intermeshing of various other forces that might be technical, aesthetic, economic, chemical –that might have to do with capacities of human bodies as affordances– and which pass between all such bodies and are composed through and among them" (p.56). In this way, it is imperative to consider the physical side of technological, social and environmental impacts.

Beth Carruthers (2006) asserts that increasingly sciences and environmental groups are looking to the arts for collaboration. Art can be a learning tool and also a social agent. It can "facilitate a process of learning through the engaged senses, bypassing conditioned patterns of thinking and allowing other ways of knowing to come forward, at times subtly, at times overwhelmingly" (p.5). Artists and creative discourses can play an important role in constructing an urgently needed environmental awareness like never before "as it is in the face of what is now obvious to all as an immediate global crisis within our sustaining and environing world. Because this crisis has been and continues to be nurtured and produced by past and current cultural practices and ideologies, artists, immersed in world and cultural practices, are ideally situated to locate and develop responses" (Carruthers, 2006, p.5).

More environmental information is being made available by means of mobile devices, which also provide the means to transmit information and participate in communal, participative processes. Non-profit organizations and communities, such as *Public Lab*[52] in Cambridge, are emerging, where individuals can learn how to investigate environmental concerns. Using inexpensive DIY[53] techniques, these groups seek to change how people see the world in environmental, social, and political terms.

From the perspective of art and creation engaging with the biosphere and with mobility, we believe the following lines of research and action could be positive in the move towards a world based on environmental justice:

1. Changing points of view. Using emergent storytelling and locative media to feel/perceive from another perspective including non-human animals.
2. Creating new cartographies to improve our perception.
3. Entertainment is also culture, and as Guattari has suggested, within every discipline and background it is possible to inspire a change of values and open our minds to new ways of perceiving, feeling, and then acting. The audience wants to choose, to decide, and so it is necessary to create participative and collaborative forms of entertainment. The *CYOD* project shows that this is possible.
4. Art and creativity can facilitate a process of learning and can operate as a social agent. Artists or creative activists can 'translate' vital information into forms that reach individuals, communities and organizations.

CONCLUSION

We cannot claim that art and/or technology can solve global problems, change the world, or change people's minds. But when art and technology act "in collaboration with social movements, scientists, governments, and communities, the sky's the limit" (Lippard, 2011, p.11). Environmental art is like a magnifying glass, says Lippard, it reveals details and depths we do not see or do not think about. Environmental art that makes use of contemporary tools (mobility, hypermedia, transmedia, and so on), as in the case of the artworks described in this article, operate like a 'liquid magnifying glass', flowing between the pillars of contemporary communication.

Art practices, and creative and cultural work are changing with the rise of new forms of storytelling: locative media-based, transmedia, multimodal, and longitudinal. This is the era of the Creative Activist. This term, coined by Kathy Eldon, refers to something other than political activism. A Creative Activist uses art and media to scale up the message to a global level, to reach a lot of people and to move them (Tercek, 2014).

Finally, we should remember Lynn Margulis' words (2001): "[w]e need honesty. We need to be freed from our species-specific arrogance, (...) No evidence exists that we are 'chosen,' the unique species for which all the others were made" (p.149). Planet Earth –or Gaia– is not human, nor does it belong to humans. Humans are not the center of life; we are a recent, rapidly growing part of an enormous ancient whole. Gaia is the place for all the living beings we know. It is also the place where our dreams are born. If we feel our lives to be something more than a succession of hedonistic distractions then it is crucial that we reconnect with the biosphere. We hope contemporary eco-art practices, let's call them Creative Artivism, will help us to resolve the simple question: where on Earth are we going?

REFERENCES

Blum, J., Blintham, M., Jacobs, R., Shipp, V., Kefalidou, G., Brown, M., & McAuley, D. (2013). The Timestreams Platform: Artist Mediated Participatory Sensing for Environmental Discourse. *Proceedings of UbiComp '13 the 2013 ACM International Joint Conference on Pervasive and Ubiquitous Computing*. Zurich, Germany: ACM.

Bogost, I. (2008). The Rhetoric of Video Games. In K. Salen (Ed.), *The Ecology of Games: Connecting Youth, Games, and Learning* (pp. 117–140). Cambridge, MA: The MIT Press.

Carruthers, B. (2006). Mapping the Terrain of Contemporary Ecoart Practice and Collaboration. *Proceedings of ART in ECOLOGY – A Think Tank on Arts And Sustainability*. Vancouver, Canada: Canadian Commission for UNESCO.

Colebrook, C. (2011). *Extinctions*. London, United Kingdom: Open Humanities Press.

Crothers, L. (2013). *Globalization and American Popular Culture* (3rd ed.). Plymouth, United Kingdom: Rowman and Littlefield.

da Costa, B., & Philip, K. (Eds.). (2008). *Tactical Biopolitics. Art, Activism, and Technoscience*. Cambridge: The MIT Press. doi:10.7551/mitpress/9780262042499.001.0001

Desblache, L. (2011). *La Plume des bêtes. Les animaux dans le roman*. Paris, France: L'Harmattan.

Drum, M. (2013). Place-based, Somatic Augmented Reality as Critical Practice. AR[t] 4, 16-20.

Farman, J. (2010). Mapping the digital empire: Google Earth and the process of postmodern cartography. *New Media & Society*, *12*(6), 869–888. doi:10.1177/1461444809350900

Farman, J. (2012). *Mobile Interface Theory*. New York: Routledge.

Fuller, M. (2005). *Media ecologies: Materialist energies in art and technoculture*. Cambridge: The MIT Press.

Gottlieb, B. (2009). *Gratitude for Tecnology*. Dresden, Germany: Atropos Press.

Gray, J. (2002). *Straw Dogs. Thoughts on humans and other animals*. London, United Kingdom: Granta Books.

Guattari, F. (1989). *Les trois écologies*. Paris, France: Galilée.

Haraway, D. (2008a). *When species meet.* Minneapolis: University of Minnesota Press.

Haraway, D. (2008b). *Haraway on Multispecies Cosmopolitics: Staying with the Trouble.* Minneapolis: University of Minnesota Press.

Jenkins, H. (P.I.) (2009). Confronting the Challenges of Participatory Culture Media Education for the 21st Century. The John D. and Catherine Foundation Reports on Digital Media and Learning. Massachusetts: MIT Press.

Jevbratt, L. (2012). With the eyes of another. Antennae, 21, 76-93.

Levine, P. (2014). On Common Ground: Here as There. In J. Farman (Ed.), *The Mobile Story: Narrative Practices with Locative Technologies* (pp. 143–158). New York: Routledge.

Lippard, L. (1981). Hot Potatoes: Art and Politics in 1980. Block, 4, 2-9.

Lovejoy, M., Paul, C., & Vesna, V. (Eds.). (2011). *Contexts Providers: Conditions of Meaning in Media Arts.* Bristol, United Kingdom: Intellect.

Margulis, L. (2001). The Symbiotic Planet. London, United Kingdom: Phoenix.

Morton, T. (2010). *The ecological thought.* Cambridge, MA: Harvard University Press.

Moyer, T., & Harper, G. (Eds.). (2011). *The New Earthwork. Art, Action, Agency.* Seattle: Isc Press.

Murray, J. (2012). *Inventing the Medium. Principles of Interaction Design as a Cultural Practice.* Cambridge: MIT Press.

Ohlenschläger, K. (2007). The Biosphere as Infosphere. In S. Himmelsbach & Y. Volkart (Eds.), *Ecomedia. Ecological Strategies in Today's Art.* Oldenburg: Edith-Ruß-Haus für Medienkunst.

Paull, J. (2011). Incredible Edible Todmorden: Eating the Street. In Farming Matters 27(3), 28-29.

Plant, S. (2001). *On the Mobile*: Motorola. Retrieved from http://www.it-c.dk/courses/MGB/F2002/on_the_mobile.pdf

Postman, N. (2000). The Humanism of Media Ecology. *Proceedings of the Media Ecology Association, 1*, 10–16.

Puleo, A. (2011). *Ecofeminismo: Para otro mundo posible.* Madrid, Spain: Cátedra.

Puleo, A. (2012). Speaking from the South of Europe. In *Deportate, esuli, profughe 20*, 78-89. Retrieved from http://www.unive.it/media/allegato/dep/n20-2012/Ricerche/Riflessione/8_Puleo_Ecofemminismo_sud.pdf

Quinn, D. (1992). *Ishmael: An Adventure of the Mind and the Spirit.* New York, London: Bantam/Turner Books.

Salen, K. (Ed.). (2008). *The Ecology of Games: Connecting Youth, Games, and Learning.* Cambridge, MA: The MIT Press.

Schoonover, K. (2013). Documentaries without documents? Ecocinema and the toxic. Necsus, 2(2), 483–507.

Scolari, C. A., & Ibrus, I. (2014). Transmedia Critical: Empirical Investigations into Multiplatform and Collaborative Storytelling. *International Journal of Communication*, *8*, 2191–2200.

Tercek, R. (2014). *Documentary Filmmakers in Surveillance Society*. TEDxHollywood. Retrieved from http://youtu.be/sh9t1UE7b7A

Transnational Temps. (2007). Transnational Temps presents: eco-scope. In J. Martín (Ed.), *Inclusiva-net. New Art Dynamics in Web 2 Mode* (pp. 195–213). Madrid, Spain: Área de las Artes, Dirección General de Promoción y Proyectos Culturales.

Zeng, M. (2011). Examining Geocaching Practices through a Mobilities Lens. [Alberta, Canada: MEA.]. *Proceedings of the Media Ecology Association*, *12*, 113–122.

KEY TERMS AND DEFINITIONS

Creative Artivism: Using art and media to scale up an activist message to a global level, to reach lot of people and to move them.

Ecoartivism: Artivism is used to describe artistic practices that are also activist (ecological, social, political). Ecoartivism are the artistic and activist practices in an ecological context.

Media: The platforms or "vehicle" for content in the global project. It may be digital (mobile phone, pad, PC, Web-based, etc.) or analog (objects, board, cards, social, etc.).

Space of Possibility: The space of all possible actions and meanings that occur in the course of a transmedia project, an alternate reality game or other type of emergent narrative.

Ubiquitous Computing: A design framework that makes possible a computing technology linked to the environment. Computation is integrated directly into objects or using mobile devices. Also referred to as ubicomp.

ENDNOTES

1. See the *Global Environment Outlook* (GEO-5), published by the United Nations Environment Programme: http://www.unep.org/geo/geo5.asp
2. "Are Toxins Astir? Release the Hounds" by Mark Glassman. In *The New York Times*, June 26, 2003. http://www.nytimes.com/2003/06/26/technology/circuits/26dogs.html
3. *Bat Billboard*: http://www.woebken.net/bat_billboard.html
4. *Bat Billboard* at the MOMA in New York. http://www.moma.org/interactives/exhibitions/2011/talktome/objects/146223/
5. *OOZ*: http://www.nyu.edu/projects/xdesign/ooz/
6. For more information about *Environmental Health Clinic* look this presentation by the artist at TED Talks: http://www.ted.com/talks/natalie_jeremijenko_the_art_of_the_eco_mindshift#t-213931
7. For more information about this look at Donna Haraway presentation as the Institute for the Humanities Research's 2013 Distinguished Lecturer: http://vimeo.com/62081248

8 *SF: String Figures, Multispecies Muddles, Staying with the Trouble*. Presentation at the University of Alberta, March 24, 2014. http://www.situsci.ca/event/sf-string-figures-multispecies-muddles-staying-trouble

9 About Crittercam: http://animals.nationalgeographic.com/animals/crittercam-about/

10 Sharks Now Tweet When Approaching the Beach. Turning transmitter data into Twitter alerts. *Adweek*, 27 December, 2013. http://www.adweek.com/adfreak/sharks-now-tweet-when-approaching-beach-154662

11 *Heartlands* is a game controlled by player's heart rates. As a player takes a walk, their journey is mapped on a smart mobile phone by GPS (global positioning system) and a heart rate monitor that tracks the player's heart rate. The goal is to keep their heart rate healthy as they explore, in order to score points. A virtual map of the journey is recreated live on the mobile phone and projected on a large screen (exhibition), reflecting the real journey as it happens: http://www.i-am-ai.net/work/heartlands-ere-be-dragons/

12 *The Predicted Machine*: www.thepredictionmachine.org/

13 *Timestream Platform*: http://timestreams.org/

14 About *Indeterminates Hikes+*: http://ecoarttech.org/projects/indeterminate-hike/

15 The original text is: "Les amimaux de la littérature d'aujourd'hui nous parlent tantôt à travers les discours raisonné le plus structuré, tantôt à travers la fantasie la plus débridée. Mais plus jamais, ils sont porteurs de sagesse, car ils font éclater nos propres limites, et nous font prendre conscience de l'étroitesse et des dangers d'une vision pyramidale des êtres dont l'être humain se trouve au sommet". Thanks to Lucile Desblache for the translation.

16 *Bear 71* is a interactive documentary created in 2012 by Leanne Allison and Jeremy Mendes. Available from

17 Blast Theory, *A Machine to See With*. http://www.blasttheory.co.uk/projects/a-machine-to-see-with/

18 *Safari Urbis*: http://transnationaltemps.net/safariurbis/

19 Highrise. The Towers in the World. The World in the Towers: http://highrise.nfb.ca/

20 Katerina Cizek: Case Study: HIGHRISE: Out My Window. http://vimeo.com/32321155

21 About *Highrise*: http://highrise.nfb.ca/about/

22 *One Millionth Tower*: http://highrise.nfb.ca/onemillionthtower/

23 Ximena Alarcón website: http://ximenaalarcon.net/

24 Listening and remembering (Ximena Alarcón) http://ximenaalarcon.net/soundingunderground/listeningtexts/listening.php

25 Futurefarmers site: http://www.futurefarmers.com/

26 Pam Warhurst: *How we can eat our landscapes* (TED Talks) http://www.ted.com/playlists/28/sustainability_by_design

27 *Earthcache*: http://www.earthcache.org/

28 *Cache In Trash Out*: http://www.geocaching.com/cito/

29 *Ecocaching*: http://ecocaching.net/

30 *Sauti Ya Wakulima*: http://sautiyawakulima.net

31 Small-scale Agriculture and Mobile Technologies: http://www.sautiyawakulima.net/research/

32 Wildlife Witness: http://taronga.org.au/support-us/take-action/wildlife-witness

33 *WildObs*: http://wildobs.com

34 *iNaturalist*: http://www.inaturalist.org/

35 *Noah*: http://www.projectnoah.org/

36 *BirdsEye*: http://www.birdseyebirding.com/

37 *SimioMobile*: http://simiomobile.com/

38 *iMine*: http://www.i-mine.org/

39 About Senseable City lab: http://senseable.mit.edu/

40 *Two Lungs*: http://senseable.mit.edu/twolungs/

41 *Marine Debris Tracker*: http://www.marinedebris.engr.uga.edu/

42 *Global Forest Watch*: http://www.globalforestwatch.org/

43 *Rainforest Connection*: https://rfcx.org/

44 June 24, 2014. Press Release from Rainforest Connection: http://goo.gl/Rmnj8Y

45 *Choose Your Own Documentary*: http://www.cyod.co.uk

46 *Choose Your Own Adventure* is a series of children's gamebooks where each story is written from a second-person point of view, with the reader assuming the role of the protagonist and making choices that determine the main character's actions and the plot's outcome. The concept was created by Edward Packard in 1976. The official site: http://www.cyoa.com/

47 Official London Theatre press: http://goo.gl/1h5zZU

48 *Ingress*: http://www.ingress.com/

49 *SimCity*: http://www.simcity.com

50 Look at the Key Terms and Definitions section.

51 More information is available at "The Future of Storytelling", a Latitude's website: http://futureofstorytellingproject.com.

52 The Public Lab: http://publiclab.org/

53 Do it yourself (DIY), the method of building, modifying, or repairing something without the aid of experts or professionals.

Chapter 18
Mobile Media and Social Interaction:
Mobile Services and Content as Drivers of Social Interaction

Manuel José Damásio
Lusófona University, Portugal

Inês Teixeira-Botelho
Lusófona University, Portugal

Sara Henriques
Lusófona University, Portugal

Patrícia Dias
Lusófona University, Portugal

ABSTRACT

This chapter discusses the new social configurations society is undergoing on the basis of media emergence. Media are embedded in the arousal of communication and information transmission becoming the form, the infrastructure and the institution for the social and culture. This chapter focuses on mobile communication, having as central goal to debate on the processes of mediatization and mediation of society, as well as on the processes of belonging and social cohesion. Data from mobile internet adoption and use will be discussed in the light of the above mentioned theoretical approaches. An empirical case study will also be approached and results will provide contributions for the understanding of this type of technology adoption processes and the increasing importance of mobility in cultural and social practices, promoting an exciting discussion on the centrality of media nowadays and the current transformation processes society is undergoing.

INTRODUCTION

This chapter deals with the different ways contemporary mobile content, services and applications mold specific individual and collective experiences that have consequences for different aspects and forms of social cohesion, involvement and belonging. These consequences mainly result from contemporary social practices related with mobility and mobile communication (mobile internet services) use in community life. Departing from the awareness that, nowadays, media influence is spread to all spheres of modern life,

DOI: 10.4018/978-1-4666-8838-4.ch018

our goal is to examine these processes of transformation on the field of mobile communication, services and trends. Thereby, this chapter intends to approach and discuss how social transformations may be supported by a mediatized environment grounded on the power and added-valueof mobile features, namely services offered via the use of mobile internet (m-internet), the so-called mobile applications (apps).

Our main assumption envisions that media come to be a central element, both for social relationships to take place, as well as for instrumental tasks and daily chores. Our intention will be to focus on mobile communication and on the value of mobility for current social arrangements, exploring the impact of this type of access, and practices performed through it, on the levels of social cohesion and sense of belonging to a group or community. Mobility and mobile communication here will be approached as a new and original social shared system that allows for original individual and social patterns of orientation and interaction and that are being adopted worldwide, compelling individuals and institutions to conform an adopt this new mobile trend and mobile media logic. Our main problem questions 1)whether the emergence of network technologies, namely the adoption and use of mobile features and services (apps) that allow for social interaction, have any impact on associated social configurations such as social cohesion, sense of belonging and integration in social groups; 2)whether contemporary social arrangements are in some way an outcome of technological processes, in this case regarding mobile technologies; 3)whether there are other intervenient forces beyond media-related aspects and what are they; and finally, 4)what is the value of mobility for current social practices of interaction and participation.

In order to further explore these questions, an empirical research project on the study of mobile internet use and adoption will be presented and discussed. This project includes a three-year investigation which explored the 1) relevance of mobility as a driving force for social involvement; 2) the role of a mobile application (app) in community life and its value for the promotion of social engagement, social cohesion, and subjective expression; and lastly, 3) the emergence of new consumption rituals and cross-media behaviors after the dissemination and adoption of a mobile app for communication and content sharing and distribution.

PART I: LITERATURE REVIEW

Media and Technology: The Mediatization of Society

Media technologies, and more particularly, digital media, are spread and embedded in all spheres and institutions of our society in a way that it seems almost impossible to think of a world without media: how would we talk to other people, how would we get the news, how would we search for information, etc. Changes in social practices, social interactions and institutional procedures are only examples of the social influence of digital media and their use. Newly emerging patterns of social and technical interaction are synonymous with social change and lead to specific dynamics and outcomes (Damásio, Henriques & Costa, 2011). The on-going interaction between digital media and their users in contemporary society results in a dynamic process of social change that leaves no social sphere or dimension untouched.

Social changes resulting from the spread of communication technology in society (Rogers, 1985) have been studied not only from the standpoint of people's adoption and uses of new communication technologies (Venkatesh, Morris, Davis & Davis, 2003; Ling, 2007; Wang & Wang, 2010), but also from the more general perspective of how the media affect society and culture (Deuze, 2006; Latour, 2011). Communication studies have relied on two influential wide-ranging concepts for understanding the broader social consequences of media, including the so called new media (Bolter et al, 2006): mediatization and mediation (Mansell, 2002; Schulz, 2004; Livingstone, 2009; Couldry, 2012). Therefore, mediatization deals mainly with the specific claim that many cultural and social processes are now constrained to take on a form suitable for media (Couldry, 2008), while mediation requires us to understand how processes of communication change the social and cultural environments that support them, as well as the relationships that individuals and institutions have with that environment and with each other (Silverstone, 2002).

Mediatization theory, although focusing on the social and cultural effects of the media, assumes that society is permeated by media in all its dimensions instead of approaching them as an independent variable. Thus, the process of mediatization results from the interaction between the technological characteristics of each medium and human agency applied to its use. The concept of mediatization points to the intensified importance of the media in society and stimulates an adequate analysis of on-going transformation processes. Within the process of mediatisation, we may distinguish between a direct (strong) and an indirect (weak) form of mediatization (Hjarvard, 2008). Trying to keep it simple, direct mediatization refers to situations where formerly non-mediated activity converts to a mediated form, i.e., the activity is performed through interaction with a medium (Jansson, 2002). Indirect mediatization is when a given activity is increasingly influenced with respect to form, content, or organisation by mediagenic symbols or mechanisms. It occurs when the cultural context surrounding an activity is affected by the presence of media, in both symbolic and economic terms. The relationship between direct and indirect mediatization can be described as a cause and effect relationship: there is social change induced by the media and their influence spreads to all spheres of society – indirect mediatisation, in addition, because the individuals and the organisations nowadays use the media to perform all types of activities – direct mediatization also occurs.

However, we can also distinguish these direct and indirect effects by mirroring the more traditional effects theories approaches. In the case of the effects' theories, direct media effects correspond to the consequences of media on the people who use them and indirect effects refers to effects that result from media-induced behaviour change on persons who do not necessarily recognise such changes in themselves and are unaware of what the media have to offer. In many cases, causal explanations of media effects are supplemented by functional ones. Social actors are considered individuals who behave according to their intentions. They adjust their behaviour to the requirements of the social environment they act in, and the media on whose resonance they depend. While media effects research is focused on individuals, in the case of mediatization, attention is focused on structures that are changed by the individuals' orientation to the media. Individual behaviour is nothing but an indicator of structural aspects. From this viewpoint, there would be no cause and effect relationship but, rather, a functional one whereby direct effects of mediatization would constitute the micro level of mediatization and indirect effects would constitute the macro level. Whatever the case, what we verify is that mediation seems to correspond to the operational features of mediatization, here understood as general processes of social and cultural realities construction (Krotz, 2007). Direct effects of mediatization correspond to concrete practices of media use for social interaction, while indirect effects correspond to the influence the media have on different sites of modern life.

Following previous work from Thompson (1995), Schulz (2004) and Krotz (2007), Hjarvard (2008), describes the mediatization of contemporary society as a double-sided process, in which the media have become an independent social institution which others have to accommodate; while at the same time the media are integrated in all social institutions, as these are increasingly performed through the media. The concept has evolved to characterise a given phase or situation in the overall development of society and culture, in which the logic of the media exerts a particularly predominant influence on other social institutions (Hjarvard, 2008). Currently, mediatization theory is concerned not only with the relationship between media, agency and social change, but also, mainly, with the fact that society is becoming increasingly subject to, or dependent on, the media and their logic.

Mediatization affects mainly social interaction within institutions, between institutions and in society at large (Hjarvard, 2008). Furthermore, Finnemann (2011), argues that mediatization as a meta-process that encompasses all dimensions of society is only fully enabled by digital media, whose mediation blurs previous social boundaries (e.g. between public and private, local and global, sender and receiver, synchrony and asynchrony, individuals and institutions) and requires new social practices. Hjarvard (2008) sums up the main changes in social interaction resulting from digital mediatization: communication takes place independently of temporal and spatial structures, several social interactions can be enacted simultaneously, actors are more capable of managing and optimising their social networks and interactions, and actors have more access to information and more tools to manage it.

On the other hand, the concept of mediation refers to the role that media technology and institutions play in communication and social interaction, mainly as enablers. Livingstone (2009) stresses that mediation has always existed in human societies, but recognises that its scale, scope and intensity in contemporary society require a deeper analysis. Schulz (2004) identifies four processes whereby the media change human interaction: 1) the media extend human communication abilities; 2) mediation substitutes unmediated forms of social interaction; 3) the media propitiate the amalgamation of activities; and, 4) users tend to accommodate to technological features and characteristics. Furthermore, Couldry (2008) argues that the media create a common framework of references, a shared context of cognitive categories and social meanings which frame social interaction. Mediation tries to capture the particularities that characterise each kind of communication technology and the society that uses it, the variety of dynamics implied in the flows of production, circulation, interpretation or reception, and recirculation, as interpretations flow back into production or outwards into general social and cultural life. Therefore, we can affirm that mediation describes the concrete act of communication by means of a medium in a specific social context (Quico, 2012) and, by contrast, mediatization refers to a longer-lasting process, whereby social and cultural institutions and modes of interaction are changed as a consequence of the growth of media influence (Haythornthwaite, 2005; Couldry, 2008). Mediatization goes beyond the concept of media as intermediaries, alluding to the pervasiveness of media in all spheres of society and to the fact that media themselves have become the main social institution of contemporaneity (Hepp, Hjarvard and Lundby, 2010).

The Value of Mobility in a Mediatized World

The consequences of technological mediation on human social forms of interaction has always been one of the concerns of researchers within the field of communication and human-computer interaction (HCI). Nowadays the internet is accessible through a multiplicity of mobile devices, such as mobile phones, tablets, mp3, GPS and others. The usage of these devices and the possibility of accessing internet features "on the go" contribute to the emergence of new practices of social networking and communication, as these devices make it easier to maintain networks of relationships and continuous patterns of interaction (Damásio, Henriques, Botelho & Dias, 2013; Ling, 2008). In this context, convergence can be found at three levels: between devices – the phone and other devices used to access the internet; between activities – talking and browsing; and between the local and the global through the permanent dynamic articulation of communication and coordination. An additional layer can be found that concerns convergence between content in different media. Nevertheless this holy grail of "produce once distribute many" still finds a lot of drawbacks in actual production contexts that while from a conceptual point of view push this specific form of convergence, from a production point of view fall into a myriad of ramifications due to the different formats and specifications in use by each specific media form.

Such convergence processes are in line with a techno-social process that signals the broader coming-together of technological networks that connect computers, and nowadays smartphones, and the social networks that connect human beings. This convergence process could be broadly summarized under "m-internet" (mobile internet). M-internet entails access to value-added applications and multimedia content via a device with mobile communication capacities that allows for physical displacement during its use. The concept is still a debatable one, far from a clear and delimited description. Some authors agree that m-internet refers to the use of internet on mobile devices (Gerpott, 2010; Wang & Wang, 2010) but the understanding of what a mobile device exactly is in our days is something of a riddle. Some consider m-internet a synonymous of wireless internet that includes both going online with a laptop using a Wi-Fi connection or broadband card or going online via a mobile phone (Smith, 2010). For others, mobile devices are considered mobile phones, with two broad categories: a) smartphones and feature phones, b) notebooks and tablets. These latter merge both mobile features and pc features, since they are mobile but they work as a PC (Townsend & Rios, 2011). To some extent, going online via wireless connectivity is one of the agreed features of m-internet. Wireless connectivity enables internet access wherever whenever (Wellman, Quan-Haase, Boase & Chen, 2002), an important aspect being to perform activities on the go.

From now on in this paper, m-internet will be considered the access to the internet and apps via wirelessly connectivity, 3G, LTE or 4G platforms using mobile devices that allow for physical displacement during use operations, which means, to perform activities on the move, anywhere and at any time.

Social interaction has been the basis for the dissemination and usage of mobile phones and m-internet (Ling, 2008; Lasen, 2004) as relationships are core aspects in human life and interaction is considered an essential human need. Mobile phones usage has changed actual relationships, patterns of interaction and organizational procedures. People often maintain a perpetual contact with their close network of interactions, relationships have become more intense as users are always in touch with their network of close relationships, people express their emotions and feelings more often when using a mobile phone (Lasen, 2004; Levinson, 2004), communication is usually more spontaneous and less formal (Plant, 2002; Intel, 2005)Using a mobile phone is a *sine qua non* condition for social integration and coordination in current society (Taylor & Harper, 2002). More recent research areas consider the technological convergence between mobile phones and the internet and a mobile lifestyle, suggesting that mobile phones are much more than phones; they are mobile devices that integrate phone, television, computer, mp3 player, game console and GPS (Ling, 2007). Studies have demonstrated that the mobile phone does not contribute significantly to the increase of the social networks of users, but rather fosters more frequent contact with those that are already part of these networks (Katz & Aakhus, 2002). A series of projects and authors have considered these matters in the context of contemporary debates regarding the ways society is being transformed by the growing use and dissemination of the internet and associated technologies. Accordingly, the consequences of internet use may vary in accordance with the type of activities the users perform online, but also with the type of connectivity they have (Ling, 2008).

Mobile devices have become so integrated in daily life that they are often associated with multi-screening, i.e. the sequential or simultaneous use of more than one screened-medium (Dias & Teixeira-Botelho, 2014). One of the most common multi-screening practices associates new digital media such as smartphones and tablets to a mass-medium, the TV. This phenomenon has been described by Haridakis and Hanson (2009) as co-viewing, describing the articulation between watching videos on YouTube and related social interactions on social networks such as Facebook. Also, the concept of social television (Ducheneaut et al., 2008; Eriksson Consumer Lab, 2012) describes a particular type of multi-screening in which there is an important component of social networking simultaneously to television viewing. This practice is relatively common, as study from Eriksson Consumer Lab (2012) indicates that 62% of US TV viewers use social media while watching TV. Thus, multi-screening behavior is often motivated by utilitarian goals, such as not wasting time or accomplishing tasks, but the main motivation is affective. Giglietto (2014) agrees, stressing that Twitter in particular is enabling an active viewing of TV content, involving voting, commenting and even some content creation. Thus, instead of drawing attention from television, simultaneous activities in other screened-devices are actually redirecting attention to the TV set and enhancing engagement with TV content, contributing to a converging and dynamic media landscape. Concerning connectivity, there is an extensive body of literature highlighting social capital and sense of belonging as key drivers for multi-screening (i.e. Xu and Yan, 2011; Riedler, Köbler, Goswami & Krcmar, 2013). The importance of social capital and the role played by smartphones and tablets in the creation of new forms of sense of belonging, involvement and social cohesion points to new opportunities in mobile content innovation, that should explore, on the one hand, the articulation with other media in multi-screening behavior, and on the other hand, the enhancement of other activities by adding a permanent social layer to everything we do, providing the desired sense of belonging and sense of permanent connectedness.

Another relevant concept is 'connected viewing' (Smith & Boyles, 2012) that is more focused on consumer behavior and aims to characterize all the different activities that are performed through mobile devices while watching television. A study from Pew Internet Research (Smith & Boyles, 2012) showed that the most frequent simultaneous to watching television activity is keeping busy during commercial breaks, followed by non-TV related activities such as checking the email, web browsing and downloading apps. Other TV-related activities that are also frequent included checking whether something they heard on TV is true or not (22%), searching what other people are saying about television content (20%) and commenting online about the content (19%) and, with a less frequent use, voting for a reality show or contest (6%). This study distinguishes between 'connected viewing' and 'distracted viewing' by arguing that the first concept is more frequent for smartphone users and requires a connection between watching TV and smartphone activities, and the second refers to mobile phone users who use the device to keep busy while the television content is not engaging their attention. In what concerns 'connected viewing', television content is usually the trigger for smartphone activities. To 'distracted viewers', the activities performed are unrelated to the television content and usually end up drawing the attention from the TV completely. On the same issue, it is also relevant the distinction from mobile devices as 'companions' or as 'enhancers', suggested by Eriksson Consumer Lab's study (2012). As 'companions', people use mobile phones simultaneously with TV viewing but there is no connection between those activities, i.e. playing a game on mobile phone while listening to news on TV. However, as 'enhancers', people watch TV while they interact with applications that are related to the channel or content they watching. This enhances the TV experience by adding it a social layer, extra information or allowing participation. Thus, emergent smartphone and tablet use practices should drive mobile content innovation, taking the users' needs, preferences and expectations as a departing point.

Sense of Belonging and Social Cohesion

Social belonging is a natural and intrinsic feature of all human beings. Being part of a group, experiencing connectedness and bonds with other people in dyadic, small or larger groups, is part of our developmental process, at social, emotional and cognitive levels (Newman, Lohman & Newman, 2007). Relationships are core aspects in human life and an essential human need. At the same time, social interaction has been the basis for the dissemination and usage of mobile phones and m-internet (Ling, 2008; Lasen, 2004).

Human beings are, first of all, social. This means they need contact with other people, they need interaction, exchange and communication, and they need to feel included in social groups, developing a sense of belonging to those groups or communities. The concept of belonging can be analyzed from a dualistic perspective, representing two sides of the same coin. On the one side, we can understand belonging as being an experience of being part of a group or community, which offers us a sense of personal continuity of self as well as an involvement in a larger social context with which we share something in common (Jansson, 2002). This perceptive focus on the development of a sense of belonging to a group or community is what supports and helps us in building up and maintaining our own identities, through which we can identify ourselves. On the other side of the coin, we can approach belonging as being a perception of possession or appropriation. Belonging means something that belongs to me. From a communitarian perspective, if I belong to a network, if I am part of it, then that network also belongs to me. As a member of a social group I will also gain advantages, whether social, cultural or economic, that derive from the position I occupy within that network (Burt, 2000). These two ways of understanding the concept of belonging should not be separated in order to fully comprehend the concept (Jansson, 2009).

Sense of belonging is a concept intimately related to that of social networks and the formulation of a network as a set of links and connections we have with others. Sense of group belonging, whether physical or virtual, is a psychological construct that relates to the affective aspect of group belonging, including feelings of being a valued group member, of being proud of one's group, of belonging to something and of having something to belong to (Newman, Lohman & Newman, 2007). The affective facet of a sense of group belonging is the most internally consistent aspect of group membership.

Within society, we are connected to each other via a web of social ties and bonds that offer us a sense of who we are and to what we belong (Kerstetter et al, 2008). These social networks offer members effective and supportive bonds that have relevant influences across individuals' lives, promoting the development of a sense of belonging.

A social network is a regular aggregate of ties within a group having some influence on the social and affective behavior of the people involved. This aggregate represents the group of people with whom one maintains contact and has a common bond – the group to which one belongs. Social networks evolve as actors develop some kind of links, either formal or informal, related to friendship, trust, influence, recognition, work, knowledge or any of the many other forms of social interaction. To be part of a social network means that dependent links exist between subjects, such as social cohesion, social preferences, pressure or hierarchies. Therefore, in order to understand a social network and the processes of belonging, one should not only regard community members as individual separate entities, but consider the overall interactions and relationships (Hanneman & Riddle, 2005; Borgatti, Everett & Freeman, 2002).

In this sense, interaction becomes a condition for the emergence of a sense of belonging. Social interaction stands first of all for the exchange of information at an interpersonal level. This basic informal form of social interaction may result in participation when there is an increase in interpersonal trust and adherence to social norms. Participation then arises as group involvement and formal social interaction.

In any case, interaction seems to precede participation. Some authors (Hampton & Wellman, 2003) claim that the diversity of network and social settings, where the interactions occur, is the key explanatory factor. If social interaction is the propeller of social relations, their settings will maximise interaction - the more diversified they are the better, thereby allowing the subject access to more information. This process defines the interplay between the social and the material realm, where the sense of belonging is constituted. If social interaction is ubiquitous, in the sense that everyone engages in social actions, the emergence of structures within network resources that facilitate such practices points to the constitution of a spaces that increasingly facilitates belonging.

Sense of belonging should also be discussed as playing a central role in the notion of community. With the development of society, in several different aspects of social and community life, the notion of community has undergone a process of transformation. According to Kollock and Smith (2002), a community is a group of people who interact together and share symbols and conventions that are common to external facts acquired via a continuous social contact. McMillan and Chavis (1986), were amongst the first authors to define something similar to a community sense of belonging, a process they named "sense of community". According to them, sense of community refers to an individual sense of being part of a group and sharing with others a common need based on membership, influence, fulfillment of needs and shared emotional connections. From a more classical perspective, the understanding of the concept of community is related to the idea of "home" – Heimat – of belonging and safeness. However, with the widespread changes in communication, physical mobility and ways of interacting, the issue of the familiar has become complex and the whole experience of a place of belonging has become more complex, since we can now find the familiar in the most distant places and the strange in very near neighborhoods.

PART II: EMPIRICAL RESEARCH

A Case Study of M-Internet Adoption: 'Lusofona Mobile'

The ComuniMEDIA Research Project

ComuniMEDIA was a three years research project coordinated by the authors of this chapter at CICANT/ULHT from 2011. All data from this project is primary empirical data collected between 2012 and 2014 through the use of different research instruments and methods for data collection, such as direct interviews, online interviews, focus group discussions and direct inquiries, all analyzed via the application of different techniques of data analysis such as qualitative or quantitative techniques and mix methods.

This project as main goals were a) to characterize mobile internet and activities carried out mainly through the mobile access; b) to identify the importance of mobility and mobile devices in promoting social engagement, interaction and subjective expression; c) to compare the fixed and mobile internet access and the consequences of the type of access in users practices and in the production and accumulation of social capital; d) to analyze the interdependencies and prospects of the industry and users regarding the adoption of this technology and consumption practices.

The central research problem addressed by the authors in this project question mainly, RQ1) how mobile communication technologies affect the activities and practices users perform online, as well as, RQ2) how social networks are created and sustained by mobile communication (the emergence of web based communities).

The ComuniMEDIA project dealt with one essential problematic - the transformation of online activities and associated social contexts as a consequence of changes in the modes of access to the internet (i.e growth in mobile based web access). This research focused on two areas of interest: a) the identification of the patterns of use and access to the internet via mobile based technologies; b) the role of mobility and uses of media content in promoting original forms of access and social interaction that increase social capital and transform the proprieties of the existing social context.

Research hypotheses synthesize the two areas of interest stated above and have as a driving force the core idea that online activities and forms of social constructs (i.e. virtual communities) will be deeply affected by the emergence of mobile forms of access and use of the internet. The perspective guiding the construction of our hypotheses is based on the concept of complementarity and convergence, suggesting that changes in access and type/levels of media content complements social outcomes. Research hypotheses are: H1a) the type of access has an impact on the activities and actions performed online, which leads to less diversified levels of web use. H1b) Individuals with mobile access have a stronger participation in virtual communities and are more satisfied with that process; H2) social activities, and the interaction outcomes arousing from them, constitute one of the main drivers for the adoption process of mobile features; H3) The prevalence of mobile technologies over other forms of access reinforces existing forms of social belonging and social capital though promoting a process of closure within the community.

Method

Project Structure

ComuniMEDIA research project was developed in three stages: 1) an exploratory phase on mobile media based on the analysis of secondary data worldwide about mobile media; 2) an empirical stage about mobile media at a national level based on the production of two major studies, the first one with the main stakeholders of Mobile Industry at a National and International level, and the second one with mobile technologies' users. The combinations of the results from these two studies originated the third stage of the research project, 3) an experimental stage in which a mobile app was produced and introduced within a particular community and its adoption processes and usage was analyzed.

In this chapter the second and third stage of ComuniMEDIA project will be described and discussed.

The second stage, as mentioned, included two studies. The first study was based on set of interviews to a number of stakeholders from the mobile industry, with the central aim of analyzing the overall picture of the use of mobile Internet access tools, the main motivations, attractions, limitations, obstacles to the spread and use of mobile technologies. The second study was based on the application of a survey to a representative sample of the population, the main objective being to analyze the prospects of Internet users on the use of mobile Internet access tools, the value of mobility and influence of the same activities practiced online, using for these goals.

The combination of these two studies offers an overall view of the process of m-internet adoption and use – the point of view from the Industry's stakeholders (people who develop and distribute the technology) and the point of view from the users (people that actually uses the technology) as two sides of a coin. This combination allows us to make practical inferences about the drivers and main obstacles to adoption of m-internet and to analyze the relevance of this type of technology on current social and cultural practices and patterns of communication, considering both the macro and micro environments.

Stakeholders and Users' Study: Second Project Stage

Regarding the first study – stakeholders study - interviews were conducted with key stakeholders in the access and use of the wireless internet process in Portugal namely: João Seabra, Key Account Manager at Sony Ericsson Portugal; Sofia Patricio, responsible for Nokia Portugal communication; Hugo Jorge, Director of Marketing at LG; Victor Head, Deputy Director for the area of Internet Marktest; Carlos Figueiredo, business group director of GfK Portugal; Tiago Cabral, partner of Netsonda; Luis Frade, Manager at TIMwe; Roberto Gomes Optimus in the Internet - Mobile Multimedia Services; Henrique Fonseca, responsible for Internet service Vodafone Portugal; and Peter Farausto director of voice product ZON. Convenience sampling was based on market quota and company size.

Each interview consisted of 18 predetermined questions developed by our research team. The questionnaire developed for this study was based on previous studies that also analyzed stakeholders' views on technology adoption (Quico, Damásio, Henriques & Veríssimo, 2012; Araújo, Cardoso & Espanha, 2008; The World Internet Project, 2012). All interviews were performed face-to-face, recorded, transcribed and analyzed with NVIVO software. The qualitative analysis was carried out in a sequential manner: all the transcribed interviews were read and each question was analyzed regarding the relevant themes approached within. After that, themes were organized and converted into a list of tree nodes. Finally, that list was improved with contributions from all researchers. The final version presents eleven tree nodes hierarchically organized.

Complementarily to this approach, a quantitative study was conducted based on the application of an inquiry about m-internet usage, with the main goal of analyzing users' perspective of m-internet services and possibilities, in order to identify the factors that contribute to its growth and the activities performed via m-internet, namely in comparison with fixed access to the internet. The inquiry was developed based on previous measures (Zickuh, Kathryn, Smith, Aaron, 2011; Smith, Aaron). A pre-test was performed with a convenience sample of 40 individuals. The final version includes a descriptive analysis of mobile phone and m-internet usage, an analysis of the type of activities and tasks performed via m-internet access and the levels of participation and satisfaction of individuals within communities in that process, and for last demographic characterization items. The inquiry has in total 20 questions and was taken first in November 2011 and then updated in May 2013. Data exploration was mainly based on quantitative techniques of analysis.

Given the length of the inquiry, two relevant questions were chosen to be discussed for this paper - the first one is focused on participants' online activities, both via mobile and via fixed internet access; the second is based on a measure developed for analyzing the levels of satisfaction and participation of m-internet users. These variables refer to satisfaction and participation when using the internet and, specifically, when using social websites or social web platforms whether via mobile or via fixed access.

This measure was previously tested and adjusted via an exploratory and confirmatory factor analysis; the final measure reached a good adjustment for the data (cronbach α=0,897, CR=0,895; VEM=0,594). The quantitative data was analyzed with IBM SPSS 20. Statistical analysis was performed with a level of confidence of 95% (α=0,05). Having in view to analyze whether the type of access has any influence on the activities performed online and to understand what kind of online activities are more conducted via mobile internet, a multi-dimensional scale analysis was also performed. The values of proximity were calculated based on the answers to this item in the inquiry, using the algorithm Proxcal (PROXimity SCALing) with the support of SPSS (v14, SPSS Inc. Chicago, IL). The selection of the minimum number of dimensions to keep in the model in order to reproduce parsimoniously the similarities / dissimilarities between the activities (via mobile and fixed access) was held respecting the screen-plot criteria and the proximity versus transformed distance graphics criteria. The quality of the model was analyzed through the STRESS-I index and DAF index using the reference values defined in Marôco (2007).

In order to analyze whether the levels of satisfaction and participation with online experience have any influence on the decision of using or not using m-internet we use the Wilcoxon-Mann-Whitney non-parametric since the conditions to use a parametric test failed. After confirming the influence of satisfaction and participation in the intention to use m-internet, a discriminant analysis was made in order to identify the variables that better differentiate between the groups and to create a model (discriminant function) that will allow us to identify, a priori, the subjects that have the intention of adopting and using m-internet and the ones that probably won't have that intention. The assumptions to perform this analysis were tested, namely normality assumptions with the test Kolmogorov-smirnov and homogeneity of variance-covariance matrices from each group with the test M-Box. Most variables do not present a normal distribution, though since the size sample is larger than 30 (N=1107) so we will assume normality assumption. Regarding M-Box, the assumption of homogeneity is valid (p>0,05).

Experimental Stage: Third Project Stage

This stage corresponded to the last phase of the project aiming to develop an added-value mobile solution for the market and, more precisely, for our institution academic life context and for our community. The mobile platform intended to meet some challenges including: 1) Gathering of information dispersed through the various channels and tailor our services to the mobility inherent in the current lifestyle; 2) Encourage the sharing of knowledge, the development of partnerships and enable communication between staff, teachers and students from all areas of knowledge; 3) Develop a unique identity and support the integration of individuals into a collective whole that is the higher education institution; 4) Provide information services to visitors to make better use of the time and resources we offer; 5) Collect interaction data that allowed us to research on the problems posed.

Android base platform version 2.3.4 - GingerBread API Level 10 was chosen. Three functional areas were identified as mandatory for this app: communication between users of the application through instant messaging (IM); Consultation of academic data, including notes, inscriptions, time and cash position; Consult all type of news and information about the University.

After launching the app for free on Google market (Play store) in September 2013 and after a month of use, our sample consisted of a group of approximately 2700 students of the institution.

On this stage, we used mixed quantitative methods, collecting and processing data regarding the use of the app, data for social network analysis with UCINET and metrics from Google Analytics software; and other qualitative methods such as conducting focus group sessions and individual interviews with users (n=30).The focus groups were conducted with groups composed of 5 students that used the mobile application. Of the 30, 15 were first-year students and the remaining 15 students from 2nd or 3rd year.

Results

Stakeholders Study

During the interviews, stakeholders pointed that mobile phones' sales are decreasing worldwide however, on the other hand, smartphones are having an exponential growth due to the price being so similar to the ordinary mobile phone or feature phone. As consequence of the smartphone widespread, stakeholders claim that the access to m-internet is growing, as smartphones naturally need internet for full extent using of their features.

According to stakeholders, the age group between 25-34 years is the one that uses m-internet the most in Portugal. The main activities are social media and email access, news reading and meteorology checking.

Vodafone believes that the use of m-internet will soon exceed the use of internet on a PC. In addition, their data shows that m-internet access is more utilitarian and frequent along the day but has a shorter duration. Also, there are peaks of m-internet usage at lunch time, between 6 p.m. and 7 p.m. and a primetime between 10 p.m. and 11 p.m. Friday is usually the day with more traffic.

According to all interviews, m-internet users are becoming multiplatform, which means that they can access it everywhere, anytime and through different devices. This characteristic allows mobility and consequently a better exploitation of the internet and its possibilities, the creation of a more appealing experience and spending less time on a normal computer. Thus, the stakeholders' perspective points to the possibilities afforded by technological convergence as the main adoption driver for smartphones and m-internet.

For the majority of the stakeholders, with exception of those from mobile networks operators, the price plans available are the main obstacle to the expansion of m-internet, as they still have high prices and very limited traffic. Other obstacles listed were the lack of information and literacy of Portuguese people regarding the existent services.

Among the motivations to access m-internet, stakeholders also referred that social networks represent a very high share of internet traffic through mobile phones. Market research companies highlighted that a high percentage of young people aged 15 to 17 access social networks on mobile phones, with the groups aged groups 18 to 24 and 25 to 34 also reaching significant values. Mobility is also one of the main reasons to use m-internet as it gives access to new services and possibilities (e.g. geolocation) and new types of consuming, sharing and communicatingGeo-location is another motivation mentioned by the stakeholders as geo-location services are among the most popular apps.

Stakeholders considered that m-internet access makes it easier to communicate and share information in real time. This fact contributes to social closeness and public life involvement. For the stakeholders', smartphone convergence is not only a technological process, but also a cultural one, that results in new social practices that "open" the users to the outside world.

M-Internet Users Study

The individuals from this sample are aged between 14 and 64 years (mean age = 38,69), 49% being male and 51% being female. The vast majority of the sample (81%) has elementary and high school education, 42,7% frequented elementary school, 38,0% frequented high school, 9% have a degree, 1,4% have a master, 0,3% have a PhD and 7,1% have less than primary school. The sampling was conducted in a random way in all regions of the country and the inquiry application was performed face-to-face in the individuals' households.

Given the length of the questionnaire, only a partial set of the results will be presented here. Results indicate that almost all participants have a mobile phone device (96,6%), however only 19,1% own smartphones. From the total sample, only 15,7% answered they use m-interned (84,3% do not use m-internet), 14,5% use it via mobile phone, 1,5% use it via tablets, 3,3% use it in their iPods, 1,4% use it via iPad and finally, 2,1% use it via portable playstation.

Figure 1. Reasons for not using m-internet

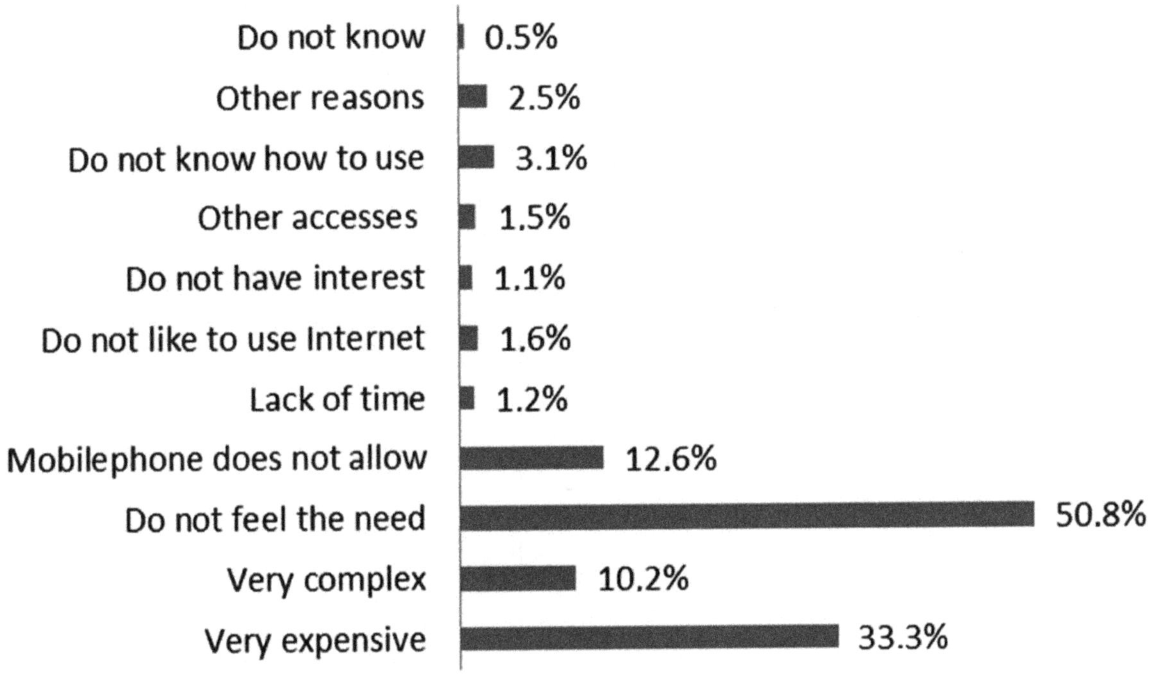

Participants that replied they do not use m-internet (84%) were asked their reasons. The main reasons presented were: they do not feel the need of using it (50,8%) or it is very expensive (33,3%). Figure 1 shows other reasons mentioned.

Regarding the activities performed via m-internet, the inquiry posted a list of activities performed online, both via mobile or via fixed access to the internet (the same online activity was always questioned twice, one regarding the mobile access and one regarding the fixed access to the internet), and asked participants their agreement with the sentences on a scale from 1 to 7, where 1 meant "I never do that" and 7 meant "I always do that". The more frequently conducted via mobile are ordered in Figure 2.

In order to explore if there is a divergence between actions performed online when fixed or mobile internet access was considered, we performed a multidimensional scaling analysis evaluating the proximity (similarity/ dissimilarity) between the activities analyzed in the inquiry, intending to examine groups or types of activities that are more frequently performed via mobile. Method was explained in previous section. Three dimensions were retained in order to reproduce properly the perceived similarities between the activities (STRESS-I=0,129, DAF=0,983). Figure 3 illustrates the three-dimensional perceptual map.

In this perceptual map the activities conducted online either via fixed or mobile access to the internet are distributed spatially according to its similarities (based on the respondent's answers) in a spatial map with three dimensions that we defined as: dimension 1 - interaction; dimension 2 – time; dimension 3 - effort. Each point in the perceptual map represents the positioning of each activity according to these three dimensions, allowing to represent perceptually and to analyze the proximity and distance between the activities, considering the dimensions mentioned.

Two main groups of activities emerge, while the remaining activities are distributed randomly within the graphic space. The two main groups are located at the superior left part and at the superior right of the perceptual map. The superior left part includes the following activities: using the e-mail via mobile

Figure 2. Activities conducted via mobile and fixed internet access

		Mean	Std. deviation
Setle a doubt	fixed internet	5,99	1,62
	mobile internet	3,91	2,29
See who's online	fixed internet	4,42	2,43
	mobile internet	2,90	2,19
Professional searches	fixed internet	5,45	2,18
	mobile internet	2,87	2,20
Download music	fixed internet	5,51	2,14
	mobile internet	2,28	1,92
Watch videos	fixed internet	5,62	2,04
	mobile internet	3,03	2,21
Weather info	fixed internet	4,56	2,34
	mobile internet	3,74	2,35
Read e-mails	fixed internet	5,88	1,81
	mobile internet	4,17	2,26
SNS	fixed internet	5,61	2,05
	mobile internet	4,24	2,31
Info about roads, paths	fixed internet	3,98	2,30
	mobile internet	3,99	2,26
Read news	fixed internet	4,93	2,19
	mobile internet	2,90	2,16
chats	fixed internet	4,61	2,32
	mobile internet	3,89	2,29
Online games	fixed internet	4,20	2,60
	mobile internet	2,44	2,08
Online services	fixed internet	5,39	1,99
	mobile internet	3,44	2,08
Surf the web	fixed internet	5,80	1,82
	mobile internet	3,41	2,11
MSN	fixed internet	5,48	2,08
	mobile internet	3,36	2,24
Geo-location	fixed internet	3,21	2,26
	mobile internet	3,64	2,33

Figure 3. Three-dimensional perceptual map of online activities (Euclidean distance model: STRESS-I=0,129; DAF=0,983)

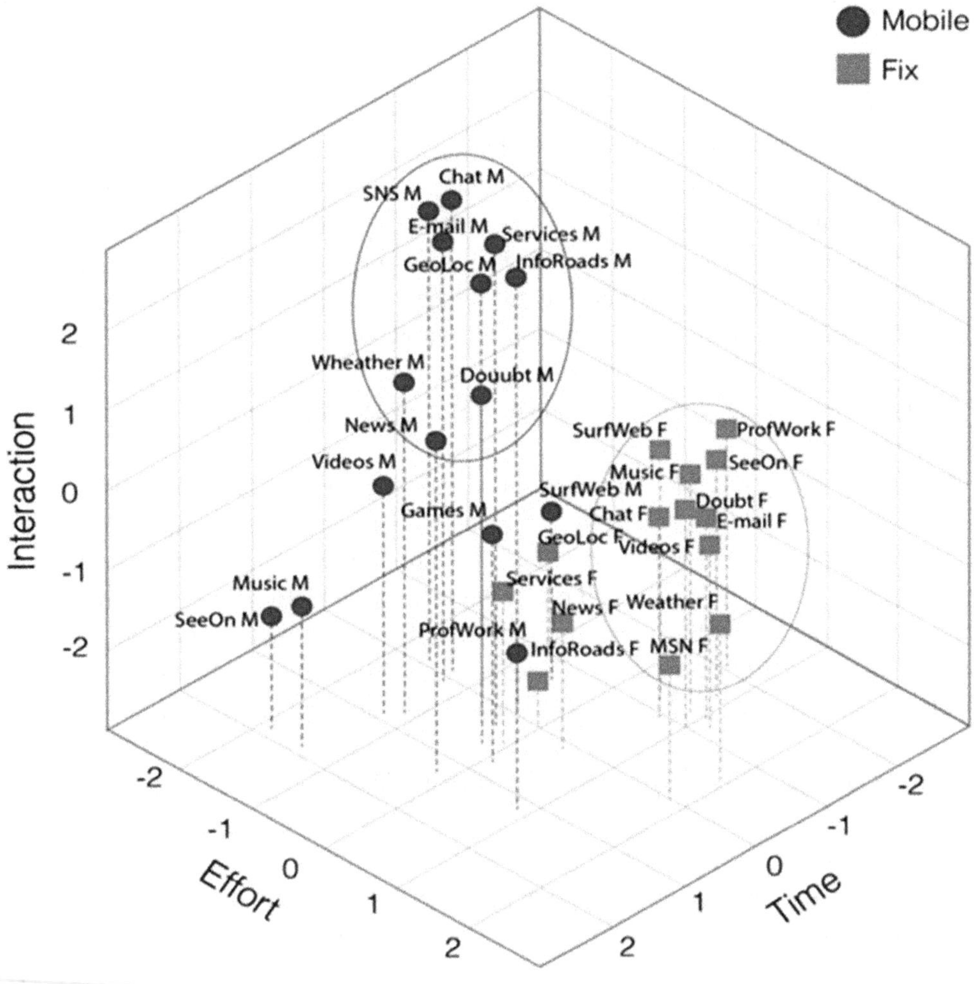

access, using social network site via mobile, using geo-location applications via mobile, using online services via mobile, searching information about maps, paths or roads, using chats via mobile. These activities score higher values on mobile access. Regarding the superior right part of the perceptual map, we can find another group of online activities, such as: downloading music, settling a doubt, looking for someone online, performing professional/ work searches online, watching videos online, which present higher values on the fixed access. The perceptual map distinguishes between two main groups of activities: one more related to the fixed access and the other one more related to the mobile access.

Interpreting this data and considering the dimensions' conceptual meaning, one can say that the activities more frequently done via mobile access commonly imply communication and participation, such as using chats, e-mailing, using social network sites, and activities that need to be done on the go, such as searching information about streets, roads or maps, used when someone is lost or is trying to find a new place. The activities more frequently done via fixed access are normally activities that take more time to be performed, imply higher levels of attention or knowledge. This distinction is thus another divergence identified in our research, this one concerning the activities performed online according to the type of access. These results are consistent with our previous argument of smartphones remediating the internet through the convergence process of these two technologies.

One last and highly relevant aspect, concerns the fact that all activities that involve the reinforcement of close ties within our dense personal network, are much stronger in mobile technologies. This shows this technology potential to increase high levels of connectivity with existing social networks – bonding type of social capital – and lower levels with outside networks – bridging type of social capital.

In May 2013 the same questionnaire was applied again to a representative sample of the population (N = 1201, aged 14 and 64, in which 49% were men and 51% women) and found that the percentage of Internet users Portugal evolved from 15.7% to 57.5%, which translates into a very representative growth.

Who does not use the mobile Internet referred as the main reasons that it is expensive (71%), no need to do so (66%) and no interest (49%). The justifications are the same when the results obtained in November 2011.

The data also show that the preferred mobile device to access the Internet continues to be the smartphone.

As regard to activities the results remained. In all online activities, only two have higher values in the use of mobile internet. These two activities are searching for information about paths and the use of geolocation services. However, despite higher values for mobile access numbers are very close to the fixed internet use, again indicating that users still prefer the use of fixed internet to perform online activities.

These data allowed us to establish that the use of mobile internet by Portuguese population is growing and tends to evolve in the coming years. The evolution of the number of users since November 2011 until May 2013 is remarkable, as currently already exceeds half of its population.

Experimental Stage: Development of a Mobile App

A social network analysis (SNA) was performed with data from chat functionality collected after the launch of the second version of the app, on September 2013. Data reveals a great number of isolated acts, which is possibly explained by students checking and experimenting chat functionality, but not really using it. However, data indicated that there were 212 users and 292 chat rooms were created. Results showed that interactions were mostly performed between students from the same year and from the same course. This indicates a preference in the use of the app to contact with close colleagues and strengthening bonding relationships.

Google metrics indicate that there are at least 1400 users, though the number is growing everyday. Most of them prefer to use wireless to access the app, but they also use 3G and 4G access. The functionality most commonly used is the week schedule. The chat appears as the less used feature. From analytics, it was also possible to analyze data regarding demographics, network operators, mobile device brand, screen resolution, OS version, app version, language, country, new users, active users in real time and their behavior, user engagement, and most accessed screens and functions.

Data from focus groups indicate that students developed a new type of behavior regarding the use of some academic services that are now only accessed via the app – week schedule, classrooms indications, and grades. This new practice, allied to a general sense of satisfaction with the development of the app, made them feel closer to the university and the academic life.

The possibility of accessing the app on the go develop a new ritual of arranging the day and preparing for the next day. Students expressed they felt an increase in their sense of commitment with university issues. Moreover, they also mentioned that the usage of features related to events and information have increased the sense of integration with academic life.

An analysis between older and new students was performed and results indicate that older students have mostly strengthened their academic sense of belonging and involvement. 1st year students reported they felt more quickly integrated in academic life and that it prevent them to feel lost and displaced as they could access academic information anytime anywhere.

Results showed a correlation between the use of a mobile app in a university context, the increase of the sense of belonging and greater levels of commitment and engagement with academic life. There was a positive relationship between the use of a mobile technology and individual and group social involvement. Results also indicated evidence that the use of that mobile app helped students to maintain close social ties with their closer colleagues (from the same year and from the same course). Close contacts are relevant as they provide support in everyday tasks as well as crisis times. Ling and Stald (2010) stated that cohesion in tighter circles of friends enhances well-being and social participation and group engagement. Moreover, results showed that students use the app for both instrumental and expressive tasks as well as for the coordination of their academic life.

PART III: CONCLUSION

Future Research Directions

Future research will continue to focus on mobile technology and user experience and the value of mobility for current society. In an ever more mediatized society, the role communication technologies has for the organization of society, from cultural to social and economic aspects, is a significant aspect that calls for constant research. We intend to continue studying problems associated with mobile technologies in different aspects of social life, from health care to sports, psychological and physical handicaps, community life, social interaction and patterns of behavior, intervention to application and dissemination. We are already developing a research project on health care based on the implementation of interventions with mobile and wearable media technologies. The potential of wearable technologies to educate and empower individuals by providing objective means to self-monitor their health and wellness or other aspects of their life, is becoming clearer as innovative devices that help people to accurately monitor parameters such as biometrics performance, blood glucose levels, activity, sleep cycles or food intake are multiplying. We intend to follow and analyze the adoption process and acceptance of these wearable and mobile technologies, although mobile devices are far from being widespread. Studies on these processes of adoption may help to support and suggest guidelines on how to better disseminate and integrate these practices in people's life, from younger people to adults and even elderly people.

Conclusion

Today, media have taken over more and more functions for communication, organization and living in actual society. More than never, society is characterized by dense networks of mediated communication (Burt, 2000), which has consequences for people's identities and social relations, as well as the ways in which institutions and organisations conduct themselves and for culture and society as a whole.

Regarding research questions put forward within the ComuniMEDIA project, results from users and stakeholders empirical studies, as well as from the experimental study with *Lusófona Mobile*, indicate that the type of access to the internet has an influence on the activities and common practices users perform online. Therefore, mobile and fixed access to the internet allow for differences in the tasks performed online, with users preferring the mobile access to perform to communicate, interact and participate via practices that require less effort and time and medium or high levels of participation. The more common activities performed via m-internet were mainly using chats, e-mailing, using social network sites, and activities that need to be done on the go, such as searching information about streets, roads or maps, used when someone is lost or is trying to find a new place. On the other side, via fixed internet access users are more likely to perform activities that take more time and higher levels of effort, such as attention or knowledge. A relevant result from these studies was that all activities related to the strengthening of close and boding ties are more commonly performed via mobile access, as interacting practices, which points out the potential and value of this type of access to increase high levels of connectivity with existing social networks – bonding type of social capital. This was also a result from the case study, indicating that mobile interaction was mainly performed with bonding ties and close friends. As well, results indicate that individuals with mobile access show a stronger participation in virtual communities and are more satisfied with that process and that social activities, and the interaction outcomes arousing from them, constitute one of the core drivers for the adoption and usage process of mobile features.

When discussing networks of belonging, we are interested in exploring whether the mediatization process is reinforced by the particular features of digital media, both technological and social, which are reflected in changes in the sense of belonging, or whether digital media are a neutral infra-structure that enables and mediates social interaction, and users are the driving force in the mediatization of belonging by their enacting it every day through their mediated uses of digital media. Thereby, we are interested in clarifying and comparing the relevance of technology itself versus the relevance of the social dimension of technology usage in the process of mediatisation, more particularly, in the process of mobile technology adoption. What is the central driving force to adopt, use and disseminate a particular technology – is it the technology itself and its features, for instances the innovative content of mobile applications, the mobility it allows, etc, or is it the social element, the relationships and interaction levels that that particular technology allows me to have with my peers that belong to the same social group or community?

The presented case study of *Lusófona Mobile* was an instrument that allowed us to better examine and understand this tension between adoption processes foster by technology features or by users intrinsic motivation. The experienced difficulties during the app distribution and appropriation processes, as well as the type of activities performed when using the app (more self-organizing and management tasks as well as interaction tools) indicate the significance of the social dimension for the adoption process. Users mostly used the mobile app as a mediation tool for their institutional daily-life, to organize and manage their lives and to interact with others, from colleagues to teachers. On one side, they are the outcome from an institutional point of view of mediatization indirect effects (as they live most of their lives immersed in a digital world), but on the other side, they depict the operational features of mediation and the direct effects of mediatization upon, in this case, younger generations. This should lead us to conceive mediatization not only as an institutional configuration, but also as a process dependent upon individual action. Mediation would then become the key facet prompting the transformations of sense of belonging.

Sense of belonging is intrinsic to our daily lives as members of communities. However, in such a highly-mediated context as that which is emerging, this need to be part of a group is evolving and communities are no longer set only within physical proximity, but also in mediated contexts made available through social media platforms. The sense of networked belonging that is developed through different new media practices may be a key element to understanding audiences. Digital media and their increasing user-friendliness are challenging audiences to become more than content consumers, but to assume active roles as consumers, producers and / or 'produsers'.

Networked belonging is therefore essential for people to develop their identities as media creators through interaction with peers and audiences that may be anywhere in the world (Jenkins, 2006; Lange & Ito, 2009), and is one of the key transformations of sense of belonging that we have isolated. In contemporary society, young peoples' interaction with the media comes through involvement, while producing content through peer-networked experiences and affiliations established through dialogue, sharing and discussions, contribute to the sense of networked belonging that we suggest. Sense of belonging can be found in the most primitive and ancient human cultures and even within animal groups. For those who lived before the Web 2.0 it is easy to remember these facets of human life, where media was already present but not in a so dominant way that affected all aspects of social or professional life. There is also a very well-known discussion of whether the Web 2.0 contribute to foster social relationships and interaction between people or, on the contrary, whether all these possibilities offered by media nowadays and dispelling and keeping people dispersed and away. As it is usually heard, in the old days, people knew each other, they knew their neighbours, they talked with others in the most different social situation and nowadays there is a common sense that people are more focused on themselves. Would it be possible that nowadays networking through social media is more about self-communication to others than about a real exchange of information or communication with others? Would the notion that there is a sense of increased belonging through social networking be more illusory than it might at first appear? These are questions that remain for further research.

REFERENCES

Araújo, V., Cardoso, G., & Espanha, R. (2008). The adoption of digital television in Portugal: a media literacy approach. In A. Urban, B. Sapio, & T. Turk (Eds.), *Digital Television Revisited - Linking Users, Markets and Policies, COST Action 298 Participation in the Broadband Society*. Hungary.

Bolter, J., McIntyre, B., Gandy, M., & Schweitzer, P. (2006). New Media and the Permanent Crisis of Aura. *Convergence (London)*, *12*(1), 21–39. doi:10.1177/1354856506061550

Borgatti, S., Everett, M., & Freeman, L. (2002). *UCINET 6 for Windows: Software for Social Network Analysis. User's Guide. Harvard*. Natick: Analytic Technologies.

Burt, R. (2000). The Network Structure of Social Capital. *Research in Organizational Behavior*, *22*, 345–423. doi:10.1016/S0191-3085(00)22009-1

Couldry, N. (2008). Does the Media have a future? *European Journal of Communication*, *24*(4), 437–449. doi:10.1177/0267323109345604

Couldry, N. (2012). *Media, society, world: social theory and digital media practice*. London: Polity.

Damásio, M., Henriques, S., Botelho, I., & Dias, P. (2013). Social activities and mobile internet diffusion: a search for the holy. Mobile Media & Communication, 1(3), 335-355. London: SAGE Communication.

Damásio, M., Henriques, S., Torres da Silva, M., Pacheco, L., & Brites, M. (2014, April 3 - 4). The diversity of European media audiences. Mass and self-media, old and new media. International Conference *(New) Audience Practices.*, Catholic University of Portugal.

Damásio, M. J., Henriques, S., & Costa, C. (2011, *September 1-2*). Mobile access to the internet: the Portuguese case. Proceedings of Transforming Audiences 3 - Online and mobile media, everyday creativity and DIY culture – *International Conference.* . University of Westminster, London, UK.

Dias, P., & Botelho, I. (2014, November 12 - 15). Is the Second Screen becoming the First? An exploratory study of second screening use practices. *Paper presented at ECREA's 5th European Communication Conference* - Lisbon.

Ducheneaut, N., Moore, R., Oehlberg, L., Thorton, J., & Nickell, E. (2008). Social TV: Designing for distributed, sociable television viewing. *International Journal of Human Computer Interaction*, 24(2). Retrieved from http://tinyurl.com/pun4dxr

TV and Video: An analysis of evolving consumer habits. Eriksson Consumer Lab. (2012). Eriksson Consumer Lab. Retrieved from http://tinyurl.com/bvgrfot

Finnemann, N. (2011). Mediatization Theory and Digital Media *Communications*, *36*(1), 67–89. doi:10.1515/comm.2011.004

Gerpott, T. (2010). Communication behaviors and perceptions of mobile internet adopters. *Emerald*, *12*(4), 54–73.

Giglietto, F., & Selva, D. (2014). Second Screen and Participation: A content analysis on a full season dataset of tweets. *Journal of Communication*, *64*(2), 260–277. doi:10.1111/jcom.12085

Hampton, K., & Wellman, B. (2003). Neighboring in Netville: How the internet supports community and social capital in a wired suburb. *City & Community Journal*, *2*(4), 277–311. doi:10.1046/j.1535-6841.2003.00057.x

Hanneman, R., & Riddle, M. (2005). *Introduction to Social Network Methods*. Riverside, CA: University of California.

Haridakis, P., & Hanson, G. (2009). Social Interaction and Co-Viewing with YouTube: Blending mass communication reception and social connection. *Journal of Broadcasting & Electronic Media*, 53(2), 317-335. Retrievedfrom http://tinyurl.com/o5t2blu

Haythornthwaite, C. (2005). Social networks and internet Connectivity effects. *Information Communication and Society*, *8*(2), 125–147. doi:10.1080/13691180500146185

Hepp, A., Hajvard, S., & Lundby, K. (2010). Mediatization – Empirical Perspectives: An introduction to a special issue. *Communications*, *35*(3), 223–228. doi:10.1515/comm.2010.012

Hjarvard, S. (2008). The Mediatization of Society. *Nordicom Review*, *29*(2), 105–134. PMID:19361120

Jansson, A. (2002). Spatial Phantasmagoria: The Mediatization of Tourism Experience. *European Journal of Communication*, *17*(4), 429–443. doi:10.1177/0267323102017004020l

Jansson, A. (2009). Mobile Belongings: Texturation and stratification in mediation processes. In K. Lundby (Ed.), *Mediatization: Concept, changes, consequences.* (pp. 243–261). Oxford, Wien.

Jenkins, H. (2006). *Convergence Culture*. New York: NYU Press.

Katz, J., & Aakhus, M. (Eds.). (2002). *Perpetual Contact, Mobile Communication, Private Talk, Public Performance*. Cambridge: Cambridge University Press. doi:10.1017/CBO9780511489471

Kerstetter, D., Yarnal, C., Son, J., Yen, I., & Baker, B. (2008). Functional support associated with belongings to the Red Hat Society, a leisure-based social network. *Journal of Leisure Research*, *40*(4), 531–555.

Kollock, P., & Smith, M. (2002).Communities in cyberspace. In Smith, & Kollock (Eds.), Communities in Cyberspace (pp. 3-26). London: Kindle Edition.

Krotz, F. (2007). *Mediatisierung: FallstudienzumWandell von Kommunikation*. Wiesbaden: VS VerlagfürSocialwissenschaften.

Lange, P., & Ito, M. (2009). Hanging out, messing around, and geeking out: kids living and learning with the new media. In P. Lange and M. Ito (Eds.), Creative production (pp. 243-256). Cambridge: The MIT Press.

Lasen, A. (2004). *Emotions and Digital Devices – Affective Computing and Mobile Phones*. Surrey: Surrey University.

Levinson, P. (2004). *Cellphone: The Story of the World's Most Mobile Medium and How it has Transformed Everything!* New York: Palgrave MacMillan.

Ling, R. (2007). *Mobile Communication and Mediated Ritual*. Norway and USA: Telenor & University of Michigan.

Ling, R. (2008). *New tech, new ties: how mobile communication is reshaping social cohesion*. New York: MIT press.

Ling, R., & Stald, G. (2010). Mobile communities: Are we talking about a village, a clan, or a small group? *America behavioral. Scientist (Philadelphia, Pa.)*, *53*, 1133–1146.

Livingstone, S. (2009). On the Mediation of Everything: ICA Presidential Address 2008. *Journal of Communication*, 59, 1-18.

Mansell, R. (2002). Conclusion. Social relations, mediating power and technologies. In R. Mansell (Ed.), Inside the communication revolution: evolving patterns of social and technical interaction. (pp. 251-270). Oxford University Press.

Marôco, J. (2007). *Análise estatística com SPSS*. Lisboa: Edições Sílabo.

McMillan & Chavis. (1986). *Sense of community: a definition and theory. Journal of Community Psychology. (14)*. NY: New York University Press.

Newman, B., Lohman, B., & Newman, P. (2007). Peer-group membership and a sense of belonging: Their relationship to adolescent behaviour problems. *Adolescence*, *42*(166), 241–282. PMID:17849935

Pew Research Internet Project. (2014). *Mobile Technology Fact Sheet*. Pew Internet & American Life Project. Retrieved from http://www.pewinternet.org/fact-sheets/mobile-technology-fact-sheet/

Plant, S. (2002). *On the mobile. Receiver*. Vodafone.

Quico, C. (2012). Digital TV switchover in Portugal: What is in it for the viewer. *International Journal of Digital Television*, *3*(1), 163–179. doi:10.1386/jdtv.3.2.163_1

Quico, C., Damásio, M.J., Baptista, A., Sequeira, A. Veríssimo, I. Cardoso, M. & Henriques, S. (2014). Using digital interactive television for the promotion of health and wellness. *International Journal of Health Promotion and Education*.

Quico, C., Damásio, M. J., Henriques, S., & Veríssimo, I. (2012). Digital terrestrial TV switchover process in Portugal: viewers between a rock and a hard place. In Sousa, H., & Denicoli, S. (Eds.), *Digital Communication Policies in the Information Society Promotion Stage* eBook.. Universidade do Minho. Retrieved from http://www.lasics.uminho.pt/ojs/index.php/digital_communication_polices/index

Riedler, K. Goswami & Krcmar (2013). Tweeting to Fells Connected: A model for social connectedness in online social networks. *International Journal of Human-Computer Interaction*, vol. 29 (10). Retrievedfrom http://tinyurl.com/ma2pprb

Schulz, W. (2004). Reconstruction mediatization as an analytical concept. [London: Sage.]. *European Journal of Communication*, *19*(1), 87–101. doi:10.1177/0267323104040696

Silverstone, R. (2002). *Why Study the Media?* London: Sage.

Smith, A. (2010, July 7). Mobile Access 2011. *Pew Internet & American Life Project*. Retrieved from http://pewinternet.org/Reports/2010/Mobile-Access-2010.aspx

Smith, A., & Boyles, J. L. (2012). The rise of the 'connected viewer. *Pew Internet & American Life Project, Pew Research Center*. Retrieved from http://tinyurl.com/nfgbj62

Taylor, A., & Harper, R. (2002). *Age-old Practices in the 'New World': a Study of Gift-Giving between Teenage Mobile Users*. Surrey: Surrey University. doi:10.1145/503376.503455

Thompson, J. B. (1995). *The Media and Modernity: A social theory of the media*. Cambridge: Polity Press.

Venkatesh, V., Morris, M., Davis, G., & Davis, F. (2003). User acceptance of information technology: Toward a unified view. *Management Information Systems Quarterly*, *27*(3), 425–478.

Wang, H., & Wang, S. (2010). User acceptance of mobile internet based on the unified theory of acceptance and use of technology: Investigating the determinants and gender differences. *Social Behavior and Personality*, *38*(3), 414–426. doi:10.2224/sbp.2010.38.3.415

Wellman, B., Quan-Haase, A., Boase, J., & Chen, W. (2002). *Examining the Internet in Everyday Life*. Nijmegen. [online] doi:10.1002/9780470774298

Xu, H., & Yan, R. N. (2011). Feeling Connected via Television Viewing: Exploring the scale and its correlates. *Communication Studies*, 62(2). Retrieved from http://tinyurl.com/mymqnp3

Zickuhr, K., & Smith, A. (2012). Digital differences. *Pew Internet & American life project*. Retrieved from http://www.pewinternet.org/files/old-media/Files/Reports/2012/PIP_Digital_differences_041312.pdf

KEY TERMS AND DEFINITIONS

Lusófona Mobile: Lusófona Mobile is a mobile application designed for students from Lusófona University of Humanities and Technologies in Portugal, available in Google Play for Android smartphones. This mobile application was develop by CICANT (Centre of Research in Applied Communications, Culture and New Technologies) with the objective of providing academic services, centralizing information access, offering communication tools in mobility and provide data to the research project ComuniMEDIA.

M-Activities: Mobile activities are considered activities carried out online via a mobile device that allows for physical displacement during the activities performance.

M-Communication: Mobile communication is considered communication and interaction, whether oral, written, verbal, non-verbal, image based or symbolic communication, via mobile devices.

Mediatization: Mediatization deals with the specific claim that many cultural and social processes are now constrained to take on a form suitable for media. Mediatization was originally conceived as a media logic associated with mass communication media and the large influence these exert over organizations and institutions.

M-Internet: Mobile internet is the use of internet via mobile devices, mobile phones, smartphones, tablets, iPad, iPhone, PSP, or other devices that allow for physical displacement during operations.

Mobile User Experience: Mobile user experience involves the user's perceptions and feelings before, during and after their interaction with mobile technology. It represents a more subjective experience of use. User experience usually encompasses the experiential, emotional and affective aspects of human-computer interaction and the person's perception of practical issues such as utility, usefulness, ease of use and efficiency.

Compilation of References

Socio-economic impact of mHealth. *Touching lives through mobile health - Assessment of the global market opportunity.* (2013).

Statistics. (2013). *ITU.int.* Retrieved from http://www.itu.int/en/ITU-D/Statistics/Pages/stat/default.aspx

2008: Population projections 2008–2060 in EU27. (2008). Eurostat.

2013 Big Data Survey Research Brief. (2013). *SAS.* Retrieved from http://www.sas.com/resources/whitepaper/wp_58466.pdf

2014. The production and consumption of mobile news. InGoggin, G., & Hjorth, L. (Eds.), *The Routledge companion to mobile media* (pp. 135–145). New York: Routledge.

51Degrees. (n. d.). Retrieved from http://51degrees.mobi/

Accessing the news in an age of mobile media: tracing displacing and complementary effects of mobile news on newspapers and online news. (2015). *Mobile Media & Communication, 3*(1), 53-74. doi:10.1177/2050157914549039

Acquisti, A. (2010). *The Economics of Personal Data and the Economics of Privacy. 30 Years after the OECD Privacy Guidelines*. Brussels: OECD.

ACSI. (2014). *ACSI e-Business Report 2014.* Retrieved from http://www.theacsi.org

Adams, R. (2007). Applying advanced concepts of cognitive overload and augmentation in practice; the future of overload. In D. Schmorrow, K. M. Stanney, and L. M. Reeves (Eds.), Foundations of Augmented Cognition (2nd Edition) (pp. 223-229).

Adams, R., Bessant, J., & Phelps, R. (2006). Innovation management measurement: A review. *International Journal of Management Reviews, 8*(1), 21–47. doi:10.1111/j.1468-2370.2006.00119.x

Adler, E. (2014a). 'Reverse showrooming': Bricks-and-Mortar retailers fight back. *Business Insider.* Retrieved from http://www.businessinsider.com/tablets-and-smartphones-are-used-completely-differently-for-shopping-and-buying-2014-4?utm_source=alerts&nr_email_referer=1

Adler, E. (2014b). Tablets and smartphones are used completely differently for shopping and buying. *Business Insider.* Retrieved from http://www.businessinsider.com/tablets-and-smartphones-are-used-completely-differently-for-shopping-and-buying-2014-4?utm_source=alerts&nr_email_referer=1

Adomavicius, G., & Tuzhilin, A. (2005). Toward the Next Generation of Recommender Systems: A Survey of the State-of-the-Art and Possible Extensions. *IEEE Transactions on Knowledge and Data Engineering, 17*(6), 734–749. doi:10.1109/TKDE.2005.99

Aggarwal, C. C., Ashish, N., & Sheth, A. (2012). The Internet of Things: A Survey from the Data-Centric Perspective. In C. C. Aggarwal (Ed.), *Managing and Mining Sensor Data* (pp. 383–428). New York City: Springer.

Agosti, M., Crivellari, F., & Di Nunzio, G. M. (2012). Web Log Analysis: A Review Of A Decade Of Studies About Information Acquisition, Inspection And Interpretation Of User Interaction. *Data Mining and Knowledge Discovery*, *24*(3), 663–696. doi:10.1007/s10618-011-0228-8

Aguado, J. M. (2013). La industria de contenido en la era post-PC: Horizontes, amenazas y oportunidades. In Notícias e Mobilidade: Jornalismo e Dispositivos Móveis (pp. 5-28). Covilha, Portugal: LabCom, Universidade de Beira Interior.

Aguado, J. M., & Castellet, A. (2013). Periodismo móvil e información ubicua [Mobile journalism and ubiquitous information]. In J.M. Aguado, C. Feijóo, & I. J. Martínez (Eds.), La comunicación móvil: Hacia un nuevo ecosistema digital. [The mobile communication: towards a new digital ecosystem] (pp. 187-217). Barcelona, Espanha: Editorial Gedisa.

Aguado, J. M., & Güere, H. N. (2013). Comunicación móvil, ecosistema digital e industrias culturales [Mobile communication, digital ecosystem and cultural industries]. In J.M. Aguado, C. Feijóo, & I. J. Martínez (Coords.), La comunicación móvil: Hacia un nuevo ecosistema digital [The mobile communication: Towards a new digital ecosystem] (pp. 57-77). Barcelona, Espanha: Editorial Gedisa.

Aguado, J.M; Feijóo, C; Martínez, I.J. (cords.). (2013). La comunicación móvil. Hacia un Nuevo ecosistema digital. Barcelona: Gedisa.

Aguado, J. M., Feijóo, C., & Martínez, I. J. (2011). Contenidos digitales y comunicación móvil: Hacia la transformación de Internet. *Derecho a Comunicar*, *2*(3), 1–15.

Aguado, J. M., Feijóo, C., Martínez, I. J., & Roel, M. (2012). Mobile Television, A Paradigmatic Case of the Uncertainties and Opportunities of the New Media Ecosystem. In A. Abruzzese, N. Barile, J. Gebhardt, J. Vincent, & L. Fortunati (Eds.), *The New Television Ecosystem* (pp. 141–158). Frankfurt am Main: Peter Lang.

Aguado, J. M., & Martínez, I. (2014). Feeding digital omnivores: The impact of mobile media in digital entertainment. In G. Goggin & L. Hjorth (Eds.), *The Routledge Companion to Mobile Media* (pp. 181–194). New York, London: Routledge.

Aguado, J. M., & Martínez, I. (2014). Feeding Digital Omnivores: The Impact Of Mobile Media In Digital Entertainment. In G. Goggin & L. Hjorth (Eds.), *The Routledge Companion to Mobile Media* (pp. 181–194). New York, London: Routledge.

Aguado, J. M., & Martínez, I. J. (2010). Liquid digital selves: Mobile Media Implicit Cultures, Social Relations and Identity Management. *Encyclopaideia: Journal of Phenomenology and Education*, *15*(1), 63–88.

Ahonen, T. (2014a). The comprehensive app economics blog 2014. *Communities dominate brands*. Retrieved from http://communities-dominate.blogs.com/brands/2014/08/the-comprehensive-app-economics-blog-2014-yes-peak-app-is-apt-name-sheer-disaster-industry-with-only.html

Ahonen, T. (2014b). Enter the iflop, what will be seen as first apple failure after Steve Jobs. *Communities dominate brands*. Retrieved from http://communities-dominate.blogs.com/brands/2014/09/enter-the-iflop-what-will-be-seen-as-first-apple-failure-after-steve-jobs-but-the-first-edition-appl.html

Ahonen, T. (2014c). Seems like my work is done: As we shift from 'hunters' in mobile industry to 'farmers.' *Communities dominate brands*. Retrieved from http://communities-dominate.blogs.com/brands/2014/07/as-we-shift-from-hunters-in-mobile-industry-to-farmers-most-of-the-cool-stuff-is-done-and-dusted.html

Ahonen, T. (2014d). Smartphone top 10 for Q2 of 2014, same ole same ole. *Communities dominate brands*. Retrieved from http://communities-dominate.blogs.com/brands/2014/08/smartphone-top-10-for-q2-of-2014-same-ole-same-ole.html

Ahonen, T. (2015). Smartphone Year 2014 Final Stats: Brand top 10, OS top 4, installed base, regional split; and bloodbath Q4 results. *Communities dominate brands*. Retrieved from communities-dominate.blogs.com: http://communities-dominate.blogs.com/brands/2015/02/smarphone-year-2014-final-stats-brand-top-10-os-top-5-installed-base-regional-split-and-bloodbath-q4.html

Ahonen, T. T., & Barrett, J. (2002). *Services for UMTS: Creating killer applications in 3G*. Singapore: John Wiley & Sons. doi:10.1002/0470014172

Akamai (2014). *The State of the Internet*. Q1 2014 Report, Vol. 7(1). Retrieved from http://www.akamai.com/stateoftheinternet/

Akerman, N. (1998). *The necessity of friction*. Heidelberg, Germany: Westwiew Press.

Åkesson, M. (2009). Digital innovation in the value networks of newspapers. *Gothenburg Studies in Informatics*, 42. Retrieved from http://hdl.handle.net/2077/20887

Alam, H., Hartono, R., Kumar, A., Rahman, F., Tarnikova, Y., & Wilcox, C. (2003). Web page summarization for handheld devices: a natural language approach.*Seventh International Conference on Document Analysis and Recognition*, (pp. 1153-1158). doi:10.1109/ICDAR.2003.1227838

Ali-Yrkkö, J., Kalm, M., Pajarinen, M., Rouvinen, P., Seppälä, T., & Tahvanainen, A.-J. (2013). Microsoft Acquires Nokia: Implications for the Two Companies and Finland. Retrieved from http://pub.etla.fi/ETLA-Muistio-Brief-16.pdf

Amen, U. (2010). Consumer attitude towards mobile advertising. *Interdisciplinary Journal of Contemporary Research in Business*, *2*(3), 75–104.

AOL-BBDO. (2012). Seven Shades of Mobile. The Hidden Motivations of Smartphone Users. Research. Retrieved from http://advertising.aol.com/sites/default/files/ content/download/ attachment/research/aol-bbdo-7-shades-mobile-abstract-final.pdf

Apple Insider. (2013). *Apple awarded patent for more accurate haptic feedback system*. Retrieved from http://appleinsider.com/articles/13/02/19/apple-awarded-patent-for-more-accurate-haptic-feedback-system

Apple Reinvents the Phone with iPhone. (2007). *Apple Press Info*.

Appnation (2013). State of the App Economy Report. Retrieved from http://appnationconference.com/main/research/

Arai, K., & Herman, T. (2010). Method for Automatic E-Comic Scene Frame Extraction for Reading Comic on Mobile Devices. *Seventh International Conference of Information Technology New Generations (ITNG)* (pp. 370-375). doi:10.1109/ITNG.2010.22

Araújo, V., Cardoso, G., & Espanha, R. (2008). The adoption of digital television in Portugal: a media literacy approach. In A. Urban, B. Sapio, & T. Turk (Eds.), *Digital Television Revisited - Linking Users, Markets and Policies, COST Action 298 Participation in the Broadband Society*. Hungary.

Archibugi, D., Filipetti, A., & Frenz, M. (2013). Economic crisis and innovation: Is destruction prevailing over accumulation? *Research Policy*, *42*(2), 303–314. doi:10.1016/j.respol.2012.07.002

Ashton, K. (2009). That 'internet of things' thing. *RFiD Journal*, *22*, 97–114.

Atelier Paribas. (2013). *Big data, big culture? The Growing Power of the Data and its Outlook for the Economy of Culture*. Retrieved from http://www.forum-avignon.org/sites/default/files/editeur/EtudeATELIER_FA_2013.pdf

Avid (2014). *Avid Everywhere, White Paper, A Vision for the Future of the Media Industry*. Retrieved from www.avid.com/static/resources/common/documents/whitepapers/Avid_Everywhere.pdf

Bacache-Beauvallet, M., Bourreau, M., Moreaux, F., (2011). *Portrait des musiciens à l'heure du numérique*. Paris: Editions rue d'Ulm.

Baeza-Yates, R., Dupret, G., & Velasco, J. (2007). A Study of Mobile Search Queries in Japan. *Proceedings of the international World Wide Web Conference*. Retrieved from http://lahuen.dcc.uchile.cl/~gdupret/pubs/banff_japan.pdf

Baeza-Yates, R., & Raghavan, P. (2010). Next Generation Web Search. In S. Ceri & M. Brambilla (Eds.), *Search computing, LNCS 5950* (pp. 11–23). Berlin, Heidelberg: Springer-Verlag.

Baeza-Yates, R., & Raghavan, P. (2010). Next Generation Web Search. In *Search Computing* (pp. 11–23). Springer. doi:10.1007/978-3-642-12310-8_2

Bain. (2013a). *Big Data: The Organization Challenge*. Retrieved from www.bain.com

Bain. (2013b). *The age of curation: From abundance to discovery*. Report for the Forum of Avignon. Retrieved from www.bain.com

Bakardjieva, M. (2011). The internet in everyday life: Exploring the tenets and contributions of diverse approaches. In M. Consalvo & C. Ess (Eds.), The handbook of internet studies (pp. 59–82). Oxford: Wiley-Blackwell.

Bakhshi, H.T. (2010). *Culture of Innovation? An economic analysis of innovation in arts and cultural organisations*. Nesta.

Balez, M. (2011). Personally yours on the go. Retrieved from http://googlemobile.blogspot.com/2010/11/personally-yours-on-go-personalized.html

Ballon, P. (2009). Platform types and gatekeeper roles: The case of the mobile communications industry (pp. 17-19). CBS- Copenhagen Business School. Retrieved from http://www2.druid.dk/conferences/viewpaper.php?id=5952&cf=32

Ballon, P. (2009). The Platformisation of the European Mobile Industry. *Communications & Stratégies*, *75*(3), 15–34.

Ballon, P. (2009). The Platformisation of the European Mobile Industry. *Communications & Stratégies*, (75): 15–33.

Ballon, P., & Van Heesvelde, E. (2011). ICT platforms and regulatory concerns in Europe. *Telecommunications Policy*, *35*(8), 702–714. doi:10.1016/j.telpol.2011.06.009

Baluja, S. (2006). Browsing on small screens: recasting web-page segmentation into an efficient machine learning framework.*Proceedings of the 15th international conference on World Wide Web (WWW '06)* (pp. 33-42). New York, NY, USA: ACM. doi:10.1145/1135777.1135788

Banks, A. (2014). *2014 Brazil Digital Future in Focus*. São Paulo: comScore. Retrieved from http://www.comscore.com

Barbosa, S., Firmino da Silva, F., & Nogueira, L. (2014). Análise da convergência de conteúdos em produtos jornalísticos com presença multiplataforma. In E.C. Pellanda, & S. Barbosa (Eds.), Jornalismo e mídias móveis no contexto da convergência (pp.95-126). Porto Alegre: EdiPUCRS.

Barbosa, S. (2013). Jornalismo convergente e continuum multimídia na quinta geração do jornalismo em redes digitais. In J.*Canavilhas (Ed.). Notícias e mobilidade: o jornalismo na era dos dispositivos digitais* (pp. 33–54). Covilhã: Livros Labcom.

Barbosa, S., Firmino da Silva, F., Nogueira, L., & Almeida, Y. (2013). A atuação jornalística em plataformas móveis: Estudo sobre produtos autóctones e a mudança no estatuto do jornalista. *Brazilian Journalism Research*, *9*(2), 10–29.

Barlow, J., Singh, D., Bayer, S., & Curry, R. (2007). A systematic review of the benefits of home telecare for frail elderly people and those with long-term conditions. *Journal of Telemedicine and Telecare*, *13*(4), 172–179. doi:10.1258/135763307780908058 PMID:17565772

Barnhurst, K. G., & Nerone, J. (2001). *The form of news: a history*. New York: Guilford Press.

Barns, M. (2014). Expect a seismic shift in video consumption. *Re/code*. Retrieved from http://recode.net/2014/02/26/expect-a-seismic-shift-in-video-consumption/

Baron, N. S. (2008). *Always on: Language in an online and mobile world*. New York, NY: Oxford University Press. doi:10.1093/acprof:oso/9780195313055.001.0001

Barrows, J. (2010). The Jane Austen Industry and long tail marketing. *Yale Economic Review*, *6*(2), 36–38.

Barsotti, A. (2014). Two screen, two paths: News production for smartphones and tablets on the brazilian newspaper scene. *#. ISOJ Journal*, *4*(1), 112–132.

Barton, B., & Barton, M. (1993). Ideology and the map: Toward a postmodern visual design practice. In N. Blyler & C. Thralls (Eds.), *Professional communication: The social perspective* (pp. 49–78). Newbury Park, CA: Sage.

Barylick, C., & Honan, M. (2007). iPhone release brings out the crowds. *MacWorld,* retrieved from http://www.macworld.com/article/1058682/iphone_crowds.html

Batiston, B., & Bello Neto, G. B. (2013). Jornalismo para tablets: a dialética entre pesquisa e prática, experiências desenvolvidas na universidade. In R. Paulino & V. Rodrigues (Eds.), *Jornalismo para tablets: pesquisa e prática*.

Bauman, Z. (2005). *Liquid Life*. Cambridge, UK: Polity Press.

Bautista, A. (2009). Relaciones interculturales en educación mediadas por narraciones audiovisuales. *Comunicar*, *33*(33), 149–156. doi:10.3916/c33-2009-03-006

Belk, R. W. (2013). Extended self in a digital world. *The Journal of Consumer Research*, *40*(3), 477–500. doi:10.1086/671052

Benford, S., & Giannachi, G. (2009). Temporal Convergence in Shared Networked Narratives: The Case of Blast Theory's Day of the Figurines. *Leonardo*, *42*(5), 443–448. doi:10.1162/leon.2009.42.5.443

Benghozi, P.-J., Salvador, E., & Simon, J.-P. (2015). *R&D and innovation in the cultural-creative sectors: the case of cinema*. Retrieved from http://is.jrc.ec.europa.eu/pages/ISG/EURIPIDIS/documents/JRC95536.pdf

Benton, J. (2014). The mobile majority. *Nieman Reports, 68*(2), 38-39. Retrieved from http://www.nieman.harvard.edu/assets/pdf/Nieman%20Reports/backissues/NRSpring2014.pdf

Bercovici, J. (2014, May 3). Flipboard buys Zite from CNN; founder Mark Johnson leaving. *Forbes*. Retrieved from http://www.forbes.com

Berman, S. J., & Kesterson-Townes, L. (2012). Connecting with the digital customer of the future. *Strategy and Leadership*, *40*(6), 29–35. doi:10.1108/10878571211278868

Bickmore, T. (1999). Web Page Filtering and Re-Authoring for Mobile Users. *The Computer Journal*, *42*(6), 534–546. doi:10.1093/comjnl/42.6.534

Blattberg, R. C., & Deighton, J. (1991). Interactive marketing: Exploiting the age of addressability. *Sloan Management Review*, *33*(1), 5–14.

Bloem, J., van Doorm, M., Duivestein, S., & Sjöström, A. (2013). *The App Effect*. Amsterdam: VINT Bariet.

Blum, J., Blintham, M., Jacobs, R., Shipp, V., Kefalidou, G., Brown, M., & McAuley, D. (2013). The Timestreams Platform: Artist Mediated Participatory Sensing for Environmental Discourse. *Proceedings ofUbiComp '13the 2013 ACM International Joint Conference on Pervasive and Ubiquitous Computing*. Zurich, Germany: ACM.

Blundell, R. (2009). *Communicating Science: a novel frame of reverence* [Master's Dissertation]. Retrieved from ProQuest Dissertations and Theses database (UMI No. 1461771).

Boczkowski, P., & Ferris, J. (2005). Multiple Media, Convergent Processes, and Divergent Products: Organizational Innovation in Digital Media Production at a European Firm. *The Annals of the American Academy of Political and Social Science*, *597*(1), 32–47. doi:10.1177/0002716204270067

Boehm, E. (2011). *Mobile Healthcare's Slow Adoption Curve*. Forrester Research, Inc.

Bogdanowicz, M., de Prato, G., Feijóo, C., Nepelski, D., & Simon, J. P. (2010). *Born digital/grown digital. Assessing the future competitiveness of the EU video games software industry*. JRC Scientific and Technical Reports. Retrieved from http://ftp.jrc.es/EURdoc/JRC60711.pdf

Bogost, I. (2008). The Rhetoric of Video Games. In K. Salen (Ed.), *The Ecology of Games: Connecting Youth, Games, and Learning* (pp. 117–140). Cambridge, MA: The MIT Press.

Bohlin, E., Olsson, T., & Westlund, O. (2007). Mobile Barometer 2007 – An Analysis of the Swedish Mobile Society. Tokyo: Mobile Society Research Institute Report, NttDoCoMo. Retrieved from: http://www.moba-ken.jp/wp-content/pdf/research07-01.pdf

Bolter, J., & Grusin, R. (1998). *Remediation: Understanding new media*. MA: MIT Press Cambridge.

Bolter, J., McIntyre, B., Gandy, M., & Schweitzer, P. (2006). New Media and the Permanent Crisis of Aura. *Convergence (London)*, *12*(1), 21–39. doi:10.1177/1354856506061550

Bonfiglio, S. (2012). The Role of ICT in a Healthcare Moving from "Clinical-Centric" to "Patient-Centric". In Impact Analysis of Solutions for Chronic Disease Prevention and Management (pp. 250-253). Springer Berlin Heidelberg.

Bontempi, M. E., & Mairesse, J. (2008). *Intangible capital and productivity. An exploration on a panel of Italian manufacturing firms*. NBER (The National Bureau of Economic Research) Working Paper No. 14108. Cambridge (USA). Retrieved from http://www.nber.org/papers/w14108

Borgatti, S., Everett, M., & Freeman, L. (2002). *UCINET 6 for Windows: Software for Social Network Analysis. User's Guide. Harvard*. Natick: Analytic Technologies.

Borodin, Y., Mahmud, J., & Ramakirshnan, I. (2007). Context browsing with mobiles - when less is more.*Proceedings of the 5th international conference on Mobile systems, applications and services (MobiSys '07)*. New York, NY, USA. doi:10.1145/1247660.1247665

Boston Consulting Group. (2012). *The Value of Our Digital Identity*. Liberty Global Policy Series. Retrieved from https://www.bcgperspectives.com/content/articles/digital_economy_consumer_insight_value_of_our_digital_identity/

Boudreau, K. J., & Jeppesen, L. B. (2014). Unpaid crowd complementors: The platform network effect mirage. *Available at SSRN 2475034*.

Bourdieu, P. (2005). *Capital cultural, escuela y espacio social*. México: Siglo XXI.

Bouwman, H., et al. (2009). Reconsidering the actual and future use of mobile services. *Information Systems and e-Business Management*, 7, 301–317.

Boyd, E. (2011). Brains And Bots Deep Inside Yahoo's CORE Grab A Billion Clicks. *Fast Company*. Retrieved from http://www.fastcompany.com/1770673/brains-and-bots-deep-inside-yahoos-core-grab-billion-clicks

Boyd, D. (2014). *It's Complicated. The Social Lifes of Networked Teens. Yale New Haven and London*. University Press.

Brake, D. R. (2014). *Sharing our Lives Online. Risks and Exposure in Social Media*. Basingstoke, UK: Palgrave Macmillan.

Bright, P. (2014). Neither Microsoft, Nokia, nor anyone else should fork Android. It's unforkable. *Ars Technica*. Retrieved from http://arstechnica.com/information-technology/2014/02/neither-microsoft-nokia-nor-anyone-else-should-fork-android-its-unforkable/

Brin, S. (2013). Why Google Glass? *TED Talks*. Retrieved from http://www.ted.com/talks/sergey_brin_why_google_glass

Broder, A. (2002). A Taxonomy of Web Search. *ACM Sigir Forum, 36*(2), 3-10. Retrieved from http://portal.acm.org/citation.cfm?id=792552

Broder, A. (2002). A Taxonomy Of Web Search. *ACM Sigir Forum, 36*(2), 3-10. Retrieved from http://portal.acm.org/citation.cfm?id=792552

Broder, A., Fontoura, M., Josifoovski, V., & Riedel, L. (2007). A semantic approach to contextual advertising. *Mobile Data Engineering Laboratory, Department of Computer Science and Engineering*. Korea University. Retrieved from http://websense.korea.ac.kr/projects/ads/a_semantic_approach_to_contextual_advertising.pdf

Bruns, A. (2011). Gatekeeping, Gatewatching, Real-Time feedback: New challenges for Journalism. *Brazilian Journalism Research, 7*(11), 117–136.

Bruns, A., & Highfield, T. (2012). Blogs, Twitter, and Breaking News: The produsage of citizen journalism. In R. A. Lind (Ed.), *Produsing theory in a digital world: The intersection of audiences and production* (pp. 15–32). New York, NY: Peter Lang.

Budiu, R., & Nielsen, J. (2012). *Mobile Usability*. San Francisco, CA: New Riders Press.

Buongiorno Annual report (2011). Retrieved from http://www.buongiorno.com/financial-information

Buongiorno company website. (n. d.). Retrieved from http://www.buongiorno.com

Burke, K. (1966). *Language as symbolic action: Essays on life, literature, and method*. Los Angeles: University of California Press.

Burke, R., Felfernig, A., & Göker, M. H. (2011). Recommender Systems: An overview. *AI Magazine, 32*(3), 13–18.

Burt, R. (2000). The Network Structure of Social Capital. *Research in Organizational Behavior, 22*, 345–423. doi:10.1016/S0191-3085(00)22009-1

Busson, A., & Landau, O. (2011). La transformation de la chaîne de la valeur de l'audiovisuel. In La richesse des réseaux numériques - Actes du colloque Médias Proceedings of Aix Marseille University conference.

Byrne, D. (2012). *How music works*. Edinburgh, London: Canongate.

Cádima, F. R. (2013). A Google, o sistema de media e a agregação de informação[Google, media system and news aggregation]. *Intercom: Revista Brasileira de Ciências da Comunicação, 36*(1), 19–37.

Cai, D., Yu, S., Wen, J., & Ma, W. (2003). VIPS: A vision-based segmentation algorithm. MSR-TR-2003-70.

Calin, M., Dellarocas, C., Palme, E., & Sutanto, J. (2013, February 16). Attention allocation in information-rich environments: The case of news. *Boston University School of Management Research Paper Series*. Retrieved from http://ssrn.com/abstract=2225359doi: 10.2139/ssrn.2225359

Cameron, A. (1998). The medium is messy. *Eye (London, England), 30*, 6–7.

Campbell, S. W. (2007). A Cross-Cultural Comparison of Perceptions And Uses Of Mobile Telephony. *New Media & Society*, *9*(2), 343–363. doi:10.1177/1461444807075016

Campbell, S. W. (2013). Mobile media and communication: A new field, or just a new journal? *Mobile Media & Communication*, *1*(1), 8–13. doi:10.1177/2050157912459495

Campbell, S. W. (2015). Mobile Communication and Network Privatism: A Literature Review of The Implications For Diverse, Weak, and New Ties. *Communication Research*, *3*(1), 1–21. Retrieved July 7, 2015, from http://www.rcommunicationr.org/index.php/component/jdownloads/finish/16/55?Itemid=0.

Canadian Hexagram-CIAM. (2015). Retrieved from http://www.hexagram.ca/

Canavilhas, J. (2013). Jornalismo Móvel e Realidade Aumentada: o contexto na palma da mão [Mobile Journalism and Augmented Reality: the context in your hand). *Verso e Reverso*, *64*, 2–8. doi: 10.4013/ver.2013.27.64.01

Canavilhas, J., & Santana, D. (2011). Jornalismo para plataformas móveis de 2008 a 2011: Da autonomia à emancipação. *Revista Líbero*, *14*(28), 53–66.

Carr, D. (2013). Giving Viewers What They Want. *The New York Times*. Retrieved from http://www.nytimes.com/2013/02/25/business/media/for-house-of-cards-using-big-data-to-guarantee-its-popularity.html?pagewanted=all&_r=0

Carruthers, B. (2006). Mapping the Terrain of Contemporary Ecoart Practice and Collaboration.*Proceedings of ART in ECOLOGY – A Think Tank on Arts And Sustainability*. Vancouver, Canada: Canadian Commission for UNESCO.

Castellet, A., & Feijóo, C. (2013). Los actores en el ecosistema móvil [The actors in mobile ecosystem]. In J.M. Aguado, C. Feijóo, & I. J. Martínez (Coords.), La comunicación móvil: Hacia un nuevo ecosistema digital [The mobile communication: towards a new digital ecosystem] (pp. 27-56). Barcelona, Espanha: Editorial Gedisa.

Castellet, A., Aguado, J. M., & Martínez, I. (2013). Los nuevos actores que cambian las reglas y el juego: Cómo Apple y Google han transformado la distribución de contenidos. In Zilles, Cuenca, & Rom (Eds.), *Breaking the media value chain* (pp. 333-42). Barcelona: Facultat de Comunicació Blanquerna, Universitat Ramon Llull. Retrieved from http://www.slideshare.net/acastellet/castellet-aguado-martinez-cicr2013

Castellet, A., Aguado, J. M., & Martínez, I. J. (2012, December 7-8). Towards a medium-less ecosystem: The impact of app stores in the digital content environment, *I Technological Innovation in a Mobile World Conference*, Murcia.

Castellet, A., & Feijóo, C. (2013). Los actores en el ecosistema móvil. In J. M. Aguado, C. Feijóo, & I. Martínez (Eds.), *La comunicación móvil. Hacia un nuevo ecosistema digital* (pp. 27–56). Barcelona: Gedisa.

Castellet, A., & Feijóo, C. (2013). Los Actores En El Ecosistema Móvil. In *La Comunicación Móvil. Hacia Un Nuevo Ecosistema Digital*. Barcelona: Gedisa.

Castells, M., Fernández-Ardèvol, M., Qiu, J. L., & Sey, A. (2006). *Mobile communication and society*. Cambridge, MA: The MIT Press.

Castells, M., Fernandez-Ardevol, M., Qiu, J. L., & Sey, A. (2009). *Mobile communication and society: A global perspective*. Cambridage, MA: MIT Press.

Casual Games Association. (2013). *Smartphone & Tablet Gaming 2013: Games Market Sector Report*. Newzoo.

Centre for Research-Creation in Media Arts and Technologies. (2015). Retrieved from http://hexagram.concordia.ca

Cerra, A., & James, C. (2012). *Identity Shift. Indianapolis.* IN: John Wiley & Sons.

Chang, W., Chang, C., & Li, Q. (2012). Customer lifetime value: A review. *Social Behavior and Personality, 40*(7), 1057–1064. doi:10.2224/sbp.2012.40.7.1057

Chan, M. (2015). *Examining The Influences Of News Use Patterns, Motivations, And Age Cohort On Mobile News Use: The case of Hong Kong, Mobile Media & Communication, 3(2).* OnlineFirst.

Chen, Y., Ma, M., & Zhang, H. (2003). Detecting web page structure for adaptive viewing on small form factor devices. *Proceedings of the 12th international conference on World Wide Web,* Budapest, Hungary (pp. 225-233).

Chen, G. D., Chang, C. K., & Wang, C. Y. (2008). Ubiquitous learning website: Scaffold learners by mobile devices with information-aware techniques. *Computers & Education, 50*(1), 77–90. doi:10.1016/j.compedu.2006.03.004

Chen, H., Chiang, R. H., & Storey, V. C. (2012). Business Intelligence and Analytics: From Big Data to Big Impact. *Management Information Systems Quarterly, 36*(4), 1165–1188.

Chen, P. T., & Hsieh, H. P. (2012). Personalized mobile advertising: Its key attributes, trends, and social impact. *Technological Forecasting and Social Change, 79*(3), 543–557. doi:10.1016/j.techfore.2011.08.011

China Internet Network Information Center (CCNIC). (2014). Statistical Report on Internet Development in China. Retrieved from http://www1.cnnic.cn/AU/MediaC/rdxw/hotnews/201401/t20140117_43849.htm

Christensen, C., Skok, D., & Allworth, J. (2012). Be the disruptor. *Nieman Reports, 66*(3). Retrieved from http://www.nieman.harvard.edu/assets/pdf/Nieman%20Reports/backissues/Fall2012.pdf

Christensen, C. M. (1997). *The innovator's dillemma.* Cambridge, MA: Harvard Business School Press.

Christian, D. (2007). *This Fleeting World, a Short History of Humanity.* Great Barrington, MA: Berkshire Publishing Group.

Christiansen, L. (2011). Personal privacy and Internet marketing: An impossible conflict or a marriage made in heaven? *Business Horizons, 54*(6), 509–514. doi:10.1016/j.bushor.2011.06.002

Chun-Rong, H., Chu-Song, C., & Pau-Choo, C. (2005). Tangible Photorealistic Virtual Museum. *IEE Computer Graphics and Applications, 25*(1), 15–17. doi:10.1109/MCG.2005.22 PMID:15691165

Church, K., & Smyth, B. (2009). Understanding the intent behind mobile information needs. *Proceedings of the 13th International Conference on Intelligent User Interfaces,* 247-256. Retrieved from https://wiki.ucop.edu/download/attachments/34668692/Understanding+the+Intent+Behind+Mobile+Information+Needs+p247-church.pdf

Church, K., Smyth, B., Bradley, K., & Cotter, P. (2008). A Large Scale Study of European Mobile Search Behaviour. *Proceedings of the 10th international conference on human-computer interaction with mobile devices and services.* ACM Press. doi:10.1145/1409240.1409243

CIDUS. (2015). Retrieved from: www.usc.es/es/servizos/cidus

CIGI-Ipsos. (2014). *83% of Global Internet Users Believe Affordable Access to the Internet Should be a Basic Human Right.* Retrieved November 18, 2014 from http://www.ipsos-na.com/knowledge-ideas/public-affairs/ipsos-global-advisory/?q=internet-basic-right

Cisco Visual Networking Index. (2013, February). *Global Mobile Data Traffic Forecast Update, 2012–2017.* Retrieved from http://www.cisco.com/en/US/solutions/collateral/ns341/ns525/ns537/ns705/ns827/white_paper_c11-520862.pdf

Cisco Visual Networking Index. (2014, February). *Global Mobile Data Traffic Forecast Update, 2013–2018.* Retrieved from http://www.cisco.com/c/en/us/solutions/collateral/service-provider/visual-networking-index-vni/white_paper_c11-520862.html

Cisco. (2014). *Cisco visual networking index: Global mobile data traffic forecast update, 2013–2018*. Retrieved from http://www.cisco.com/c/en/us/solutions/collateral/service-provider/visual-networking-index-vni/white_paper_c11-520862.html

Cisco. (2015). *Visual Networking Index: Global Mobile Data Traffic Forecast Update 2014–2019* [White Paper].

CITIUS. (2015). Retrieved from: https://citius.usc.es/Coiro

COADEC. (2014, September). The Startup Manifesto.

Coates, K., & Ellison, A. (2014). *Introducción al diseño de la información*. Badalona: Parramón.

Cocosila, M., Coursaris, C., & Yuan, Y. (2004). M-healthcare for patient self-management: A case for diabetics. *International Journal of Electronic Healthcare*, *1*(2), 221–241. doi:10.1504/IJEH.2004.005869 PMID:18048222

Codagnone, C. (2009). Reconstructing the whole: present and future of Personal Health Systems.

Cohen, W. (2010). Fifty Years of Empirical Studies of Innovative Activity and Performance. Handbook of the Economics of Innovation (Vol. 1, pp. 129–213).

Colebrook, C. (2011). *Extinctions*. London, United Kingdom: Open Humanities Press.

Colombo, F. (1974). *Televisione: la realtà come spettacolo*. Milano: Bompiani.

Comic Life. (2015). Retrieved from http://plasq.com/products/comiclife3

ComScore Device Essentials. (2013b). Retrieved from http://direct.comscore.com/Default.aspx

ComScore. (2013, March). *Europe digital future in focus 2013: Key insights from 2012 and what they mean for the coming year*. Retrieved from http://www.comscore.com

ComScore. (2013a). *Mobile in Future 2013*. Retrieved from http://fr.slideshare.net/Briancrotty/comscore-mobile-futureinfocusreport2013

ComScore. (2014). *The U.S. Mobile App Report*. Retrieved from https://www.comscore.com/fre/Insights/Presentations-and-Whitepapers/2014/The-US-Mobile-App-Report?&elqCampaignId=631

Consortium, G. E. M. (2014). *Global entrepreneurship monitor 2013 global report*. Recovered on February.

Constantinou, A., Vakulenko, M., & Kapetanakis, M. (2011). *Mobile Megatrends 2011*. Vision Mobile. Retrieved from http://www.visionmobile.com/rsc/researchreports/mobile-megatrends-2011-(visionmobile).pdf

Conti, M., & Giordano, S. (2014). Mobile ad hoc networking: Milestones, challenges, and new research directions. *Communications Magazine, IEEE*, *52*(1), 85–96. doi:10.1109/MCOM.2014.6710069

Cormen, T. H. (2013). *Algorithms unlocked*. Cambridge, MA: The MIT Press.

Correa, E. S. (Ed.). (2012). *Curadoria digital e o campo da comunicação* [Digital curation and the communication field]. São Paulo, Brazil: ECA – USP.

Costello, S. (2014a). IDC lowers tablet forecast on phablet growth. Retrieved from http://www.mobileworldlive.com/idc-lowers-tablet-forecast-phablet-growth?utm_campaign=MWL-D-20140604&utm_medium=email&utm_source=Eloqua&elq=0261f07a8f5044f4a5229868728f0b3a&elqCampaignId=1830

Costello, S. (2014b). Asia leading the way for apps. Retrieved from http://www.mobileworldlive.com/asia-leading-way-apps

Costello, S. (2014c). Take 2: LTE Broadcast learns from mobile TV missteps. Retrieved from http://www.mobileworldlive.com/take-2-lte-broadcast-learns-mobile-tv-missteps?utm_campaign=MWL_20140724&utm_medium=email&utm_source=Eloqua&elq=f275168051b04f6d85e33d64c63b79fc&elqCampaignId=2043

Couldry, N. (2008). Does the Media have a future? *European Journal of Communication*, *24*(4), 437–449. doi:10.1177/0267323109345604

Couldry, N. (2012). *Media, society, world: social theory and digital media practice*. London: Polity.

Crossan, M., & Apaydin, M. (2010). A Multi-Dimensional Framework of Organizational Innovation: A Systematic Review of the Literature. *Journal of Management Studies*, *47*(6), 1154–1190. doi:10.1111/j.1467-6486.2009.00880.x

Crothers, L. (2013). *Globalization and American Popular Culture* (3rd ed.). Plymouth, United Kingdom: Rowman and Littlefield.

CTIA. (2011). Wireless Quick Facts. Retrieved from http://www.ctia.org/media/industry_info/index.cfm/AID/10323

da Costa, B., & Philip, K. (Eds.). (2008). *Tactical Biopolitics. Art, Activism, and Technoscience*. Cambridge: The MIT Press. doi:10.7551/mitpress/9780262042499.001.0001

Damásio, M. J., Henriques, S., & Costa, C. (2011, *September 1-2*). Mobile access to the internet: the Portuguese case. Proceedings of Transforming Audiences 3 - Online and mobile media, everyday creativity and DIY culture – *International Conference*. . University of Westminster, London, UK.

Damásio, M., Henriques, S., Botelho, I., & Dias, P. (2013). Social activities and mobile internet diffusion: a search for the holy. Mobile Media & Communication, 1(3), 335-355. London: SAGE Communication.

Damásio, M., Henriques, S., Torres da Silva, M., Pacheco, L., & Brites, M. (2014, April 3 - 4). The diversity of European media audiences. Mass and self-media, old and new media. International Conference *(New) Audience Practices.*, Catholic University of Portugal.

Dandekar, K., Raju, B. I., & Srinivasan, M. A. (2003). 3-D finite-element models of human and monkey fingertips to investigate the mechanics of tactile sense. *Journal of Biomechanical Engineering*, *125*(5), 682–691. doi:10.1115/1.1613673 PMID:14618927

Danova, T. (2014). Smart watches will lead wearable computing to mass market status. *Business insider*. Retrieved from http://www.businessinsider.com/global-smartwatch-sales-set-to-explode-2014-3?utm_source=alerts&nr_email_referer=1

Davenport, T. H., & Dyché, J. (2013). *Big Data in Big Companies*. International Institute for Analytics (iianalytics.com). Retrieved from http://www.sas.com/resources/asset/Big-Data-in-Big-Companies.pdf

Dawkins, R. (2006). *The Selfish Gene*. Oxford, United Kingdom: Oxford University Press.

De Prato, G. (2014). The video games industry. In G. De Prato, E. Sanz, & J. P. Simon (Eds.). *Digital Media Worlds; The new media economy*. Oxford: Palgrave. doi:10.1057/9781137344250.0019

De Prato, G., Feijoo, C., & Simon, J. P. (2014). Innovations in the Video Games Industry: Changing global markets. *Communications & Strategies,* 94, 17–38.

De Prato, G., Sanz, E., & Simon, J. P. (2014). *Digital Media Worlds. The New Economy of Media*. New York: Palgrave-Macmillan. doi:10.1057/9781137344250

De Prato, G., & Simon, J. P. (2015). "The next wave: Big Data? *Communications & Strategies*, 97, 15–39.

De Souza e Silva, A., & Sheller, M. (Eds.). (2014). *Mobility and Locative Media: Mobile Communication in Hybrid Spaces*. New York: Routledge.

de Souza e Silva, A., & Sutko, D. M.de Souza e Silva. (2008). Playing Life and Living Play: How Hybrid Reality Games Reframe Space, Play, and the Ordinary. *Critical Studies in Media Communication*, *25*(5), 447–465. doi:10.1080/15295030802468081

de Souza e Silva, A., & Sutko, D. M.de Souza e Silva. (2011). Theorizing Locative Technologies Through Philosophies of the Virtual. *Communication Theory*, *21*(1), 23–42. doi:10.1111/j.1468-2885.2010.01374.x

De Wit, M., Mateus, M., Moitry, J., & Acharys, A. (2009). Dumb pipe vs. Smart pipe? Better to design a valuable pipe. *Oliver Wyman*. Retrieved from http://www.oliverwyman.com/ow/pdf_files/CMT07-DPipeSPipe-final.pdf

Deciphering Trends in Mobile Search. (2007.). *Computer*, *40*(8), 58–62. doi:10.1109/MC.2007.270

Deegan, R. (2014). Mobile HCI: Issues Surrounding Cognition, Distraction, Usability and Performance. *International Journal of Mobile Human Computer Interaction*, *6*(1), 1–14.

Desblache, L. (2011). *La Plume des bêtes. Les animaux dans le roman*. Paris, France: L'Harmattan.

Deuze, M. (2012). *Media life*. Cambridge, United Kingdom: Polity Press.

Deuze, M. (2012). *Media Life*. Cambridge: Polity Press.

Dewar, C. (2014). Rebalancing the value from voice and SMS to data. Retrieved from http://www.mobileworldlive.com/rebalancing-value-voice-sms-data?utm_campaign=MWL_20140902&utm_medium=email&utm_source=Eloqua&elq=698c8f2d564f402380d40b1fec23acb1&elqCampaignId=2215

Dewey, J. (2007). Experience and Education (Original edition 1938). New York, NY: Simon and Schuster.

Dharmasiri, C., Jayendranath, R., Ariyarathna, A. L., Perera, P. M., & Weerawarana, S. M. (2013) Web browsers on smart mobile devices: A gap analysis on the state of the art. Proceedings of the Advances in ICT for Emerging Regions (ICTer) 2013 International Conference, 75(79), 11-15. doi:10.1109/ICTer.2013.6761158

Dhar, V. (2013). Data science and prediction. *Communications of the ACM*, *56*(12), 64–73. doi:10.1145/2500499

Dias, P., & Botelho, I. (2014, November 12 - 15). Is the Second Screen becoming the First? An exploratory study of second screening use practices. *Paper presented at ECREA's 5th European Communication Conference* - Lisbon.

Díez, L. G., & Cuadrado, P. P. (2001). *Principios básicos sobre diseño periodístico*. Madrid: Universitas.

Dimmick, J. W., Feaster, J. C., & Hoplamazian, G. J. (2011). News in the interstices: the niches of mobile media in space and time. *New Media & Society*, *13*(1), 23–39. doi:10.1177/1461444810363452

Directive 95/46/EC. (1995, October 24). European Parliament.

Doyle, K. (2014). Smartphone obsolescence: How the personal cloud and IoT will disrupt the handset. *Gigaom research*. Retrieved from http://research.gigaom.com/report/smartphone-obsolescence-how-the-personal-cloud-and-iot-will-disrupt-the-handset/?utm_source=GeneralUsers&utm_campaign=15c55a60e5-v%3Awhatsnewd%3A08-11&utm_medium=email&utm_term=0_1dd83065c6-15c55a60e5-99640621

Drum, M. (2013). Place-based, Somatic Augmented Reality as Critical Practice. AR[t] 4, 16-20.

Dubois, C., Avignon, M., & Escudier, P. (2012). *Observer la terre depuis l'espace, enjeu des données spatiales pour la société*. Paris, France: Dunod.

Ducheneaut, N., Moore, R., Oehlberg, L., Thorton, J., & Nickell, E. (2008). Social TV: Designing for distributed, sociable television viewing. *International Journal of Human Computer Interaction*, 24(2). Retrieved from http://tinyurl.com/pun4dxr

Dudenhoeffer, C. (2014). The existing wearable technology landscape. *Wearable world news*. Retrieved from http://wearableworldnews.com/2014/05/06/wearable-world-taxonomy/

Durrani, A. (2014, February, 24). Mobile now driving 80% of Twitter traffic. *Media Week*. Retrieved from http://www.mediaweek.co.uk/

Dutta, S., & Bilbao-Osorio, B. (2012). The Global Information Technology Report 2012. Living. In *A Hyperconnected World, Insight Report*. Geneva: RO-Kundig.

EC. (2008). EU crackdown of ringtone scams. Frequently asked questions. E. C. DG for Health and Consumers. Retrieved from http://europa.eu/rapid/pressReleasesAction.do?reference=MEMO/08/516&format=HTML&aged=0&language=EN&guiLanguage=en

El-Hussein, M. O. M., & Cronje, J. C. (2010). Defining Mobile Learning in the Higher Education Landscape. *Journal of Educational Technology & Society*, *13*(3), 12–21.

Ellis, D. (2006). Extracting information from music audio. *Communications of the ACM*, *49*(8), 32–37. doi:10.1145/1145287.1145310

EMC Privacy Index. (2014). Retrieved from http://portugal.emc.com/campaign/privacy-index/index.htm?pid=home-emcprivacyindex-120614?fromGlobalSelector

Emmanouilidis, C., Koutsiamanis, R.-A., & Tasidou, A. (2013). Mobile guides: Taxonomy of architectures, context awareness, technologies and applications. *Journal of Network and Computer Applications*, *36*(1), 103–125. doi:10.1016/j.jnca.2012.04.007

Epstein, M. (2009). *Moving Story.* Paper presented at the Media in Transition 6: Stone and Papyrus, Cambridge, Mass.

Erickson, K., Kretschmer, M., & Dinusha Mendis, D. (2013, January). *Copyright and the Economic Effects of Parody: An Empirical Study of Music Videos on the YouTube Platform and an Assessment of the Regulatory Options Parody and Pastiche.* Study III. Retrieved from http://zenodo.org/record/8374/files/CREATe-Working-Paper-2013-04.pdf

Ericsson Mobility Report. (2013). *On the Pulse of the Networked Society*. Retrieved from www.ericsson.com

Ericsson Mobility Report. (2014). Retrieved from www.ericsson.com/mobility-report

ETSE. (2015). Retrieved from http://www.usc.es/etse/en

Euromonitor International. (2014). *WTM Global trends report.*

European Banking Authority. (2014). *Eba Consumer Trends Report 2014.*

European Commission. (2013). *Finance for Growth. Report of the high level expert group on SME and infrastructure financing*. doi:10.3917/reco.596.1075

European Commission. (2014). Green Paper on mobile health ("mHealth").

European Commission. (2014). *Sizing the EU app economy*. Luxembourg; doi:10.2759/43252

European Commission. (2015). eCall: Time saved = lives saved. *Digital Agenda for Europe*.

Eye Film Institute. (2015). Retrieved from https://www.eyefilm.nl/en

Eysenbach, G. (2001). What is e-health? *Journal of Medical Internet Research, 3*(2), e20. doi:10.2196/jmir.3.2.e20 PMID:11720962

Fagerberg, J. (2005). Innovation: A guide to the Literature. In J. Fagerberg & D. C. Mowery (Eds.), *The Oxford Handbook of Innovation*. Oxford: Oxford University Press.

Farman, J. (2010). Mapping the digital empire: Google Earth and the process of postmodern cartography. *New Media & Society, 12*(6), 869–888. doi:10.1177/1461444809350900

Farman, J. (2012). *Mobile Interface Theory*. New York: Routledge.

Farman, J. (2012). *Mobile Interface Theory: Embodied Space and Locative Media*. New York, NY: Routledge.

FCCC. (2015). Retrieved from http://www.usc.es/en/centros/cc_comunicacion/index.html

Featherstone, M. (2009). Ubiquitous media: An introduction. Theory, Culture & Society, 26 (2-3). Nottingham, SAGE & TCS Nottingham Trent University, 1–22.

Feijoo, C., Gelabert, J., Lindmark, S., Matía, B., Tarín, C., & Villar, J. P. (2013). Public and Commercial Models of Access in the Digital Era. *European Parliamen*t. Retrieved from http://www.europarl.europa.eu/studies

Feijóo, C., Gómez Barroso, J. L., Marín-Andreu, Á., & Rojo-Alonso, D. (2006). An overview of contents exploitation scenarios in the new digital media (pp. 72-95). ENCIP. i2010: Comments and Contributions. Eurocpr 2006: Selected Papers.

Feijoo, C., Gómez-Barroso, J. L., & Aggarwal, S. (2015). Economics of Big Data. In B. &. Latzer, Internet Economics (p. 27). Cheltenham Glos: Edward Elgar Publishing Limited.

Feijóo, C., Gómez-Barroso, J.L. & Martínez, I. J. (2010). Nuevas vías para la comunicación empresarial: publicidad en el móvil. *El profesional de la información, 19*(2), 140-148.

Feijóo, C., Maghiros, I., & Gómez-Barroso, J. L. (2008). *Techno-Economic models for mobile content.* Proceedings of the 17th Biennial Conference of the International Telecommunications Society, Montréal, Canada.

Feijóo, C. (2013). Soportes digitales y transformación de las industrias de contenidos. *El Profesional de La Informacion, 22*(1), 5–9. doi:10.3145/epi.2013.ene.01

Feijóo, C., & Gómez Barroso, J. L. (2013). Hacia una economía de la información personal. In J. M. Aguado (Ed.), *Feijóo, C. & Martínez, I. (2013), La Comunicación Móvil: Hacia un Nuevo Ecosistema Digital* (pp. 305–322). Barcelona: Gedisa.

Feijóo, C., Gómez Barroso, J. L., Aguado, J. M., & Barrero, S. (2012). Mobile gaming: Industry challenges and policy implications. *Telecommunications Policy, 36*(3), 212–221. doi:10.1016/j.telpol.2011.12.004

Feijóo, C., Gómez-Barroso, J. L., & Martínez Martínez, I. J. (2010). Nuevas vías para la comunicación empresarial: Publicidad en el móvil. *El Profesional de la Información, 19*(2), 140–148. doi:10.3145/epi.2010.mar.04

Feijóo, C., Gómez-Barroso, J. L., & Voigt, P. (2014). Exploring the economic value of personal information from firms' financial statements. *International Journal of Information Management, 34*(2), 248–256. doi:10.1016/j.ijinfomgt.2013.12.005

Feijóo, C., Lindmark, S., Villar, J. P., Tarín, C., Gelabert, J., & Matía, B. (2013). *Public and commercial modes of access to content and information in the digital era. Culture and Education. IP/B/CULT/IC/2012-018*. Brussels: European Parliament.

Feijóo, C., Maghiros, I., Abadie, F., & Gomez-Barroso, J. L. (2009). Exploring a heterogeneous and fragmented digital ecosystem: Mobile content. *Telematics and Informatics, 26*(3). 282–292. doi:10.1016/j.tele.2008.11.009

Feijóo, C., Pascu, C., Misuraca, G., & Lusoli, W. (2009). The next paradigm shift in the mobile ecosystem: Mobile social computing and the increasing relevance of users. *Communications & Stratégies*, *75*(3), 57–77.

Feijóo, C., & Ramos, S. (2013). La música en el móvil o el dispositivo como soporte. In J. M. Aguado (Ed.), *Feijóo, C. & Martínez, I. (2013), La Comunicación Móvil: Hacia un Nuevo Ecosistema Digital* (pp. 157–186). Barcelona: Gedisa.

Feinleib, D. (2012). Big-Data-Trends. *Forbes*.

Feldman, V. (2005). *Leveraging Mobile Media*: Cross-Media Strategy and Innovation Policy for Mobile Media Communication. Heildeberg: Physica-Verlag.

Feldmann, V. (2005). *Information Age Economy: Leveraging mobile media: Cross-media strategy and innovation policy for mobile media communication*. Heidelberg, Germany: Physica-Verlag.

Feldmann, V. (2005). *Leveraging mobile media: Cross-media strategy and innovation policy for mobile media communication*. Heidelberg: Physica-Verlag.

Fenn, J., & Linden, A. (2005). Gartner's Hype Cycle Special Report for 2005. Retrieved from http://www.gartner.com/DisplayDocument?doc_cd=130115

Ferguson, T. (2014). Phablets carving out their own app niche. Retrieved from http://www.mobileworldlive.com/phablets-carving-app-niche-flurry?utm_campaign=MWL-A-20140911&utm_medium=email&utm_source=Eloqua&elq=9ec8057650a74b4099530ab35c7ff091&elqCampaignId=2286

Fidalgo, A., Serrano Tellería, A., Carvalheiro, J. R., Canavilhas, J., & Correia, J. C. (2013). Human Being as a Communication Portal: The construction of the Profile on Mobile Phones. *Revista Latina de Comunicación Social,* 68, 545-565. Retrieved November 8, 2014 from http://www.revistalatinacs.org/068/paper/989_Covilha /23_Telleriaen.html

Fidalgo, A. (2009). Pushed news: When the news comes to the cellphone. *Brazilian Journalism Research*, *5*(2), 113–124.

Filloux, F. (2013) Facebook Home: Another Android Lock Pick. *Monday Note*. Retrieved from http://www.mondaynote.com/2013/04/07/facebook-home-another-android-lock-pick/

Filloux, F. (2014). Mobile Facts to Keep in Mind. Part 1. *Mondaynote*. Retrieved from http://www.mondaynote.com/2014/07/06/mobile-facts-to-keep-in-mind-part-1/

Finnemann, N. (2011). Mediatization Theory and Digital Media *Communications*, *36*(1), 67–89. doi:10.1515/comm.2011.004

Fiordelli, M., Diviani, N., & Schulz, P. J. (2013). Mapping mHealth research: A decade of evolution. *Journal of Medical Internet Research*, *15*(5), e95. doi:10.2196/jmir.2430 PMID:23697600

Flichy, P. (2010). *Le sacre de l'amateur*. Paris, Le Seuil: La République des idées.

Flurry. (2014). Apps Solidify Leadership Six Years into the Mobile... *Flurry Insights Blog*.

Flusser, V. (1988). *On writing, complexity and the technical revolutions. Interview in Onasbrück, European Media Art Festival.* September 1988. Retrieved November 8, 2014 from https://www.youtube.com/watch?v=lyfOcAAcoH8&app=desktop

Fonkych, K., & Taylor, R. (2005). *The state and pattern of health information technology adoption*. Rand Corporation.

Foridi, L. (2014). The Facebook-ification of everything! Sex, authenticity and reality for the status update era. *Salon*. Sunday, Aug 31. Retrieved from http://www.salon.com/2014/08/31/the_facebook_ification_of_everything_sex_authenticity_and_reality_for_the_status_update_era/?utm_source=twitter&utm_medium=socialflow

Fortes, F., Lv, R., Dehghan, D., & Moraveji, N. (2007). Taxonomies for News Aggregator Interfaces and User Patterns. Microsoft Research Technical Report. Retrieved from http://research.microsoft.com/pubs/70396/tr-2007-02.pdf

Fortunati, L. (2002). The Mobile Phone: Towards New Categories and Social Relations. *Information Communication and Society*, *5*(4), 514–528. doi:10.1080/13691180208538803

Foth, M. (Ed.). (2011). *From social butterfly to engaged citizen: Urban informatics, social media, ubiquitous computing, and mobile technology to support citizen engagement*. MIT Press.

Foucault, M., & Miskowiec, J. (1986). Of Other Spaces. *Diacritics*, *16*(1), 22–27. doi:10.2307/464648

Fowler, C., & Voci, P. (2011). *Brief Encounters: Theorizing Screen Attachments Outside the Movie Theatre*. Retrieved from http://www.screeningthepast.com/2011/11/brief-encounters-theorizing-screen-attachments-outside-the-movie-theatre/

Franciscato, C. E. (2010). Uma proposta de incorporação dos estudos sobre inovação nas pesquisas em jornalismo. *Estudos em Jornalismo e Mídia*, *7*(1), 8–18.

Friedman, L. (2013). The App Store turns five: A look back and forward. *MacWorld*, retrieved from http://www.macworld.com/article/2043841/the-app-store-turns-five-a-look-back-and-forward.html

Frier, S. (2014). Facebook test lets users search old posts by keyword. *Bloomberg* [Web page]. Retrieved from: http://www.bloomberg.com/news/2014-08-28/facebook-test-lets-users-search-old-posts-by-keyword.html

Frost & Sullivan. (2008). *Wireless Healthcare Technologies in Europe*. Mobile.

Fuchs, C. (2014). *Social Media: A Critical Introduction*. London, UK: Sage. doi:10.4135/9781446270066.n2

Fudzee, M., & Abawajy, J. (2008). *A classification for content adaptation system*. Paper presented at 0th International Conference on Information Integration and Web-based Applications & Services. New York, NY, USA.

Fuller, M. (2005). *Media ecologies: Materialist energies in art and technoculture*. Cambridge: The MIT Press.

Fundación Pedro Barrié de la Maza. (2015). Retrieved from http://www.programacatedral.com

Fu, W.-T., & Pirolli, P. (2007). SNIF-ACT: A cognitive model of user navigation on the world wide web. *Human-Computer Interaction*, *22*(4), 355–412.

Gagnon, D. (2010). *Mobile Learning Environments*. Retrieved from http://www.educause.edu/ero/article/mobile-learning-environments

Gallant, L. M., Boone, G., & LaRoche, C. S. (2014). Mobile Usability: State of the Art and Implications. *Interdisciplinary Mobile Media and Communications: Social, Political, and Economic Implications, 344*.

Gans, H. J. (2004). *Deciding What's News: A Study of CBS Evening News, NBC Nightly News, Newsweek, and Time* (25 th ed.). Evanston, IL: Northwestern University Press.

García Jiménez, L. (2012, December 7-8). The use, consumption and perception of smartphones by digital natives. Proceedings of *I Technological Innovation in a Mobile World Conference*, Murcia.

García, M. (2012). iPad design lab: storytelling in the age of the tablet. (Kindle Edition).

Garcia, M., & Calantone, R. (2002). A critical look at technological innovation typology and innovativeness terminology: A literature review. *Journal of Product Innovation Management*, *19*(2), 110–132. doi:10.1016/S0737-6782(01)00132-1

Gardner, H., & Davis, K. (2013). *The app generation: how today's youth navigate identity, intimacy, and imagination in a digital world*. New Haven, CT: Yale University Press.

Gartner. (2013). Gartner Says Mobile App Stores Will See Annual Downloads Reach 102 Billion in 2013.

Gassée, J. L. (2014). Tim Cook needs to hire human curators to fix the Apple App Store. *Quartz*. Retrieved from http://qz.com/247370/apple-tim-cook-desperately-needs-to-fix-the-app-store/

Gavalas, D., Konstantopoulos, C., Mastakas, K., & Pantziou, G. (2014). Mobile Recommender Systems in Tourism. *Journal of Network and Computer Applications*, *39*, 319–333. doi:10.1016/j.jnca.2013.04.006

Gawer, A., & Cusumano, M. A. (2014). Industry platforms and ecosystem innovation. *Journal of Product Innovation Management*, *31*(3), 417–433. doi:10.1111/jpim.12105

Gay, G. (2009). Context-aware mobile computing: Affordances of space, social awareness, and social influence. *Synthesis Lectures on Human-Centered Informatics*, *2*(1), 1–62. doi:10.2200/S00135ED1V01Y200905HCI004

Gellert, W., Gottwald, S., Hellwich, M., Kästner, H., & Küstner, H. (1989). *La Enciclopedia Concisa de las Matemáticas* (Vol. 2). VNR.

George, D., Stryjak, J., Meloán, M., Castells, P. (2015, July). The Mobile Economy 2015. *GSMAIntelligence*. Retrieved from www.gsmaintelligence.com

Germany's consumers not sold on wearables. (2014). eMarketer. Retrieved from http://www.emarketer.com/Article/Germanys-Consumers-Not-Sold-on-Wearables/1011141/2

Geron, T. (2013, November 11). LinkedIn buys pulse newsreader for $90 million. *Forbes*. Retrieved from http://www.forbes.com

Gerpott, T. (2010). Communication behaviors and perceptions of mobile internet adopters. *Emerald*, *12*(4), 54–73.

Giardina, M. (2012). Emerging media channels: Mobile tablet powered news dissemination. *f the Technology & Emerging Media Track–Annual Conference of the Canadian Communication Association*. Waterloo, Canada: Canadian Communication Association.

Gibson, J. J. (1979a). The Theory of Affordances. In J. Giesekink, & W. Mangold (Eds) (2014), The People, Place and Space Reader. New York: Routledge.

Gibson, J. J. (1979). *The ecological approach to visual perception*. Hillsdale, NJ: Lawrence Erlbaum Associates, Inc.

Giglietto, F., & Selva, D. (2014). Second Screen and Participation: A content analysis on a full season dataset of tweets. *Journal of Communication*, *64*(2), 260–277. doi:10.1111/jcom.12085

Gillet, J. (2013). Top-20 Global Mobile Operator Groups by Connections and Revenue. *GSMA Intelligence*. Retrieved from http://www.mobileworldlive.com/top-20-global-mobile-operator-groups-by-connections-and-revenue-q1-2013

Gillet, J. (2014). Measuring mobile penetration. *GSMA Intelligence*. Retrieved from https://gsmaintelligence.com/analysis/2014/5/measuring-mobile-penetration/430/

Glickman, L., & Kim, A. (2012). Product placement and technology: Developments, opportunities, and challenges. *The Entertainment and Sports Lawyer*, *30*(1), 30–41.

Global mobile statistics . (2012). Retrieved from http://mobithinking.com/mobile-marketing-tools/latest-mobile-stats/d#mobilebehavior

Goffey, A. (2008). Algorithm. In M. Fuller (Ed.), *Software studies: a lexicon* (pp. 15–20). Cambridge, MA: The MIT Press. doi:10.7551/mitpress/9780262062749.003.0002

Goffman, E. (1959). *The presentation of self in everyday life*. New York, NY: Doubleday.

Goggin, G. (2011). Ubiquitous apps: Politics of openness in global mobile cultures. *Digital Creativity*, *22*(3), 148–159. doi:10.1080/14626268.2011.603733

Goggin, G., Dwyer, T., & Martin, F. (2014). Moveable Media: Mobile internet and new policy modes. In M. Glowacki & L. Jackson (Eds.), *Creativity, Innovation and Interaction: Public Media Management Fit for the 21st Century*. New York: Routledge.

Goggin, G., Martin, F., & Dwyer, T. (2014). Locative News. *Journalism Studies*. doi: 10.1080/1461670X.2014.890329

Goldfarb, A., & Tucker, C. (2008). Economic and business dimensions: Search engine advertising. *Communications of the ACM*, *51*(11), 22–24. doi:10.1145/1400214.1400222

Gold, J. E., Driban, J. B., Thomas, N., Chakravarty, T., Channell, V., & Komaroff, E. (2012). Postures, typing strategies, and gender differences in mobile device usage: An observational study. *Applied Ergonomics*, *43*(2), 408–412. doi:10.1016/j.apergo.2011.06.015 PMID:21764031

Gómez-Barroso, J. L., Compañó, R., Feijóo, C., Bacigalupo, M., Westlund, O., Ramos, S., & García-Jiménez, M. C. (2010). Prospects of mobile search. Seville: Institute for Prospective Technological Studies. Retrieved from http://ipts.jrc.ec.europa.eu/publications/pub.cfm?id=3159

Gómez-Barroso, J. L., Compañó, R., Feijóo, C., Bacigalupo, M., Westlund, O., Ramos, S., et al. (2010). *Prospects of Mobile Search*, European Commission, JRC, Institute for Prospective Technological Studies (IPTS), EUR 24148 EN 2010. (http://ipts.jrc.ec.europa.eu/publications/pub.cfm?id=3159)

Gómez-Barroso, J. L., Feijóo, C., & Compañó, R. (2011). Opportunities in the mobile search market. *Computer*, *44*(11), 93–95. doi:10.1109/MC.2011.347

Gómez-Barroso, J. L., Feijóo, C., & Compañó, R. (2011). Promising prospects in mobile search: Business as usual or techno-economic disruptions? *IEEE Signal Processing Magazine*, *28*(4), 131–135. doi:10.1109/MSP.2011.941135

Google & Ypsos. (2012). The new multi-screen world. Retrieved from http://www.thinkwithgoogle.com/insights/library/studies/the-new-multi-screen-world-study/

Google. (2012). Project Glass: One day [YouTube video]. Retrieved from https://www.youtube.com/watch?v=9c6W4CCU9M4&feature=youtu.be

Google. (2012, August). The new multiscreen world. Understanding cross-platform consumer behavior. Retrieved from http://www.google.com/think/research-studies/the-new-multi-screen-world-study.html#

Gordon, E., & de Souza e Silva, A. (2011). Net Locality: Why Location Matters in a Networked World. West Sussex, United Kingdom: Wiley-Blackwell.

Gordon, E., & Souza e Silva, A. (2011). *Net locality: Why location matters in a networked world.* West Sussex, United Kingdom: Wiley-Blackwell. doi:10.1002/9781444340679

Gottlieb, B. (2009). *Gratitude for Tecnology*. Dresden, Germany: Atropos Press.

Gray, J. (2002). *Straw Dogs. Thoughts on humans and other animals.* London, United Kingdom: Granta Books.

Greengard, S. (2012). Advertising gets personal. *Communications of the ACM*, *55*(8), 18–20. doi:10.1145/2240236.2240243

GRETEL. (2005). La explotación de los contenidos audiovisuales en la transición a los nuevos medios digitales. In C. Feijóo (Ed.), El sector audiovisual y su evolución. Madrid: Colegio Oficial de Ingenieros de Telecomunicación (COIT).

Grosser, B. (2014). What Do Metrics Want? How Quantification Prescribes Social Interaction on Facebook. *Computational Culture: A Journal of Software Studies*. Retrieved November 17, 2014 from http://computationalculture.net/article/what-do-metrics-want

Grubenmann, S. (2013). Innovation in Journalism – Towards a Holistic Approach. Retrieved from http://www.researchgate.net/publication/235799323_Innovation_in_Journalism__Towards_a_Holistic_Approach

Grusin, R. (2010). *Premediation: Affect and mediality after 9/11*. New York, NY: Palgrave Macmillan. doi:10.1057/9780230275270

GSM Association. (2011). *Mobile Education Landscape Report*.

GSMA Intelligence. (2015). Definitive data and analysis for the mobile industry. Retrieved from https://gsmaintelligence.com/

GSMA Intelligence. (2014a). *The Mobile Economy. Asia Pacific*. Retrieved from www.gsmaintelligence.com

GSMA Intelligence. (2014b). *The Mobile Economy 2014*. Retrieved from www.gsmaintelligence.com

GSMA Intelligence. (2014c). Half a billion Chinese citizens have subscribed to the mobile internet. Retrieved from www.gsmaintelligence.com

GSMA Intelligence. (2014d). Chinese carriers dominate global operator ranking as M&A deals shake up US market. Retrieved from http://www.mobileworldlive.com/chinese-carriers-dominate-global-operator-ranking-ma-deals-shake-us-market?utm_campaign=MWL_20140924&utm_medium=email&utm_source=Eloqua&elq=e31fd99c468d440995cbeebe6fee5f83&elqCampaignId=2366

Guattari, F. (1989). *Les trois écologies*. Paris, France: Galilée.

Gubbins, M. (2014). *Audience in the mind*, Sampomedia. Report commissioned by Cine regio. Retrieved from www.SampoMedia.com

Guirguis, S. K., & Hassan, M. A. (2010). A smart framework for web content and resources adaptation in mobile devices. *Proceedings of the 12th international conference on Advanced communication technology*, Piscataway, NJ, USA (Vol. 1, pp. 487-492).

Haddon, L., & Vincent, J. (2008). Children's broadening use of mobile phones. In G. Goggin & L. Hjorth (Eds.), *Mobile technologies: From telecommunications to media* (pp. 37–49). Abingdon: Routledge.

Ham, D.-H., Heo, J., Fossick, P., Wong, W., Park, S., Song, C., & Bradley, C. (2006). Conceptual framework and models for identifying and organizing usability impact factors of mobile phones.*Proceedings of the 18th Australia conference on Computer-Human Interaction: Design: Activities, Artefacts and Environments (OZCHI '06)* (pp. 261-268). New York, NY, USA: ACM. doi:10.1145/1228175.1228221

Hampton, K., & Wellman, B. (2003). Neighboring in Netville: How the internet supports community and social capital in a wired suburb. *City & Community Journal*, *2*(4), 277–311. doi:10.1046/j.1535-6841.2003.00057.x

Handal, B., El-Khoury, J., Campbell, C., & Cavanagh, M. (2013). *A framework for categorising mobile applications in mathematics education*. Paper presented at the Australian Conference on Science and Mathematics Education: Students in transition – The learners' journey, Sydney, Australia.

Handford, R. (2014b). ESPN, Hulu hail mobile video traffic but questions remain over advertising, network capacity. Retrieved from http://www.mobileworldlive.com/espn-hulu-hail-video-first-world-questions-remain-advertising-network-capacity?utm_campaign=MWL_20140910&utm_medium=email&utm_source=Eloqua&elq=1de7cc88861646dbab7a1eaca0392bc5&elqCampaignId=2274

Hanford, R. (2014a). China Mobile highlight radical changes with 5G. Retrieved from http://www.mobileworldlive.com/china-mobile-top-scientist-offers-string-5g-pearls?utm_campaign=MWL_20140923&utm_medium=email&utm_source=Eloqua&elq=c79149105ec54bb28ba1d429fefa4568&elqCampaignId=2360

Hanneman, R., & Riddle, M. (2005). *Introduction to Social Network Methods*. Riverside, CA: University of California.

Haraway, D. (2008a). *When species meet*. Minneapolis: University of Minnesota Press.

Haraway, D. (2008b). *Haraway on Multispecies Cosmopolitics: Staying with the Trouble*. Minneapolis: University of Minnesota Press.

Haridakis, P., & Hanson, G. (2009). Social Interaction and Co-Viewing with YouTube: Blending mass communication reception and social connection. *Journal of Broadcasting & Electronic Media*, 53(2), 317-335. Retrievedfrom http://tinyurl.com/o5t2blu

Haumont, S., & Siren, R. (2007). *dotMobi, a Key Enabler for the Mobile Internet*. Paper presented at Mobile Internet User Experience Workshop.

Hayles, N. K. (1999). *How we became posthuman: Virtual bodies in cybernetics, literature, and informatics*. Chicago, IL: University of Chicago Press. doi:10.7208/chicago/9780226321394.001.0001

Haythornthwaite, C. (2005). Social networks and internet Connectivity effects. *Information Communication and Society*, *8*(2), 125–147. doi:10.1080/13691180500146185

Heerden, A. V., Tomlinson, M., & Swartz, L. (2012). Point of care in your pocket: A research agenda for the field of m-health. *Bulletin of the World Health Organization*, *90*(5), 393–394. doi:10.2471/BLT.11.099788 PMID:22589575

Heggestuen, J. (2014a). Carrier billing hasn't gone away. *Business Insider*. Retrieved from http://www.businessinsider.com/carrier-billing-hasnt-gone-away--here-is-why-it-has-got-developers-app-stores-and-mobile-carriers-so-excited-2014-2?utm_source=alerts&nr_email_referer=1

Heggestuen, J. (2014b). Carrier billing is another way to pay on phones. *Business Insider*. Retrieved from http://www.businessinsider.com/carrier-billing-is-another-way-to-pay-on-phones--and-its-creating-a-big-opportunity-to-monetize-mobile-users-2014-3?utm_source=alerts&nr_email_referer=1

Heggestuen, J. (2014c). Google is cutting deals with wireless carriers everywhere... *Business Insider*. Retrieved from http://www.businessinsider.com/google-play-and-carrier-billing-2014-4?utm_source=alerts&nr_email_referer=1

Henley, J. (2014). L'exemple de The Guardian. In J.,M. Charon, J. Papet, (Eds.). *Le journalisme en questions* (pp 169-171).

Henry, S. L., Abou-Zahra, S., & Brewer, J. (2014, April). The role of accessibility in a universal web.*Proceedings of the 11th Web for All Conference* (p. 17). ACM. doi:10.1145/2596695.2596719

Hepp, A., Hajvard, S., & Lundby, K. (2010). Mediatization – Empirical Perspectives: An introduction to a special issue. *Communications*, *35*(3), 223–228. doi:10.1515/comm.2010.012

Hermida, A. (2014). *Tell Everyone: Why We Share and Why it Matters*. Toronto: Doubleday Canada.

Hesmondhalgh, D. (2007). *The Cultural Industries*. London, U.K.: Sage Publications Ltd.

Hexagram UQAM. (2015). Retrieved from http://www.hexagram.uqam.ca/

Hight, J. (2003). Narrative Archeology. Retrieved from http://www.xcp.bfn.org/hight.html

Hirst, M., Harrison, J., & Mazepa, P. (2014). *Communication and new media: From broadcast to narrowcast*. Don Mills: Oxford University Press.

Hjarvard, S. (2008). The Mediatization of Society. *Nordicom Review*, *29*(2), 105–134. PMID:19361120

Hofstede, G. (2001). *Culture's consequences: Comparing values, behaviors, institutions, and organizations across nations*. Thousand Oaks, USA: Sage Publications.

Holden, W. (2008). Making music with mobile. Basingstoke: Juniper Research. Retrieved from http://www.juniper-research.com/shop/viewwhitepaper.php?id=98&whitepaper=56

Holden, R. J., & Karsh, B. T. (2010). The technology acceptance model: Its past and its future in health care. *Journal of Biomedical Informatics*, *43*(1), 159–172. doi:10.1016/j.jbi.2009.07.002 PMID:19615467

Horning, R. (2014). Social Media Is Not Self-Expression. The New Inquiry, November 14. Retrieved November, 17, 2014, from http://thenewinquiry.com/blogs/marginal-utility/social-media-is-not-self-expression/

Horozov, T., Narasimhan, N., & Vasudevan, V. (2006). Using location for personalized POI recommendations in mobile environments. Saint. *International Symposium on Applications and the Internet (SAINT'06)*, 124-129. doi:10.1109/SAINT.2006.55

Hou, H.-T., Wu, S.-Y., Lin, P.-C., Sung, Y.-T., Lin, J.-W., & Chang, K.-E. (2014). A Blended Mobile Learning Environment for Museum Learning. *Journal of Educational Technology & Society*, *17*(2), 207–218.

Hsi, S. (2008). Designing for Mobile Visitor Engagement. In L. Tallon (Ed.), *Digital technologies and the museum experience: handheld guides and other media*. Landham, MD: AltaMira Press.

Huang, M.-J., & Lee, T.-L. (2008). *An integrated software processor with autofilling out web forms*. Paper presented at Computer Systems Architecture Conference.

Humphreys, L., Von Pape, T., & Karnowski, V. (2013). Evolving Mobile Media: Uses And Conceptualizations Of The Mobile Internet. *Journal of Computer-Mediated Communication*, *18*(4), 491–507. doi:10.1111/jcc4.12019

IDC (2011). *Worldwide and U.S. Mobile Applications, Storefronts, Developer, and In-App Advertising 2011-2015 Forecast: Emergence of Postdownload Business Models*. International Data Corporation - IDC Corporate USA.

IEG World Bank. (2013). *World Bank Group Support for Innovation and Entrepreneurship. An Independent evaluation.*

IFPI. (2013). IFPI Digital Music Report 2013. Retrieved from http://www.promusicae.es/files/imagenes/file/DMR2013fullv9.pdf

IFPI. (2014). Digital music report. Retrieved from http://www.ifpi.org/downloads/Digital-Music-Report-2014.pdf

IIT. (2015). Retrieved from https://www.usc.es/en/institutos/iit/index.html

Industry associations, trade press, consultancies and research institutes

INE Portugal. (2014, November 6). *Sociedade da Informação e do Conhecimento. Inquérito à Utilização de Tecnologias da Informação e da Comunicação pelas Famílias*. Retrieved from http://www.ine.pt/xportal/xmain?xpid=INE&xpgid=ine_destaques&DESTAQUESdest_boui=211422735&DESTAQUESmodo=2

Ingram, M. (2013). Did Facebook miss a massive opportunity by building a walled garden instead of a truly open platform? *Giga OM*. Retrieved from http://gigaom.com/2013/07/24/did-facebook-miss-a-massive-opportunity-by-building-a-walled-garden-instead-of-a-truly-open-platform/

Innovation report. (2014, March 24). *The New York Times*. Retrieved from http://visualdays.no/files/2014/05/224608514-The-Full-New-York-Times-Innovation-Report.pdf

Insight, C. C. S. (2014). *Global smartphone market analysis and outlook: Disruption in a changing market.* Retrieved from http://www.lenovo.com/transactions/pdf/CCS-Insight-Smartphone-Market-Analysis-Full-Report-07-2014.pdf

Internet live statistics. (2015). Retrieved from http://www.internetlivestats.com/internet-users/

Ipsos Media, C. T. (2013). *Our Mobile Planet: Sweden*. Retrieved from http://services.google.com/fh/files/misc/omp-2013-sw-en.pdf

Isaacson, W. (2011). *Steve Jobs*. New York: Simon & Schuster.

Isbell, K. (2010, August 30). The rise of the news aggregator: legal implications and best practices. *Berkman Center Research Publication.* Retrieved from http://ssrn.com/abstract=1670339. doi:10.2139/ssrn.1670339

Istepanian, R., Laxminarayan, S., & Pattichis, C. S. (2006). *M-health*. Springer Science+ Business Media, Incorporated.

ITU. (2010). *The World in 2010. ICT Facts and Figures*. Retrieved from http://www.itu.int/en/ITU-D/Statistics/Documents/facts/ICTFactsFigures2010.pdf

ITU. (2013). ICT Facts and Figures, International telecommunication Union, Geneva, Switzerland. Retrieved from http://www.itu.int/ITU-D/ict/facts/material/ICTFactsFigures2013.pdf

ITU. (2013). ICT Facts and Figures. Retrieved from http://www.itu.int/en/ITU-D/Statistics/Documents/facts/ICTFactsFigures2013-e.pdf

Jacobs, A. (2009). The pathologies of big data. *Communications of the ACM*, *52*(8), 36–44. doi:10.1145/1536616.1536632

Jansson, A. (2002). Spatial Phantasmagoria: The Mediatization of Tourism Experience. *European Journal of Communication*, *17*(4), 429–443. doi:10.1177/0267323102017004020l

Jansson, A. (2009). Mobile Belongings: Texturation and stratification in mediation processes. In K. Lundby (Ed.), *Mediatization: Concept, changes, consequences.* (pp. 243–261). Oxford, Wien.

Jaokar, A., & Fish, T. (2006). *Mobile Web 2.0*. London: Futuretext.

Jaokar, A., & Fish, T. (2006). *Mobile Web 2.0: The innovator's guide to developing and marketing next generation wireless/mobile applications*. London: Futuretext.

Jasiewicz, M. I. (2012, December 20). Copyright protection in an opt-out world: Implied license doctrine and news aggregators. *The Yale Law Journal*. Retrieved from http://ssrn.com/abstract=2207489

Jayawardhena, C., Kuckertz, A., Karjaluoto, H., & Kautonen, T. (2009). Antecedents to permission based mobile marketing: An initial examination. *European Journal of Marketing*, *43*(3/4), 473–499. doi:10.1108/03090560910935541

Jenkins, H. (P.I.) (2009). Confronting the Challenges of Participatory Culture Media Education for the 21st Century. The John D. and Catherine Foundation Reports on Digital Media and Learning. Massachusetts: MIT Press.

Jenkins, H. (2006). *Convergence Culture*. New York: NYU Press.

Jenkins, H. (2009). *Confronting the Challenges of Participatory Culture: Media Education for the 21st Century. The John D. and Catherine T. MacArthur Foundation Reports on Digital Media and Learning*. Cambridge, MA: MIT Press.

Jenkins, H., Ford, S., & Green, J. (2013). *Spreadable Media. Creating Value and Meaning in a Networked Culture*. New York, NY: New York University Press.

Jenkins, H., Ford, S., & Green, J. (2013). *Spreadable media: Creating value and meaning in a networked culture*. New York, NY: New York University Press.

Jevbratt, L. (2012). With the eyes of another. Antennae, 21, 76-93.

Jones, S., Karnowski, V., Ling, R., & von Pape, T. (2013). Welcome to Mobile Media & Communication. *Mobile Media & Communication, 1*(1), 3–7. doi:10.1177/2050157912471456

Joshi, P. M., & Liu, S. (2009). Web document text and images extraction using DOM analysis and natural language processing.*Proceedings of the 9th ACM symposium on Document engineering (DocEng '09)*, (pp. 218-221). New York, NY, USA. doi:10.1145/1600193.1600241

Julier, G. (2010). *La cultura del diseño*. Barcelona: Gustavo Gili.

Juniper Research. (2015). Apps & App Stores Mobile Market Research - Market Trends & Forecasts.

Kaikkonen, A. (2008). Full or tailored Mobile Web- where and how do people browse on their mobiles?*Proceedings of the International Conference on Mobile Technology, Applications, and Systems (Mobility '08)*. New York, NY, USA: ACM. doi:10.1145/1506270.1506307

KakaoTalk. (n. d.). Retrieved from www.kakao.com/

Kamvar, M., & Baluja, S. (2006 April 22–27). A Large Scale Study of Wireless Search Behavior: Google Mobile Search. Proceedings of the SIGCHI conference on human factors in computing systems, Montreal (pp. 701–709). Retrieved from http://www.esprockets.com/papers/kamvar-baluja.chi06.pdf

Karahanna, E., Straub, D. W., & Chervany, N. L. (1999). Information Technology Adoption Across Time: A Cross-Sectional Comparison Of Pre-Adoption And Post-Adoption Beliefs. *Management Information Systems Quarterly, 23*(2), 183–213. doi:10.2307/249751

Katz, E., & Aarhus, M. (2002). *Perpetual Contact: Mobile Communication, Private Talk, Public Performance*. Cambridge: Cambridge University Press. doi:10.1017/CBO9780511489471

Keil, J., Zoellner, M., Engelke, T., Wientapper, F., & Schmitt, M. (2013). Controlling and filtering information density with spatial interaction techniques via handheld Augmented Reality. In *Virtual augmented and mixed reality. Designing and developing augmented and virtual environments* (pp. 49–57). Springer Berlin Heidelberg.

Kelly, S. M. (2014, July 10). Report: Nearly 80% of apps in Apple's app store are basically lifeless. *Mashable*. Retrieved from http://mashable.com

Kerr, A. (2014, September 3-5). The Occupational Implications of Digital Games as a Service [Presentation] Proceedings of the Cost Conference, *The Dynamics of Virtual Work: the Transformation of Labour in a digital Economy,* Hatfield.

Kerris, N., & Dowling, Steve. (2007). Apple Reinvents the Phone with iPhone (press release). *Apple.com*. Retrieved from http://www.apple.com/pr/library/2007/01/09iphone.html

Kerstetter, D., Yarnal, C., Son, J., Yen, I., & Baker, B. (2008). Functional support associated with belongings to the Red Hat Society, a leisure-based social network. *Journal of Leisure Research, 40*(4), 531–555.

Khalaf, S. (2013). Flurry five-year report: it's an app world. The web just lives in it. *Flurry Analytics*. Retrieved from http://blog.flurry.com/bid/95723/Flurry-Five-Year-Report-It-s-an-App-World-The-Just-Web-Lives-in-It

Kim, J., Lee, S., Geum, Y., & Park, Y. (2012). Patterns of innovation in digital content services: The case of App Store applications. *Innovation: Management. Policy & Practice*, *14*(4), 540–556. doi:10.5172/impp.2012.14.4.540

Kitamura, S. (2013). The Relationship between use of the internet and traditional information sources: an empirical study in Japan, *SAGE Open*, April-June, 1–9.

Kitchin, R., & Dodge, M. (2011). *Code/space: software and everyday life*. Cambridge, MA: The MIT Press. doi:10.7551/mitpress/9780262042482.001.0001

Klemens, G. (2010). *The cellphone: The history and technology of the gadget that changed the world*. Jefferson, NC: McFarland & Company.

Knobel, J., Lankshear, M., & Leu, C. D. J. (Eds.). (2014). *Handbook of research on new literacies*. New York: Routledge.

Kohlborn, T., Korthaus, A., Riedl, C., & Krcmar, H. (2009). Service aggregators in business networks. In Enterprise Distributed Object Computing Conference Workshops, 2009. EDOCW 2009. 13th (pp. 195-202). doi:10.1109/EDOCW.2009.5331992

Kollock, P., & Smith, M. (2002).Communities in cyberspace. In Smith, & Kollock (Eds.), Communities in Cyberspace (pp. 3-26). London: Kindle Edition.

Koudounas, V., & Iqbal, O. (1999). *Mobile Computing: Past, present and future*. London: Department of Computing, Imperial College London.

Kramer, A. D. I., Guillory, J. E., & Hancock, J. T. (2014). Experimental evidence of massive-scale emotional contagion through social networks. *Proceedings of the National Academy of Sciences of the United States of America*, *111*(24), 8788–8790. doi:10.1073/pnas.1320040111 PMID:24889601

Kremhelmer, S., & Zenger, H. (2008). Advertising and the screening role of mass media. *Information Economics and Policy*, *20*(2), 107–119. doi:10.1016/j.infoecopol.2007.11.003

Krippendorf, K. (2004). *Content Analysis: An Introducing to its Methodology*. Beverly Hills: Sage Publications.

Krotz, F. (2007). *Mediatisierung: FallstudienzumWandell von Kommunikation*. Wiesbaden: VS VerlagfürSocialwissenschaften.

Kurzweil, R. (2005). *The singularity is near: When humans transcend biology*. New York, NY: Viking Adult.

Labs, E. (2013). Intel capital, Samsung venture investment corporation, and Telefónica Digital make strategic investment. Retrieved from http://www.prnewswire.com/news-releases/intel-capital-samsung-venture-investment-corporation-and-telefonica-digital-make-strategic-investment-in-expect-labs-205361051.html

Lam, H., & Baudisch, P. (2005). Summary thumbnails: readable overviews for small screen web browsers.*Proceedings of the SIGCHI conference on Human factors in computing systems (CHI '05)* (pp. 681-690). New York, NY, USA: ACM. doi:10.1145/1054972.1055066

Laney, D. (2001). 3D data management: Controlling data volume, velocity and variety. *META Group Research Note, 6*.

Lange, P., & Ito, M. (2009). Hanging out, messing around, and geeking out: kids living and learning with the new media. In P. Lange and M. Ito (Eds.), Creative production (pp. 243-256). Cambridge: The MIT Press.

Laouris, Y., & Eteokleous, N. (2005). *We need an educationally relevant definition of mobile learning*. Paper presented at the 4th World Conference on Mobile Learning, Cape Town, South Africa.

Larequi, J. C. (1994). *El diseño periodístico en prensa diária*. Barcelona: Bosch.

Lasen, A. (2004). *Emotions and Digital Devices – Affective Computing and Mobile Phones*. Surrey: Surrey University.

Laurila, J., Gatica-Pérez, D., Aad, I., Blom, J., Bonet, O., Do, T., et al. (2013). The mobile data challenge: Big data for mobile computing research. *Nokia Research*. Retrieved from http://research.nokia.com/files/public/MDC2012_Overview_LaurilaGaticaPerezEtAl.pdf

Le Diberder, A. (2012). Le modèle économique des jeux vidéo, un colosse en péril. In J. P. Lafrance (Ed.). *Les jeux vidéo. Quand jouer c'est communiquer, Hermes (Vol. 62,* pp. 136–143). Paris: CNRS.

Leber, J. (2013, April 12). How Wireless Carriers Are Monetizing Your Movements. *MIT Technology Review*.

LedoAndión, M. &Castelló Mayo, E. (2012). Identification of barriers and motivations in the circulation of audiovisual media contents in the Lusophone space: Communication, interculturality, citizenship, and interactivity. *Communicatio Socialis*, *XXV*(2), 85–116.

LedoAndión, M. &Castelló Mayo, E. (2013). Cultural Diversity across the Networks: The Case of National Cinema. *Comunicar. Scientific Journal of Media Education*, *XX*(40), 183–191.

LedoAndión, M., López Gómez, A., & Castelló Mayo, E. (2014). Creative Destruction as Economic Pattern of Europe's Small-Scale Audiovisual Media Markets and Minority-Language Film Industries. *China-USA Business Review*, XIII(1), 12-19.

Lee, E. (2012). Apple's ITunes Would Be One of World's Biggest Media Companies, Retrieved from http://go.bloomberg.com/tech-blog/2012-12-03-apple%E2%80%99s-itunes-would-be-one-of-world%E2%80%99s-biggest-media-companies/

Lehtonen, T., Benamar, S., Laamenen, V., Luoma, I., Ruotsalainen, O., Salonen, J., & Mikkonen, T. (2006). Towards user-friendly mobile browsing.*Proceedings of the 2nd international workshop on Advanced architectures and algorithms for internet delivery and applications (AAA-IDEA '06)*. New York, NY, USA. doi:10.1145/1190183.1190190

Lessig, L. (2006). Code: Version 2.0. New York: Basic Books. Retrieved from http://codev2.cc/download+remix/Lessig-Codev2.pdf

Leurdijk, A., & Nieuwenhuis, O. (2012). *Statistical, Ecosystems and Competitiveness Analysis of the Media and Content Industries: The Music Industry*. Luxembourg: Publication Office of the European Union. Retrieved from http://ipts.jrc.ec.europa.eu/publications/pub.cfm?id=5022

Leurdijk, A., & Nieuwenhuis, O. (2012). The Music Industry: Statistical Ecosystems and Competitiveness Analysis of the Media and Content Industries. *JRC Technical Reports*. Retrieved from ftp://ftp.jrc.es/pub/EURdoc/JRC69816.pdf

Levine, P. (2014). On Common Ground: Here as There. In J. Farman (Ed.), *The Mobile Story: Narrative Practices with Locative Technologies* (pp. 143–158). New York: Routledge.

Levinson, P. (2004). *Cellphone: The Story of the World's Most Mobile Medium and How it has Transformed Everything!* New York: Palgrave MacMillan.

Levy, D. A. L., & Newman, N. (Eds.). (2014). *Reuters Institute Digital News Report. Tracking the Future of News*. Retrieved from http://reutersinstitute.politics.ox.ac.uk/publications/risj.html

Lewis, S. C., & Westlund, O. (2015). Actors, Actants, Audiences, and Activities in Cross-Media News Work. *Digital Journalism*, *3*(1), 19–37. doi:10.1080/21670811.2014.927986

Li, F., & Whalley, J. (2002). Deconstruction of the telecommunications industry: From value chains to value networks. *Telecommunications Policy*, *26*(9/10), 451–472. doi:10.1016/S0308-5961(02)00056-3

Lind, M. R., & Zmud, R. W. (1991). The influence of a convergence in understanding between technology providers and users on information technology innovativeness. *Organization Science*, *2*(2), 195–217. doi:10.1287/orsc.2.2.195

Ling, R. (2007). *Mobile Communication and Mediated Ritual*. Norway and USA: Telenor & University of Michigan.

Ling, R. (2008). *New tech, new ties: how mobile communication is reshaping social cohesion*. New York: MIT press.

Ling, R. (2008). The Mediation of Ritual Interaction via the Mobile Telephon. In J. Katz (Ed.), *Handbook of Mobile Communication*. Cambridge, MA: MIT Press. doi:10.7551/mitpress/9780262113120.003.0013

Ling, R. (2012). *Taken for grantedness*. Cambridge, MA: MIT Press.

Ling, R. (2012). *Taken for grantedness: The embedding of mobile communication into society*. Cambridge, MA: MIT Press.

Ling, R., & Stald, G. (2010). Mobile communities: Are we talking about a village, a clan, or a small group? *America behavioral. Scientist (Philadelphia, Pa.)*, *53*, 1133–1146.

Lin, H., Yeo, B., & Chen, Y. (2013). Mobilizing timely location-based advertising: A study of effectiveness on persuasion. *Australian Journal of Business and Management Research*, *3*(6), 15–29.

Lippard, L. (1981). Hot Potatoes: Art and Politics in 1980. Block, 4, 2-9.

Lira, W., Ferreira, R., de Souza, C., & Carvalho, S. (2014, September). Experimenting on the cognitive walkthrough with users. *Proceedings of the 16th international conference on Human-computer interaction with mobile devices & services* (pp. 613-618). ACM. doi:10.1145/2628363.2628428

Liu, C. L., Sinkovics, R. R., Pezderka, N., & Haghirian, P. (2012). Determinants of consumer perceptions toward mobile advertising: A comparison between Japan and Austria. *Journal of Interactive Marketing*, *26*(1), 21–32. doi:10.1016/j.intmar.2011.07.002

Livingstone, S. (2009). On the Mediation of Everything: ICA Presidential Address 2008. *Journal of Communication*, 59, 1-18.

Li, X. (2013). Innovativeness, Personal Initiative, News Affinity And News Utility As Predictors Of The Use Of Mobile Phones As News Devices. *Chinese Journal of Communication*, *6*(3), 350–373. doi:10.1080/17544750.2013.789429

Lohr, S. (2012). The age of big data. *New York Times*.

Long, C., & Wong, R. C. W. (2014). Viral marketing for dedicated customers. *Information Systems*, *46*, 1–23. doi:10.1016/j.is.2014.05.003

Louçã, F. (2014). The elusive concept of innovation for Schumpeter, Marschak and the early econometricins. *Research Policy*, *43*(8), 1442–1449. doi:10.1016/j.respol.2014.02.002

Lovejoy, M., Paul, C., & Vesna, V. (Eds.). (2011). *Contexts Providers: Conditions of Meaning in Media Arts*. Bristol, United Kingdom: Intellect.

Lovelock, J. (1995). *The Ages of Gaia: A Biography of Our Living Earth*. Oxford, United Kingdom: Oxford University Press.

Luchetta, G. (2013). The Law and Economics of Intermediaries of Personal Information. Centre for European Policy Studies. Retrieved from http://ssrn.com/abstract=2228358

Lü, L., Medo, M., Yeung, C. H., Zhang, Y. C., Zhang, Z. K., & Zhou, T. (2012). Recommender Systems. *Physics Reports*, *159*(1), 1–50. doi:10.1016/j.physrep.2012.02.006

Lunden, I. (2014, April 23). Facebook passes 1B mobile users, 200M messenger users in Q1. *TechCrunch*. Retrieved from http://techcrunch.com

Lusoli, W., Bacigalupo, M., Lupiañez, F., Andrade, N., Monteleone, S., & Maghiros, I. I. (2012 June 4). EU27 survey of electronic identity, privacy and data protection [Final Report]. *IPTS of the EU Commission*. Retrieved from http://is.jrc.ec.europa.eu/pages/TFS/eidsurvey.html

Lynch, C. (2008). Big data: How do your data grow? *Nature*, *455*(7209), 28–29. doi:10.1038/455028a PMID:18769419

Mace, M. (2014). The real meaning of the Fire Phone. *Mobile opportunity*. Retrieved from http://www.mobileopportunity.blogspot.com.es/2014/07/the-real-meaning-of-fire-phone.html

Machado, E. (2000). *La estructura de la noticia en las redes digitales: un estudio de las consecuencias de las metamorfosis tecnológicas en el periodismo* [Doctoral thesis]. Universidad Autónoma de Barcelona, Spain.

Machado, E. (2010). Creatividad e innovación en el periodismo digital. *Proceedings of II Congreso Internacional de Ciberperiodismo y Web 2.0*, Bilbao, Universidad del País Vasco (pp. 64-72).

MacMillan, D., Burrows, P., & Ante, S. E. (2009). Inside the App Economy. *Business Week*, *4153*, 347–357.

Macri, P. (2011). Per una teoria copernicana delle informazioni. *European Journalism Observatory*. Retrieved from http://it.ejo.ch

Madden, M. (2014). Public Perceptions of Privacy and Security in the Post-Snowden Era. Pew Research. November 12, 2014. Retrieved November, 18, 2014, from http://www.pewinternet.org/2014/11/12/public-privacy-perceptions/

Madrigal, A. C. (2014). What is Pinterest? A Database Of Intentions. *The Atlantic* [Web . Retrieved from: http://www.theatlantic.com/technology/archive/2014/07/what-is-pinterest-a-database-of-intentions/375365/?single_page=true

Maghiros, I., Abadie, F., & Pascu, C. (2007). *Introduction to the State of the Art of the Creative Content Sector*. Seville: Institute of Prospective Technological Studies.

Mai, T., Hua, C., Bo, T. J., & Wei, L. (2010). A Study of Image Compression Technique Based on Wavelet Transform. *Proceedings of Genetic and Evolutionary Computing (ICGEC)2010 Fourth International Conference* (pp. 826-828).

Mair, F. S., May, C., Finch, T., Murray, E., Anderson, G., & Sullivan, F. et al. (2007). Understanding the implementation and integration of e-health services. *Journal of Telemedicine and Telecare*, *13*(5suppl 1), 36–37. doi:10.1258/135763307781645112

Malik, P., Vignesvara Ilavarasan, P. (2013). Information and Communications Technology Sector in India. In N. De Prato, & S.P. Simon (Eds), Asia in the Global ICT Innovation Network. Dancing with the tigers (pp. 19-43). Chandos, Whitney Oxforshire.

Manjoo, F. (2011). The Great Tech War of 2012. *Fast company*. Retrieved from http://www.fastcompany.com/magazine/160/tech-wars-2012-amazon-apple-google-facebook

Manjoo, F. (2011, November). The Great Tech War of 2012. *Fast Company*. Retrieved from http://www.fastcompany.com/magazine/160/tech-wars-2012-amazon-apple-google-facebook

Manovich, L. (2013). *Software takes command: extending the language of new media*. New York, NY: Bloomsbury Academic.

Mansell, R. (2002). Conclusion. Social relations, mediating power and technologies. In R. Mansell (Ed.), Inside the communication revolution: evolving patterns of social and technical interaction. (pp. 251-270). Oxford University Press.

Manyika, J. (2013). The impact of disruptive technology: a conversation with Eric Schmidt. *Mckinsey & Company*. Retrieved from http://www.mckinsey.com/insights/high_tech_telecoms_internet/disruptive_technologies

Manyika, J., Chui, M., Brown, B., Bughin, J., Dobbs, R., Roxburgh, C., & Byers, A. (2011). *Big data: The next frontier for innovation, competition, and productivity*. McKinsey Global Institute.

Marcotte, E. (2011). Responsive Web Design. New York, NY: A Book Apart.

Margtullis, L. (1998). *Symbiotic Planet: A New Look at Evolution*. New York, NY: Basic Books.

Margulis, L. (2001). The Symbiotic Planet. London, United Kingdom: Phoenix.

María Galdón, P., Ignacio Madrid, R., de la Rubia-Cuestas, E. J., Diaz-Estrella, A., & Gonzalez, L. (2013). Enhancing mobile phones for people with visual impairments through haptic icons: the effect of learning processes. *Assistive Technology*, *25*(2), 80–87. doi:10.1080/10400435.2012.715112 PMID:23923690

Marichal, J. (2012). *Facebook Democracy. The Architecture of Disclosure and the Threat to Public Life*. Farnham, Surrey; Burlington, VT: Ashgate.

Marketshare, N. (2014). Search engine market share. Retrieved from http://www.netmarketshare.com/search-engine-market-share.aspx?qprid=4&qpcustomd=1&qptimeframe=Y

Markham, A. (2013). Undermining "Data": A Critical Examination of a Core Term in Scientific Inquiry. First Monday, 7 October, 18 (10). Retrieved November, 8, 2014, from http://uncommonculture.org/ojs/index.php/fm/article/view/4868/3749

Marôco, J. (2007). *Análise estatística com SPSS*. Lisboa: Edições Sílabo.

Marquez, I. V. (2010). Música en la nube. Experiencias musicales interactivas en la red. *Telos*, (83): 139–148.

Martí Parreño, J., Sanz-Blas, S., Ruiz-Mafé, C., & Aldás-Manzano, J. (2013). Key factors of teenagers' mobile advertising acceptance. *Industrial Management & Data Systems*, *113*(5), 732–749. doi:10.1108/02635571311324179

Martin, M. (2012). Retrieved from http://searchengineland.com/responsive-design-alone-is-not-mobile-seo-124202

Martínez-Conde, S., & Macknick, S. L. (2013). *Los engaños de la mente*. Barcelona: Destino.

Martin, J. (2015). *Mobile News Use and Participation in Elections: A Bridge for the Democratic Divide? Mobile Media & Communication, 3(2)*. OnlineFirst.

Marton, A., Avital, M., & Blegind Jensen, T. (2013). Reframing Open Big Data. *ECIS 2013 Completed Research*, 146.

Marwick, A. (2013). *Status Update: Celebrity, Publicity, and Branding in the Social Media Age*. New Haven, CT: Yale University Press.

Mashable. (2014). *Mobile-Minded*. Retrieved November 18, 2014 from http://mashable.com/2014/11/17/human-smartphone-relationship/#share-action:eyJzIjoidCIsImkiOiJfejQ5OXB4eDY0bnExZ3kwMyJ9

Matsa, K. E., & Mitchell, A. (2014). 8 Key Takeaways about Social Media and News. *Pew State*. Retrieved from http://www.journalism.org/2014/03/26/8-key-takeaways-about-social-media-and-news/

May, C., Mort, M., Mair, F., & Williams, T. (2001). Factors affecting the adoption of telehealthcare in the United Kingdom: The policy context and the problem of evidence. *Health Informatics Journal*, *7*(3-4), 131–134. doi:10.1177/146045820100700304

McAfee, A., Brynjolfsson, E., Davenport, T. H., Patil, D., & Barton, D. (2012). Big Data: The management revolution. *Harvard Business Review*, *90*(10), 61–67. PMID:23074865

McClelland, S. (2012). The big picture/The rise of Asia. *Intermedia*, *40*(4). 16–17.

McConville, D. (2011). Reflexive Ecologies: Visualizing Priorities. In M. Lima (Ed.), *Visual Complexity: A visual exploration on mapping complex networks*. New York, NY: Princeton Architectural Press.

McDonald, A. (2014). Good companions – the second screen and multiscreen monetisation. *Digital TV Europe net*. Retrieved from http://www.digitaltveurope.net/221941/good-companions-the-second-screen-and-multiscreen-monetisation/

McKinsey. (2011). *Big Data: The next frontier for innovation, competition and productivity*. Retrieved from http://www.mckinsey.com/~/media/McKinsey/dotcom/Insights%20and%20pubs/MGI/Research/Technology%20and%20Innovation/Big%20Data/MGI_big_data_full_report.ashx

McLuhan, M. (1962). *The Gutenberg galaxy: The making of typographic man*. Toronto, ON: University of Toronto Press.

McLuhan, M. (1997). *Essential McLuhan*. New York, NY: Basic Books.

McMillan & Chavis. (1986). *Sense of community: a definition and theory. Journal of Community Psychology. (14)*. NY: New York University Press.

McQuail, D. (2006). *Mass Communication Theory*. London, UK: Sage.

Measuring the Information Society 2012. (2012). Retrieved from http://www.itu.int/en/ITU-D/Statistics/Documents/publications/mis2012/MIS2012_without_Annex_4.pdf

Measuring the Information Society 2013. (2013). Retrieved from http://www.itu.int/en/ITU-D/Statistics/Documents/publications/mis2013/MIS2013_without_Annex_4.pdf

Meeker, M. (2014). Internet Trends 2014 – Code Conference. KPCB. 28 May 2014. Retrieved November, 8, 2014, from http://www.kpcb.com/internet-trends

MEF & AVG. (2014). *Global Consumer Trust Report 2014*. Retrieved November 18, 2014 from http://www.mefmobile.org/activities-and-analytics/analytics/global-consumer-trust-report

Méndez, R., Otero, A., Jarque, S., & Flores, J. (2012). Exploración en tiempo real de la reconstrucción virtual de los instrumentos del Pórtico de la Gloria. *VAR. Virtual Archaeology Review*, *3*, 49–53.

Metz, R. (2014). This Company Thinks Your Car Wants Google Glass. *MIT Technology Review*. Retrieved from http://www.technologyreview.com/news/529646/this-company-thinks-your-car-wants-google-glass/

Meyer, D. (2013). Want Android and Firefox OS in one phone? Geeksphone has something lined up for you. *Giga OM*. Retrieved from http://gigaom.com/2013/11/28/want-android-and-firefox-os-in-one-phone-geeksphone-has-something-lined-up-for-you/?utm_source=Triggermail&utm_medium=email&utm_term=Mobile%20Insights&utm_campaign=Post%20Blast%20%28sai%29%3A%20Netflix%20Is%20Deepening%20Its%20Personalization%20Research

Meyrowitz, J. (1985). *No sense of place: The impact of electronic media on social behavior*. Oxford University Press.

Michaud, L. (2013, December 4). *Evolution of video game industry ecosystem(s). Business models and trends in the evolving market for video games*. Proceedings of the Evolution of video games industry's ecosystems, innovation and regulation chair/mosaic conference, Paris. Retrieved from http://innovation-regulation2.telecom-paristech.fr/wp-content/uploads/2013/12/2013_GameCo_Evolution-of-business-models_Light.pdf

Milchen, J., & Woods, C. (2014). Seven ways businesses and communities can counter "showrooming". *AMIBA*. Retrieved from http://www.amiba.net/resources/showrooming

MindMeld. (2014). The MindMeld API: Introducing multilingual support. Retrieved from https://www.youtube.com/watch?v=14xQK2h9yGY&utm_source=Expect+Labs+Mailing+List&utm_campaign=d32c75f15a-Multilingual_Video&utm_medium=email&utm_term=0_bcc914a475-d32c75f15a-99209801

Minges, M. (2013). *Key Trends in the Development of the Mobile Sector*. Part 2. Statistical Appendix. Pp.115-134. World Bank. Retrieved from http://siteresources.worldbank.org/EXTINFORMATIONANDCOMMUNICATIONANDTECHNOLOGIES/Resources/IC4D_Part2_.pdf?utm_source=Research&utm_campaign=69d2fb0ede-snapshot-2014-05-22&utm_medium=email&utm_term=0_1791988daa-69d2fb0ede-294293285

Mitchelstein, E., & Boczkowski, P. (2009). Between tradition and change: A review of recent research on online news production. *Journalism, 10*(5), 562–586. doi:10.1177/1464884909106533

Mobile Accessibility. Targeting the mobile browser. (n. d.). Retrieved from http://www.reflektions.com/miniml/template_permalink.asp?id=508

Mobile news: A review and model of journalism in an age of mobile media. (2013). *Digital Journalism, 1,* 6–26.

Mobile tracker 2015: http://www.mobileworldlive.com/mhealth-tracker. Retrieved on January 2015.

Mobile Web Best Practices 1.0, W3C Recommendation. (2008). Retrieved from http://www.w3.org/TR/mobile-bp/#d0e253

MobiThinking. (2014). Retrieved from http://mobithinking.com/mobile-marketing-tools/latest-mobile-stats/e#smartphoneactivities

Mohomed, I. (2007). Correlation-based content adaptation for Mobile Web browsing.*Proceedings of the ACM/IFIP/USENIX 2007 International Conference on Middleware (Middleware '07)*, New York, NY, USA (pp. 101-120). doi:10.1007/978-3-540-76778-7_6

Möller, A., Diewald, S., Roalter, L., & Kranz, M. (2012). MobiMed: Comparing object identification techniques on smartphones.*Proceedings of the 7th Nordic Conference on Human-Computer Interaction: Making sense through design* (pp. 31–40). doi:10.1145/2399016.2399022

Montessori, M. (2004). The Discovery of the Child (Original Edition 1948). Delhi, India: Aakar Books.

Moreno Sánchez, I. (2002). *Musas y nuevas tecnologías: el relato hipermedia*. Barcelona: Paidós.

Morgan Stanley. (2012). *Tablet Landscape Evolution. Window(s) of Opportunities.* Retrieved from http://www.morganstanley.com/views/perspectives/index.html

Morgan, J. (2002). *Born with a Bang: The Universe Tells Our Cosmic Story*. Nevada City, CA: Dawn Publications.

Morton, T. (2010). *The ecological thought*. Cambridge, MA: Harvard University Press.

Moyer, T., & Harper, G. (Eds.). (2011). *The New Earthwork. Art, Action, Agency*. Seattle: Isc Press.

Murray, J. H. (2012). Inventing the medium: principles of interaction design as a cultural practice (Kindle Edition). Cambridge: MIT Press.

Murray, J. (2012). *Inventing the Medium. Principles of Interaction Design as a Cultural Practice*. Cambridge: MIT Press.

Nash, T. (2014). Apple, developers think you are just one inconsistent loop. *Tapdaq Blog*. Retrieved from http://blog.tapdaq.com/insight/apple-developers-think-you-are-just-one-inconsistent-loop/

Netflix's war on Mass Culture. (2013). *New republic*. Retrieved from http://www.newrepublic.com/article/115687/netflixs-war-mass-culture

Network, U. W. (2011). *Wireless infrastructure: Can our current system handle bandwidth demands*. Phoenix, USA: College of Information Systems and Technology.

Neuman, W. R., & Guggenheim, L. (2011, May). The Evolution of Media Effects Theory: A Six Stage Model of Cumulative Research. *Communication Theory*, *21*(2), 169–196. doi:10.1111/j.1468-2885.2011.01381.x

New(S) Functions for the Mobile. (2010). *New Media & Society*, *12*, 91–108. doi:10.1177/1461444809355116

Newman, N., & Levy, D. A. L. (2014). Reuters Institute Digital News Report 2014: tracking the future of news. Retrieved from http://www.digitalnewsreport.org

Newman, N., & Levy, D. A. L. (2014). *Reuters institute digital news report 2014: Tracking the future of news*. Retrieved from https://reutersinstitute.politics.ox.ac.uk/sites/default/files/Reuters%20Institute%20Digital%20News%20Report%202014.pdf

Newman, B., Lohman, B., & Newman, P. (2007). Peer-group membership and a sense of belonging: Their relationship to adolescent behaviour problems. *Adolescence*, *42*(166), 241–282. PMID:17849935

Newman, N., & Levy, D. A. L. (2013). *Reuters Institute Digital News Report 2013: Tracking the future of news*. Reuters Institute for the Study of Journalism, University of Oxford.

News Consumption in an Age of Mobile Media: Patterns, People, Place and Participation. (2015). *Mobile Media & Communication, 3(2)*. doi: 10.1177/2050157914563369

News-Video-on-the-Web. (2014). *Pew*. Retrieved from http://www.journalism.org/files/2014/03/News-Video-on-the-Web.pdf

Newswire, N. (2014a). Smartphones: So many apps, so much time. Retrieved from http://www.nielsen.com/us/en/insights/news/2014/smartphones-so-many-apps--so-much-time.html

Newswire, N. (2014b). Wearable Bluetooth users, constantly connected consumers. Retrieved from http://www.nielsen.com/us/en/insights/news/2014/wearable-bluetooth-users-constantly-connected-consumers.html

Nicholas, D., Dobrowolski, T., Withey, R., Russell, C., Huntington, P., & Williams, P. (2003). Digital information consumers, players and purchasers: Information seeking behaviour in the new digital interactive environment. *Aslib Proceedings*, *55*(1/2), 23–31. doi:10.1108/00012530310462689

Nichols, J., Hua, Z., & Barton, J. (2008). Highlight: A System for Creating and Deploying Mobile Web Applications. *Proceedings of UIST'2008*. Monterey, CA. doi:10.1145/1449715.1449757

Nickerson, R., Muntermann, J., Varshney, U., & Isaac, H. (2009). *Taxonomy development in information systems: Developing a taxonomy of mobile applications*. Paper presented at the European Conference in Information Systems, Verona, Italy.

Nicolaou, A. (2013). Best practices on the move: Building web apps for mobile devices. *Communications of the ACM*, *56*(8), 45–51. doi:10.1145/2492007.2492023

Nielsen, J. (1994). *Usability Engineering*. San Diego, CA: Academic Press.

Noguera, J. M., Villi, M., Nyiro, N., De Blasio, E., & Bourdaa, M. (2013). The role of the media industry when participation is a product. In N. Carpentier, K. Schroder, & L. Hallett (Eds.), *Audience Transformations. Shifting Audience Positions in Late Modernity* (pp. 174–192). London: Routledge.

Nordfors, D. (2003). *The concept of innovation journalism and a programme for developing it.* Stockholm: Vinnova. Retrieved from http://www.innovationjournalism.org

Norman, D. A. (1988). *The design of everyday things.* New York: Doubleday.

Norman, D. A. (1988). *The psychology of everyday things.* New York, NY: Basic books.

Norman, D. A. (2010). *Living with complexity.* Cambridge, MA: MIT Press.

Nyre, L., Bjornestad, S., Tessem, B., & Oie, K. V. (2012). Locative Journalism: Designing A Location-Dependent News Medium For Smartphones. *Convergence (London), 18*(3), 297–314. doi:10.1177/1354856512441151

Odin, R. (1995). *Le film de famille: usage privé, usage public.* Paris: Éditions Méridiens Klincksieck.

Odin, R. (1999). *Le cinéma en amateur », Communications, n° 68.* Seuil.

OECD. (2009). *The Impact of the Global Crisis on SME and Entrepreneurship Financing and Policy Responses. Policy.*

OECD. (2013). *The app economy. OECD Digital Economy Papers, no. 230.* Paris, France: OECD Publishing.

OECD. (2014). *Financing SMEs and Entrepreneurs 2014.* Paris: OECD Publishing; doi:10.1787/fin_sme_ent-2014-en

Ofcom (2014). *Ofcom's Communications Market Report 2013.* Retrieved from http://stakeholders.ofcom.org.uk/market-data-research/market-data/communications-market-reports/cmr13/uk/

Ohlenschläger, K. (2007). The Biosphere as Infosphere. In S. Himmelsbach & Y. Volkart (Eds.), *Ecomedia. Ecological Strategies in Today's Art.* Oldenburg: Edith-Ruß-Haus für Medienkunst.

Okazaki, S., Molina, F. J., & Hirose, M. (2012). Mobile advertising avoidance: Exploring the role of ubiquity. *Electronic Markets, 22*(3), 169–183. doi:10.1007/s12525-012-0087-1

Oliver, M. B., & Raney, A. A. (Eds.). (2014). *Media and Social Life.* New York: Routledge.

Ongena, G., Bouwman, H., & Gillebaard, H. (2012). Displacement And Supplemental Effects Of The Mobile Internet On Fixed Internet Use. *International Journal on Media Management, 14*(4), 279–299. doi:10.1080/14241277.2012.684192

ONTSI. (2012). Las TIC en los hogares españoles. Estudio de demanda y uso de Servicios de Telecomunicaciones y Sociedad de la Información. XXXVI Oleada (Abril – Junio 2012). Madrid. Retrieved from http://www.ontsi.red.es/ontsi/es/estudios-informes/xxxvi-oleada-del-panel-de-hogares

Opera Mini & Opera Mobile browsers . (n. d.). Retrieved from http://www.opera.com/mobile/

Oppegaard, B., & Shine, G. (2014). Going Mobile: Changing the Face of Interpretation in the National Park Service. *Federal History Journal, 6*(1), 123–142.

O'Reilly. Tim. (2011). Some context for thinking about the future of media. Madrid: Ficod. Retrieved from http://www.slideshare.net/timoreilly/ficod-2011-pdf-with-notes/download

Organisation for Economic Co-operation and Development (2013, December 16). *The App Economy.* Publishing. doi: 10.1787/5k3ttftlv95k-en

Otero, A., Méndez, R., & Flores, J. (2012). Pórtico de la Gloria Virtual, aplicación interactiva en tiempo real para la exploración virtual del Pórtico de la Gloria. *VAR. Virtual Archaeology Review, 3,* 18–25.

Oulasvirta, A., Rattenbury, T., Ma, L., & Raita, E. (2012). Habits Make Smartphone Use More Pervasive. *Personal and Ubiquitous Computing, 16*(1), 105–114. doi:10.1007/s00779-011-0412-2

Pagani, M. (2004). Determinants Of Adoption Of Third Generation Mobile Multimedia Services. *Journal of Interactive Marketing*, *18*(3), 46–59. doi:10.1002/dir.20011

Palacios, M. (2013). Prefácio - O mundo no bolso e o contexto na palma da mão. [The world in the pocket and the context in the hand]. In S. Barbosa & L. Mielniczuk (Eds.), *Jornalismo e tecnologias móveis* [Journalism and mobile technologies] (pp. 1–5). Covilhã, Portugal: Livros Labcom.

Palacios, M. S., & Da Cunha, R. (2012). A tactilidade em dispositivos móveis: Primeiras reflexões e ensaio de tipologias. *Contemporanea (Bologna, Italy)*, *10*(3), 668–685.

Palacios, M. S., & Da Cunha, R. E. S. (2012). A tactilidade em dispositivos móveis: Primeiras reflexões e ensaio de tipologias[Tactility and mobile devices: First approximations and a typology]. *Contemporanea (Bologna, Italy)*, *10*(3), 668–685.

Palacios, M., Barbosa, S., Firmino da Silva, F., & Da Cunha, R. (2014). Aplicativos jornalísticos vespertinos para tablets. Cartografia do fenômeno ante o desafio de uma produção original e inovadora. *Sur le Journalisme*, *3*(2), 40–55.

Pariser, E. (2011). The Filter Bubble: What The Internet Is Hiding From You. Penguin UK.

Park, M., Hong, J., & Cho, S. (2007). Location-Based Recommendation System Using Bayesian User's Preference Model in Mobile Devices. In Ubiquitous Intelligence and Computing (pp. 1130-1139). Springer Berlin Heidelberg. doi:10.1007/978-3-540-73549-6_110

Park, Y. J., & Mo Jang, S. (2014). Understanding privacy knowledge and skill in mobile communication. *Computers in Human Behavior*, *38*, 296–303. doi:10.1016/j.chb.2014.05.041

Parry, R. (2008). Afterword: The future in our hands? Putting potential into practice. In L. Tallon (Ed.), *Digital technologies and the museum experience: handheld guides and other media*. Landham, MD: AltaMira Press.

Pasternack, J., & Roth, D. (2009). Extracting article text from the web with maximum subsequence segmentation. *Proceedings of WWW '09 the 18th international conference on World Wide Web*, New York, NY, USA (pp. 971-980). doi:10.1145/1526709.1526840

PatientView. (2012). A review by patient groups and empowered consumers. Retrieved from http://myhealthapps.net/

Paull, J. (2011). Incredible Edible Todmorden: Eating the Street. In Farming Matters 27(3), 28-29.

Pauwels, C., & Picone, I. (2012). The tussle with trust: Trust in the News media ecology. *Computer Law & Security Report*, *28*(5), 542–550. doi:10.1016/j.clsr.2012.07.003

PCs in Decline: TV Screen Takes Over As Americans' First Choice For Internet TV and Movies. (2013). *Strategy Analytics blog*. Retrieved from http://blogs.strategyanalytics.com/CM/post/2013/03/22/PCs-In-Decline-TV-Screen-Takes-Over-As-Americans'-First-Choice-For-Internet-TV-and-Movies.aspx

Pellanda, E. (2009). Comunicação móvel no contexto brasileiro. In A. Lemos (Ed.), *Josgrilberg. F. Comunicação e mobilidade aspectos socioculturais das tecnologias móveis de comunicação no Brasil*. Salvador: EDUFBA.

Pentland, A. P. (1998). Wearable Intelligence. *Scientific American*, 90–95.

Pentland, A., Lazer, D., Brewer, D., & Heibeck, T. (2009). Improving Public Health and Medicine by use of Reality Mining. *Studies in Health Technology and Informatics*, *149*, 93–102. PMID:19745474

Pérez, J., & Feijóo, C. (2000). Convergencia, competencia y regulación en los mercados de las telecomunicaciones, el audiovisual e Internet. GRETEL. Madrid: Colegio Oficial de Ingenieros de Telecomunicación (COIT).

Pérez, J., Feijóo, C., & García, F. (2002). Nuevo diseño europeo de las telecomunicaciones, el audiovisual e Internet. GRETEL. Madrid: Colegio Oficial de Ingenieros de Telecomunicación (COIT).

Pernice, K., Nielsen, J., Farrell, S., Mizobuchi, S., Ishida, N., Stover, U. A., & Richardson, A. (2014). *Usability Guidelines for Accessible Web Design. Evidence--Based User Experience Research, Training*. Consulting.

Perry, M. (2012). Newspaper Ad Revenues Fall to 60-Yr. Low in 2011. Retrieved form http://mjperry.blogspot.com.es/2012/02/newspaper-ad-revenues-fall-to-50year.html

Perse, E. M., & Dunn, D. G. (1998). Utility of Home Computers and Media Use: Implications of Multimedia and Connectivity. *The Journal of Broadcasting & Electronic Media*, *42*(4), 435–456. doi:10.1080/08838159809364461

Pescher, C., Reichhart, P., & Spann, M. (2014). Consumer decision-making processes in mobile viral marketing campaigns. *Journal of Interactive Marketing*, *28*(1), 43–54. doi:10.1016/j.intmar.2013.08.001

Peters, C. (2012). Journalism to go: The changing spaces of news consumption. *Journalism Studies*, *13*(5-6), 5–6. doi:10.1080/1461670X.2012.662405

Pew Research Center. (2012, October 1). *The future of mobile news: The explosion in mobile audiences and a close look at what it means for news*. Retrieved from http://www.journalism.org/files/legacy/Futureofmobilenews%20_final1.pdf

Pew Research Center. (2014). Key Indicators in Media and News. *State of the News Media 2014*. Retrieved from http://www.journalism.org/files/2014/03/Key-Indicators-in-Media-and-News-2014.pdf

Pew Research Centre. (2012). One in ten 'dual-screened' the Presidential Debate. Retrieved from http://www.people-press.org/2012/10/11/one-in-ten-dual-screened-the-presidential-debate/1/

Pew Research Centre. (2013). *The State of the News Media 2013*. Retrieved from http://stateofthemedia.org/2013/overview-5/

Pew Research Internet Project. (2014). *Mobile Technology Fact Sheet*. Pew Internet & American Life Project. Retrieved from http://www.pewinternet.org/fact-sheets/mobile-technology-fact-sheet/

Pew Research. (2014). *Social Networking Fact Sheet*. Retrieved November 18, 2014 from http://www.pewinternet.org/fact-sheets/social-networking-fact-sheet/

Plant, S. (2001). *On the Mobile*: Motorola. Retrieved from http://www.it-c.dk/courses/MGB/F2002/on_the_mobile.pdf

Plant, S. (2002). *On the mobile. Receiver*. Vodafone.

Podesta,J., & al, (2014, May). Big Data: Seizing Opportunities, Preserving Values. *Executive Office of the President*. Retrieved from http://www.whitehouse.gov/blog/2014/05/01/findings-big-data-and-privacy-working-group-review

Porter, M. E. (2008). *Competitive advantage: Creating and sustaining superior performance* (Original work published 1985). New York: Simon and Schuster.

Portugal, I. N. E. (2014). *Sociedade da Informação e do Conhecimento. Inquérito à Utilização de Tecnologias da Informação e da Comunicação pelas Famílias*. 6 de Novembro de 2014. Retrieved November, 18, 2014, from http://www.ine.pt/xportal/xmain?xpid=INE&xpgid=ine_destaques&DESTAQUESdest_boui=211422735&DESTAQUESmodo=2

Postman, N. (2000). The Humanism of Media Ecology. *Proceedings of the Media Ecology Association*, *1*, 10–16.

Preston, A. (2014). The Death of Privacy. The Guardian. Sunday 3 August 2014. Retrieved September, 11, 2014, from http://www.theguardian.com/world/2014/aug/03/internet-death-privacy-google-facebook-alex-preston?CMP=twt_gu

Preston, P., & Rogers, J. (2011). Social networks, legal innovations and the "new" music industry. *EuroCPR*. Retrieved from http://www.eurocpr.org/data/2011/2_Preston_Rogers.pdf

Privacy Index, E. M. C. (2014). Retrieved November, 18, 2014, from http://portugal.emc.com/campaign/privacy-index/index.htm?pid=home-emcprivacyindex-120614?fromGlobalSelector

Puleo, A. (2012). Speaking from the South of Europe. In *Deportate, esuli, profughe 20*, 78-89. Retrieved from http://www.unive.it/media/allegato/dep/n20-2012/Ricerche/Riflessione/8_Puleo_Ecofemminismo_sud.pdf

Puleo, A. (2011). *Ecofeminismo: Para otro mundo posible*. Madrid, Spain: Cátedra.

Pulman, S. (2011). Facebook as a mobile media platform. *Storytelling*. Retrieved from http://www.storytelling.es/facebook-as-a-mobile-media-platform/

Punie, Y., Lusoli, W., Centeno, C., Misuraca, G., & Broster, D. (2009). *The Impact of Social Computing in the EU Information Society and Economy*. JCR Scientifical and Technical Reports. European Commission. Retrieved from ftp://ftp.jrc.es/pub/EURdoc/JRC54327.pdf

Purcell, K., Brenner, J., & Rainie, L. (2012). *Search engine use 2012*. Pew Internet & American Life Project Search. Retrieved from http://pewinternet.org/Reports/2012/Search-Engine-Use-2012.aspx

Purcell, K., Brenner, J., & Rainie, L. (2012). *Search Engine Use 2012*. Pew Research Center's Internet & American Life Project. Retrieved from http://pewinternet.org/Reports/2012/Search-Engine-Use-2012.aspx

Purcell, K., Rainie, L., Mitchell, A., Rosenstiel, T., & Olmstead, K. (2010, March 1). *Understanding the participatory news consumer: How internet and cell phone users have turned news into a social experience*. Retrieved from http://www.pewinternet.org/files/old-media//Files/Reports/2010/PIP_Understanding_the_Participatory_News_Consumer.pdf

Purcell, K., Rainie, L., Mitchell, A., Rosentiel, T., & Olmstead, K. (2010). Understanding the participatory news consumer. How internet and cellphone users have turned news into a social experience. *Pew Research Centre. Project for Excellence in Journalism*. Retrieved from http://www.pewinternet.org/~/media//Files/Reports/2010/PIP_Understanding_the_ParticipatoryNews_Consumer.pdf

PWC. (2005). *HealthCast 2020: Creating a sustainable future*. Price Waterhouse Coopers.

PWC. (2012). *Emerging mHealth: Paths for growth*. Price Waterhouse Coopers.

Quico, C., Damásio, M. J., Henriques, S., & Veríssimo, I. (2012). Digital terrestrial TV switchover process in Portugal: viewers between a rock and a hard place. In Sousa, H., & Denicoli, S. (Eds.), *Digital Communication Policies in the Information Society Promotion Stage* eBook.. Universidade do Minho. Retrieved from http://www.lasics.uminho.pt/ojs/index.php/digital_communication_polices/index

Quico, C., Damásio, M.J., Baptista, A., Sequeira, A. Veríssimo, I. Cardoso, M. & Henriques, S. (2014). Using digital interactive television for the promotion of health and wellness. *International Journal of Health Promotion and Education.*

Quico, C. (2012). Digital TV switchover in Portugal: What is in it for the viewer. *International Journal of Digital Television, 3*(1), 163–179. doi:10.1386/jdtv.3.2.163_1

Quinn, D. (1992). *Ishmael: An Adventure of the Mind and the Spirit*. New York, London: Bantam/Turner Books.

Quinn, S. (2005). *Convergent journalism: the fundamentals of multimedia reporting*. New York: Peter Lang Publishing.

Quinton, F. (2014). Game of Thrones, le transmédia couronné de succès. *INA Global*. Retrieved from t: www.inaglobal.fr/television/article/game-thrones-le-transmedia-couronne-de-succes-7542?tq=8

Quotidiania Delirante. (2015). Retrieved from http://itunes.apple.com/es/app/quotidiania-delirante-1-imotion/id356356294?mt=8

Ramos, S., & Feijóo, C., González, a, Rojo, D., & Gómez-Barroso, J. L. (2004). Barriers to widespread use of mobile Internet in Europe. An overview of the new regulatory framework market competition analysis. *The Journal of the Communications Networks*, *3*(3), 76–83.

Ranjini Mei Hua, S., & Melissa, E. (2012). *Investigating the potential of mobile phones for e-governance in Indonesia.* Paper presented at the CPRAfrica2012/CPR7 conference. Retrieved from www.cprsouth.org

Rau, P. L. P., Zhou, J., Chen, D., & Lu, T. P. (2014). The influence of repetition and time pressure on effectiveness of mobile advertising messages. *Telematics and Informatics*, *31*(3), 63–476.

Rawolle, J., & Hess, T. (2000). New digital media and devices: An analysis for the media industry. *International Journal on Media Management*, *2*(2), 89–99. doi:10.1080/14241270009389926

Reding, V. (2008, November 13). Digital Europe: The Internet mega-trends that will shape tomorrow's Europe. Speech at the European Internet Foundation Special Event. A view of the Digital World in 2025, Brussels. Retrieved from http://europa.eu/rapid/press-release_SPEECH-08-616_en.htm

Regalado, A. (2014). Why did Google pay $400 million for DeepMind? *MIT Technology Review*. Retrieved from www.technologyreview.com

Reports consumer 2012. (2013). *Nielsen*. Retrieved from http://www.nielsen.com/libs/cq/core/content/login.html?resource=%2Fcontent%2Fdam%2Fcorporate%2Fus%2Fen%2Freports-download%2F2013&$$login$$=%24%24login%24%24

Research2Guidance (2013). *The mobile health global market report 2013-2017: the commercialisation of mHealth apps* (Vol. 3).

Research2Guidance. (2014). *mHealth App Developer Economics 2014. The State of the Art of mHealth App Publishing.*

Rheingold, H. (2004). Mobile media and political collective action. In J. Katz (Ed.), *The handbook of mobile communication studies.* Cambridge, MA: MIT Press.

Rheingold, H. (2012). *Net smart. How to thrive online.* Cambridge: The MIT Press.

Riedler, K. Goswami & Krcmar (2013). Tweeting to Fells Connected: A model for social connectedness in online social networks. *International Journal of Human-Computer Interaction*, vol. 29 (10). Retrievedfrom http://tinyurl.com/ma2pprb

Ries, E. (2011). *The Lean Startup: How Today's Entrepreneurs Use Continuous Innovation to Create Radically Successful Businesses.* New York: Crown Publishing Group.

Riot Games. (n. d.). Retrieved from http://www.riotgames.com/about

Rius, À., Clarisó, R., & Masip, D. (2014). Student projects empowering mobile learning in higher education. *RUSC. Universities and Knowledge Society Journal*, *11*(1), 192–207. doi:10.7238/rusc.v11i1.1901

RNCOS. (2011). *Global Mobile TV Forecast to 2013.* Retrieved from http://www.rncos.com/Report/IM194.htm

Roccetti, M., Marfia, G., Varni, A., & Zanichelli, M. (2013). How to Outreach the External World from a Museum: The Case of the Marsili's Spirit App. *Arts and Technology Lecture Notes of the Institute for Computer Sciences. Social Informatics and Telecommunications Engineering*, *116*, 25–32.

Rochet, J.-C., & Tirole, J. (2006). Two-Sided Markets: A Progress Report. *The Rand Journal of Economics*, *37*(3), 645–667. http://www.jstor.org/stable/25046265 doi:10.1111/j.1756-2171.2006.tb00036.x

Rodríguez, M. P. (2011). Imágenes del otro. Identidad e inmigración en la literatura y el cine. *Journal of Spanish Cultural Studies*, *3*(3), 375–377. doi:10.1080/14636204.2011.669256

Rogers, E. M. (2003). *Diffusion Of Innovations. New York*. New York: Free Press.

Rogers, E. M. (2003). *Diffusion of innovations*. New York: Free Press.

Roschelle, J. (2003). Keynote paper: Unlocking the learning value of wireless mobile devices. *Journal of Computer Assisted Learning*, *19*(3), 260–272. doi:10.1046/j.0266-4909.2003.00028.x

Rosen, L. (2013). Always On, All the Time: Are We Suffering From FoMO? *Psychologytoday*. Retrieved November 17, 2014 from http://www.psychologytoday.com/blog/rewired-the-psychology-technology/201305/always-all-the-time-are-we-suffering-fomo

Rosenbaum, S. (2011). *Curation nation: How to win in a world where consumers are creators*. New York, NY: McGraw-Hill.

Rossetti, R. (2013). Categorias de inovação para os estudos em comunicação. *Comunicação & Inovação*, *14*(27), 63–72.

Rovio Entertainment. (2014). *Annual Report 2013*. Retrieved from www.rovio.com

Russom, P. (2011). *Big data analytics*. TDWI Best Practices Report, Fourth Quarter.

Sagan, C. (1994). Pale Blue Dot: A Vision of the Human Future in Space. Ann Arbor, MI: University of Michigan.

Sahtouris, E. (2000). Earthdance: Living Systems in Evolution. Bloomington, IN: iUniverse.

Salaverría, R. (2012, May 8). Se ve… y se toca: hacia un nuevo lenguaje periodística multimedia. Retrieved February 15, 2014, from http://www.salaverria.es

Salaverría, R., & Negredo, S. (2008). *Periodismo integrado: convergencia de medios y reorganización de redacciones*. Barcelona: Editorial Sol90.

Salaverría, R., García Avilés, J. A., & Masip, P. M. (2010). Concepto de convergencia periodística. In: López García, X. & Pereira Fariña, X. (Orgs.), Convergencia digital: reconfiguración de los medios de comunicación en España. Santiago de Compostela: USC, 41-64.

Salen, K. (Ed.). (2008). *The Ecology of Games: Connecting Youth, Games, and Learning*. Cambridge, MA: The MIT Press.

Salmon, K. (2013). Creators, producers, distributors, consumers, public authorities...Who really has overall control? *Forum d'Avignon*. Retrieved from avignon.org.

Salmon, K., (2012). The manufacturing of innovation Management & creation, perspectives for economic growth. *Forum d'Avignon*. Retrieved from avignon.org.

Samarajiva, R. (2014). The potential of "big data" for improving public-service delivery in developing countries. Retrieved from http://lirneasia.net/search/

Santaella, L. (2011). *Percepção: fenomenologia, ecologia, semiótica*. São Paulo: Cengage.

Santiago(e)Tapas Contest. (2011). Retrieved from: http://www.santiagoetapas.com

Sasseen, J., Olmstead, K., & Mitchell, A. (2013). Digital: As mobile grows rapidly, the pressures on news intensify. The state of the news media 2013: An annual report on American journalism. Pew Research Center's Project for Excellence in Journalism, Washington, D.C., 2012.

Satuf Rezende, I. (2013). Jornalismo para tablets e a emergência de contratos comunicacionais, paper present at *XIII Congreso Internacional Ibercom*. Retrieved from https://www.academia.edu/7848328/Jornalismo_para_tablet_e_a_emerg%C3%AAncia_de_contratos_comunicacionais

Sauro, J., & Kindlund, E. (2005). A method to standardize usability metrics into a single score.*Proceedings of the SIGCHI conference on Human factors in computing systems (CHI '05)*. New York, NY, USA. doi:10.1145/1054972.1055028

Scaglione, M., & Hamoudia, M. (2010). Online and mobile social networking: present and perspectives.*21st European Regional International Telecommunications Society Conference*. Copenhagen.

Schmidt, E. (2010). Mobile World Congress keynote address [YouTube video]. Retrieved from http://www.youtube.com/watch?v=YuqiE2lukDM

Schmidt, E., & Cohen, J. (2013). *The new digital age: transforming nations, businesses, and our lives*. New York, NY: Alfred A. Knopf.

Schoonover, K. (2013). Documentaries without documents? Ecocinema and the toxic. Necsus, 2(2), 483–507.

Schuermans, S. (2013). The naked android. *VisionMobile*. Retrieved from http://www.visionmobile.com/blog/2013/11/the-naked-android/?utm_source=VisionMobile+subscriber+list&utm_campaign=2248ef3bba-2013_06+Newsletter&utm_medium=email&utm_term=0_efe0e34ffa-2248ef3bba-394601118

Schulz, W. (2004). Reconstruction mediatization as an analytical concept.[London: Sage.]. *European Journal of Communication*, *19*(1), 87–101. doi:10.1177/0267323104040696

Scoble, R., & Israel, S. (2013). *Age of Context: mobile sensors, data and the future of privacy*. United States: Self Published.

Scolari, C., Aguado, J., & Feijóo, C. (2012). Mobile Media: towards a definition and taxonomy of contents and applications. *iJIM*, 6 (2), 29-38.

Scolari, C. A. (2009). Transmedia Storytelling, Narrative Worlds, and Branding in Contemporary Media Production. *International Journal of Communication*, *3*, 586–606.

Scolari, C. A., Aguado, J. M., & Feijóo, C. (2012). Mobile Media: Towards a Definition and Taxonomy of Contents and Applications. *International Journal of Interactive Mobile Technologies*, *6*(2).

Scolari, C. A., Aguado, J. M., & Feijóo, C. (2012). Mobile Media: Towards a definition and taxonomy of contents and applications. *International Journal of Interactive Mobile Technologies*, *6*(2), 29–38. doi: 10.3991/ijim.v6i2.1880

Scolari, C. A., Aguado, J. M., & Feijoo, C. (2012). Mobile media: Towards a definition and taxonomy of contents and applications. *International Journal of Interactive Mobile Technologies*, *6*, 29–38.

Scolari, C. A., & Ibrus, I. (2014). Transmedia Critical: Empirical Investigations into Multiplatform and Collaborative Storytelling. *International Journal of Communication*, *8*, 2191–2200.

Scolari, C., Aguado, J. M., & Feijóo, C. (2013). Una ecología del medio móvil: Contenidos y aplicaciones. In J. M. Aguado, C. Feijóo, & I. Martínez (Eds.), *La Comunicación Móvil: Hacia un Nuevo Ecosistema Digital* (pp. 79–106). Barcelona: Gedisa.

Serrano Tellería, A. (2014). Interface Design on Mobile Phones: The Delimitation of the Public and Private Spheres. In Paiva, F; Moura, C. (Orgs.) DESIGNA: INTERFACE International Conference on Design Research (pp.87-108) Covilhã, Portugal: Labcom Books, University of Beira Interior. ISBN 978-989-654-141-5. Retrieved December 12, 2014, from http://www.livroslabcom.ubi.pt/book/111.

Serrano Tellería, A. (2015). Liquid Spheres or Constellations: Reflections Towards Mobile Devices. In Carvalheiro J.R., Serrano Tellería A. (eds.) Mobile and Digital Communication: Approaches to Public and Private. 173-198. Covilhã, Portugal: Labcom Books, University of Beira Interior. ISBN: 978-989-654-235-1 (paper) 978-989-654-237-5 (pdf) 978-989-654-236-8 (epub). Retrieved July 17, 2015, from http://www.livroslabcom.ubi.pt/book/141.

Serrano Tellería, A., & Oliveira, M. (2015). Liquid Spheres on Smartphones: The Personal Information Policies. *International Journal of Interactive Mobile Technologies*, *9*(1). Retrieved July 17, 2015, from http://online-journals.org/index.php/i-jim/article/view/4065

Shankar, V., & Batra, R. (2009). The growing influence of online marketing communications. *Journal of Interactive Marketing*, *23*(4), 285–287. doi:10.1016/j.intmar.2009.07.006

Shoemaker, P. J., Vos, T. P., & Reese, S. D. (2008). Journalists as Gatekeepers. In K. Wahl-Jorgensen & T. Hanitzsch (Eds.), *The handbook of journalism studies* (pp. 73–87). New York, NY: Routledge.

Sidhe. (n. d.). Retrieved from www.sidhe.co.nz

Silberschatz, A., & Galvin, P. (1991). *Operating system concepts*. New York: Addison-Wesley.

Silcoff, S., McNish, J., & Erman, B. (2013, November 4). BlackBerry financing aims for a new lease on life. *The Globe and Mail*. Retrieved from http://www.theglobeandmail.com/report-on-business/blackberry-fairfax-deal-dies-thorsten-heins-out/article15240310/

Silva, F. F. (2008). Jornalismo reconfigurado: tecnologias móveis e conexões sem fio na reportagem de campo, paper presented at *Intercom – Sociedade Brasileira de Estudos Interdisciplinares da Comunicação, XXXI Congresso Brasileiro de Ciências da Comunicação*. Retrieved from http://rtietz.com/posimagemfeevale/page0/files/Fernando_firmino_intercom.pdf

Silver, J. (2013, November 8). Panel 2. Stake-holders' view of evidence for copyright policy. In M. Kretschmer, & R. Towse (Eds). Proceedings of Digital Conference, Bournemouth University (pp. 58-62). Retrieved from www.copyright-evidence.org/create/esrc_evidence_symposium

Silverstone, R. (2002). *Why Study the Media?* London: Sage.

Simon, J. P. (2012). *The Dynamics of the Media and Content Industries: A Synthesis*. Retrieved from http://is.jrc.ec.europa.eu/pages/ISG/MCI.html

Simon, J. P. (2015). Le Big Data: un enjeu pour les industries creatives. *INA Global*. Retrieved from http://www.inaglobal.fr/numerique/article/le-big-data-un-enjeu-pour-les-industries-creatives-8065

Simon, J. P. (2014). Media in the changing media-IT- telecom ecosystem. In G. De Prato, E. Sanz, & J. P. Simon (Eds.). *Digital Media Worlds; The new media economy*. Oxford: Palgrave. doi:10.1057/9781137344250.0011

Singh, K. (2014). 10 Rising Social Networks you should explore. *Mashable*. Retrieved from http://mashable.com/2014/07/28/social-networks-on-the-rise/?utm_cid=mash-com-fb-socmed-link

Smith, A. (2010, July 7). Mobile Access 2011. *Pew Internet & American Life Project*. Retrieved from http://pewinternet.org/Reports/2010/Mobile-Access-2010.aspx

Smith, A., & Boyles, J. L. (2012). The rise of the 'connected viewer. *Pew Internet & American Life Project, Pew Research Center*. Retrieved from http://tinyurl.com/nfgbj62

Smith, C. (2014). *Project Ara development boards release date set for late July. BGR*. Retrieved from http://bgr.com/2014/07/15/project-ara-development-boards-release-date/

Smith, G. (2010). *What Media Classes Really Want to Discuss: A Student Guide*. New York, NY: Routledge.

Snijders, C., Matzat, U., & Reips, U.-D. (2012). Big Data: Big gaps of knowledge in the field of Internet. *International Journal of Internet Science*, *7*, 1–5.

Soba, M., & Aydin, M. (2013). Product placement efficiency in marketing communication strategy. *International Journal of Business and Management*, *8*(12), 111–116. doi:10.5539/ijbm.v8n12p111

Soderlan, S. (1999). Learning information extraction rules for semi-structured and free text. *Machine Learning*, *34*(1), 233–272. doi:10.1023/A:1007562322031

Sohn, T., Chun, F., Li, Y., Battestini, A., Setlur, V., & Mori, K. (2011). Myngle: unifying and filtering web content for unplanned access between multiple personal devices.*Proceedings of the 13th international conference on Ubiquitous computing (UbiComp '11)* (pp. 257-266). New York, NY, USA: ACM. doi:10.1145/2030112.2030147

Sonnenberg, C. (2014) Maintaining Thematic Consistency within Mobile Content Adaptation.*Proceedings of the International Conference: Industry, Engineering & Management Systems*, Cocoa Beach, FL.

Spier, F. (2011). *Big History and the Future of Humanity*. New York, NY: John Wiley & Sons.

Stald, G. (2008). Mobile Identity: Youth, Identity, and Mobile Communication Media. In Buckingham D. Youth, Identity, and Digital Media (pp. 143–164). The John D. and Catherine T. MacArthur Foundation Series on Digital Media and Learning, Cambridge, MA: The MIT Press.

Statista. (2014). *Global market share of search engines 2010-2014*. Retrieved from http://www.statista.com/statistics/216573/worldwide-market-share-of-search-engines/

Steventon, A., Bardsley, M., Billings, J., Dixon, J., Doll, H., Hirani, S., ... & Newman, S. (2012). Effect of telehealth on use of secondary care and mortality: findings from the Whole System Demonstrator cluster randomised trial. *BMJ: British Medical Journal, 344*.

Stone, L. (2014). *Continuous Partial Attention*. Retrieved November 8, 2014 from http://lindastone.net/qa/continuous-partial-attention/

Strip Generator. (2015). Retrieved from http://stripgenerator.com

Stroetmann, K. A., Kubitschke, L., Robinson, S., et al. (2010). How can telehealth help in the provision of integrated care? World Health Organization. Regional Office for Europe, 2010.

Subramaniam, N. (2012). Android One is Google's bet to sweep Indian smartphone market. *Tech2*. Retrieved from http://m.tech.firstpost.com/news-analysis/android-one-is-googles-bet-to-sweep-indian-smartphone-market-232006.html

Sultan, F., & Rohm, A. J. (2008). How to market to generation M(obile). *MIT Sloan Management Review*, *49*(4), 34–41.

Surowiecki, J. (2012). Blackberry Season. *The New Yorker*. Retrieved from http://www.newyorker.com/talk/financial/2012/02/13/120213ta_talk_surowiecki

Syndicat National du Jeu Vidéo (SNJV). (2012). *Les chiffres des marchés du jeu vidéo dans le monde et en France*. Retrieved from http://www.snjv.org/fr/industrie-francaise-jeu-video/

Tallon, L. (2008). *Digital technologies and the museum experience: handheld guides and other media*. Landham, MD: Altamira Press.

Tallon, P. P. (2013). Corporate governance of big data: Perspectives on value, risk, and cost. *Computer*, *46*(6), 32–38. doi:10.1109/MC.2013.155

Taneja, H., Webster, J. G., Malthouse, E. C., & Ksiazek, T. B. (2012). Media Consumption Across Platforms: Identifying User-Defined Repertoires. *New Media & Society*, *14*(6), 951–968. doi:10.1177/1461444811436146

Tatzgern, M., Kalkofen, D., & Schmalstieg, D. (2013). Dynamic compact visualizations for augmented reality. In Virtual Reality (VR), 2013 IEEE (pp. 3-6). IEEE. doi:10.1109/VR.2013.6549347

Taylor, A., & Harper, R. (2002). *Age-old Practices in the 'New World': a Study of Gift-Giving between Teenage Mobile Users*. Surrey: Surrey University. doi:10.1145/503376.503455

Taylor, C. A., Samuels, N., & Ramey, J. A. (2011). How people really use the mobile web: A framework for understanding motivations, behaviors, and contexts. In J. Lumsden (Ed.), *Human-Computer interaction and innovation in handheld, mobile and wearable technologies* (pp. 195–214). Hershey, PA: Information Science Reference; doi:10.4018/978-1-60960-499-8.ch012

Tehrani, K., & Michael, A. (2014). Wearable technology and wearable devices: Everything you need to know. *Wearable Devices Magazine*. Retrieved from http/:www.wearabledevices.com:what-is-a-wearable-device

Teilhard de Chardin, P. (1959). *The Phenomenon of Man*. Paris, France: Éditions du Seuil.

Telenor Group. (2011). Customer loyalty lies with the operating system. Retrieved from http://telenor.com/en/news-and-media/articles/2011/customer-loyalty-lies-with-the-operating-system

Tellabs. (2011). *Tellabs "End of Profit" study executive summary*. Retrieved from http://www.tellabs.com/markets/tlab_end-of-profit_study.pdf

Tencent website. (n. d.). Retrieved from www.tencent.com/en-us/at/abouttencent.shtml

Tercek, R. (2014). *Documentary Filmmakers in Surveillance Society*. TEDxHollywood. Retrieved from http://youtu.be/sh9t1UE7b7A

The State of the News Media 2013. (2013). *Pew*. Retrieved from //stateofthemedia.org/2013/

Thierer, A. (2004). Are "dumb pipe" mandates smart public policy? Vertical integration, net neutrality and the network layers model. *Journal on Telecommunications & High Technology Law*, *3*(2), 275.

Thomas, D. (2014). Telecoms groups blast EU net neutrality plans. *FT.Com*. Retrieved from http://www.ft.com/intl/cms/s/0/089476a4-9ca3-11e3-9360-00144feab7de.html?utm_content=buffer7eb3d&utm_medium=social&utm_source=twitter.com&utm_campaign=buffer&siteedition=intl

Thompson, J. B. (1995). *The Media and Modernity: A social theory of the media*. Cambridge: Polity Press.

Thomson, A., & Orihuela, R. (2014). Phone carriers call for EU rules to hit facebook and google. *Bloomberg*. Retrieved from http://www.bloomberg.com/news/2014-08-31/vodafone-heads-to-spain-to-protest-eu-network-traffic-proposals.html

Thorson, E. (2008). Changing patterns of news consumption and participation. *Information Communication and Society*, *11*(4), 473–489. doi:10.1080/13691180801999027

TIME Mobility Poll. (2013). *TIMEMagazine*. Retrieved from http://www.qualcomm.com/media/documents/time-mobility-poll-cooperation-qualcomm

Toffler, A. (1980). *The Third Wave: Democratization in the Late Twentieth Century*. New York: Bantam Book.

Toondoo. (2015). Retrieved from http://www.toondoo.com

Traffic and market data report. (2011, November). *Ericsson*.

Transnational Temps. (2007). Transnational Temps presents: eco-scope. In J. Martín (Ed.), *Inclusiva-net.New Art Dynamics in Web 2 Mode* (pp. 195–213). Madrid, Spain: Área de las Artes, Dirección General de Promoción y Proyectos Culturales.

Transparency Market Research. (2013). *Big Data Market - Global Scenario, Trends, Industry Analysis, Size, Share And Forecast 2012 – 2018.* Retrieved from http://www.slideshare.net/wayne31jan/big-data-market-38557173

Trends In U.S. Video Gaming – The Rise Of Cross-Platform. (2012). *Nielsen.* Retrieved from http://www.gamesindustryblog.com/2012/03/trends-in-u-s-video-gaming-the-rise-of-cross-platform/

Turkle, S. (2008). Always-on/always-on-you: The tethered self. In J. Katz (Ed.), *Handbook of Mobile Communication Studies* (pp. 121–138). Cambridge, MA: MIT Press. doi:10.7551/mitpress/9780262113120.003.0010

Turkle, S. (2011). *Alone together: why we expect more from technology and less from each other.* New York, NY: Basic Books.

Turkle, S. (2011). *Alone together: why we expect more from technology and less from each other?* New York, NY: Basic Books.

TV and Video: An analysis of evolving consumer habits. Eriksson Consumer Lab. (2012). Eriksson Consumer Lab. Retrieved from http://tinyurl.com/bvgrfot

Uhlmann, S., & Lugmayr, A. (2012). Portable personality and its personalization algorithms: An overview and directions. In A. Lugmayr, H. Franssila, P. Näränen, O. Sotamaa, J. Vanhala, & Z. Yu (Eds.), *Media in the ubiquitous era: Ambient, social and gaming media* (pp. 66–93). Hershey, PA: Information Science Reference. doi:10.4018/978-1-60960-774-6.ch004

UK Online Measurement Company (UKOM). Tablets now account for 8% of website traffic: UKOM and comScore announce early data release date. Retrieved from http://www.ukom.uk.net/news/tablets-now-account-for-8-of-website-traffic-ukom-and-comscore-announce-ear

UNCTAD. (2010). *Creative Economy Report 2010 – Creative Economy: A Feasible Development Option.* Geneva. Retrieved from http://unctad.org/en/Docs/ditctab20103_en.pdf

United Nations Conference on Trade and Development. (2011). *Information Economy Report 2011. ICTs as an Enabler for Private Sector Development.* New York and Geneva: United Nations publication.

United Nations University. (2011). *Entrepreneurship and Economic Development* (W. Naudé, Ed.). London: Palgrave Macmillan.

Vaidhyanathan, S. (2011). *The Googlization of Everything (and Why We Should Worry).* Berkeley, CA: University of California Press.

Vaidhyanathan, S. (2011). *The Googlization Of Everything: And Why We Should Worry.* Berkeley & Los Angeles, CA: University of California Press.

Vakulenko, M., Schuermans, S., Constantinou, A., & Kapetanakis, M. (2011). *Mobile platforms: The clash of ecosystems.* VisionMobile. Retrieved from http://www.visionmobile.com/rsc/researchreports/VisionMobile-Clash-of-Ecosystems_v1.pdf

Van Damme, K., Courtois, C., Verbrugge, K., & de Marez, L. (2015). *What's happening to News? A Mixed-Method Audience Centred Study on Mobile News Consumption. Mobile Media & Communication, 3(2).* OnlineFirst.

Van de Ven, A. H. (1986). Central Problems in the Management of Innovation. *Management Science, 32*(5), 590–607. doi:10.1287/mnsc.32.5.590

Van Dijck, J. (2013). You have one identity: performing the self on Facebook and LinkedIn. *Media, Culture & Society,* 35(2), 199-215. Retrieved November 14, 2014 from http://mcs.sagepub.com/content/35/2/199

Van Dijck, J. (2007). *Mediated Memories in the Digital Age.* Stanford, CA: Stanford UP.

Van Dijck, J. (2014). Datafication, Dataism and Dataveillance: Big Data between Scientific Paradigm and Ideology. *Surveillance & Society*, *2*(2), 197–208.

Varian, H. (2009). Online ad auctions. *The American Economic Review*, *99*(2), 430–434. doi:10.1257/aer.99.2.430

Varian, H. (2014). Big data: New tricks for econometrics. *The Journal of Economic Perspectives*, *28*(2), 3–28. doi:10.1257/jep.28.2.3

Varnali, K., Yilmaz, C., & Toker, A. (2012). Predictors of attitudinal and behavioral outcomes in mobile advertising: A field experiment. *Electronic Commerce Research and Applications*, *11*(6), 570–581. doi:10.1016/j.elerap.2012.08.002

Venkatesh, V., Morris, M. G., Davis, G. B., & Davis, F. D. (2003). User Acceptance Of Information Technology: Toward A Unified View. *Management Information Systems Quarterly*, *27*(3), 425–478.

Venkatesh, V., Morris, M., Davis, G., & Davis, F. (2003). User acceptance of information technology: Toward a unified view. *Management Information Systems Quarterly*, *27*(3), 425–478.

Verbrugge, K., Stevens, I., De Marez, L., Veeckman, C., Coppens, P., Van Hamme, L., & Lievens, B. (2013). Mobimeter 2013: A Report On The Adoption And Use Of Mobile Media Technologies In Flanders. Gent, Belgium.

Vernadsky, V. (1998) The Biosphere (Original edition 1926). Berlin, Germany: A Peter N. Nevraumont book Copernicus Series. Springer Science & Business Media.

Villalba, E., Casas, I., Abadie, F., & Lluch, M. (2013). Integrated personal health and care services deployment: Experiences in eight European countries. *International Journal of Medical Informatics*, *82*(7), 626–635. doi:10.1016/j.ijmedinf.2013.03.002 PMID:23587432

Villalba, E., Salvi, D., Ottaviano, M., Peinado, I., Arredondo, M. T., & Akay, A. (2009). Wearable and mobile system to manage remotely heart failure. Information Technology in Biomedicine. *IEEE Transactions on*, *13*(6), 990–996.

Virgin Mobile. (2014). Custom: make it yours. Retrieved from http://www.virginmobileusa.com/custom/#/

Virpi, R., Popescu, A., Koivisto, A., & Vartianien, E. (2006). Minimap: A Web Page Visualization Method for Mobile Phones. *Proceedings of SIGCHI*, *06*, 35–44.

Vishwanath, S., Vaidya, K., Nawal, R., Kumar, A., Parthasarathy, S., & Verma, S. (2012). *Touching lives through mobile health-Assessment of the global market opportunity*. Pricewaterhouse Coopers.

Vision Mobile. (2014a). *App Economy Forecasts 2013-2016*. Retrieved from http://www.visionmobile.com/product/app-economy-forecasts-2013-2016/

Vision Mobile. (2014b). *App Profits and Costs. The ingredients of a successful app business and how to replicate them*. Retrieved from http://www.visionmobile.com/product/app-economy-profits/

Visionmobile. (2014). *Mobile Megatrends 2014*.

Voci, P. (2010). *China on video: Smaller-screen realities, Routledge Studies in Asia's Transformations*. London, New York: Routledge.

W3C Authoring Challenges for Device Independence . (n. d.). Retrieved from http://www.w3.org/TR/2003/NOTE-acdi-20030901/

W3C mobileOK Basic Tests . (n. d.). Retrieved from http://www.w3.org/TR/mobileOK-basic10-tests/

W3C Web Accessibility Initiative Standards Committee . (n. d.). Retrieved from http://www.w3.org/WAI/mobile/

Walker Rettberg, J. (2014). *Seeing Ourselves Through Technology*. Basingstoke, UK: Palgrave Macmillan. Retrieved November 8, 2014 from http://www.academia.edu/8482366/Seeing_Ourselves_Through_Technology_How_We_Use_Selfies_Blogs_and_Wearable_Devices_to_See_and_Shape_Ourselves

Wang, A. (2006). The Shazam music recognition service. *Communications of the ACM*, *49*(8), 44–48. doi:10.1145/1145287.1145312

Wang, H., & Wang, S. (2010). User acceptance of mobile internet based on the unified theory of acceptance and use of technology: Investigating the determinants and gender differences. *Social Behavior and Personality*, *38*(3), 414–426. doi:10.2224/sbp.2010.38.3.415

Wangsness, R. (1986). *Electromagnetic Fields*. Wiley.

Waring, J. (2014). China to continue to drive iPhone sales – despite subsidy cuts. Retrieved from http://www.mobileworldlive.com/china-continue-drive-iphone-sales-despite-subsidy-cuts?utm_campaign=MWL_20140820&utm_medium=email&utm_source=Eloqua&elq=c995dcfc0f634901998b4f6b791c4791&elqCampaignId=2149

Warman, P. (2013b). Mobile Gaming. A Global View on Impact and Opportunities. Retrieved from http://fr.slideshare.net/Markies/newzoo-mobile-first4junev1

Watson, C., McCarthy, J., & Rowley, J. (2013). Consumer attitudes towards mobile marketing in the smart phone era. *International Journal of Information Management*, *33*(5), 840–849. doi:10.1016/j.ijinfomgt.2013.06.004

Wee, W. (2014a). KakaoTalk reveals it has 140 million users, made $203 million in revenue last year". *Technostorm*. Retrieved from https://sg.news.yahoo.com/kakaotalk-reveals-140-million-users-060150446.html

Wee, W. (2014b). How KakaoTalk Grew From a Startup to a Giant Mobile Platform. Retrieved from *Techniasia*. http://www.techinasia.com/kakao-founding-story/

Wei, R., Lo, V., Xu, X., Chen, Y. K., & Zhang, G. (2014). Predicting mobile news use among college students: the role of press freedom in four Asian cities. *New Media & Society*, *14*(4), 637–654. doi:10.1177/1461444813487963

Wellman, B. (2001). Physical Place and Cyberplace: The Rise of Personalized Networking. *International Journal of Urban and Regional Research*, *25*(2), 227–252. doi:10.1111/1468-2427.00309

Wellman, B., Quan-Haase, A., Boase, J., & Chen, W. (2002). *Examining the Internet in Everyday Life*. Nijmegen. [online] doi:10.1002/9780470774298

West, J., & Mace, M. (2010). Browsing as the killer app: Explaining the rapid success of Apple's iPhone. Retrieved from www.joelwest.org/Papers/WestMace2010-WP.pdf

West, J., & Mace, M. (2010). Browsing as the killer app: Explaining the rapid success of apple's iphone. *Telecommunications Policy*, *34*(5-6), 270–286. doi:10.1016/j.telpol.2009.12.002

Westlund, O. (2008). From mobile to phone to mobile device: news consumption on the go. *Canadian Journal of Communication*, *33*, 443–463.

Westlund, O. (2012). Producer-centric vs. Participation-centric: On the sahping of mobile media. *Northern Lights: Film & Media Studies Yearbook*, *10*(1), 107–121. doi:10.1386/nl.10.1.107_1

Westlund, O. (2013). Mobile news: A review and model of journalism in an age of mobile media. *Digital Journalism*, *1*(1), 6–26. doi:10.1080/21670811.2012.740273

Westlund, O., & Färdigh, M. A. (2012). Conceptualizing Media Generations: The Print-, Online- And Individualized Generations. *Observatorio Journal*, *6*, 181–213.

Westlund, O., Gómez-Barroso, J.-L., Compañó, R., & Feijóo, C. (2011). Exploring the logic of mobile search. *Behaviour & Information Technology, 30*(5), 691–703. doi:10.1080/0144929X.2010.516020

Westlund, O., & Lewis, S. (2014). Agents of media innovation: actors, actants, and audiences. *The Journal of Media Innovations, 1*(2). doi:10.5617/jmi.v1i2.856

Westlund, O., & Weibull, L. (2013). Generation, life course and news media use in Sweden 1986–2011. *Northern Lights, 11*(1), 147–173. doi:10.1386/nl.11.1.147_1

White, D. M. (1950). The "Gate Keeper": A case study in the selection of news. *The Journalism Quarterly, 27*(4), 383–390.

Whitney, L. (2014). KitKat chews up more than 20% of android devices. *CNET*. Retrieved from http://www.cnet.com/news/kitkat-chews-up-more-than-20-percent-of-android-devices/

Wicker, S. B. (2012). The loss of location privacy in the cellular age. *Communications of the ACM, 55*(8), 60–68. doi:10.1145/2240236.2240255

Wilcox, M., & Voskoglou, C. (2014). *Developer economics. State of the developer nation Q3 2014*. Vision Mobile. Retrieved from http://visionmobile.com

Wilcox, M., & Voskoglou, C. (2014). *State of the Developer Nation, Q3*, 2014.

Wilson, S., & McCarthy, G. (2010). The mobile university: From the library to the campus. *RSR. Reference Services Review, 38*(2), 214–232. doi:10.1108/00907321011044990

Wingfield, N. (2008, July 10). IPhone Calls On Software Developers. *The Wall Street Journal*. Retrieved from http://online.wsj.com/article/SB121565491776341571.html?

Wireless Intelligence. (2013a). BRIC markets generating almost a quarter of global mobile sales. Retrieved from https://wirelessintelligence.com/analysis/2013/02/bric-markets-generating-almost-a-quarter-of-global-mobile-sales/368/

Wireless Intelligence. (2013b). Global and regional mobile revenue trends. Retrieved from https://wirelessintelligence.com/analysis/2013/01/global-and-regional-mobile-revenue-trends/367/

Wirth, W., Von Pape, T., & Karnowski, V. (2008). An integrative model of mobile phone appropriation. *Journal of Computer-Mediated Communication, 13*(3), 593–617. doi:10.1111/j.1083-6101.2008.00412.x

Wit, M. d., Mateus, M., Moitry, J., & Acharys, A. (2009). *Dumb pipe vs. Smart pipe? Better to design a valuable pipe*. Oliver Wyman. Retrieved from http://www.oliverwyman.com/ow/pdf_files/CMT07-DPipeSPipe-final.pdf

Wolf, C., & Hohlfeld, R. (2012). Revolution in Journalism? Mobile Devices as a New Means of Publishing. In C. Martin & T. von Pape (Eds.), *Images in Mobile Communication: New content, new uses, new perspectives* (pp. 81–99). Germany: Springer VS Verlag für Sozialwissenschaften. doi:10.1007/978-3-531-93190-6_5

Wolff, M. (2013, January 4). This tipping-point for paywalls does not fix newspapers' larger crisis. *The Guardian*. Retrieved from http://m.guardian.co.uk/commentisfree/2013/apr/01/tipping-point-paywalls-newspapers-crisis

World Economic Forum. (2009). *Educating the Next Wave of Entrepreneurs.Unlocking entrepreneurial capabilities to meet the global challenges of the 21st Century.*

World Health Organisation. (2011). mHealth (Vol. 3). Geneva, Switzerland: WHO Press.

World Health Organization. (2011). mHealth: New horizons for health through mobile technologies. Global Observatory for eHealth series (Vol. 3).0

WURFL - Wireless Universal Resource FiLe. (n. d.). Retrieved from http://wurfl.sourceforge.net/

Wu, T. (2011). *The Master Switch: The Rise and Fall of Information Empires*. London: Atlantic Books.

Xiang, P., Yang, X., & Shi, Y. (2006). Effective Page Segmentation Combining Pattern Analysis and Visual Separators for Browsing on Small Screens.*IEEE/WIC/ACM International Conference on Web Intelligence (WI '06)*, (pp. 831-840). Washington, DC, USA. doi:10.1109/WI.2006.67

Xiao, Y., Tao, Y., Li, Q., & Li, W. (2008). Web Page Adaptation for Small Screen Mobile Device: A New P2P Collaborative Deployment Approach. *Intelligent Networks and Intelligent Systems, 08*, 191–196.

Xiong, Y., Wang, H., & Li, Y. (2014). Research on Mobile Communication Based on Multimedia Technology.*Proceedings of the 9th International Symposium on Linear Drives for Industry Applications* (Vol. 4, pp. 677-684). doi:10.1007/978-3-642-40640-9_85

Xu, H., & Yan, R. N. (2011). Feeling Connected via Television Viewing: Exploring the scale and its correlates. *Communication Studies*, 62(2). Retrieved from http://tinyurl.com/mymqnp3

Xu, H., Rossom, M. B., Gupta, S., & Carroll, J. M. (2012). Measuring Mobile User's Concern for Information Privacy. Proceedings of the *33rd International Conference on Information Systems*. Orlando, FLA. Retrieved from http://faculty.ist.psu.edu/xu/papers/Xu_etal_ICIS_2012a.pdf

Yang, B., Kim, Y., & Yoo, C. (2013). The integrated mobile advertising model: The effects of technology- and emotion-based evaluations. *Journal of Business Research*, *66*(9), 1345–1352. doi:10.1016/j.jbusres.2012.02.035

Yang, F., & Wang, Z. M. (2009). A mobile location-based information recommendation system based on GPS and WEB2.0 services. *Journal WSEAS Transactions on Computers*, *8*(4), 725–734.

Yeung, K. (2013, September 11). Social reader app Flipboard now has 85 million registered users. *The Next Web*. Retrieved from http://thenextweb.com

Ynduráin, E., Feijóo, C., Ramos, S., & Campo, C. (2010). Context-aware mobile applications design: Implications and challenges for a new industry. *The Journal of the International Telecommunications Professional*, *4*(4), 16–28.

Young, E. (2014). 90 million wearables will be sold this 2014, analyst says. *TechnologyTell*. Retrieved from http://www.technologytell.com/gadgets/138712/90-million-wearables-will-be-sold-this-2014-analyst-says/

Youngblood, G. (1970). *Expanded cinema*. New York, NY: Dutton.

Zeimpekis, V., Giaglis, G. M., & Lekakos, G. (2002). A taxonomy of indoor and outdoor positioning techniques for mobile location services. *ACM SIGecom Exchanges*, *3*(4), 19–27. doi:10.1145/844351.844355

Zeng, M. (2011). Examining Geocaching Practices through a Mobilities Lens.[Alberta, Canada: MEA.]. *Proceedings of the Media Ecology Association*, *12*, 113–122.

Zhang, D. (2007). Web content adaptation for mobile handheld devices. *Communications of the ACM*, *50*(2), 75–79. doi:10.1145/1216016.1216024

Zhou, X., & Dong, C. (2014). Study on the impact of network interactivity on purchase intention in the social media era. *International Journal of Business and Social Science*, *5*(6), 259–265.

Zhu, H., Cao, H., Chen, E., Xiong, H., & Tian, J. (2014). Mobile App Classification with Enriched Contextual Information. *IEEE Transactions on Mobile Computing*, *13*(7), 1550–1563. doi:10.1109/TMC.2013.113

Zickuhr, K., & Smith, A. (2012). Digital differences. *Pew Internet & American life project*. Retrieved from http://www.pewinternet.org/files/old-media/Files/Reports/2012/PIP_Digital_differences_041312.pdf

About the Contributors

Juan Miguel Aguado teaches Communication Theory at the School of Communication and Information Studies, University of Murcia (Spain). His publications include 'Mobile Society: Culture, Identity and Technology' (Edited with Inmaculada J. Martínez, 2008), Movilizad@s: Women and mobile Communication in the Information Society' (with Inmaculada J. Martinez and Iolanda Tortajada, Spanish Women Studies Institute, 2009), 'Technology and Social Complexity (Edited with Eva Buchinger and Bernard Scott, Editum, 2010) and 'Mobile Communication: Towards a New Media Ecosystem' (Gedisa, 2013). Juan Miguel's research includes mobile communication's ecosystem and its social and cultural implications. Together with Inmaculada J. Martínez he is currently coordinating an international four year research project on the impact of personal information management as an economic asset revolutionizing both the digital media and advertising systems.

Claudio Feijóo holds an MSc and PhD in Telecommunication Engineering and an MSc in Quantitative Economics. He is professor at the Technical University of Madrid (UPM) where he researches on the future socio-economic impact of emerging information society technologies, in particular, from an ultra-broadband, mobile and content perspective. Currently he serves as Co-Director of the Sino-Spanish Campus at Tongji University in Shanghai. He is also the co-founder and coordinator of the ICT area in the international master programme on city sciences and main responsible of the organizing committee of the International Conference in City Sciences. He spent two years at the Institute for Prospective Technological Studies of the European Commission researching on the future prospects of mobile content and applications. He also directed the Chair in Telecommunications Regulation and Information Society Public Policies at UPM. He participated in the information society development plans and broadband deployment strategies while being adviser for the Spanish State Secretary on Telecommunications and Information Society. On the entrepreneurial side, for three years he was dedicated to launch a university spin-off devoted to the transfer of know-how in technology, media and telecommunications. Since then he has participated in numerous start-up companies. He has also been involved in numerous research, development and consulting projects, both public and private, in Europe, Latin America, North of Africa and Asia. He lectures regularly in international seminars and postgraduate courses and has authored more than 200 publications in books, journals and conferences. He is guest lecturer at IE Business School on digital business and disruptive ICT, and member of the board of the International Telecommunications Society.

Inmaculada J. Martinez is an Assistant Professor at the School of Communication and Information Studies, University of Murcia (Spain) where she teaches Advertising Ecosystem. Her publications include 'Mobile Society: Culture, Identity & Technology ' (2008), 'Mobilized: Women & Mobile in the Information Society ' (2010) and 'Mobile Communications: Towards a New Digital Ecosystem' (2013). Her research interests include mobile advertising and marketing, mobile merchandising, the new media impact on consumer behaviour and mobile communications from a gender perspective. Together with Juan Miguel Aguado, she is currently coordinating an international four year research project on the impact of personal information management as an economic asset revolutionizing both the digital media and advertising systems.

* * *

Frederic Adam is a freelance Art director and an expert in Mobile Learning and spatial narratives. He's currently working with scientists in Australia and the United States. Between 2011 and 2013 he was as Associate Teacher at the Fine Arts School, at the University of Murcia and an artist member of the Art Collective Transnational Temps and Founder of the GPS Museum web portal (gpsmuseum.eu) about Locative Media and creativity.

Shivom Aggarwal is a Research Associate in CeDInt (Centro domótica Integral), Universidad Politécnica de Madrid, Spain and doctoral candidate in Operations and Technology department at IE Business School, Madrid, Spain. His main research areas include Big Data, Smart Cities, Cloud Computing, Knowledge Management and Future Technologies. He also works with the European Union under FP7 framework and other governmental organizations on policy making for future technologies such as Big Data and Artificial Intelligence. Prior to joining academia, he worked in mergers and acquisitions of high-tech companies including Big Data Analytics, Information Security and Mobile Computing with an investment bank based in Munich, Germany. He is a Computer Engineer (with a specialization in Information Security) by training with a Masters in international management. He also possesses international certifications such as a CEH (Certified Ethical Hacker), an RHCE (Red Hat Certified Engineer) and Network Management. He has been an entrepreneur and management consultant with technology start-ups based in Europe and Asia.

Alberto Arenal has a MSc in Telecommunications Engineering and a MSc in City Sciences from the Technical University of Madrid and a Business Administrations degree from the National Distance Learning University (UNED). He is currently a doctoral candidate at the Technical University of Madrid in the Entrepreneurship field and his main research include the qualitative and quantitative analysis of the entrepreneurial ecosystems, with focus on cities. Since 2009, he has been working in several positions in REDTEL, the Spanish Association of Telecommunications Operators, and in AMETIC, the Association of Electronics, Information and Communications Technologies, Telecommunications and Digital Content Companies in Spain, focusing on the study of the regulatory framework of the Digital Economy and the relationship between the different stakeholders.

Cristina Armuña has both a MSc in Telecommunications Engineering and MSc in City Sciences by the Technical University of Madrid. Currently, she is a PhD candidate at the Technical University of Madrid with focus on entrepreneurial ecosystems and obtaining a Law degree from the National Distance Learning University (UNED). She has two years of experience as financial consulting at Management Solutions and afeter this she has been acting for five years as Manager of the Telecommunications Infrastructures Department at the Spanish Association of Telecommunications Operators REDTEL, focused on the mobile regulatory framework. She currently combines the research field with strategic consulting at Noon Venture (a consultancy firm specialized in strategic advice in the domain of Technology, Media and Telecommunications), and startup creation and mentoring at Inténtalo Startups (educational platform for accelerating potential entrepreneurs).

Victoria Bleda holds a degree in Law and a master's in Public Administration. She has been working in and around the music industry for the last 20 years. She has been part of record labels, both traditional and online, led a music editorial initiative, and worked for intellectual property rights collecting societies.

João Canavilhas is an associate professor and has done his PhD at the Universidad de Salamanca (Spain) with a thesis entitled Webnoticia: propuesta de modelo periodístico para la WWW. He teaches at the Universidade da Beira Interior (Covilhã PORTUGAL), where he is Vice-Rector and researcher at Labcom Laboratory of Online Communication He is the author or co-author of 4 books, 20 chapter books and 30 papers in national and international scientific journals. His most important contribution to online journalism is the "Tumbled Pyramid theory". His research work focuses on various aspects of "Communication and New Technologies", particularly in the fields of online journalism, e-politics, social media and journalism for portable devices.

Andreu Castellet has a PhD Degree in Social and Law Sciences from the University of Murcia (Spain), where he teaches in the Department of Information and Documentation. He has also been in the Media Management and Transformation Centre (MMTC) of Jönköping International Business School (Sweden) as a Visiting Researcher. He holds a Bachelor in Journalism from the Autonomous University of Barcelona (Spain). Also, he has a worked as a journalist in a number of media groups in Spain and in the UK. His research interests include digital media management, innovation, structure and regulation; and mobile ecosystem strategies.

Enrique Castelló-Mayo holds a PhD from the Complutense University of Madrid, and he was Assistant Professor in this University and University of Valladolid. Actually he is Associated Professor and Researcher at University of Santiago de Compostela, member of the Audiovisual Studies Group and Audiovisual Coordinator of CIDUS (Platform for Innovation and Experimentation and Innovation in Digital Contents). His research lines include: audiovisual formats and applied production technologies, transmedia narratives. His background includes working in Antena 3 TV, Agencia EFE, Information and Communication New Technologies Program and as external collaborator of Televisión Española and ATEI.

Rodrigo da Cunha holds a Doctorate in Communication and Contemporary Culture from the Federal University of Bahia (UFBA). Adjunct Professor of News Design, at Department of Communication, Federal University of Pernambuco (UFPE), Brazil. Research member in Online Journalism Research Group (GJOL) and in the Convergent Journalism Laboratory Project. Your research is about news apps interfaces for mobile devices, news design, infographics, history of design, digital journalism, and cyberculture.

Fernando Firmino da Silva holds a Doctorate in Communication and Contemporary Culture from the Federal University of Bahia (UFBA). Professor in the Department of Communication at the State University of Paraíba (UEPB) and professor of the PosGraduate Program in Journalism at the Federal University of Paraíba.

Manuel José Damásio holds a PhD in communications/Multimedia Systems from the Universidade Nova de Lisboa and an MSc in digital media by Napier University (UK). He is an Associate Professor and the Head of the Film and Media Arts department at Universidade Lusófona in Lisob, Portugal. He holds vast experience in consulting and production concerns of audiovisual and multimedia production. He is a member of the R&D research unit CICANT - Centre for Research in Applied Communication, Culture and New Technologies and a Member of the board of GEECT the European Association of Film and Media schools.

Giuditta De Prato joined the Institute for Prospective Technologies as a post-doctoral researcher in January 2009 to contribute to projects on the economic aspects of the Information Society and on the impacts of Information Society Technologies, mainly focusing on ICT R&D, the software sector, patents and innovation. She has a PhD in Economics and Institutions from the University of Bologna (Italy). Before joining IPTS, she was a software developer and IT consultant from 1992 to 2005. From 2005 to 2009, she was a contract research assistant, focussing on research activities on local development, evaluation, ICT and open source at the University of Bologna, where she also lectured on macroeconomics and environmental economics. She co-edited together with E. Sanz and J.P Simon, Digital Media Worlds; The new media economy, Oxford, Palgrave, 2014.

Patrícia Dias is the author of "The Mobile Phone and Daily Life", a book about the social impact of the mobile phone. She is a lecturer and a researcher at the Catholic University of Lisbon (UCP) and at Lusófona University (ULHT), specializing in social media and digital marketing. She coordinates the Advanced Training in Social Media at UCP. She holds a PhD in Communication Studies from the Catholic University of Portugal with Summa Cum Laude distinction.

Jesús Ferrandis has a MSc in Telecommunications Engineering from the Technical University of Madrid and a MSc in Economics from National University of Distance Education in Spain. His career has been linked to Telefonica along different positions in the field of Technology, R+D and Corporate Strategy. He is an expert on ICT based entrepreneurship and advisor to different educational programs in this field.

Julian Flores González is an associate professor at the University of Santiago de Compostela (Spain) in the Department of Electronics and Computer Science. Julian got his degree in Physical Sciences, specializing in electronics in 1991, a PhD in 1996 and afterwards a post-doc in the Graphics Visualization and Usability Center of the Georgia Institute of Technology (USA). His main research activity has focused on the development of interactive graphical applications in real time. These applications have covered a range of areas from applications, to the revaluation of assets, to the rehabilitation of people with different types of diseases.

José Luis Gómez-Barroso is an Associate Professor in the Department of Applied Economics and Economic History at Universidad Nacional de Educación a Distancia (UNED). He holds a PhD and a degree in Economics from the Universidad Nacional de Educación a Distancia (UNED). He also holds a degree in Telecommunication Engineering from the Universidad Politécnica de Madrid as well as a degree in Law from the Universidad Complutense. His research interests lie on the economics and regulation of ICT markets, and on the factors hindering the development of an information society. Gómez-Barroso has participated in different research projects, some of them for the European Commission. He has published more than sixty academic papers and chapters in books.

Sara Henriques is a PhD student in Communication Sciences at the Universidade Lusófona de Humanidades e Tecnologias, she has a master's degree in Educational Psychology from the Instituto Superior de Psicologia Aplicada – Instituto Universitário (ISPA). She also started a PhD in Developmental Psychology which she opted to reassign to Communication Sciences due to the work she is currently performing at the Centre for Research in Applied Communications, Culture and New Technologies at ULHT. She has developed academic research in the developmental and cognitive psychology field and more recently in the communication sciences field. Her academic interests and recent academic activity is based on the influence of online social media in social relationships and patterns of individual and collective behavior and on the transformations of social practices and social relationships via the dissemination and use of new social technologies.

Antía López-Gómez is a Professor in the Department of Communication Sciences and a Researcher in the Audiovisual Studies Group of the University of Santiago de Compostela (Spain). Since 1997, she holds a Ph.D. from the Universidad Complutense in Madrid (Spain) in Information Sciences. Her research lines include: textual, content and format analysis.

Roi Méndez Fernández is a Computer Systems engineer from the University of Santiago de Compostela (Spain) since 2007 and got the title of Master in Computing, specializing in Computer Graphics, within the Universitat Politecnica de Catalunya (Spain). In 2009, developing his master's thesis at the University of Manchester (UK). His fields of interest focus on virtual and augmented reality, immersive environments and human-computer interaction. Currently, he is a pre-doctoral student at CITIUS, focusing his research on the use of virtual sets in live television program broadcasting.

Abhishek Nayak is currently a Doctoral Candidate and a research scholar in the Marketing department at the IE Business School, Madrid. His key research areas include consumer analytic (understanding and predicting consumer purchase intentions) and advertising effectiveness during recessions (optimizing resources through effective media allocation). Before pursuing his Phd, he completed his MBA in International Business and Bachelors in Electronics Engineering from India. He has industry experience in the Sales and Business Analytic domain in the logistics and the IT industry.

Suzana Oliveira Barbosa holds a Doctorate in Communication and Contemporary Culture from the Federal University of Bahia (UFBA). Professor in the Department of Communication and Communication and Contemporary Culture Program, Faculty of Communication, at the Federal University of Bahia (UFBA). Investigator-coordinator of the Research Group on Online Journalism (GJOL). Coordinator of the Convergent Journalism Laboratory Project. Her research is about journalism, digital journalism, database journalism, convergence journalism and mobile journalism.

Brett Oppegaard was the individual recipient of the regional and national 2012 George and Helen Hartzog Award for his research into mobile app development and media delivery systems within the National Park Service, as well as the national 2013 John Wesley Powell Prize winner for outstanding achievement in the field of historical displays.

Marcos Palacios has a B.A First Class Honours (Sociology) University of Liverpool (1975); Ph.D. (Sociology) Center for Latin American Studies, University of Liverpool (1979). Full professor in the Department of Communication and Communication and Contemporary Culture Program, Faculty of Communication, at the Federal University of Bahia (UFBA). Visiting Professor at University of Beira Interior (Portugal). Investigator-coordinator of the Research Group on Online Journalism (GJOL). Research member in the Convergent Journalism Laboratory Project. Research experience in Communication Studies, with special reference to Cyberculture, New Media and Journalism and sociological aspects of Communication.

Ignacio Peinado holds an MsC in Telecommunication Engineering by the Technical University of Madrid. Currently, he is working as an Innovation Manager in the University Hospital of Getafe, Spain. He has been working more than 10 years in the design, development of validation of ICT solutions for healthcare and inclusion in the context of different international research projects. In total, he has collaborated on more than 10 projects funded by the European Commission, in both technical and management aspects. Moreover, he has worked as a user experience and accessibility consultant for different start-ups and SMEs.

Verónica Perales (European PhD) is a Professor of Multimedia and Transmedia Narratives at the Fine Arts School and also at the School of Communication, University of Murcia (Spain). Co-founder of international art collective Transnational Temps (2001), questioning different aspects of the underlying ideology of technological progress. The mass media´s trivialisation of certain environmental problems, perception, dialogue, citizen action and the strategies of superfluous green cleaning are some of the issues addressed by the group, whose ultimate aim is to reinforce encourage genuine respect for the biosphere. Her artistic practice is also related to environmental activism and ecofeminism.

Sergio Ramos gained his PhD and Master in Telecommunications Engineering from the Technical University of Madrid - UPM. He holds an MBA from the Stockholm School of Economics and a Degree in Advanced Management from the IE Business School. His career has been linked to two main areas of activity: ICT and Information Society Sector, and Innovation Management and Entrepreneurship. His starting research interest was related to the study of the mobile Internet phenomenon in Europe from the regulatory, market and business perspectives, being his Ph.D. Thesis (2005) one of the first studies in Europe in the field. Currently he is focused on practical research and experimentation on entrepreneurial ecosystems, particularly in the characterization of the trigger factors that stimulate entrepreneurship and innovation. Among his most relevant positions, he worked for the European Commission as Resident Twinning Adviser of an EC Twinning Project for the Public Utilities Commission of Latvia, to design and monitor the transposition process of EU electronic communications regulatory framework into Latvian legislation. In addition, he worked for the Spanish Association of Telecom Network Operators (REDTEL) as a Director of Network Infrastructure Deployment, and served as Innovation and Entrepreneurship Manager at CeDInt-UPM (Research Centre for Smart Buildings and Energy Efficiency at UPM). Currently based in Madrid, he combines lecturing, research and managerial activities at UNED, UPM and IE Business School, with strategic consulting for the last eight years at Noon Venture (a consultancy firm specialized in strategic advice in the domain of Technology, Media and Telecommunications), and startup creation and mentoring at Inténtalo Startups (educational platform for accelerating potential entrepreneurs).

Leocadio Rodriguez Mañas holds a PhD in Medicine. He is the Head of the Geriatric Department at the University Hospital of Getafe and Director of the Spanish Network on Aging and Frailty (RETICEF) of the Ministry of Economy and Competitiveness, Co-Director of the Toledo Study on Healthy Aging-TSHA. He has been the Principal Investigator in 36 research projects, 4 of them funded by the European Commission focused on the field of frailty in elderly people. He has authored more than 130 papers in reputed journals.

Juan Angel Ruiz is a manager in a European leading IT company. At the same time, he is a Ph.D. student at the Applied Economy Department, in the UNED (Spain). He is currently researching on the economic value of personal information for companies in the digital economy. His interests also include global public goods, global public private partnerships, public policy evaluation, privacy policies, and the role of IT technologies in the production functions. Mr. Ruiz holds an MSc in Forest Engineering-manufacturing, from the Universidad Politécnica de Madrid (Spain); a BS in Economics-quantitative modelling, and a MS on economic growth and sustainable development by the UNED; a MS in Financial Management by the ESIC Business & Marketing School (Spain), and an Executive degree in Business Management by HEC-School of Management (Paris-Mumbai). His career has been developed in the ICT sector –in both, big and niche companies—where he has carried out consulting and program management primarily those related with business processes and supply chain management, and large contract sales in Banking, Manufacturing and the public sector (national and European Institutions). He has also assumed strategic and business development responsibilities in the public sector market, portfolio management roles, and strategic transformation program leadership within his own company for different geographies. He currently lives in Paris.

Eduardo M. Sánchez Vila was born in Barcelona, Spain in 1970. He received a B.S. degree in physics from the University of Santiago de Compostela in 1993, a Ms. degree in Neuroscience at the International University of Andalucía in 1996, and a Ms. degree in Computer Science at the University of Southern California in 2001. He also received a PhD in Physics in 2000 from the University of Santiago de Compostela. Nowadays, he is working as an Associate Professor in the Department of Electronics and Computer Science at the University and as researcher at the Research Center for Information Technologies (CITIUS) of Santiago de Compostela. His research interests are in recommender systems, satisfaction models, and computational models of somatosensory and visual systems.

Ivan Satuf is a PhD student in Communication Sciences at University of Beira Interior (Portugal), a researcher trainee at the Laboratory of Online Communication (LabCom) and a CAPES/Brazil doctoral fellow. His research interests are the fields of journalism and new media.

Ana Serrano Tellería is an Assistant Professor Ph.D (Accredited by ANECA, Spanish National Official Agency, June 2013). Postdoc, coordinator of 5 Ph.D candidates in the European FEDER project 'Public and Private in Mobile Communications' at LabCom, Beira Interior University, Portugal (April 2013 - 2015), European Union and MAIS CENTRO - Portuguese Government. Lecturer of Cyberculture (2013-15). Full grant (EEUU Embassy in Spain) for Seminar 'Entrepreneurial Journalism: A Renewed Hope' (2014), University of Castilla La Mancha, Spain. Extraordinary Ph.D Award (Creativity, innovation, impact and relevance; 2012), Ph.D. ('Initial Node Design on Cybermedia: A Comparative Study', 2010) and Bachelor in Journalism (2002) by the University of The Basque Country, Spain. Full grant (2006-08) and research contract (2008-10) to develop Ph.D by the University of the Basque Country. First cycle of English Translation and Interpretation, University of Alicante, Spain (2000). Management of European projects and International Cooperation ones in Documenta, Spain (2012-13). Responsible of the Communication Department in the Modern Art Museum of Santander (2015) and the Territorial Development Agency "Campoo Los Valles", Spain (2011). Master in Innovation Management, University of Cantabria, Spain (Full grant, 2010-11). Visiting researcher at the Faculty of Communication in the Federal University of Salvador de Bahia, Brazil (Full grant of the Spanish Ministry of Science and Innovation, 2009). Cultural Manager for the University of the Basque Country, Spain (2005), Master in Theatre and Performing Arts (2004-05), University of the Basque Country. Theatre teacher, Spain (2003-04). Online journalist in Elcorreo.com, Vocento, Spain (2002-03). Research interest: Digital / Mobile / Online Communication and Design, Entrepreneurial Journalism, Media Studies, Performing & Stage Arts. Full grants received from: the Spanish Confederation of Young Entrepreneurs with the Spanish Ministry of Employment and Social Security, University of Cantabria, SODERCAN – Government of Cantabria, University of the Basque Country, Federal University of Bahia (Brazil), Ministry of Science and Innovation (Government of Spain), European Union, Marcelino Botin Foundation, EEUU Embassy in Spain.

Jean Paul Simon is the founder of JPS Public Policy Consulting, a consulting firm specialised in media/ telecom law regulation and strategy. He has held various position in the telecom industry, worked as a senior scientist at the Institute for Prospective Technological Studies (IPTS), European Commission, Directorate-General JRC. He holds a PhD in Philosophy (1975) and is a graduate (MBA) from the Ecole des Hautes Etudes Commerciales (HEC) (MBA, econometrics, 1971). He has written several books and articles on communications and public policy. He is a frequent speaker on telecommunications and media in Asia, Europe and the USA. He co-edited with G. De Prato and E. Sanz, the book "Digital Media Worlds; The new media economy," Oxford, Palgrave, 2014.

Christian Sonnenberg is an Assistant Professor of Information Systems and the Academic Chair of Management Information Systems and Information Technology at Florida Institute of Technology. His research focuses on mobile and handheld usability, modeling and simulation techniques, online learning formats, data visualization, and web accessibility.

Inês Teixeira-Botelho is the author of "Extreme Generation," a Vodafone sponsored book about the impact of new technologies both in digital natives and emigrants. She has worked as an author for several publications, books and screenplays, as a speaker, consultant and advisor on communication, media, technology and entrepreneurship, as a researcher and project manager at Lusófona University (ULHT) and also is an invited specialist in the Advanced Training in Social Media at the Catholic University of Lisbon (UCP). She holds an Msc from Catholic University of Portugal with a Summa Cum Laude distinction.

Elena Villalba Mora holds a PhD in Biomedical Engineering from the Technical University of Madrid, where she is an Assistant Professor. She is also senior researcher at the University Hospital of Getafe, Spain. With 10 years' experience, she has participated in 11 international research projects focused on the modeling, design and validation of Health Information Technologies for long-term care. She also investigates the technological and socio-economic value of eHealth products and services across Europe. She has published 9 papers in reputed journals, and more than 25 conference papers and book chapters.

Oscar Westlund, PhD, is Associate Professor at the University of Gothenburg, currently working as Research Leader for "Medieutredningen" at the Government Offices of Sweden, Ministry of Culture. Westlund has researched the production, distribution and consumption of news through various methods for a decade, focusing especially on the shift towards mobile media. He has also conducted research into mobile search, including taking part in a project for the EU-commission based at the Institute for Prospective Technology Studies in Seville. Westlund has published more than 60 book chapters and reports, and articles in approximately 20 different international journals. He serves on the editorial boards of *Digital Journalism*, *Journal of Media Business Studies*, *Journal of Media Innovations*, and *Mobile Media & Communication*.

Index

N

O

P

R

S

T

U

V

W

Z

www.ingramcontent.com/pod-product-compliance
Lightning Source LLC
LaVergne TN
LVHW060356300926
849162LV00006B/1282